Student Solutions Manual

Cindy Trimble & Associates

Prealgebra
FIFTH EDITION

Elayn Martin-Gay

PEARSON

Prentice Hall

Upper Saddle River, NJ 07458

Vice President and Editorial Director, Mathematics: Christine Hoag
Executive Editor: Paul Murphy
Project Manager, Editorial: Mary Beckwith
Editorial Assistant: Georgina Brown
Senior Managing Editor: Linda Behrens
Project Manager, Production: Robert Merenoff
Supplement Cover Manager: Paul Gourhan
Supplement Cover Designer: Victoria Colotta
Operations Specialist: Ilene Kahn
Senior Operations Supervisor: Diane Peirano

© 2008 Pearson Education, Inc.

Pearson Prentice Hall

Pearson Education, Inc.

Upper Saddle River, NJ 07458

Pearson Prentice Hall™ is a trademark of Pearson Education, Inc.

The author and publisher of this book have used their best efforts in preparing this book. These efforts include the development, research, and testing of the theories and programs to determine their effectiveness. The author and publisher make no warranty of any kind, expressed or implied, with regard to these programs or the documentation contained in this book. The author and publisher shall not be liable in any event for incidental or consequential damages in connection with, or arising out of, the furnishing, performance, or use of these programs.

Printed in the United States of America

10   9   8   7   6   5   4   3   2

ISBN 13:   978-0-13-157638-4 Standalone

ISBN 10:   0-13-157638-0 Standalone

ISBN 13:   978-0-13-157639-1 Value Pack

ISBN 10:   0-13-157639-9 Value Pack

Pearson Education Ltd., *London*
Pearson Education Australia Pty. Ltd., *Sydney*
Pearson Education Singapore, Pte. Ltd.
Pearson Education North Asia Ltd., *Hong Kong*
Pearson Education Canada, Inc., *Toronto*
Pearson Educación de Mexico, S.A. de C.V.
Pearson Education—Japan, *Tokyo*
Pearson Education Malaysia, Pte. Ltd.

Contents

Chapter 1

Section 1.2

Practice Problems

1. The place value of the 8 in 38,760,005 is millions.

2. The place value of the 8 in 67,890 is hundreds.

3. The place value of the 8 in 481,922 is ten-thousands.

4. 54 is written as fifty-four.

5. 678 is written as six hundred seventy-eight.

6. 93,205 is written as ninety-three thousand, two hundred five.

7. 679,430,105 is written as six hundred seventy-nine million, four hundred thirty thousand, one hundred five.

8. Thirty-seven in standard form is 37.

9. Two hundred twelve in standard form is 212.

10. Eight thousand, two hundred seventy-four in standard form is 8,274 or 8274.

11. Five million, fifty-seven thousand, twenty-six in standard form is 5,057,026.

12. 4,026,301
    $= 4,000,000 + 20,000 + 6000 + 300 + 1$

13. a. Find Norway in the left-hand column. Read from left to right until the "gold" column is reached. Norway won 96 gold medals.

    b. Find the countries for which the entry in the first column (gold) is greater than 100. Germany and Russia have won more than 100 gold medals.

Vocabulary and Readiness Check

1. The numbers 0, 1, 2, 3, 4, 5, 6, 7, 8, 9, 10, 11, 12, ... are called <u>whole</u> numbers.

3. The number "twenty-one" is written in <u>words</u>.

5. In a whole number, each group of 3 digits is called a <u>period</u>.

Exercise Set 1.2

1. The place value of the 5 in 657 is tens.

3. The place value of the 5 in 5423 is thousands.

5. The place value of the 5 in 43,526,000 is hundred-thousands.

7. The place value of the 5 in 5,408,092 is millions.

9. 354 is written as three hundred fifty-four.

11. 8279 is written as eight thousand, two hundred seventy-nine.

13. 26,990 is written as twenty-six thousand, nine hundred ninety.

15. 2,388,000 is written as two million, three hundred eighty-eight thousand.

17. 24,350,185 is written as twenty-four million, three hundred fifty thousand, one hundred eighty-five.

19. 65,773 is written as sixty-five thousand, seven hundred seventy-three.

21. 1679 is written as one thousand, six hundred seventy-nine.

23. 2,800,000 is written as two million, eight hundred thousand.

1

25. 11,239 is written as eleven thousand, two hundred thirty-nine.

27. 202,700 is written as two hundred two thousand, seven hundred.

29. Six thousand, five hundred eighty-seven in standard form is 6587.

31. Fifty-nine thousand, eight hundred in standard form is 59,800.

33. Thirteen million, six hundred one thousand, eleven in standard form is 13,601,011.

35. Seven million, seventeen in standard form is 7,000,017.

37. Two hundred sixty thousand, nine hundred ninety-seven in standard form is 260,997.

39. Three hundred fifty-three in standard form is 353.

41. Four hundred eighty-four thousand, two hundred thirty-five in standard form is 484,235.

43. Fifty-four million, five hundred thousand in standard form is 54,500,000.

45. Seven hundred fourteen in standard form is 714.

47. $209 = 200 + 9$

49. $3470 = 3000 + 400 + 70$

51. $80,774 = 80,000 + 700 + 70 + 4$

53. $66,049 = 60,000 + 6000 + 40 + 9$

55. 39,680,000
$= 30,000,000 + 9,000,000 + 600,000 + 80,000$

57. The elevation of Mt. Clay is 5532 which is five thousand, five hundred thirty-two in words.

59. The height of Boott Spur is 5492 which is $5000 + 400 + 90 + 2$ in expanded form.

61. Mt. Washington is the tallest mountain in New England.

63. Chihuahua has fewer dogs registered than Golden retriever.

65. Labrador retrievers have the most registrations; 146,714 is written as one hundred forty-six thousand, seven hundred fourteen.

67. The maximum weight of an average-size Dachshund is 25 pounds.

69. The largest number is 9861.

71. No; 105.00 should be written as one hundred five.

73. answers may vary

75. 135.5 trillion in standard form is 135,500,000,000,000.

Section 1.3

Practice Problems

1. 4135
  $+ 252$
  ───────
  4387

3. Notice $12 + 8 = 20$ and $4 + 6 = 10$.
$12 + 4 + 8 + 6 + 5 = 20 + 10 + 5 = 35$

5. **a.** $14 - 6 = 8$ because $8 + 6 = 14$.

   b. $20 - 8 = 12$ because $12 + 8 = 20$

   c. $93 - 93 = 0$ because $0 + 93 = 93$.

   d. $42 - 0 = 42$ because $42 + 0 = 42$.

7. a.
```
  8 17
  6̷9̷ 7̷
  - 4 9
  -----
  64 8
```
Check:
```
  648
 + 49
 -----
  697
```

b.
```
  2 12
  3̷2̷6
  - 245
  -----
    81
```
Check:
```
   81
 + 245
 -----
  326
```

c.
```
  1234
 - 822
 -----
   412
```
Check:
```
   412
 + 822
 -----
  1234
```

9. 2 cm + 8 cm + 15 cm + 5 cm = 30 cm
The perimeter is 30 centimeters.

11.
```
  15,759
 -    458
 -------
  15,301
```
The radius of Neptune is 15,301 miles.

Calculator Explorations

1. 89 + 45 = 134

3. 285 + 55 = 340

5. 985 + 1210 + 562 + 77 = 2834

7. 865 − 95 = 770

9. 147 − 38 = 109

11. 9625 − 647 = 8978

Vocabulary and Readiness Check

1. The sum of 0 and any number is the same number.

3. The difference of any number and that same number is 0.

5. In 37 − 19 = 18, the number 37 is the minuend, the 19 is the subtrahend, and the 18 is the difference.

7. Since 7 + 10 = 10 + 7, we say that changing the order in addition does not change the sum. This property is called the commutative property of addition.

Exercise Set 1.3

1.
```
  14
 + 22
 ----
  36
```

3.
```
   62
 + 230
 -----
  292
```

5.
```
  12
  13
 + 24
 ----
  49
```

7.
```
  5267
 + 132
 -----
  5399
```

9.
```
   1  1 1
   22,781
 + 186,297
 --------
  209,078
```

11.
```
   8
   9
   2
   5
 + 1
 ----
  25
```

13.
```
  2 2
   81
   17
   23
   79
 + 12
 ————
  212
```

15.
```
  1  1 2
      24
    9006
     489
  + 2407
  ——————
  11,926
```

17.
```
  1 1  1
   6 820
   4 271
 + 5 626
 ———————
  16,717
```

19.
```
  1 1 2 2
       49
      628
    5 762
 + 29,462
 ————————
   35,901
```

21.
```
  1 2 2 2 1
   121,742
    57,279
    26,586
 + 426,782
 —————————
   632,389
```

23.
```
   62              Check:      1
  -37                        25
  ———                      + 37
   25                      ————
                             62
```

25.
```
   749            Check:    600
 - 149                    + 149
 —————                    —————
   600                      749
```

27.
```
   922            Check:    1 1
  -634                     288
  ————                   + 634
   288                   —————
                           922
```

29.
```
   600            Check:    1 1
  - 432                    168
  —————                  + 432
   168                   —————
                           600
```

31.
```
  6283            Check:      1
  - 560                    5723
  ——————                  + 560
   5723                   ——————
                            6283
```

33.
```
   533            Check:      1
  - 29                      504
  —————                    + 29
   504                    —————
                            533
```

35.
```
  1983            Check:       1
  - 1904                      79
  ——————                  + 1904
    79                    ——————
                            1983
```

37.
```
  50,000          Check:   1 1 1 1
  - 17,289                 32,711
  ———————                + 17,289
  32,711                 ————————
                           50,000
```

39.
```
  7020            Check:    1 1 1
  - 1979                    5041
  ——————                  + 1979
   5041                   ——————
                            7020
```

41.
```
  51,111          Check:   1 1 1 1
  - 19,898                 31,213
  ———————                + 19,898
  31,213                 ————————
                           51,111
```

43.
```
   986
  + 48
  —————
  1034
```

4

45.
$$\begin{array}{r} 76 \\ -\ 67 \\ \hline 9 \end{array}$$

47.
$$\begin{array}{r} 9000 \\ -\ 482 \\ \hline 8518 \end{array}$$

49.
$$\begin{array}{r} {\scriptstyle 1\ 1\ 1} \\ 10,962 \\ 4\ 851 \\ +\ 7\ 063 \\ \hline 22,876 \end{array}$$

51. $7 + 8 + 10 = 25$
The perimeter is 25 feet.

53. Opposite sides of a rectangle have the same length.
$4 + 8 + 4 + 8 = 12 + 12 = 24$
The perimeter is 24 inches.

55. $8 + 3 + 5 + 7 + 5 + 1 = 29$
The perimeter is 29 inches.

57. The unknown vertical side has length
$12 - 5 = 7$ meters. The unknown horizontal side has length $10 - 5 = 5$ meters.
$10 + 12 + 5 + 7 + 5 + 5 = 44$
The perimeter is 44 meters.

59. "Find the sum" indicates addition.
$$\begin{array}{r} {\scriptstyle 1\ 1\ 1} \\ 297 \\ +\ 1796 \\ \hline 2093 \end{array}$$
The sum of 297 and 1796 is 2093.

61. "Find the total" indicates addition.
$$\begin{array}{r} {\scriptstyle 1\ 3} \\ 76 \\ 39 \\ 8 \\ 17 \\ +\ 126 \\ \hline 266 \end{array}$$
The total of 76, 39, 8, 17, and 126 is 266.

63. "Find the difference" indicates subtraction.
$$\begin{array}{r} 41 \\ -\ 21 \\ \hline 20 \end{array}$$
The difference of 41 and 21 is 20.

65. "Increased by" indicates addition.
$$\begin{array}{r} {\scriptstyle 1} \\ 452 \\ +\ 92 \\ \hline 544 \end{array}$$
452 increased by 92 is 544.

67. "Less" indicates subtraction.
$$\begin{array}{r} 108 \\ -\ 36 \\ \hline 72 \end{array}$$
108 less 36 is 72.

69. "Subtracted from" indicates subtraction.
$$\begin{array}{r} 100 \\ -\ 12 \\ \hline 88 \end{array}$$
12 subtracted from 100 is 88.

71. Subtract 36,039 thousand from 38,067 thousand.
$$\begin{array}{r} 38,067 \\ -\ 36,039 \\ \hline 2\ 028 \end{array}$$
California's projected population increase is 2028 thousand.

73. Subtract the cost of the DVD player from the amount in her savings account.
$$\begin{array}{r} 914 \\ -\ 295 \\ \hline 619 \end{array}$$
She will have $619 left.

75.
$$\begin{array}{r} 189,000 \\ +\ 75,000 \\ \hline 264,000 \end{array}$$
The total U.S. land area drained by the Upper Mississippi and Lower Mississippi sub-basins is 264,000 square miles.

77. 530,000
     − 247,000
     ─────────
     283,000

The Missouri sub-basin drains 283,000 square miles more than the Arkansas Red-White sub-basin.

79. 70 + 78 + 90 + 102 = 340
The homeowner needs 340 feet of fencing.

81. 503
     − 239
     ────
     264

She must read 264 more pages.

83. 386,119
     + 426,990
     ─────────
     813,109

The total number sold in the U.S. in 2005 was 813,109.

85. Live rock music has a decibel level of 100 dB.

87. 88
     − 30
     ───
     58

The sound of snoring is 58 dB louder than normal conversation.

89.
     111
     5696
     + 3346
     ──────
     9042

There were 9042 stores worldwide.

91. Each side of a square has the same length.
31 + 31 + 31 + 31 = 124
The perimeter of the playing board is 124 feet.

93. California has the most Target stores.

95.
     11
     205
     121
     + 95
     ────
     421

The total number of Target stores in California, Texas, and Florida is 421 stores.

97. Florida and Georgia:
     1
     95
     + 44
     ────
     139

Michigan and Ohio:
     11
     53
     + 49
     ────
     102

Florida and Georgia have more Target stores.

99.
     1
     2029
     + 3865
     ──────
     5894

The total highway mileage in Delaware is 5894 miles.

101. The minuend is 48 and the subtrahend is 1.

103. The minuend is 70 and the subtrahend is 7.

105. answers may vary

107.
     1
     566
     932
     + 871
     ─────
     2369

The given sum is correct.

109.
```
      2 2
       14
      173
       86
    + 257
    -----
      530
```
The given sum is incorrect, the correct sum is 530.

111.
```
     1 1
     675
    +  56
    -----
     731
```
The given difference is incorrect.
```
     741
    −  56
    -----
     685
```

113.
```
     141
    + 888
    -----
    1029
```
The given difference is correct.

115.
```
     5269
    − 2385
    ------
     2884
```

117. answers may vary

119.
```
     1 2 1 3 2
    289,462
    369,477
    218,287
  + 121,685
  ---------
    998,911
```
Since 998,911 is less than one million, they did not reach their goal.
```
    1,000,000
    − 998,911
    ---------
        1 089
```
They need to read 1089 more pages.

Section 1.4

Practice Problems

1. a. To round 57 to the nearest ten, observe that the digit in the ones place is 7. Since the digit is at least 5, we add 1 to the digit in the tens place. The number 57 rounded to the nearest ten is 60.

    b. To round 641 to the nearest ten, observe that the digit in the ones place is 1. Since the digit is less than 5, we do not add 1 to the digit in the tens place. The number 641 rounded to the nearest ten is 640.

    c. To round 325 to the nearest ten observe that the digit in the ones place is 5. Since the digit is at least 5, we add 1 to the digit in the tens place. The number 325 rounded to the nearest ten is 330.

2. a. To round 72,304 to the nearest thousand, observe that the digit in the hundreds place is 3. Since the digit is less than 5, we do not add 1 to the digit in the thousands place. The number 72,304 rounded to the nearest thousand is 72,000.

    b. To round 9222 to the nearest thousand, observe that the digit in the hundreds place is 2. Since the digit is less than 5, we do not add 1 to the digit in the thousands place. The number 9222 rounded to the nearest thousand is 9000.

    c. To round 671,800 to the nearest thousand, observe that the digit in the hundreds place is 8. Since this digit is at least 5, we add 1 to the digit in the thousands place. The number 671,800 rounded to the nearest thousand is 672,000.

3. a. To round 3474 to the nearest hundred, observe that the digit in the tens place is 7. Since this digit is at least 5, we add 1 to the digit in the hundreds place. The number 3474 rounded to the nearest hundred is 3500.

b. To round 76,243 to the nearest hundred, observe that the digit in the tens place is 4. Since this digit is less than 5, we do not add 1 to the digit in the hundreds place. The number 76,243 rounded to the nearest hundred is 76,200.

c. To round 978,865 to the nearest hundred, observe that the digit in the tens place is 6. Since this digit is at least 5, we add 1 to the digit in the hundreds place. The number 978,865 rounded to the nearest hundred is 978,900.

4.

49	rounds to	50
25	rounds to	30
32	rounds to	30
51	rounds to	50
98	rounds to	+ 100
		260

5.

3785	rounds to	4000
− 2479	rounds to	− 2000
		2000

6.

11	rounds to	10
16	rounds to	20
19	rounds to	20
+ 31	rounds to	+ 30
		80

The total distance is approximately 80 miles.

7.

120,710	rounds to	120,000
22,878	rounds to	20,000
+ 45,974	rounds to	+ 50,000
		190,000

The total number of cases is approximately 190,000.

Vocabulary and Readiness Check

1. To graph a number on a number line, darken the point representing the location of the number.

3. The number 65 rounded to the nearest ten is 70 but the number 61 rounded to the nearest ten is 60.

Exercise Set 1.4

1. To round 423 to the nearest ten, observe that the digit in the ones place is 3. Since this digit is less than 5, we do not add 1 to the digit in the tens place. The number 423 rounded to the nearest ten is 420.

3. To round 635 to the nearest ten, observe that the digit in the ones place is 5. Since this digit is at least 5, we add 1 to the digit in the tens place. The number 635 rounded to the nearest ten is 640.

5. To round 2791 to the nearest hundred, observe that the digit in the tens place is 9. Since this digit is at least 5, we add 1 to the digit in the hundreds place. The number 2791 rounded to the nearest hundred is 2800.

7. To round 495 to the nearest ten, observe that the digit in the ones place is 5. Since this digit is at least 5, we add 1 to the digit in the tens place. The number 495 rounded to the nearest ten is 500.

9. To round 21,094 to the nearest thousand, observe that the digit in the hundreds place is 0. Since this digit is less than 5, we do not add 1 to the digit in the thousands place. The number 21,094 rounded to the nearest thousand is 21,000.

11. To round 33,762 to the nearest thousand, observe that the digit in the hundreds place is 7. Since this digit is at least 5, we add 1 to the digit in the thousands place. The number 33,762 rounded to the nearest thousand is 34,000.

13. To round 328,495 to the nearest hundred, observe that the digit in the tens place is 9. Since this digit is at least 5, we add 1 to the digit in the hundreds place. The number 328,495 rounded to the nearest hundred is 328,500.

15. To round 36,499 to the nearest thousand, observe that the digit in the hundreds place is 4. Since this digit is less than 5, we do not add 1 to the digit in the thousands place. The number 36,499 rounded to the nearest thousand is 36,000.

17. To round 39,994 to the nearest ten, observe that the digit in the ones place is 4. Since this digit is less than 5, we do not add 1 to the digit in the tens place. The number 39,994 rounded to the nearest ten is 39,990.

19. To round 29,834,235 to the nearest ten-million, observe that the digit in the millions place is 9. Since this digit is at least 5, we add 1 to the digit in the ten-millions place. The number 29,834,235 rounded to the nearest ten-million is 30,000,000.

21. Estimate 5281 to a given place value by rounding it to that place value. 5281 rounded to the tens place is 5280, to the hundreds place is 5300, and to the thousands place is 5000.

23. Estimate 9444 to a given place value by rounding it to that place value. 9444 rounded to the tens place is 9440, to the hundreds place is 9400, and to the thousands place is 9000.

25. Estimate 14,876 to a given place value by rounding it to that place value. 14,876 rounded to the tens place is 14,880, to the hundreds place is 14,900, and to the thousands place is 15,000.

27. To round 19,264 to the nearest thousand, observe that the digit in the hundreds place is 2. Since this digit is less than 5, we do not add 1 to the digit in the thousands place. Therefore, 19,264 rounded to the nearest thousand is 19,000.

29. To round 38,387 to the nearest thousand, observe that the digit in the hundreds place is 3. Since this digit is less than 5, we do not add 1 to the digit in the thousands place. Therefore, 38,387 points rounded to the nearest thousand is 38,000 points.

31. To round 67,520,000,000 to the nearest billion, observe that the digit in the hundred-millions place is 5. Since this digit is at least 5, we add 1 to the digit in the billions place. Therefore, $67,520,000,000 rounded to the nearest billion is $68,000,000,000.

33. To round 3,023,894 to the nearest hundred-thousand, observe that the digit in the ten-thousands place is 2. Since this digit is less than 5, we do not add 1 to the digit in the hundred-thousands place. Therefore, $3,023,894 rounded to the nearest hundred-thousand is $3,000,000.

35. U.S.: To round 207,897,000 to the nearest million, observe that the digit in the hundred-thousands place is 8. Since this digit is at least 5, we add 1 to the digit in the millions place. The number 207,897,000 rounded to the nearest million is 208,000,000.
India: To round 69,193,300 to the nearest million, observe that the digit in the hundred-thousands place is 1. Since this digit is less than 5, we do not add 1 to the digit in the millions place. The number 69,193,300 rounded to the nearest million is 69,000,000.

37.

39	rounds to	40
45	rounds to	50
22	rounds to	20
+ 17	rounds to	+ 20
		130

39.

449	rounds to	450
− 373	rounds to	− 370
		80

41.

1913	rounds to	1900
1886	rounds to	1900
+ 1925	rounds to	+ 1900
		5700

43.

1774	rounds to	1800
− 1492	rounds to	− 1500
		300

45.

3995	rounds to	4000
2549	rounds to	2500
+ 4944	rounds to	+ 4900
		11,400

47. $463 + 219$ is approximately
$460 + 220 = 680$.
The answer of 680 is incorrect because
$463 + 219 = 682$.

49. $229 + 443 + 606$ is approximately
$230 + 440 + 600 = 1270$.
The answer of 1278 is correct.

51. $7806 + 5150$ is approximately
$7800 + 5200 = 13,000$.
The answer of 12,956 is correct.

53.

899	rounds to	900
1499	rounds to	1500
+ 999	rounds to	+ 1000
		3400

The total cost is approximately $3400.

55.

1429	rounds to	1400
− 530	rounds to	− 500
		900

Boston in approximately 900 miles farther
from Kansas City than Chicago is.

57.

20,320	rounds to	20,000
− 14,410	rounds to	− 14,000
		6 000

The difference in elevation is approximately
6000 feet.

59.

782,623	rounds to	780,000
− 382,894	rounds to	− 380,000
		400,000

Jacksonville was approximately 400,000
larger than Miami.

61.

909,608	rounds to	910,000
− 906,993	rounds to	− 907,000
		3 000

The decrease in enrollment is approximately
3000 children.

63. $339 million is $339,000,000 in standard
form. $339,000,000 rounded to the nearest
ten-million is $340,000,000.
$339,000,000 rounded to the nearest
hundred-million is $300,000,000.

65. $179 million is $179,000,000 in standard
form.
$179,000,000 rounded to the nearest ten-
million is $180,000,000.
$179,000,000 rounded to the nearest
hundred-million is $200,000,000.

67. 5723, for example, rounded to the nearest
hundred is 5700.

69. a. The smallest possible number that
rounds to 8600 is 8550.

    b. The largest possible number that rounds
to 8600 is 8649.

71. answers may vary

73. 54 rounds to 50
17 rounds to 20
$50 + 20 + 50 + 20 = 140$
The perimeter is approximately 140 meters.

Section 1.5

Practice Problems

1. a. $6 \times 0 = 0$

    b. $(1)8 = 8$

c. $(50)(0) = 0$

d. $75 \cdot 1 = 75$

2. a. $6(4 + 5) = 6 \cdot 4 + 6 \cdot 5$

b. $30(2 + 3) = 30 \cdot 2 + 30 \cdot 3$

c. $7(2 + 8) = 7 \cdot 2 + 7 \cdot 8$

3. a.
$$\begin{array}{r} 29 \\ \times\ 6 \\ \hline 54 \\ 120 \\ \hline 174 \end{array}$$

b.
$$\begin{array}{r} 648 \\ \times\ 5 \\ \hline 40 \\ 200 \\ 3000 \\ \hline 3240 \end{array}$$

4.
$$\begin{array}{r} 306 \\ \times\ 81 \\ \hline 306 \\ 24\ 480 \\ \hline 24,786 \end{array}$$

5.
$$\begin{array}{r} 726 \\ \times\ 142 \\ \hline 1\ 452 \\ 29\ 040 \\ 72\ 600 \\ \hline 103,092 \end{array}$$

6. Area = length $\cdot$ width
  = (360 miles)(280 miles)
  = 100,800 square miles
The area of Wyoming is 100,800 square miles.

7.
$$\begin{array}{r} 16 \\ \times\ 45 \\ \hline 80 \\ 640 \\ \hline 720 \end{array}$$
The printer can print 720 pages in 45 minutes.

8. $8 \times 11 = 88$
$5 \times 9 = 45$

$$\begin{array}{r} 88 \\ +\ 45 \\ \hline 133 \end{array}$$

The total cost is \$133.

9.
$$\begin{array}{rl} 163 & \text{rounds to} \\ \times\ 391 & \text{rounds to} \end{array}$$
$$\begin{array}{r} 200 \\ \times\ 400 \\ \hline 80,000 \end{array}$$

There are approximately 80,000 words on 391 pages.

Calculator Explorations

1. $72 \times 48 = 3456$

3. $163 \cdot 94 = 15,322$

5. $983(277) = 272,291$

Vocabulary and Readiness Check

1. The product of 0 and any number is <u>0</u>.

3. In $8 \cdot 12 = 96$, the 96 is called the <u>product</u> and 8 and 12 are each called a <u>factor</u>.

5. Since $(3 \cdot 4) \cdot 6 = 3 \cdot (4 \cdot 6)$, we say that changing the <u>grouping</u> in multiplication does not change the product. This property is called the <u>associative</u> property of multiplication.

7. Area of a rectangle = <u>length</u> $\cdot$ width.

Exercise Set 1.5

1. $1 \cdot 24 = 24$

3. $0 \cdot 19 = 0$

5. $8 \cdot 0 \cdot 9 = 0$

7. $87 \cdot 1 = 87$

9. $6(3 + 8) = 6 \cdot 3 + 6 \cdot 8$

11. $4(3 + 9) = 4 \cdot 3 + 4 \cdot 9$

13. $20(14 + 6) = 20 \cdot 14 + 20 \cdot 6$

15.
$$
\begin{array}{r}
64 \\
\times\ 8 \\
\hline
512
\end{array}
$$

17.
$$
\begin{array}{r}
613 \\
\times\ 6 \\
\hline
3678
\end{array}
$$

19.
$$
\begin{array}{r}
277 \\
\times\ 6 \\
\hline
1662
\end{array}
$$

21.
$$
\begin{array}{r}
1074 \\
\times\ 6 \\
\hline
6444
\end{array}
$$

23.
$$
\begin{array}{r}
89 \\
\times\ 13 \\
\hline
267 \\
890 \\
\hline
1157
\end{array}
$$

25.
$$
\begin{array}{r}
421 \\
\times\ 58 \\
\hline
3\ 368 \\
21\ 050 \\
\hline
24,418
\end{array}
$$

27.
$$
\begin{array}{r}
306 \\
\times\ 81 \\
\hline
306 \\
24\ 480 \\
\hline
24,786
\end{array}
$$

29.
$$
\begin{array}{r}
780 \\
\times\ 20 \\
\hline
15,600
\end{array}
$$

31. $(495)(13)(0) = 0$

33. $(640)(1)(10) = (640)(10) = 6400$

35.
$$
\begin{array}{r}
1234 \\
\times\ 39 \\
\hline
11\ 106 \\
37\ 020 \\
\hline
48,126
\end{array}
$$

37.
$$
\begin{array}{r}
609 \\
\times\ 234 \\
\hline
2\ 436 \\
18\ 270 \\
121\ 800 \\
\hline
142,506
\end{array}
$$

39.
$$
\begin{array}{r}
8649 \\
\times\ 274 \\
\hline
34\ 596 \\
605\ 430 \\
1\ 729\ 800 \\
\hline
2,369,826
\end{array}
$$

41.
$$
\begin{array}{r}
589 \\
\times\ 110 \\
\hline
5\ 890 \\
58\ 900 \\
\hline
64,790
\end{array}
$$

43.
$$
\begin{array}{r}
1941 \\
\times\ 2035 \\
\hline
9\ 705 \\
58\ 230 \\
3\ 882\ 000 \\
\hline
3,949,935
\end{array}
$$

45. Area = (length)(width)
         = (9 meters)(7 meters)
         = 63 square meters

47. Area = (length)(width)
         = (40 feet)(17 feet)
         = 680 square feet

49.   576   rounds to   600
         × 354   rounds to   ×   400
                                  ─────────
                                  240,000

51.   604   rounds to   600
         × 451   rounds to   ×   500
                                  ─────────
                                  300,000

53. 38 × 42 is approximately 40 × 40, which is 1600. The best estimate is c.

55. 612 × 29 is approximately 600 × 30, which is 18,000.
         The best estimate is c.

57. $80 \times 11 = (8 \times 10) \times 11$
         $= 8 \times (10 \times 11)$
         $= 8 \times 110$
         $= 880$

59. $6 \times 700 = 4200$

61.   2240
         ×   2
         ──────
          4480

63.   125
         ×   3
         ──────
           375
         There are 375 calories in 3 tablespoons of olive oil.

65.   94
         × 35
         ──────
          470
         2820
         ──────
         3290
         The total cost is $3290.

67. a. $4 \times 5 = 20$
            There are 20 boxes in one layer.

    b.   20
             × 5
             ──────
              100
            There are 100 boxes on the pallet.

    c.   100
            × 20
            ──────
             2000
            The weight of the cheese on the pallet is 2000 pounds.

69. Area = (length)(width)
         = (110 feet)(80 feet)
         = 8800 square feet
         The area is 8800 square feet.

71. Area = (length)(width)
         = (350 feet)(160 feet)
         = 56,000 square feet
         The area is 56,000 square feet.

73.   94
         × 62
         ──────
          188
         5640
         ──────
         5828
         There are 5828 pixels on the screen.

75.   60
         × 35
         ──────
          300
         1 800
         ──────
         2 100
         There are 2100 characters in 35 lines.

77.
$$\begin{array}{r} 160 \\ \times\ \ 8 \\ \hline 1280 \end{array}$$
There are 1280 calories in 8 ounces.

79.

T-Shirt Size	Number of Shirts Ordered	Cost per Shirt	Cost per Size Ordered
S	4	$10	$40
M	6	$10	$60
L	20	$10	$200
XL	3	$12	$36
XXL	3	$12	$36
Total Cost			$372

81. There are 31 days in March.
$$\begin{aligned} 31 \times 700,000 &= 31 \times 7 \times 100,000 \\ &= 217 \times 100,000 \\ &= 21,700,000 \end{aligned}$$
They would use 21,700,000 quarts in March.

83.
$$\begin{array}{r} 128 \\ +\ \ 7 \\ \hline 135 \end{array}$$

85.
$$\begin{array}{r} 134 \\ \times 16 \\ \hline 804 \\ 1340 \\ \hline 2144 \end{array}$$

87.
$$\begin{array}{r} 19 \\ +\ 4 \\ \hline 23 \end{array}$$
The sum of 19 and 4 is 23.

89.
$$\begin{array}{r} 19 \\ -\ 4 \\ \hline 15 \end{array}$$
The difference of 19 and 4 is 15.

91. $6 + 6 + 6 + 6 + 6 = 5 \cdot 6$ or $6 \cdot 5$

93. **a.** $3 \cdot 5 = 5 + 5 + 5 = 3 + 3 + 3 + 3 + 3$

   b. answers may vary

95.
$$\begin{array}{r} 203 \\ \times\ 14 \\ \hline 812 \\ 2030 \\ \hline 2842 \end{array}$$

97. $42 \times 3 = 126$
$42 \times 9 = 378$
The problem is
$$\begin{array}{r} 42 \\ \times 93 \end{array}$$

99. answers may vary

101. On a side with 7 windows per row, there are $7 \times 23 = 161$ windows. On a side with 4 windows per row, there are $4 \times 23 = 92$ windows.
$161 + 161 + 92 + 92 = 506$
There are 506 windows on the building.

Section 1.6

Practice Problems

1. **a.** $9\overline{)72}$ with 8 on top, because $8 \cdot 9 = 72$.

   b. $40 \div 5 = 8$ because $8 \cdot 5 = 40$.

   c. $\dfrac{24}{6} = 4$ because $4 \cdot 6 = 24$.

2. **a.** $\dfrac{7}{7} = 1$ because $1 \cdot 7 = 7$.

b. $5 \div 1 = 5$ because $5 \cdot 1 = 5$.

c. $\overset{11}{1\overline{)11}}$ because $11 \cdot 1 = 11$.

d. $4 \div 1 = 4$ because $4 \cdot 1 = 4$.

e. $\dfrac{10}{1} = 10$ because $10 \cdot 1 = 10$.

f. $21 \div 21 = 1$ because $1 \cdot 21 = 21$.

3. a. $\dfrac{0}{5} = 0$ because $0 \cdot 5 = 0$.

b. $\overset{0}{8\overline{)0}}$ because $0 \cdot 8 = 0$.

c. $7 \div 0$ is undefined because if $7 \div 0$ is a number, then the number times 0 would be 7.

d. $0 \div 14 = 0$ because $0 \cdot 14 = 0$.

4. a.
$$\begin{array}{r} 818 \\ 6\overline{)4908} \\ -48 \\ \hline 10 \\ -6 \\ \hline 48 \\ -48 \\ \hline 0 \end{array}$$

Check:
$$\begin{array}{r} 818 \\ \times \quad 6 \\ \hline 4908 \end{array}$$

b.
$$\begin{array}{r} 553 \\ 4\overline{)2212} \\ -20 \\ \hline 21 \\ -20 \\ \hline 12 \\ -12 \\ \hline 0 \end{array}$$

Check:
$$\begin{array}{r} 553 \\ \times \quad 4 \\ \hline 2212 \end{array}$$

c.
$$\begin{array}{r} 251 \\ 3\overline{)753} \\ -6 \\ \hline 15 \\ -15 \\ \hline 03 \\ -3 \\ \hline 0 \end{array}$$

Check:
$$\begin{array}{r} 251 \\ \times \quad 3 \\ \hline 753 \end{array}$$

5. a.
$$\begin{array}{r} 304 \\ 7\overline{)2128} \\ -21 \\ \hline 02 \\ -0 \\ \hline 28 \\ -28 \\ \hline 0 \end{array}$$

Check: $304 \times 7 = 2128$

b.
$$\begin{array}{r} 5\,100 \\ 9\overline{)45{,}900} \\ -45 \\ \hline 0\,9 \\ -9 \\ \hline 000 \end{array}$$

Check: $5100 \times 9 = 45{,}900$

6. a.
$$
\begin{array}{r}
234 \text{ R } 3 \\
4\overline{)\,939} \\
\underline{-8} \\
13 \\
\underline{-12} \\
19 \\
\underline{-16} \\
3
\end{array}
$$
Check: $234 \cdot 4 + 3 = 939$

b.
$$
\begin{array}{r}
657 \text{ R } 2 \\
5\overline{)\,3287} \\
\underline{-30} \\
28 \\
\underline{-25} \\
37 \\
\underline{-35} \\
2
\end{array}
$$
Check: $657 \cdot 5 + 2 = 3287$

7. a.
$$
\begin{array}{r}
9067 \text{ R } 2 \\
9\overline{)\,81,605} \\
\underline{-81} \\
0\,6 \\
\underline{-0} \\
60 \\
\underline{-54} \\
65 \\
\underline{-63} \\
2
\end{array}
$$
Check: $9067 \cdot 9 + 2 = 81,605$

b.
$$
\begin{array}{r}
5827 \text{ R } 2 \\
4\overline{)\,23,310} \\
\underline{-20} \\
3\,3 \\
\underline{-3\,2} \\
11 \\
\underline{-8} \\
30 \\
\underline{-28} \\
2
\end{array}
$$
Check: $5827 \cdot 4 + 2 = 23,310$

8.
$$
\begin{array}{r}
524 \text{ R } 12 \\
17\overline{)\,8920} \\
\underline{-85} \\
42 \\
\underline{-34} \\
80 \\
\underline{-68} \\
12
\end{array}
$$

9.
$$
\begin{array}{r}
49 \text{ R } 60 \\
678\overline{)\,33,282} \\
\underline{-27\,12} \\
6\,162 \\
\underline{-6\,102} \\
60
\end{array}
$$

10.
$$
\begin{array}{r}
57 \\
3\overline{)\,171} \\
\underline{-15} \\
21 \\
\underline{-21} \\
0
\end{array}
$$
Each student got 57 CDs.

11.
$$
\begin{array}{r}
44 \\
12\overline{)\,532} \\
\underline{-48} \\
52 \\
\underline{-48} \\
4
\end{array}
$$
There will be 44 full boxes and 4 printers left over.

12. Find the sum and divide by 7.

$$
\begin{array}{r}
4 \\
7 \\
35 \\
16 \\
9 \\
3 \\
\underline{+\,52} \\
126
\end{array}
\qquad
\begin{array}{r}
18 \\
7\overline{)\,126} \\
\underline{-7} \\
56 \\
\underline{-56} \\
0
\end{array}
$$
The average time is 18 minutes.

Calculator Explorations

1. $848 \div 16 = 53$

3. $5890 \div 95 = 62$

5. $\dfrac{32,886}{126} = 261$

7. $0 \div 315 = 0$

Vocabulary and Readiness Check

1. In $90 \div 2 = 45$, the answer 45 is called the quotient, 90 is called the dividend, and 2 is called the divisor.

3. The quotient of any number (except 0) and the same number is 1.

5. The quotient of any number and 0 is undefined.

Exercise Set 1.6

1. $54 \div 9 = 6$

3. $36 \div 3 = 12$

5. $0 \div 8 = 0$

7. $31 \div 1 = 31$

9. $\dfrac{18}{18} = 1$

11. $\dfrac{24}{3} = 8$

13. $26 \div 0 =$ undefined.

15. $26 \div 26 = 1$

17. $0 \div 14 = 0$

19. $18 \div 2 = 9$

21.
$$
\begin{array}{r}
29 \\
3{\overline{\smash{\big)}\,87}} \\
\underline{-6} \\
27 \\
\underline{-27} \\
0
\end{array}
$$
Check: $3 \cdot 29 = 87$

23.
$$
\begin{array}{r}
74 \\
3{\overline{\smash{\big)}\,222}} \\
\underline{-21} \\
12 \\
\underline{-12} \\
0
\end{array}
$$
Check: $74 \cdot 3 = 222$

25.
$$
\begin{array}{r}
338 \\
3{\overline{\smash{\big)}\,1014}} \\
\underline{-9} \\
11 \\
\underline{-9} \\
24 \\
\underline{-24} \\
0
\end{array}
$$
Check: $3 \cdot 338 = 1014$

27. $\dfrac{30}{0}$ is undefined.

29.
$$
\begin{array}{r}
9 \\
7{\overline{\smash{\big)}\,63}} \\
\underline{-63} \\
0
\end{array}
$$
Check: $7 \cdot 9 = 63$

31.
$$
\begin{array}{r}
25 \\
6{\overline{\smash{\big)}\,150}} \\
\underline{-12} \\
30 \\
\underline{-30} \\
0
\end{array}
$$
Check: $25 \cdot 6 = 150$

33.
$$
\begin{array}{r}
68 \text{ R } 3 \\
7{\overline{)479}} \\
\underline{-42} \\
59 \\
\underline{-56} \\
3
\end{array}
$$
Check: $7 \cdot 68 + 3 = 479$

35.
$$
\begin{array}{r}
236 \text{ R } 5 \\
6{\overline{)1421}} \\
\underline{-12} \\
22 \\
\underline{-18} \\
41 \\
\underline{-36} \\
5
\end{array}
$$
Check: $236 \cdot 6 + 5 = 1421$

37.
$$
\begin{array}{r}
38 \text{ R } 1 \\
8{\overline{)305}} \\
\underline{-24} \\
65 \\
\underline{-64} \\
1
\end{array}
$$
Check: $8 \cdot 38 + 1 = 305$

39.
$$
\begin{array}{r}
326 \text{ R } 4 \\
7{\overline{)2286}} \\
\underline{-21} \\
18 \\
\underline{-14} \\
46 \\
\underline{-42} \\
4
\end{array}
$$
Check: $326 \cdot 7 + 4 = 2286$

41.
$$
\begin{array}{r}
13 \\
55{\overline{)715}} \\
\underline{-55} \\
165 \\
\underline{-165} \\
0
\end{array}
$$
Check: $55 \cdot 13 = 715$

43.
$$
\begin{array}{r}
49 \\
23{\overline{)1127}} \\
\underline{-92} \\
207 \\
\underline{-207} \\
0
\end{array}
$$
Check: $49 \cdot 23 = 1127$

45.
$$
\begin{array}{r}
97 \text{ R } 8 \\
97{\overline{)9417}} \\
\underline{-873} \\
687 \\
\underline{-679} \\
8
\end{array}
$$
Check: $97 \cdot 97 + 8 = 9417$

47.
$$
\begin{array}{r}
209 \text{ R } 11 \\
15{\overline{)3146}} \\
\underline{-30} \\
14 \\
\underline{-0} \\
146 \\
\underline{-135} \\
11
\end{array}
$$
Check: $209 \cdot 15 + 11 = 3146$

49.
$$
\begin{array}{r}
506 \\
13{\overline{)6578}} \\
\underline{-65} \\
07 \\
\underline{-0} \\
78 \\
\underline{-78} \\
0
\end{array}
$$
Check: $13 \cdot 506 = 6578$

51.
$$
\begin{array}{r}
202 \text{ R } 7 \\
46\overline{)9299} \\
-92 \\
\overline{09} \\
-0 \\
\overline{99} \\
-92 \\
\overline{7}
\end{array}
$$

Check: $202 \cdot 46 + 7 = 9299$

53.
$$
\begin{array}{r}
54 \\
236\overline{)12744} \\
-1180 \\
\overline{944} \\
-944 \\
\overline{0}
\end{array}
$$

Check: $236 \cdot 54 = 12{,}744$

55.
$$
\begin{array}{r}
99 \text{ R } 100 \\
103\overline{)10{,}297} \\
-9\,27 \\
\overline{1\,027} \\
-927 \\
\overline{100}
\end{array}
$$

Check: $99 \cdot 103 + 100 = 10{,}297$

57.
$$
\begin{array}{r}
202 \text{ R } 15 \\
102\overline{)20619} \\
-204 \\
\overline{21} \\
-0 \\
\overline{219} \\
-204 \\
\overline{15}
\end{array}
$$

Check: $102 \cdot 202 + 15 = 20{,}619$

59.
$$
\begin{array}{r}
579 \text{ R } 72 \\
423\overline{)244{,}989} \\
-211\,5 \\
\overline{33\,48} \\
-29\,61 \\
\overline{3\,879} \\
-3\,807 \\
\overline{72}
\end{array}
$$

Check: $579 \cdot 423 + 72 = 244{,}989$

61.
$$
\begin{array}{r}
17 \\
7\overline{)119} \\
-7 \\
\overline{49} \\
-49 \\
\overline{0}
\end{array}
$$

63.
$$
\begin{array}{r}
511 \text{ R } 3 \\
7\overline{)3580} \\
-35 \\
\overline{08} \\
-7 \\
\overline{10} \\
-7 \\
\overline{3}
\end{array}
$$

65.
$$
\begin{array}{r}
2132 \text{ R } 32 \\
40\overline{)85312} \\
-80 \\
\overline{53} \\
-40 \\
\overline{131} \\
-120 \\
\overline{112} \\
-80 \\
\overline{32}
\end{array}
$$

67.
$$
\begin{array}{r}
6\ 080 \\
142\overline{)863{,}360} \\
-852 \\
\hline
11\ 3 \\
-0 \\
\hline
11\ 36 \\
-11\ 36 \\
\hline
00 \\
-0 \\
\hline
0
\end{array}
$$

69.
$$
\begin{array}{r}
23\ R\ 2 \\
5\overline{)117} \\
-10 \\
\hline
17 \\
-15 \\
\hline
2
\end{array}
$$

The quotient is 23 R 2.

71.
$$
\begin{array}{r}
5\ R\ 25 \\
35\overline{)200} \\
-175 \\
\hline
25
\end{array}
$$

200 divided by 35 is 5 R 25.

73.
$$
\begin{array}{r}
20\ R\ 2 \\
3\overline{)62} \\
-6 \\
\hline
02 \\
-0 \\
\hline
2
\end{array}
$$

The quotient is 20 R 2.

75.
$$
\begin{array}{r}
33 \\
65\overline{)2145} \\
-195 \\
\hline
195 \\
-195 \\
\hline
0
\end{array}
$$

There are 33 students in the group.

77.
$$
\begin{array}{r}
165 \\
318\overline{)52470} \\
-318 \\
\hline
2067 \\
-1908 \\
\hline
1590 \\
-1590 \\
\hline
0
\end{array}
$$

The person weighs 165 pounds on Earth.

79.
$$
\begin{array}{r}
310 \\
18\overline{)5580} \\
-54 \\
\hline
18 \\
-18 \\
\hline
0
\end{array}
$$

The distance is 310 yards.

81.
$$
\begin{array}{r}
88\ R\ 1 \\
3\overline{)265} \\
-24 \\
\hline
25 \\
-24 \\
\hline
1
\end{array}
$$

There are 88 bridges every 3 miles over the 265 miles, plus the first bridge, for a total of 89 bridges.

83.
$$
\begin{array}{r}
10 \\
492\overline{)5280} \\
-492 \\
\hline
360
\end{array}
$$

There should be 10 poles, plus the first pole for a total of 11 light poles.

85.
$$
\begin{array}{r}
5 \\
5280\overline{)26400} \\
-26400 \\
\hline
0
\end{array}
$$

Broad Peak is 5 miles tall.

87.
$$\begin{array}{r} 1760 \\ 3\overline{)\,5280} \\ \underline{-3} \\ 22 \\ \underline{-21} \\ 18 \\ \underline{-18} \\ 0 \end{array}$$

There are 1760 yards in 1 mile.

89.
$$\begin{array}{r} 2 \\ 10 \\ 24 \\ 35 \\ 22 \\ 17 \\ \underline{+12} \\ 120 \end{array}
\qquad
\begin{array}{r} 20 \\ 6\overline{)\,120} \\ \underline{-12} \\ 00 \end{array}$$

Average $= \dfrac{120}{6} = 20$

91.
$$\begin{array}{r} 1 \\ 205 \\ 972 \\ 210 \\ \underline{+161} \\ 1548 \end{array}
\qquad
\begin{array}{r} 387 \\ 4\overline{)\,1548} \\ \underline{-12} \\ 34 \\ \underline{-32} \\ 28 \\ \underline{-28} \\ 0 \end{array}$$

Average $= \dfrac{1548}{4} = 387$

93.
$$\begin{array}{r} 2 \\ 86 \\ 79 \\ 81 \\ 69 \\ \underline{+80} \\ 395 \end{array}
\qquad
\begin{array}{r} 79 \\ 5\overline{)\,395} \\ \underline{-35} \\ 45 \\ \underline{-45} \\ 0 \end{array}$$

Average $= \dfrac{395}{5} = 79$

95.
$$\begin{array}{r} 2 \\ 69 \\ 77 \\ \underline{+76} \\ 222 \end{array}
\qquad
\begin{array}{r} 74 \\ 3\overline{)\,222} \\ \underline{-21} \\ 12 \\ \underline{-12} \\ 0 \end{array}$$

The average temperature is 74°.

97.
$$\begin{array}{r} 1\,1\,1 \\ 82 \\ 463 \\ 29 \\ \underline{+8704} \\ 9278 \end{array}$$

99.
$$\begin{array}{r} 546 \\ \times\quad 28 \\ \hline 4\,368 \\ 10\,920 \\ \hline 15,288 \end{array}$$

101.
$$\begin{array}{r} 722 \\ -\ 43 \\ \hline 679 \end{array}$$

103. $\dfrac{45}{0}$ is undefined.

105.
$$\begin{array}{r} 9 \text{ R } 12 \\ 24\overline{)\,228} \\ \underline{-216} \\ 12 \end{array}$$

107. The quotient of 40 and 8 is $40 \div 8$, which is choice c.

109. 200 divided by 20 is $200 \div 20$, which is choice b.

111.
$$
\begin{array}{r}
797,000,000 \\
+\ 667,000,000 \\
\hline
1,464,000,000
\end{array}
$$

$$
\begin{array}{r}
732,000,000 \\
2\overline{)\ 1,464,000,000} \\
-1\ 4 \\
\hline
06 \\
-6 \\
\hline
04 \\
-4 \\
\hline
0
\end{array}
$$

The average amount spent is $732,000,000.

113. The average will increase; answers may vary.

115. No; answers may vary
Possible answer: The average cannot be less than each of the four numbers.

117. answers may vary

The Bigger Picture

1.
$$
\begin{array}{r}
\overset{1}{8}2 \\
+\ 39 \\
\hline
121
\end{array}
$$

2.
$$
\begin{array}{r}
82 \\
-\ 39 \\
\hline
43
\end{array}
$$

3.
$$
\begin{array}{r}
82 \\
\times\ 39 \\
\hline
738 \\
2460 \\
\hline
3198
\end{array}
$$

4.
$$
\begin{array}{r}
89\ R\ 11 \\
29\overline{)\ 2592} \\
-232 \\
\hline
272 \\
-261 \\
\hline
11
\end{array}
$$

5. $0 \cdot 15 = 0$

6. $0 \div 11 = 0$

7. $26 \cdot 1 = 26$

8. $36 \div 0$ is undefined.

9. $82 \div 1 = 82$

10.
$$
\begin{array}{r}
2000 \\
-\ 156 \\
\hline
1844
\end{array}
$$

Integrated Review

1.
$$
\begin{array}{r}
\overset{1}{4}2 \\
63 \\
+\ 89 \\
\hline
194
\end{array}
$$

2.
$$
\begin{array}{r}
7006 \\
-\ 451 \\
\hline
6555
\end{array}
$$

3.
$$
\begin{array}{r}
87 \\
\times\ 52 \\
\hline
174 \\
4350 \\
\hline
4524
\end{array}
$$

4.
$$
\begin{array}{r}
562 \\
8\overline{)\ 4496} \\
-40 \\
\hline
49 \\
-48 \\
\hline
16 \\
-16 \\
\hline
0
\end{array}
$$

5. $1 \cdot 67 = 67$

6. $\dfrac{36}{0}$ is undefined.

7. $16 \div 16 = 1$

8. $5 \div 1 = 5$

9. $0 \cdot 21 = 0$

10. $7 \cdot 0 \cdot 8 = 0$

11. $0 \div 7 = 0$

12. $12 \div 4 = 3$

13. $9 \cdot 7 = 63$

14. $45 \div 5 = 9$

15.
$$\begin{array}{r} 207 \\ -\ 69 \\ \hline 138 \end{array}$$

16.
$$\begin{array}{r} {\scriptstyle 1} \\ 207 \\ +\ 69 \\ \hline 276 \end{array}$$

17.
$$\begin{array}{r} 3718 \\ -2549 \\ \hline 1169 \end{array}$$

18.
$$\begin{array}{r} {\scriptstyle 11} \\ 1861 \\ +7965 \\ \hline 9826 \end{array}$$

19.
$$\begin{array}{r} 182 \text{ R } 4 \\ 7{\overline{\smash{)}\,1278}} \\ \underline{-7} \\ 57 \\ \underline{-56} \\ 18 \\ \underline{-14} \\ 4 \end{array}$$

20.
$$\begin{array}{r} 1259 \\ \times\ \ \ 63 \\ \hline 3\ 777 \\ 75\ 540 \\ \hline 79{,}317 \end{array}$$

21.
$$\begin{array}{r} 1099 \text{ R } 2 \\ 7{\overline{\smash{)}\,7695}} \\ \underline{-7} \\ 06 \\ \underline{-0} \\ 69 \\ \underline{-63} \\ 65 \\ \underline{-63} \\ 2 \end{array}$$

22.
$$\begin{array}{r} 111 \text{ R } 1 \\ 9{\overline{\smash{)}\,1000}} \\ \underline{-9} \\ 10 \\ \underline{-9} \\ 10 \\ \underline{-9} \\ 1 \end{array}$$

23.
$$\begin{array}{r} 663 \text{ R } 24 \\ 32{\overline{\smash{)}\,21{,}240}} \\ \underline{-19\ 2} \\ 2\ 04 \\ \underline{-1\ 92} \\ 120 \\ \underline{-96} \\ 24 \end{array}$$

24.
$$\begin{array}{r} 1\ 076 \text{ R } 60 \\ 65{\overline{\smash{)}\,70{,}000}} \\ \underline{-65} \\ 5\ 0 \\ \underline{-\ 0} \\ 5\ 00 \\ \underline{-4\ 55} \\ 450 \\ \underline{-390} \\ 60 \end{array}$$

25.
$$\begin{array}{r} 4000 \\ -\ 2963 \\ \hline 1037 \end{array}$$

26.
$$\begin{array}{r} 10,000 \\ -\ \ \ 101 \\ \hline 9\ 899 \end{array}$$

27.
$$\begin{array}{r} 303 \\ \times\ \ 101 \\ \hline 303 \\ 30\ 300 \\ \hline 30,603 \end{array}$$

28. $(475)(100) = 47,500$

29.
$$\begin{array}{r} 1 \\ 62 \\ +\ \ 9 \\ \hline 71 \end{array}$$
The total of 62 and 9 is 71.

30.
$$\begin{array}{r} 62 \\ \times\ \ 9 \\ \hline 558 \end{array}$$
The product of 62 and 9 is 558.

31.
$$\begin{array}{r} 6\ R\ 8 \\ 9)\overline{\ 62} \\ \underline{-54} \\ 8 \end{array}$$
The quotient of 62 and 9 is 6 R 8.

32.
$$\begin{array}{r} 62 \\ -\ \ 9 \\ \hline 53 \end{array}$$
The difference of 62 and 9 is 53.

33.
$$\begin{array}{r} 200 \\ -\ \ 17 \\ \hline 183 \end{array}$$
17 subtracted from 200 is 183.

34.
$$\begin{array}{r} 432 \\ -\ 201 \\ \hline 231 \end{array}$$
The difference of 432 and 201 is 231.

35. 9735 rounded to the nearest ten is 9740.
9735 rounded to the nearest hundred is 9700.
9735 rounded to the nearest thousand is 10,000.

36. 1429 rounded to the nearest ten is 1430.
1429 rounded to the nearest hundred is 1400.
1429 rounded to the nearest thousand is 1000.

37. 20,801 rounded to the nearest ten is 20,800.
20,801 rounded to the nearest hundred is 20,800.
20,801 rounded to the nearest thousand is 21,000.

38. 432,198 rounded to the nearest ten is 432,200.
432,198 rounded to the nearest hundred is 432,200.
432,198 rounded to the nearest thousand is 432,000.

39. $6 + 6 + 6 + 6 = 24$
$6 \times 6 = 36$
The perimeter is 24 feet and the area is 36 square feet.

40. $14 + 7 + 14 + 7 = 42$
$$\begin{array}{r} 14 \\ \times\ \ 7 \\ \hline 98 \end{array}$$
The perimeter is 42 inches and the area is 98 square inches.

41.
$$\begin{array}{r} 13 \\ 9 \\ +\ \ 6 \\ \hline 28 \end{array}$$
The perimeter is 28 miles.

42. The unknown vertical side has length
$4 + 3 = 7$ meters. The unknown horizontal
side has length $3 + 3 = 6$ meters.

$$
\begin{array}{r}
3 \\
4 \\
3 \\
7 \\
6 \\
+\,3 \\
\hline
26
\end{array}
$$

The perimeter is 26 meters.

43.
$$
\begin{array}{r}
3 \\
19 \\
15 \\
25 \\
37 \\
+\,24 \\
\hline
120
\end{array}
\qquad
\begin{array}{r}
24 \\
5\,\overline{)\,120} \\
-10 \\
\hline
20 \\
-20 \\
\hline
0
\end{array}
$$

$$\text{Average} = \frac{120}{5} = 24$$

44.
$$
\begin{array}{r}
1\,2 \\
108 \\
131 \\
98 \\
+\,159 \\
\hline
496
\end{array}
\qquad
\begin{array}{r}
124 \\
4\,\overline{)\,496} \\
-4 \\
\hline
09 \\
-8 \\
\hline
16 \\
-16 \\
\hline
0
\end{array}
$$

$$\text{Average} = \frac{496}{4} = 124$$

45.
$$
\begin{array}{r}
28,547 \\
-\,26,372 \\
\hline
2\,175
\end{array}
$$

The Lake Pontchartrain Bridge is longer by
2175 feet.

46.
$$
\begin{array}{r}
347 \\
\times\ 18 \\
\hline
2776 \\
3470 \\
\hline
6246
\end{array}
$$

The amount spent on toys is $6246.

Section 1.7

Practice Problems

1. $8 \cdot 8 \cdot 8 \cdot 8 = 8^4$

2. $3 \cdot 3 \cdot 3 = 3^3$

3. $10 \cdot 10 \cdot 10 \cdot 10 \cdot 10 = 10^5$

4. $5 \cdot 5 \cdot 4 \cdot 4 \cdot 4 \cdot 4 \cdot 4 \cdot 4 = 5^2 \cdot 4^6$

5. $4^2 = 4 \cdot 4 = 16$

6. $8^3 = 8 \cdot 8 \cdot 8 = 512$

7. $11^1 = 11$

8. $2 \cdot 3^2 = 2 \cdot 3 \cdot 3 = 18$

9. $9 \cdot 3 - 8 \div 4 = 27 - 8 \div 4 = 27 - 2 = 25$

10. $48 \div 3 \cdot 2^2 = 48 \div 3 \cdot 4 = 16 \cdot 4 = 64$

11. $(10 - 7)^4 + 2 \cdot 3^2 = 3^4 + 2 \cdot 3^2$
$$= 81 + 2 \cdot 9$$
$$= 81 + 18$$
$$= 99$$

12. $36 \div [20 - (4 \cdot 2)] + 4^3 - 6$
$$= 36 \div [20 - 8] + 4^3 - 6$$
$$= 36 \div 12 + 4^3 - 6$$
$$= 36 \div 12 + 64 - 6$$
$$= 3 + 64 - 6$$
$$= 61$$

13. $\dfrac{25+8\cdot2-3^3}{2(3-2)} = \dfrac{25+8\cdot2-27}{2(1)}$

$\qquad\qquad\quad = \dfrac{25+16-27}{2}$

$\qquad\qquad\quad = \dfrac{14}{2}$

$\qquad\qquad\quad = 7$

14. $36\div6\cdot3+5 = 6\cdot3+5 = 18+5 = 23$

15. Area $=(\text{side})^2$

$\qquad\quad = (12 \text{ centimeters})^2$

$\qquad\quad = 144 \text{ square centimeters}$

The area of the square is 144 square centimeters.

Calculator Explorations

1. $3^6 = 729$

3. $4^5 = 1024$

5. $2^{11} = 2048$

7. $7^4 + 5^3 = 2526$

9. $63\cdot75 - 43\cdot10 = 4295$

11. $4(15\div3+2) - 10\cdot2 = 8$

Vocabulary and Readiness Check

1. In $2^5 = 32$, the 2 is called the <u>base</u> and the 5 is called the <u>exponent</u>.

3. To simplify $(8+2)\cdot6$, which operation should be performed first? <u>addition</u>

5. To simplify $8\div2\cdot6$, which operation should be performed first? <u>division</u>

Exercise Set 1.7

1. $4\cdot4\cdot4 = 4^3$

3. $7\cdot7\cdot7\cdot7\cdot7\cdot7 = 7^6$

5. $12\cdot12\cdot12 = 12^3$

7. $6\cdot6\cdot5\cdot5\cdot5 = 6^2\cdot5^3$

9. $9\cdot8\cdot8 = 9\cdot8^2$

11. $3\cdot2\cdot2\cdot2\cdot2 = 3\cdot2^4$

13. $3\cdot2\cdot2\cdot2\cdot2\cdot5\cdot5\cdot5\cdot5\cdot5 = 3\cdot2^4\cdot5^5$

15. $8^2 = 8\cdot8 = 64$

17. $5^3 = 5\cdot5\cdot5 = 125$

19. $2^5 = 2\cdot2\cdot2\cdot2\cdot2 = 32$

21. $1^{10} = 1\cdot1\cdot1\cdot1\cdot1\cdot1\cdot1\cdot1\cdot1\cdot1 = 1$

23. $7^1 = 7$

25. $2^7 = 2\cdot2\cdot2\cdot2\cdot2\cdot2\cdot2 = 128$

27. $2^8 = 2\cdot2\cdot2\cdot2\cdot2\cdot2\cdot2\cdot2 = 256$

29. $4^4 = 4\cdot4\cdot4\cdot4 = 256$

31. $9^3 = 9\cdot9\cdot9 = 729$

33. $12^2 = 12\cdot12 = 144$

35. $10^2 = 10\cdot10 = 100$

37. $20^1 = 20$

39. $3^6 = 3\cdot3\cdot3\cdot3\cdot3\cdot3 = 729$

41. $3\cdot2^6 = 3\cdot2\cdot2\cdot2\cdot2\cdot2\cdot2 = 192$

43. $2\cdot3^4 = 2\cdot3\cdot3\cdot3\cdot3 = 162$

45. $15+3\cdot2 = 15+6 = 21$

47. $28 \div 7 \cdot 1 + 3 = 4 \cdot 1 + 3 = 4 + 3 = 7$

49. $32 \div 4 - 3 = 8 - 3 = 5$

51. $13 + \dfrac{24}{8} = 13 + 3 = 16$

53. $6 \cdot 5 + 8 \cdot 2 = 30 + 16 = 46$

55. $\dfrac{5 + 12 \div 4}{1^7} = \dfrac{5+3}{1} = \dfrac{8}{1} = 8$

57. $(7 + 5^2) \div 4 \cdot 2^3 = (7 + 25) \div 4 \cdot 2^3$
$= 32 \div 4 \cdot 2^3$
$= 32 \div 4 \cdot 8$
$= 8 \cdot 8$
$= 64$

59. $5^2 \cdot (10 - 8) + 2^3 + 5^2 = 5^2 \cdot 2 + 2^3 + 5^2$
$= 25 \cdot 2 + 8 + 25$
$= 50 + 8 + 25$
$= 83$

61. $\dfrac{18 + 6}{2^4 - 2^2} = \dfrac{24}{16 - 4} = \dfrac{24}{12} = 2$

63. $(3 + 5) \cdot (9 - 3) = 8 \cdot 6 = 48$

65. $\dfrac{7(9 - 6) + 3}{3^2 - 3} = \dfrac{7(3) + 3}{9 - 3} = \dfrac{21 + 3}{6} = \dfrac{24}{6} = 4$

67. $8 \div 0 + 37 =$ undefined

69. $2^4 \cdot 4 - (25 \div 5) = 2^4 \cdot 4 - 5$
$= 16 \cdot 4 - 5$
$= 64 - 5$
$= 59$

71. $3^4 - [35 - (12 - 6)] = 3^4 - [35 - 6]$
$= 3^4 - 29$
$= 81 - 29$
$= 52$

73. $(7 \cdot 5) + [9 \div (3 \div 3)] = (7 \cdot 5) + [9 \div (1)]$
$= 35 + 9$
$= 44$

75. $8 \cdot [2^2 + (6 - 1) \cdot 2] - 50 \cdot 2$
$= 8 \cdot (2^2 + 5 \cdot 2) - 50 \cdot 2$
$= 8 \cdot (4 + 5 \cdot 2) - 50 \cdot 2$
$= 8 \cdot (4 + 10) - 50 \cdot 2$
$= 8 \cdot 14 - 50 \cdot 2$
$= 112 - 50 \cdot 2$
$= 112 - 100$
$= 12$

77. $\dfrac{9^2 + 2^2 - 1^2}{8 \div 2 \cdot 3 \cdot 1 \div 3} = \dfrac{81 + 4 - 1}{4 \cdot 3 \cdot 1 \div 3}$
$= \dfrac{85 - 1}{12 \cdot 1 \div 3}$
$= \dfrac{84}{12 \div 3}$
$= \dfrac{84}{4}$
$= 21$

79. $\dfrac{2 + 4^2}{5(20 - 16) - 3^2 - 5} = \dfrac{2 + 16}{5(4) - 3^2 - 5}$
$= \dfrac{18}{5(4) - 9 - 5}$
$= \dfrac{18}{20 - 9 - 5}$
$= \dfrac{18}{11 - 5}$
$= \dfrac{18}{6}$
$= 3$

81. $9 \div 3 + 5^2 \cdot 2 - 10 = 9 \div 3 + 25 \cdot 2 - 10$
$= 3 + 25 \cdot 2 - 10$
$= 3 + 50 - 10$
$= 43$

83. $[13 \div (20 - 7) + 2^5] - (2 + 3)^2$

$= [13 \div 13 + 2^5] - 5^2$

$= [13 \div 13 + 32] - 5^2$

$= [1 + 32] - 5^2$

$= 33 - 5^2$

$= 33 - 25$

$= 8$

85. $7^2 - \{18 - [40 \div (5 \cdot 1) + 2] + 5^2\}$

$= 7^2 - \{18 - [40 \div 5 + 2] + 5^2\}$

$= 7^2 - \{18 - [8 + 2] + 5^2\}$

$= 7^2 - \{18 - 10 + 5^2\}$

$= 7^2 - \{18 - 10 + 25\}$

$= 7^2 - 33$

$= 49 - 33$

$= 16$

87. Area of a square $= (\text{side})^2$

$= (7 \text{ meters})^2$

$= 49$ square meters

89. Area of a square $= (\text{side})^2$

$= (23 \text{ miles})^2$

$= 529$ square miles

91. The statement is true.

93. $2^5 = 2 \cdot 2 \cdot 2 \cdot 2 \cdot 2$

The statement is false.

95. $(2 + 3) \cdot 6 - 2 = 5 \cdot 6 - 2 = 30 - 2 = 28$

97. $24 \div (3 \cdot 2) + 2 \cdot 5 = 24 \div 6 + 2 \cdot 5 = 4 + 10 = 14$

99. The unknown vertical length is
$30 - 12 = 18$ feet. The unknown horizontal
length is $60 - 40 = 20$ feet.
Perimeter $= 60 + 30 + 40 + 18 + 20 + 12$
$= 180$
The total perimeter of seven homes is
$7(180) = 1260$ feet.

101. $(7 + 2^4)^5 - (3^5 - 2^4)^2$

$= (7 + 16)^5 - (243 - 16)^2$

$= 23^5 - 227^2$

$= 6,436,343 - 51,529$

$= 6,384,814$

103. answers may vary; possible answer:
$(20 - 10) \cdot 5 \div 25 + 3 = 10 \cdot 5 \div 25 + 3$
$= 50 \div 25 + 3$
$= 2 + 3$
$= 5$

The Bigger Picture

1. $6^3 = 6 \cdot 6 \cdot 6 = 216$

2. $2^3 \cdot 6^1 = 2 \cdot 2 \cdot 2 \cdot 6 = 48$

3. $8 \cdot 5^2 = 8 \cdot 5 \cdot 5 = 200$

4. $1 + 2(3 + 4) = 1 + 2(7) = 1 + 14 = 15$

5. $2 + 5(10 - 3) = 2 + 5(7) = 2 + 35 = 37$

6. $200 \div 2 \cdot 2 = 100 \cdot 2 = 200$

7.
$$\begin{array}{r} 978 \\ -179 \\ \hline 799 \end{array}$$

8.
$$\begin{array}{r} 72 \\ \times\ 30 \\ \hline 2160 \end{array}$$

9.
$$\begin{array}{r} 10 \text{ R } 34 \\ 58\overline{)614} \\ \underline{-58} \\ 34 \\ \underline{-0} \\ 34 \end{array}$$

10. $3[(7-3)^2 - (25-22)^2] + 6$
$= 3[4^2 - 3^2] + 6$
$= 3(16 - 9) + 6$
$= 3(7) + 6$
$= 21 + 6$
$= 27$

Section 1.8

Practice Problems

1. $x - 2 = 7 - 2 = 5$

2. $y(x - 3) = 4(8 - 3) = 4(5) = 20$

3. $\dfrac{y+6}{x} = \dfrac{18+6}{6} = \dfrac{24}{6} = 4$

4. $25 - z^3 + x = 25 - 2^3 + 1 = 25 - 8 + 1 = 18$

5. $\dfrac{5(F-32)}{9} = \dfrac{5(41-32)}{9} = \dfrac{5(9)}{9} = \dfrac{45}{9} = 5$

6. $3(y - 6) = 6$
$3(8 - 6) \stackrel{?}{=} 6$
$3(2) \stackrel{?}{=} 6$
$6 = 6$ True
Yes, 8 is a solution.

7. $5n + 4 = 34$
Let n be 10.
$5(10) + 4 \stackrel{?}{=} 34$
$50 + 4 \stackrel{?}{=} 34$
$54 = 34$ False
No, 10 is not a solution.
Let n be 6.
$5(6) + 4 \stackrel{?}{=} 34$
$30 + 4 \stackrel{?}{=} 34$
$34 = 34$ True
Yes, 6 is a solution.
Let n be 8.
$5(8) + 4 \stackrel{?}{=} 34$
$40 + 4 \stackrel{?}{=} 34$
$44 = 34$ False
No, 8 is not a solution.

8. a. Twice a number is $2x$.

   b. 8 increased by a number is $8 + x$ or $x + 8$.

   c. 10 minus a number is $10 - x$.

   d. 10 subtracted from a number is $x - 10$.

   e. The quotient of 6 and a number is $6 \div x$ or $\dfrac{6}{x}$.

Vocabulary and Readiness Check

   1. A combination of operations on letters (variables) and numbers is an <u>expression</u>.

   3. $3x - 2y$ is called an <u>expression</u> and the letters x and y are <u>variables</u>.

   5. A statement of the form "expression = expression" is called an <u>equation</u>.

Exercise Set 1.8

1.

a	b	$a + b$	$a - b$	$a \cdot b$	$a \div b$
21	7	$21 + 7 = 28$	$21 - 7 = 14$	$21 \cdot 7 = 147$	$21 \div 7 = 3$

3.

a	b	$a + b$	$a - b$	$a \cdot b$	$a \div b$
152	0	$152 + 0 = 152$	$152 - 0 = 152$	$152 \cdot 0 = 0$	$152 \div 0$ is undefined.

5.

a	b	$a + b$	$a - b$	$a \cdot b$	$a \div b$
56	1	$56 + 1 = 57$	$56 - 1 = 55$	$56 \cdot 1 = 56$	$56 \div 1 = 56$

7. $3 + 2z = 3 + 2(3) = 3 + 6 = 9$

9. $3xz - 5x = 3(2)(3) - 5(2) = 18 - 10 = 8$

11. $z - x + y = 3 - 2 + 5 = 1 + 5 = 6$

13. $4x - z = 4(2) - 3 = 8 - 3 = 5$

15. $y^3 - 4x = 5^3 - 4(2) = 125 - 4(2) = 125 - 8 = 117$

17. $\begin{aligned} 2xy^2 - 6 &= 2(2)(5)^2 - 6 \\ &= 2 \cdot 2 \cdot 25 - 6 \\ &= 100 - 6 \\ &= 94 \end{aligned}$

19. $8 - (y - x) = 8 - (5 - 2) = 8 - 3 = 5$

21. $x^5 + (y - z) = 2^5 + (5 - 3)$
$$= 2^5 + 2$$
$$= 32 + 2$$
$$= 34$$

23. $\dfrac{6xy}{z} = \dfrac{6 \cdot 2 \cdot 5}{3} = \dfrac{60}{3} = 20$

25. $\dfrac{2y - 2}{x} = \dfrac{2(5) - 2}{2} = \dfrac{10 - 2}{2} = \dfrac{8}{2} = 4$

27. $\dfrac{x + 2y}{z} = \dfrac{2 + 2 \cdot 5}{3} = \dfrac{2 + 10}{3} = \dfrac{12}{3} = 4$

29. $\dfrac{5x}{y} - \dfrac{10}{y} = \dfrac{5(2)}{5} - \dfrac{10}{5} = \dfrac{10}{5} - 2 = 2 - 2 = 0$

31. $2y^2 - 4y + 3 = 2 \cdot 5^2 - 4 \cdot 5 + 3$
$$= 2 \cdot 25 - 4 \cdot 5 + 3$$
$$= 50 - 20 + 3$$
$$= 33$$

33. $(4y - 5z)^3 = (4 \cdot 5 - 5 \cdot 3)^3$
$$= (20 - 15)^3$$
$$= (5)^3$$
$$= 125$$

35. $(xy + 1)^2 = (2 \cdot 5 + 1)^2 = (10 + 1)^2 = 11^2 = 121$

37. $2y(4z - x) = 2 \cdot 5(4 \cdot 3 - 2)$
$$= 2 \cdot 5(12 - 2)$$
$$= 2 \cdot 5(10)$$
$$= 10(10)$$
$$= 100$$

39. $xy(5 + z - x) = 2 \cdot 5(5 + 3 - 2)$
$$= 2 \cdot 5(6)$$
$$= 10(6)$$
$$= 60$$

41. $\dfrac{7x+2y}{3x}=\dfrac{7(2)+2(5)}{3(2)}=\dfrac{14+10}{6}=\dfrac{24}{6}=4$

43.

t	1	2	3	4
$16t^2$	$16\cdot1^2=16\cdot1=16$	$16\cdot2^2=16\cdot4=64$	$16\cdot3^2=16\cdot9=144$	$16\cdot4^2=16\cdot16=256$

45. Let n be 10.

$n-8=2$

$10-8\overset{?}{=}2$

$\qquad 2=2$ True

Yes, 10 is a solution.

47. Let n be 3.

$24=80n$

$24\overset{?}{=}80\cdot3$

$24=240$ False

No, 3 is not a solution.

49. Let n be 7.

$3n-5=10$

$3(7)-5\overset{?}{=}10$

$21-5\overset{?}{=}10$

$\qquad\ \ 16=10$ False

No, 7 is not a solution.

51. Let n be 20.

$2(n-17)=6$

$2(20-17)\overset{?}{=}6$

$\qquad\ 2(3)\overset{?}{=}6$

$\qquad\qquad 6=6$ True

Yes, 20 is a solution.

53. Let x be 0.

$5x+3=4x+13$

$5(0)+3\overset{?}{=}4(0)+13$

$0+3\overset{?}{=}0+13$

$\qquad\ 3=13$ False

No, 0 is not a solution.

55. Let f be 8.

$7f=64-f$

$7(8)\overset{?}{=}64-8$

$\ \ 56=56$ True

Yes, 8 is a solution.

57. $n - 2 = 10$
Let n be 10.
$10 - 2 \stackrel{?}{=} 10$
$\quad\quad 8 = 10$ False
Let n be 12.
$12 - 2 \stackrel{?}{=} 10$
$\quad\quad 10 = 10$ True
Let n be 14.
$14 - 2 \stackrel{?}{=} 10$
$\quad\quad 12 = 10$ False
12 is a solution.

59. $5n = 30$
Let n be 6.
$5 \cdot 6 \stackrel{?}{=} 30$
$\quad 30 = 30$ True
Let n be 25.
$5 \cdot 25 \stackrel{?}{=} 30$
$\quad 125 = 30$ False
Let n be 30.
$5 \cdot 30 \stackrel{?}{=} 30$
$\quad 150 = 30$ False
6 is a solution.

61. $6n + 2 = 26$
Let n be 0.
$6(0) + 2 \stackrel{?}{=} 26$
$\quad 0 + 2 \stackrel{?}{=} 26$
$\quad\quad 2 = 26$ False
Let n be 2.
$6(2) + 2 \stackrel{?}{=} 26$
$\quad 12 + 2 \stackrel{?}{=} 26$
$\quad\quad 14 = 26$ False
Let n be 4.
$6(4) + 2 \stackrel{?}{=} 26$
$\quad 24 + 2 \stackrel{?}{=} 26$
$\quad\quad 26 = 26$ True
4 is a solution.

63. $3(n - 4) = 10$
Let n be 5.
$3(5 - 4) \stackrel{?}{=} 10$
$\quad 3(1) \stackrel{?}{=} 10$
$\quad\quad 3 = 10$ False
Let n be 7.

$3(7 - 4) \stackrel{?}{=} 10$
$\quad 3(3) \stackrel{?}{=} 10$
$\quad\quad 9 = 10$ False
Let n be 10.
$3(10 - 4) \stackrel{?}{=} 10$
$\quad 3(6) \stackrel{?}{=} 10$
$\quad\quad 18 = 10$ False
None are solutions.

65. $7x - 9 = 5x + 13$
Let x be 3.
$7(3) - 9 \stackrel{?}{=} 5(3) + 13$
$\quad 21 - 9 \stackrel{?}{=} 15 + 13$
$\quad\quad 12 = 28$ False
Let x be 7.
$7(7) - 9 \stackrel{?}{=} 5(7) + 13$
$\quad 49 - 9 \stackrel{?}{=} 35 + 13$
$\quad\quad 40 = 48$ False
Let x be 11.
$7(11) - 9 \stackrel{?}{=} 5(11) + 13$
$\quad 77 - 9 \stackrel{?}{=} 55 + 13$
$\quad\quad 68 = 68$ True
11 is a solution.

67. Eight more than a number is $x + 8$.

69. The total of a number and 8 is $x + 8$.

71. Twenty decreased by a number is $20 - x$.

73. The product of 512 and a number is $512x$.

75. The quotient of six and a number is $\dfrac{6}{x}$.

77. The sum of seventeen and a number added to the product of five and the number is $5x + (17 + x)$.

79. The product of five and a number is $5x$.

81. A number subtracted from 11 is $11 - x$.

83. A number less 5 is $x - 5$.

85. 6 divided by a number is $6 \div x$ or $\dfrac{6}{x}$.

87. Fifty decreased by eight times a number is $50 - 8x$.

89. $x^4 - y^2 = 23^4 - 72^2$
$$= 279,841 - 5184$$
$$= 274,657$$

91. $x^2 + 5y - 112 = 23^2 + 5(72) - 112$
$$= 529 + 360 - 112$$
$$= 777$$

93. $5x$ is the largest; answers may vary.

95. As t gets larger, $16t^2$ gets larger.

Chapter 1 Vocabulary Check

1. The whole numbers are 0, 1, 2, 3, ...

2. The perimeter of a polygon is its distance around or the sum of the lengths of its sides.

3. The position of each digit in a number determines its place value.

4. An exponent is a shorthand notation for repeated multiplication of the same factor.

5. To find the area of a rectangle, multiply length times width.

6. The digits used to write numbers are 0, 1, 2, 3, 4, 5, 6, 7, 8, and 9.

7. A letter used to represent a number is called a variable.

8. An equation can be written in the form "expression = expression."

9. A combination of operations on variables and numbers is called an expression.

10. A solution of an equation is a value of the variable that makes the equation a true statement.

11. A collection of numbers (or objects) enclosed by braces is called a set.

12. The 21 above is called the sum.

13. The 5 above is called the divisor.

14. The 35 above is called the dividend.

15. The 7 above is called the quotient.

16. The 3 above is called a factor.

17. The 6 above is called the product.

18. The 20 above is called the minuend.

19. The 9 above is called the subtrahend.

20. The 11 above is called the difference.

21. The 4 above is called an addend.

Chapter 1 Review

1. The place value of 4 in 7640 is tens.

2. The place value of 4 in 46,200,120 is ten-millions.

3. 7640 is written as seven thousand, six hundred forty.

4. 46,200,120 is written as forty-six million, two hundred thousand, one hundred twenty.

5. $3158 = 3000 + 100 + 50 + 8$

6. $403,225,000 = 400,000,000 + 3,000,000$
$$+ 200,000 + 20,000 + 5000$$

7. Eighty-one thousand, nine hundred in standard form is 81,900.

8. Six billion, three hundred four million in standard form is 6,304,000,000.

9. Locate Houston, Texas in the city column and read across to the number in the 2000 column. The population was 1,953,631.

10. Locate Los Angeles, California in the city column and read across to the number in the 2005 column. the population was 3,844,829.

11. Locate the smallest number in the 2005 column. San Jose, CA had the smallest population in 2005.

12. Locate the largest number in the 2005 column. New York, NY had the largest population in 2005.

13.
```
  1
  18
+ 49
────
  67
```

14.
```
  1
  28
+ 39
────
  67
```

15.
```
  462
- 397
─────
   65
```

16.
```
  583
- 279
─────
  304
```

17.
```
  428
+  21
─────
  449
```

18.
```
   1
  819
+  21
─────
  840
```

19.
```
  4000
-   86
──────
  3914
```

20.
```
  8000
-   92
──────
  7908
```

21.
```
  1 2 1
     91
   3623
+   497
───────
   4211
```

22.
```
   1 1
      82
    1647
+    238
────────
    1967
```

23.
```
   1 1
      74
     342
+    918
────────
    1334
```
The sum of 74, 342, and 918 is 1334.

24.
```
     2
     49
    529
+   308
───────
    886
```
The sum of 49, 529, and 308 is 886.

25.
```
  25,862
-  7 965
────────
  17,897
```
7965 subtracted from 25,862 is 17,897.

26.
```
  39,007
-  4 349
────────
  34,658
```
4349 subtracted from 39,007 is 34,658.

27.
```
     1
     205
+   7318
────────
    7523
```
The total distance is 7523 miles.

28.
$$\begin{array}{r} \overset{1\ 1\ 1}{62,589} \\ 65,340 \\ +\ 69,770 \\ \hline 197,699 \end{array}$$
Her total earnings were $197,699.

29. $40 + 52 + 52 + 72 = 216$
The perimeter is 216 feet.

30. $11 + 20 + 35 = 66$
The perimeter is 66 kilometers.

31.
$$\begin{array}{r} 1,256,509 \\ -\ 1,144,646 \\ \hline 111,863 \end{array}$$
The population increased by 111,863.

32.
$$\begin{array}{r} 1,517,550 \\ -\ 1,463,281 \\ \hline 54,269 \end{array}$$
The population decreased by 54,269.

33. Find the shortest bar. The balance was the least in May.

34. Find the tallest bar. The balance was the greatest in August.

35.
$$\begin{array}{r} 280 \\ -\ 170 \\ \hline 110 \end{array}$$
The balance decreased by $110 from February to April.

36.
$$\begin{array}{r} 490 \\ -\ 250 \\ \hline 240 \end{array}$$
The balance increased by $240 from June to August.

37. To round 43 to the nearest ten, observe that the digit in the ones place is 3. Since this digit is less than 5, we do not add 1 to the digit in the tens place. The number 43 rounded to the nearest ten is 40.

38. To round 45 to the nearest ten, observe that the digit in the ones place is 5. Since this digit is at least 5, we add 1 to the digit in the tens place. The number 45 rounded to the nearest ten is 50.

39. To round 876 to the nearest ten, observe that the digit in the ones place is 6. Since this digit is at least 5, we add 1 to the digit in the tens place. The number 876 rounded to the nearest ten is 880.

40. To round 493 to the nearest hundred, observe that the digit in the tens place is 9. Since this digit is at least 5, we add 1 to the digit in the hundreds place. The number 493 rounded to the nearest hundred is 500.

41. To round 3829 to the nearest hundred, observe that the digit in the tens place is 2. Since this digit is less than 5, we do not add 1 to the digit in the hundreds place. The number 3829 rounded to the nearest hundred is 3800.

42. To round 57,534 to the nearest thousand, observe that the digit in the hundreds place is 5. Since this digit is at least 5, we add 1 to the digit in the thousands place. The number 57,534 rounded to the nearest thousand is 58,000.

43. To round 39,583,819 to the nearest million, observe that the digit in the hundred-thousands place is 5. Since this digit is at least 5, we add 1 to the digit in the millions place. The number 39,583,819 rounded to the nearest million is 40,000,000.

44. To round 768,542 to the nearest hundred-thousand, observe that the digit in the ten-thousands place is 6. Since this digit is at least 5, we add 1 to the digit in the hundred-thousands place. The number 768,542 rounded to the nearest hundred-thousand is 800,000.

45.

		2
3785	rounds to	3800
648	rounds to	600
+ 2866	rounds to	+ 2900
		7300

46.

5925	rounds to	5900
− 1787	rounds to	− 1800
		4100

47.

630	rounds to	600
192	rounds to	200
271	rounds to	300
56	rounds to	100
703	rounds to	700
454	rounds to	500
+ 329	rounds to	+ 300
		2700

They traveled approximately 2700 miles.

48.

1,461,575	rounds to	1,500,000
− 1,213,825	rounds to	− 1,200,000
		300,000

The population of Phoenix is approximately 300,000 larger than that of Dallas.

49.

$$\begin{array}{r} 276 \\ \times\ \ 8 \\ \hline 2208 \end{array}$$

50.

$$\begin{array}{r} 349 \\ \times\ \ 4 \\ \hline 1396 \end{array}$$

51.

$$\begin{array}{r} 57 \\ \times\ 40 \\ \hline 2280 \end{array}$$

52.

$$\begin{array}{r} 69 \\ \times\ 42 \\ \hline 138 \\ 2760 \\ \hline 2898 \end{array}$$

53. $20(7)(4) = 140(4) = 560$

54. $25(9)(4) = 225(4) = 900$
or
$25(4)(9) = 100(9) = 900$

55. $26 \cdot 34 \cdot 0 = 0$

56. $62 \cdot 88 \cdot 0 = 0$

57.

$$\begin{array}{r} 586 \\ \times\ \ \ 29 \\ \hline 5\ 274 \\ 11\ 720 \\ \hline 16,994 \end{array}$$

58.

$$\begin{array}{r} 242 \\ \times\ \ 37 \\ \hline 1694 \\ 7260 \\ \hline 8954 \end{array}$$

59.

$$\begin{array}{r} 642 \\ \times\ \ \ 177 \\ \hline 4\ 494 \\ 44\ 940 \\ 64\ 200 \\ \hline 113,634 \end{array}$$

60.

$$\begin{array}{r} 347 \\ \times\ \ 129 \\ \hline 3\ 123 \\ 6\ 940 \\ 34\ 700 \\ \hline 44,763 \end{array}$$

61.

$$\begin{array}{r} 1026 \\ \times\ \ \ 401 \\ \hline 1\ 026 \\ 410\ 400 \\ \hline 411,426 \end{array}$$

62.
$$\begin{array}{r} 2107 \\ \times\ \ \ 302 \\ \hline 4\ 214 \\ 632\ 100 \\ \hline 636,314 \end{array}$$

63. "Product" indicates multiplication.
$$\begin{array}{r} 250 \\ \times\ \ \ 6 \\ \hline 1500 \end{array}$$
The product of 6 and 250 is 1500.

64. "Product" indicates multiplication.
$$\begin{array}{r} 820 \\ \times\ \ \ 6 \\ \hline 4920 \end{array}$$
The product of 6 and 820 is 4920.

65.
$$\begin{array}{r} 32 \\ \times\ 15 \\ \hline 160 \\ 320 \\ \hline 480 \end{array} \qquad \begin{array}{r} 38 \\ \times 11 \\ \hline 38 \\ 380 \\ \hline 418 \end{array}$$

$$\begin{array}{r} 480 \\ +\ 418 \\ \hline 898 \end{array}$$
The total cost is $898.

66.
$$\begin{array}{r} 4820 \\ \times\ \ \ 20 \\ \hline 96,400 \end{array}$$
The total cost is $96,400.

67. Area = (length)(width)
= (13 miles)(7 miles)
= 91 square miles

68. Area = (length)(width)
= (25 centimeters)(20 centimeters)
= 500 square centimeters

69. $\dfrac{49}{7} = 7$ Check: $\begin{array}{r} 7 \\ \times\ 7 \\ \hline 49 \end{array}$

70. $\dfrac{36}{9} = 4$ Check: $\begin{array}{r} 9 \\ \times\ 4 \\ \hline 36 \end{array}$

71. $\begin{array}{r} 5\text{ R }2 \\ 5{\overline{\smash{)}\,27}} \\ \underline{-25} \\ 2 \end{array}$

Check: $5 \times 5 + 2 = 27$

72. $\begin{array}{r} 4\text{ R }2 \\ 4{\overline{\smash{)}\,18}} \\ \underline{-16} \\ 2 \end{array}$

Check: $4 \times 4 + 2 = 18$

73. $918 \div 0$ is undefined.

74. $0 \div 668 = 0$ Check: $0 \cdot 668 = 0$

75. $\begin{array}{r} 33\text{ R }2 \\ 5{\overline{\smash{)}\,167}} \\ \underline{-15} \\ 17 \\ \underline{-15} \\ 2 \end{array}$

Check: $33 \times 5 + 2 = 167$

76. $\begin{array}{r} 19\text{ R }7 \\ 8{\overline{\smash{)}\,159}} \\ \underline{-8} \\ 79 \\ \underline{-72} \\ 7 \end{array}$

Check: $19 \times 8 + 7 = 159$

77.
$$
\begin{array}{r}
24\text{ R }2\\
26\overline{)626}\\
\underline{-52}\\
106\\
\underline{-104}\\
2
\end{array}
$$
Check: $24 \times 26 + 2 = 626$

78.
$$
\begin{array}{r}
35\text{ R }15\\
19\overline{)680}\\
\underline{-57}\\
110\\
\underline{-95}\\
15
\end{array}
$$
Check: $35 \times 19 + 15 = 680$

79.
$$
\begin{array}{r}
506\text{ R }10\\
47\overline{)23{,}792}\\
\underline{-23\ 5}\\
29\\
\underline{-0}\\
292\\
\underline{-282}\\
10
\end{array}
$$
Check: $506 \times 47 + 10 = 23{,}792$

80.
$$
\begin{array}{r}
907\text{ R }40\\
53\overline{)48{,}111}\\
\underline{-47\ 7}\\
41\\
\underline{-0}\\
411\\
\underline{-371}\\
40
\end{array}
$$
Check: $907 \times 53 + 40 = 48{,}111$

81.
$$
\begin{array}{r}
2793\text{ R }140\\
207\overline{)578{,}291}\\
\underline{-414}\\
164\ 2\\
\underline{-144\ 9}\\
19\ 39\\
\underline{-18\ 63}\\
761\\
\underline{-621}\\
140
\end{array}
$$
Check: $2793 \times 207 + 140 = 578{,}291$

82.
$$
\begin{array}{r}
2012\text{ R }60\\
306\overline{)615{,}732}\\
\underline{-612}\\
3\ 7\\
\underline{-0}\\
3\ 73\\
\underline{-3\ 06}\\
672\\
\underline{-612}\\
60
\end{array}
$$
Check: $2012 \times 306 + 60 = 615{,}732$

83.
$$
\begin{array}{r}
18\text{ R }2\\
5\overline{)92}\\
\underline{-5}\\
42\\
\underline{-40}\\
2
\end{array}
$$
The quotient of 92 and 5 is 18 R 2.

84.
$$
\begin{array}{r}
21\text{ R }2\\
4\overline{)86}\\
\underline{-8}\\
06\\
\underline{-4}\\
2
\end{array}
$$
The quotient of 86 and 4 is 21 R 2.

85.
$$\begin{array}{r} 27 \\ 24\overline{)\,648} \\ \underline{-48} \\ 168 \\ \underline{-168} \\ 0 \end{array}$$

27 boxes can be filled with cans of corn.

86.
$$\begin{array}{r} 13 \\ 1760\overline{)\,22{,}880} \\ \underline{-17\,60} \\ 5\,280 \\ \underline{-5\,280} \\ 0 \end{array}$$

There are 13 miles in 22,880 yards.

87. Divide the sum by 4.

$$\begin{array}{r} 76 \\ 49 \\ 32 \\ +\,47 \\ \hline 204 \end{array} \qquad \begin{array}{r} 51 \\ 4\overline{)\,204} \\ \underline{-20} \\ 04 \\ \underline{-4} \\ 0 \end{array}$$

The average is 51.

88. Divide the sum by 4.

$$\begin{array}{r} 23 \\ 85 \\ 62 \\ +\,66 \\ \hline 236 \end{array} \qquad \begin{array}{r} 59 \\ 4\overline{)\,236} \\ \underline{-20} \\ 36 \\ \underline{-36} \\ 0 \end{array}$$

The average is 59.

89. $8^2 = 8 \cdot 8 = 64$

90. $5^3 = 5 \cdot 5 \cdot 5 = 125$

91. $5 \cdot 9^2 = 5 \cdot 9 \cdot 9 = 405$

92. $4 \cdot 10^2 = 4 \cdot 10 \cdot 10 = 400$

93. $18 \div 2 + 7 = 9 + 7 = 16$

94. $12 - 8 \div 4 = 12 - 2 = 10$

95. $\dfrac{5(6^2 - 3)}{3^2 + 2} = \dfrac{5(36 - 3)}{9 + 2} = \dfrac{5(33)}{11} = \dfrac{165}{11} = 15$

96. $\dfrac{7(16 - 8)}{2^3} = \dfrac{7(8)}{8} = \dfrac{56}{8} = 7$

97. $48 \div 8 \cdot 2 = 6 \cdot 2 = 12$

98. $27 \div 9 \cdot 3 = 3 \cdot 3 = 9$

99.
$$\begin{aligned} 2 + 3[1^5 + (20 - 17) \cdot 3] + 5 \cdot 2 \\ = 2 + 3[1^5 + 3 \cdot 3] + 5 \cdot 2 \\ = 2 + 3[1 + 3 \cdot 3] + 5 \cdot 2 \\ = 2 + 3[1 + 9] + 5 \cdot 2 \\ = 2 + 3 \cdot 10 + 5 \cdot 2 \\ = 2 + 30 + 10 \\ = 42 \end{aligned}$$

100.
$$\begin{aligned} 21 - [2^4 - (7 - 5) - 10] + 8 \cdot 2 \\ = 21 - [2^4 - 2 - 10] + 8 \cdot 2 \\ = 21 - [16 - 2 - 10] + 8 \cdot 2 \\ = 21 - 4 + 8 \cdot 2 \\ = 21 - 4 + 16 \\ = 33 \end{aligned}$$

101.
$$\begin{aligned} 19 - 2(3^2 - 2^2) &= 19 - 2(9 - 4) \\ &= 19 - 2(5) \\ &= 19 - 10 \\ &= 9 \end{aligned}$$

102.
$$\begin{aligned} 16 - 2(4^2 - 3^2) &= 16 - 2(16 - 9) \\ &= 16 - 2(7) \\ &= 16 - 14 \\ &= 2 \end{aligned}$$

103. $4 \cdot 5 - 2 \cdot 7 = 10 - 14 = 6$

104. $8 \cdot 7 - 3 \cdot 9 = 56 - 27 = 29$

105. $(6-4)^3 \cdot [10^2 \div (3+17)]$
$= (6-4)^3 \cdot [10^2 \div 20]$
$= (6-4)^3 \cdot [100 \div 20]$
$= 2^3 \cdot 5$
$= 8 \cdot 5$
$= 40$

106. $(7-5)^3 \cdot [9^2 \div (2+7)] = (7-5)^3 \cdot [9^2 \div 9]$
$= (7-5)^3 \cdot [81 \div 9]$
$= 2^3 \cdot 9$
$= 8 \cdot 9$
$= 72$

107. $\dfrac{5 \cdot 7 - 3 \cdot 5}{2(11-3^2)} = \dfrac{35-15}{2(11-9)} = \dfrac{20}{2(2)} = \dfrac{20}{4} = 5$

108. $\dfrac{4 \cdot 8 - 1 \cdot 11}{3(9-2^3)} = \dfrac{32-11}{3(9-8)} = \dfrac{21}{3(1)} = \dfrac{21}{3} = 7$

109. Area $= (\text{side})^2$
$= (7 \text{ meters})^2$
$= 49$ square meters

110. Area $= (\text{side})^2$
$= (3 \text{ inches})^2$
$= 9$ square inches

111. $\dfrac{2x}{z} = \dfrac{2 \cdot 5}{2} = \dfrac{10}{2} = 5$

112. $4x - 3 = 4 \cdot 5 - 3 = 20 - 3 = 17$

113. $\dfrac{x+7}{y} = \dfrac{5+7}{0}$ is undefined.

114. $\dfrac{y}{5x} = \dfrac{0}{5 \cdot 5} = \dfrac{0}{25} = 0$

115. $x^3 - 2z = 5^3 - 2 \cdot 2$
$= 125 - 2 \cdot 2$
$= 125 - 4$
$= 121$

116. $\dfrac{7+x}{3z} = \dfrac{7+5}{3 \cdot 2} = \dfrac{12}{6} = 2$

117. $(y+z)^2 = (0+2)^2 = 2^2 = 4$

118. $\dfrac{100}{x} + \dfrac{y}{3} = \dfrac{100}{5} + \dfrac{0}{3} = 20 + 0 = 20$

119. Five subtracted from a number is $x - 5$.

120. Seven more than a number is $x + 7$.

121. Ten divided by a number is $10 \div x$ or $\dfrac{10}{x}$.

122. The product of 5 and a number is $5x$.

123. Let n be 5.
$n + 12 = 20 - 3$
$5 + 12 \overset{?}{=} 20 - 3$
$17 = 17$ True
Yes, 5 is a solution.

124. Let n be 23.
$n - 8 = 10 + 6$
$23 - 8 \overset{?}{=} 10 + 6$
$15 = 16$ False
No, 23 is not a solution.

125. Let $n = 14$.
$30 = 3(n-3)$
$30 \overset{?}{=} 3(14-3)$
$30 \overset{?}{=} 3(11)$
$30 = 33$ False
No, 14 is not a solution.

126. Let n be 20.
$$5(n-7) = 65$$
$$5(20-7) \stackrel{?}{=} 65$$
$$5(13) \stackrel{?}{=} 65$$
$$65 = 65 \quad \text{True}$$
Yes, 20 is a solution.

127. $7n = 77$
Let n be 6.
$$7 \cdot 6 \stackrel{?}{=} 77$$
$$42 = 77 \quad \text{False}$$
Let n be 11.
$$7 \cdot 11 \stackrel{?}{=} 77$$
$$77 = 77 \quad \text{True}$$
Let n be 20.
$$7 \cdot 20 \stackrel{?}{=} 77$$
$$140 = 77 \quad \text{False}$$
11 is a solution.

128. $n - 25 = 150$
Let n be 125.
$$125 - 25 \stackrel{?}{=} 150$$
$$100 = 150 \quad \text{False}$$
Let n be 145.
$$145 - 25 \stackrel{?}{=} 150$$
$$120 = 150 \quad \text{False}$$
Let n be 175.
$$175 - 25 \stackrel{?}{=} 150$$
$$150 = 150 \quad \text{True}$$
175 is a solution.

129. $5(n+4) = 90$
Let n be 14.
$$5(14+4) \stackrel{?}{=} 90$$
$$5(18) \stackrel{?}{=} 90$$
$$90 = 90 \quad \text{True}$$
Let n be 16.
$$5(16+4) \stackrel{?}{=} 90$$
$$5(20) \stackrel{?}{=} 90$$
$$100 = 90 \quad \text{False}$$
Let n be 26.
$$5(26+4) \stackrel{?}{=} 90$$
$$5(30) \stackrel{?}{=} 90$$
$$150 = 90 \quad \text{False}$$
14 is a solution.

130. $3n - 8 = 28$
Let n be 3.
$$3(3) - 8 \stackrel{?}{=} 28$$
$$9 - 8 \stackrel{?}{=} 28$$
$$1 = 28 \quad \text{False}$$
Let n be 7.
$$3(7) - 8 \stackrel{?}{=} 28$$
$$21 - 8 \stackrel{?}{=} 28$$
$$13 = 28 \quad \text{False}$$
Let n be 15.
$$3(15) - 8 \stackrel{?}{=} 28$$
$$45 - 8 \stackrel{?}{=} 28$$
$$37 = 28 \quad \text{False}$$
None are solutions.

131.
$$\begin{array}{r} 485 \\ -\ 68 \\ \hline 417 \end{array}$$

132.
$$\begin{array}{r} 729 \\ -\ 47 \\ \hline 682 \end{array}$$

133.
$$\begin{array}{r} 732 \\ \times\ 3 \\ \hline 2196 \end{array}$$

134.
$$\begin{array}{r} 629 \\ \times\ 4 \\ \hline 2516 \end{array}$$

135.
$$\begin{array}{r} {\scriptstyle 2\,2} \\ 374 \\ 29 \\ +\ 698 \\ \hline 1101 \end{array}$$

136.
$$\begin{array}{r} {\scriptstyle 2\,1} \\ 593 \\ 52 \\ +\ 766 \\ \hline 1411 \end{array}$$

137.

$$\begin{array}{r} 458 \text{ R } 8 \\ 13\overline{)\ 5962} \\ \underline{-52} \\ 76 \\ \underline{-65} \\ 112 \\ \underline{-104} \\ 8 \end{array}$$

138.

$$\begin{array}{r} 237 \text{ R } 1 \\ 18\overline{)\ 4267} \\ \underline{-36} \\ 66 \\ \underline{-54} \\ 127 \\ \underline{-126} \\ 1 \end{array}$$

139.

$$\begin{array}{r} 1968 \\ \times\ \ 36 \\ \hline 11\,808 \\ 59\,040 \\ \hline 70,848 \end{array}$$

140.

$$\begin{array}{r} 5324 \\ \times\ \ 18 \\ \hline 42\,592 \\ 53\,240 \\ \hline 95,832 \end{array}$$

141.

$$\begin{array}{r} 2000 \\ -\ \ 356 \\ \hline 1644 \end{array}$$

142.

$$\begin{array}{r} 9000 \\ -\ 519 \\ \hline 8481 \end{array}$$

143. To round 842 to the nearest ten, observe that the digit in the ones place is 2. Since this digit is less than 5, we do not add 1 to the digit in the tens place. The number 842 rounded to the nearest ten is 840.

144. To round 258,371 to the nearest hundred-thousand, observe that the digit in the ten-thousands place is 5. Since this digit is at least 5, we add 1 to the digit in the hundred-thousands place. The number 258,371 rounded to the nearest hundred-thousand is 300,000.

145. $24 \div 4 \cdot 2 = 6 \cdot 2 = 12$

146. $\dfrac{(15+3)\cdot(8-5)}{2^3+1} = \dfrac{(18)(3)}{8+1} = \dfrac{54}{9} = 6$

147. Let n be 9.
$$5n - 6 = 40$$
$$5 \cdot 9 - 6 \stackrel{?}{=} 40$$
$$45 - 6 \stackrel{?}{=} 40$$
$$39 = 40 \quad \text{False}$$
No, 9 is not a solution.

148. Let n be 3.
$$2n - 6 = 5n - 15$$
$$2(3) - 6 \stackrel{?}{=} 5(3) - 15$$
$$6 - 6 \stackrel{?}{=} 15 - 15$$
$$0 = 0 \quad \text{True}$$
Yes, 3 is a solution.

149.

$$\begin{array}{r} 53 \\ 32\overline{)\ 1714} \\ \underline{-160} \\ 114 \\ \underline{-96} \\ 18 \end{array}$$

There are 53 full boxes with 18 left over.

150.

$$\begin{array}{r} 27 \\ \times\ 2 \\ \hline 54 \end{array} \qquad \begin{array}{r} 8 \\ \times\ 4 \\ \hline 32 \end{array}$$

$$\begin{array}{r} 54 \\ +\ 32 \\ \hline 86 \end{array}$$

The total bill before taxes is $86.

Chapter 1 Test

1. 82,426 in words is eighty-two thousand, four hundred twenty-six.

2. Four hundred two thousand, five hundred fifty in standard form is 402,550.

3.
$$\begin{array}{r} 1 \\ 59 \\ + 82 \\ \hline 141 \end{array}$$

4.
$$\begin{array}{r} 600 \\ - 487 \\ \hline 113 \end{array}$$

5.
$$\begin{array}{r} 496 \\ \times \quad 30 \\ \hline 14{,}880 \end{array}$$

6.
$$\begin{array}{r} 766 \text{ R } 42 \\ 69\overline{)\,52{,}896} \\ -48\,3 \\ \hline 4\,59 \\ -4\,14 \\ \hline 456 \\ -414 \\ \hline 42 \end{array}$$

7. $2^3 \cdot 5^2 = 2 \cdot 2 \cdot 2 \cdot 5 \cdot 5 = 200$

8. $98 \div 1 = 98$

9. $0 \div 49 = 0$

10. $62 \div 0$ is undefined.

11. $(2^4 - 5) \cdot 3 = (16 - 5) \cdot 3 = 11 \cdot 3 = 33$

12. $16 + 9 \div 3 \cdot 4 - 7 = 16 + 3 \cdot 4 - 7$
$$= 16 + 12 - 7$$
$$= 28 - 7$$
$$= 21$$

13. $6^1 \cdot 2^3 = 6 \cdot 2 \cdot 2 \cdot 2 = 48$

14. $2[(6-4)^2 + (22-19)^2] + 10$
$$= 2[2^2 + 3^2] + 10$$
$$= 2[4 + 9] + 10$$
$$= 2[13] + 10$$
$$= 26 + 10$$
$$= 36$$

15. $5698 \cdot 1000 = 5{,}698{,}000$

16. Divide the sum by 5.

$$\begin{array}{r} 2 \\ 62 \\ 79 \\ 84 \\ 90 \\ + 95 \\ \hline 410 \end{array} \qquad \begin{array}{r} 82 \\ 5\overline{)\,410} \\ -40 \\ \hline 10 \\ -10 \\ \hline 0 \end{array}$$

The average is 82.

17. To round 52,369 to the nearest thousand, observe that the digit in the hundreds place is 3. Since this digit is less than 5, we do not add 1 to the digit in the thousands place. The number 52,369 rounded to the nearest thousand is 52,000.

18.
6289	rounds to	6 300
5403	rounds to	5 400
+ 1957	rounds to	+ 2 000
		13,700

19.
4267	rounds to	4300
− 2738	rounds to	− 2700
		1600

20.
$$\begin{array}{r} 107 \\ - \quad 15 \\ \hline 92 \end{array}$$

21.
$$\begin{array}{r} 15 \\ + 107 \\ \hline 122 \end{array}$$

22.
$$\begin{array}{r} 107 \\ \times\ 15 \\ \hline 535 \\ 1070 \\ \hline 1605 \end{array}$$

23.
$$\begin{array}{r} 7\ \text{R}\ 2 \\ 15{\overline{\smash{\big)}\,107}} \\ \underline{-105} \\ 2 \end{array}$$

24.
$$\begin{array}{r} 17 \\ 29{\overline{\smash{\big)}\,493}} \\ \underline{-29} \\ 203 \\ \underline{-203} \\ 0 \end{array}$$
Each can cost $17.

25.
$$\begin{array}{r} 725 \\ -\ 599 \\ \hline 126 \end{array}$$
The higher-priced one is $126 more.

26.
$$\begin{array}{r} 45 \\ \times\ 8 \\ \hline 360 \end{array}$$
There are 360 calories in 8 tablespoons of white granulated sugar.

27.
$$\begin{array}{r} 430 \\ \times\ 16 \\ \hline 2580 \\ 4300 \\ \hline 6880 \end{array} \qquad \begin{array}{r} 205 \\ \times\ 5 \\ \hline 1025 \end{array}$$

$$\begin{array}{r} 6880 \\ +\ 1025 \\ \hline 7905 \end{array}$$
The total cost is $7905.

28. Perimeter $= (5+5+5+5)$ centimeters
$$= 20 \text{ centimeters}$$

Area $= (\text{side})^2$
$$= (5 \text{ centimeters})^2$$
$$= 25 \text{ square centimeters}$$

29. Perimeter $= (20+10+20+10)$ yards
$$= 60 \text{ yards}$$
Area $= (\text{length})(\text{width})$
$$= (20 \text{ yards})(10 \text{ yards})$$
$$= 200 \text{ square yards}$$

30. Let x be 2.
$$5(x^3 - 2) = 5(2^3 - 2) = 5(8 - 2) = 5(6) = 30$$

31. Let x be 7 and y be 8.
$$\frac{3x-5}{2y} = \frac{3(7)-5}{2 \cdot 8} = \frac{21-5}{16} = \frac{16}{16} = 1$$

32. a. The quotient of a number and 17 is
$$x \div 17 \text{ or } \frac{x}{17}.$$

b. Twice a number, decreased by 20 is $2x - 20$.

33. Let n be 6.
$$5n - 11 = 19$$
$$5(6) - 11 \stackrel{?}{=} 19$$
$$30 - 11 \stackrel{?}{=} 19$$
$$19 = 19 \quad \text{True}$$
6 is a solution.

34. $n + 20 = 4n - 10$
Let n be 0.
$$0 + 20 \stackrel{?}{=} 4 \cdot 0 - 10$$
$$20 \stackrel{?}{=} 0 - 10$$
$$20 = -10 \quad \text{False}$$
Let n be 10.
$$10 + 20 \stackrel{?}{=} 4 \cdot 10 - 10$$
$$30 \stackrel{?}{=} 40 - 10$$
$$30 = 30 \quad \text{True}$$
Let n be 20.
$$20 + 20 \stackrel{?}{=} 4 \cdot 20 - 10$$
$$40 \stackrel{?}{=} 80 - 10$$
$$40 = 70 \quad \text{False}$$
10 is a solution.

Chapter 2

Practice Problems

1. **a.** If 0 represents the surface of the ocean, then 836 below the surface of the ocean is −836.

    b. If 0 represents a loss of $0, then a loss of $1 million is −1 million.

2.

3. **a.** $0 > -5$ since 0 is to the right of −5 on a number line.

    b. $-3 < 3$ since −3 is to the left of 3 on a number line.

    c. $-7 > -12$ since −7 is to the right of −12 on a number line.

4. **a.** $|-6| = 6$ because −6 is 6 units from 0.

    b. $|4| = 4$ because 4 is 4 units from 0.

    c. $|-12| = 12$ because −12 is 12 units from 0.

5. **a.** The opposite of 14 is −14.

    b. The opposite of −9 is −(−9) or 9.

6. **a.** $-|-7| = -7$

    b. $-|4| = -4$

    c. $-(-12) = 12$

7. $-|x| = -|-6| = -6$

8. The planet with the highest average temperature is the one that corresponds to the bar that extends the furthest in the positive direction (upward). Venus has the highest average temperature of 867°F.

Vocabulary and Readiness Check

1. The numbers ...−3, −2, −1, 0, 1, 2, 3, ... are called <u>integers</u>.

3. The symbols "<" and ">" are called <u>inequality symbols</u>.

5. The sign "<" means <u>is less than</u> and ">" means <u>is greater than</u>.

7. A number's distance from 0 on the number line is the number's <u>absolute value</u>.

Exercise Set 2.1

1. If 0 represents ground level, then 1235 feet underground is −1235.

3. If 0 represents sea level, then 14,433 feet above sea level is +14,433.

5. If 0 represents zero degrees Fahrenheit, then 118 degrees above zero is +118.

7. If 0 represents the surface of the ocean, then 13,000 feet below the surface of the ocean is −13,000.

9. If 0 represents a loss of $0, then a loss of $10,458 million is −10,458 million.

11. If 0 represents the surface of the ocean, then 160 feet below the surface is −160 and 147 feet below the surface is −147. Since −160 extends further in the negative direction, Guillermo is deeper.

13. If 0 represents a loss of 0%, then an 81 percent loss is −81.

15.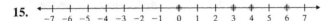

17.

19.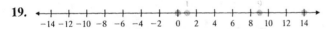

21.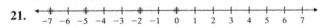

23. $0 > -7$ since 0 is to the right of −7 on a number line.

25. $-7 < -5$ since −7 is to the left of −5 on a number line.

27. $-30 > -35$ since −30 is to the right of −35 on a number line.

29. $-26 < 26$ since −26 is to the left of 26 on a number line.

31. $|5| = 5$ since 5 is 5 units from 0 on a number line.

33. $|-8| = 8$ since −8 is 8 units from 0 on a number line.

35. $|0| = 0$ since 0 is 0 units from 0 on a number line.

37. $|-55| = 55$ since −55 is 55 units from 0 on a number line.

39. The opposite of 5 is negative 5.
    $-(5) = -5$

41. The opposite of negative 4 is 4.
    $-(-4) = 4$

43. The opposite of 23 is negative 23.
$-(23) = -23$

45. The opposite of negative 85 is 85.
$-(-85) = 85$

47. $|-7| = 7$

49. $-|20| = -20$

51. $-|-3| = -3$

53. $-(-43) = 43$

55. $|-15| = 15$

57. $-(-33) = 33$

59. $|-x| = |-(-6)| = |6| = 6$

61. $-|-x| = -|-2| = -2$

63. $|x| = |-32| = 32$

65. $-|x| = -|7| = -7$

67. $-12 < -6$ since -12 is to the left of -6 on a number line.

69. $|-8| = 8$
$|-11| = 11$
Since $8 < 11$, $|-8| < |-11|$.

71. $|-47| = 47$
$-(-47) = 47$
Since $47 = 47$, $|-47| = -(-47)$.

73. $-|-12| = -12$
$-(-12) = 12$
Since $-12 < 12$, $-|-12| < -(-12)$.

75. $0 > -9$ since 0 is to the right of -9 on a number line.

77. $|0| = 0$
$|-9| = 9$
Since $0 < 9$, $|0| < |-9|$.

79. $-|-2| = -2$
$-|-10| = -10$
Since $-2 > -10$, $-|-2| > -|-10|$.

81. $-(-12) = 12$
$-(-18) = 18$
Since $12 < 18$, $-(-12) < -(-18)$.

83. If the number is 31, then the absolute value of 31 is 31 and the opposite of 31 is -31.

85. If the opposite of a number is -28, then the number is 28, and its absolute value is 28.

87. The bar that extends the farthest in the negative direction corresponds to the Caspian Sea, so the Caspian Sea has the lowest elevation.

89. The tallest bar on the graph corresponds to Lake Superior, so Lake Superior has the highest elevation.

91. The negative number on the graph closest to $0°F$ is $-81°F$, which corresponds to Mars.

93. The number on the graph closest to $-300°F$ is $-323°F$, which corresponds to the planet Uranus.

95. $0 + 13 = 13$

97.
$$\begin{array}{r} 15 \\ +\ 20 \\ \hline 35 \end{array}$$

99.
$$\begin{array}{r} {\scriptstyle 1\ 2} \\ 47 \\ 236 \\ +\ 77 \\ \hline 360 \end{array}$$

101. $2^2 = 4$, $-|3| = -3$, $-(-5) = 5$, and $-|-8| = -8$, so the numbers in order from least to greatest are $-|-8|$, $-|3|$, 2^2, $-(-5)$.

48

103. $|-1| = 1$, $-|-6| = -6$, $-(-6) = 6$, and $-|1| = -1$, so the numbers in order from least to greatest are $-|-6|$, $-|1|$, $|-1|$, $-(-6)$.

105. $-(-2) = 2$, $5^2 = 25$, $-10 = -10$, $-|-9| = -9$, and $|-12| = 12$, so the numbers in order from least to greatest are -10, $-|-9|$, $-(-2)$, $|-12|$, 5^2.

107. a. $|-9| = 9$; since $9 > 8$, then $|-9| > 8$ is true.

     b. $|-5| = 5$; since $5 < 8$, then $|-5| > 8$ is false.

     c. $|8| = 8$; since $8 = 8$, then $|8| > 8$ is false.

     d. $|-12| = 12$; since $12 > 8$, then $|-12| > 8$ is true.

109. $-(-|-8|) = -(-8) = 8$

111. False; consider $a = -2$ and $b = -3$, then $-2 > -3$.

113. True; a positive number will always be to the right of a negative number on a number line.

115. False; consider $a = -5$, then the opposite of -5 is 5, which is a positive number.

117. answers may vary

119. no; answers may vary

Section 2.2

Practice Problems

1.

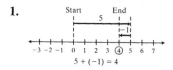

$5 + (-1) = 4$

2.

End   Start

$-6 + (-2) = -8$

3.

End   Start

$-8 + 3 = -5$

4. $|-3| + |-19| = 3 + 19 = 22$
The common sign is negative, so $(-3) + (-19) = -22$.

5. $-12 + (-30) = -42$

6. $9 + 4 = 13$

7. $|-1| = 1$, $|6| = 6$, and $6 - 1 = 5$
$6 > 1$, so the answer is positive.
$-1 + 6 = 5$

8. $|2| = 2$, $|-8| = 8$, and $8 - 2 = 6$
$8 > 2$, so the answer is negative.
$2 + (-8) = -6$

9. $-54 + 20 = -34$

10. $7 + (-2) = 5$

11. $-3 + 0 = -3$

12. $18 + (-18) = 0$

13. $-64 + 64 = 0$

14. $6 + (-2) + (-15) = 4 + (-15) = -11$

15. $5 + (-3) + 12 + (-14) = 2 + 12 + (-14)$
$= 14 + (-14)$
$= 0$

16. $x + 3y = -6 + 3(2) = -6 + 6 = 0$

17. $x + y = -13 + (-9) = -22$

18. Temperature at 8 a.m. $= -7 + (+4) + (+7)$
$$= -3 + (+7)$$
$$= 4$$
The temperature was 4°F at 8 a.m.

Calculator Explorations

  1. $-256 + 97 = -159$

  3. $6(15) + (-46) = 44$

  5. $-108,650 + (-786,205) = -894,855$

Vocabulary and Readiness Check

  1. If n is a number, then $-n + n = \underline{0}$.

  3. If a is a number, then $-(-a) = \underline{a}$.

Exercise Set 2.2

  1.

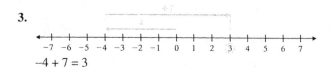

$-1 + (-6) = -7$

  3.

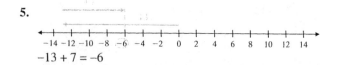

$-4 + 7 = 3$

  5.

$-13 + 7 = -6$

  7. $46 + 21 = 67$

  9. $|-8| + |-2| = 8 + 2 = 10$
      The common sign is negative, so $(-8) + (-2) = -10$.

  11. $-43 + 43 = 0$

  13. $|6| - |-2| = 6 - 2 = 4$
      $6 > 2$, so the answer is positive.
      $6 + (-2) = 4$

  15. $-6 + 0 = -6$

17. $|-5| - |3| = 5 - 3 = 2$
$5 > 3$, so the answer is negative.
$3 + (-5) = -2$

19. $|-2| + |-7| = 2 + 7 = 9$
The common sign is negative, so
$-2 + (-7) = -9$.

21. $|-12| + |-12| = 12 + 12 = 24$
The common sign is negative, so
$-12 + (-12) = -24$.

23. $|-640| + |-200| = 640 + 200 = 840$
The common sign is negative, so
$-640 + (-200) = -840$.

25. $|12| - |-5| = 12 - 5 = 7$
$12 > 5$, so the answer is positive.
$12 + (-5) = 7$

27. $|-6| - |3| = 6 - 3 = 3$
$6 > 3$, so the answer is negative.
$-6 + 3 = -3$

29. $|-56| - |26| = 56 - 26 = 30$
$56 > 26$, so the answer is negative.
$-56 + 26 = -30$

31. $|85| - |-45| = 85 - 45 = 40$
$85 > 45$, so the answer is positive.
$-45 + 85 = 40$

33. $|-144| - |124| = 144 - 124 = 20$
$144 > 124$, so the answer is negative.
$124 + (-144) = -20$

35. $|-82| + |-43| = 82 + 43 = 125$
The common sign is negative, so
$-82 + (-43) = -125$.

37. $-4 + 2 + (-5) = -2 + (-5) = -7$

39. $-52 + (-77) + (-117) = -129 + (-117)$
$= -246$

41. $12 + (-4) + (-4) + 12 = 8 + (-4) + 12$
$= 4 + 12$
$= 16$

43. $(-10) + 14 + 25 + (-16) = 4 + 25 + (-16)$
$= 29 + (-16)$
$= 13$

45. $-6 + (-15) + (-7) = -21 + (-7) = -28$

47. $-26 + 15 = -11$

49. $5 + (-2) + 17 = 3 + 17 = 20$

51. $-13 + (-21) = -34$

53. $3 + 14 + (-18) = 17 + (-18) = -1$

55. $-92 + 92 = 0$

57. $-13 + 8 + (-10) + (-27) = -5 + (-10) + (-27)$
$= -15 + (-27)$
$= -42$

59. $x + y = -20 + (-50) = -70$

61. $3x + y = 3(2) + (-3) = 6 + (-3) = 3$

63. $3x + y = 3(3) + (-30) = 9 + (-30) = -21$

65. The sum of -6 and 25 is $-6 + 25 = 19$.

67. The sum of -31, -9, and 30 is
$-31 + (-9) + 30 = -40 + 30 = -10$.

69. $0 + (-215) + (-16) = -215 + (-16) = -231$
The diver's final depth is 231 feet below the surface.

71. Sorenstam:
$-1 + 0 + (-1) + 0 + 0 + 1 + 0 + 0 + 0 + 0 + 0$
$\quad + (-1) + 1 + 0 + 0 + 0 + 0 + 0 = -1$
Hurst:
$1 + 0 + 0 + 0 + 0 + 2 + 0 + 0 + 1 + 0 + 0 + 0$
$\quad + 0 + 0 + 0 + 0 + 0 + (-1) = +3$

73. The bar for 2002 has a height of 65, so the net income in 2002 was \$65,000,000.

75. $-25 + 65 = 40$
The total net income for years 2001 and 2002 was \$40,000,000.

77. $-10 + 12 = 2$
The temperature at 11 p.m. was 2°C.

79. $-1786 + 15{,}395 = 13{,}609$
The sum of the net incomes for 2004 and 2005 is $13,609.

81. $-55 + 8 = -47$
West Virginia's record low temperature is $-47°F$.

83. $-10{,}924 + 3245 = -7679$
The depth of the Aleutian Trench is -7679 meters.

85. $44 - 0 = 44$

87.
$$\begin{array}{r} 200 \\ -\ 59 \\ \hline 141 \end{array}$$

89. answers may vary

91. $7 + (-10) = -3$

93. $-4 + 14 = 10$

95. True

97. False; for example, $4 + (-2) = 2 > 0$.

99. answers may vary

Section 2.3

Practice Problems

1. $13 - 4 = 13 + (-4) = 9$

2. $-8 - 2 = -8 + (-2) = -10$

3. $11 - (-15) = 11 + 15 = 26$

4. $-9 - (-1) = -9 + 1 = -8$

5. $6 - 9 = 6 + (-9) = -3$

6. $-14 - 5 = -14 + (-5) = -19$

7. $-3 - (-4) = -3 + 4 = 1$

8. $-15 - 6 = -15 + (-6) = -21$

9. $\begin{aligned} -6 - 5 - 2 - (-3) &= -6 + (-5) + (-2) + 3 \\ &= -11 + (-2) + 3 \\ &= -13 + 3 \\ &= -10 \end{aligned}$

10. $\begin{aligned} 8 + (-2) - 9 - (-7) &= 8 + (-2) + (-9) + 7 \\ &= 6 + (-9) + 7 \\ &= -3 + 7 \\ &= 4 \end{aligned}$

11. $x - y = -5 - 13 = -5 + (-13) = -18$

12. $3y - z = 3(9) - (-4) = 27 + 4 = 31$

13. $29{,}028 - (-1312) = 29{,}028 + 1312 = 30{,}340$
Mount Everest is 30,340 feet higher than the Dead Sea.

Vocabulary and Readiness Check

1. It is true that $a - b = \underline{a + (-b)}$. b

3. To evaluate $x - y$ for $x = -10$ and $y = -14$, we replace x with -10 and y with -14 and evaluate $\underline{-10 - (-14)}$. d

Exercise Set 2.3

1. $-8 - (-8) = -8 + 8 = 0$

3. $19 - 16 = 19 + (-16) = 3$

5. $3 - 8 = 3 + (-8) = -5$

7. $11 - (-11) = 11 + 11 = 22$

9. $-4 - (-7) = -4 + 7 = 3$

11. $-16 - 4 = -16 + (-4) = -20$

13. $3 - 15 = 3 + (-15) = -12$

15. $42 - 55 = 42 + (-55) = -13$

17. $478 - (-30) = 478 + 30 = 508$

19. $-4 - 10 = -4 + (-10) = -14$

21. $-7 - (-3) = -7 + 3 = -4$

23. $17 - 29 = 17 + (-29) = -12$

25. $-25 - 17 = -25 + (-17) = -42$

27. $-22 - (-3) = -22 + 3 = -19$

29. $2 - (-12) = 2 + 12 = 14$

31. $-37 + (-19) = -56$

33. $8 - 13 = 8 + (-13) = -5$

35. $-56 - 89 = -56 + (-89) = -145$

37. $30 - 67 = 30 + (-67) = -37$

39. $8 - 3 - 2 = 8 + (-3) + (-2) = 5 + (-2) = 3$

41. $13 - 5 - 7 = 13 + (-5) + (-7) = 8 + (-7) = 1$

43. $\begin{aligned} -5 - 8 - (-12) &= -5 + (-8) + 12 \\ &= -13 + 12 \\ &= -1 \end{aligned}$

45. $\begin{aligned} -11 + (-6) - 14 &= -11 + (-6) + (-14) \\ &= -17 + (-14) \\ &= -31 \end{aligned}$

47. $\begin{aligned} 18 - (-32) + (-6) &= 18 + 32 + (-6) \\ &= 50 + (-6) \\ &= 44 \end{aligned}$

49. $\begin{aligned} -(-5) - 21 + (-16) &= 5 + (-21) + (-16) \\ &= -16 + (-16) \\ &= -32 \end{aligned}$

51. $\begin{aligned} -10 &- (-12) + (-7) - 4 \\ &= -10 + 12 + (-7) + (-4) \\ &= 2 + (-7) + (-4) \\ &= -5 + (-4) \\ &= -9 \end{aligned}$

53. $\begin{aligned} -3 + 4 - (-23) - 10 &= -3 + 4 + 23 + (-10) \\ &= 1 + 23 + (-10) \\ &= 24 + (-10) \\ &= 14 \end{aligned}$

55. $x - y = -4 - 7 = -4 + (-7) = -11$

57. $x - y = 8 - (-23) = 8 + 23 = 31$

59. $2x - y = 2(4) - (-4) = 8 + 4 = 12$

61. $2x - y = 2(1) - (-18) = 2 + 18 = 20$

63. The temperature in March is 11°F and in February is −4°F.
$11 - (-4) = 11 + 4 = 15$
The difference is 15°F.

65. The two months with the lowest temperatures are January, −10°F, and December, −6°F.
$-6 - (-10) = -6 + 10 = 4$
The difference is 4°F.

67. $136 - (-129) = 136 + 129 = 265$
Therefore, 136°F is 265°F warmer than −129°F.

69. $\begin{aligned} 93 - 18 - 26 &= 93 + (-18) + (-26) \\ &= 75 + (-26) \\ &= 49 \end{aligned}$
She owes $49 on her account.

71. $\begin{aligned} -4 - 3 + 4 - 7 &= -4 + (-3) + 4 + (-7) \\ &= -7 + 4 + (-7) \\ &= -3 + (-7) \\ &= -10 \end{aligned}$
The temperature at 9 a.m. is −10°C.

73. $-282 - (-436) = -282 + 436 = 154$
The difference in elevation is 154 feet.

75. $-436 - (-505) = -436 + 505 = 69$
The difference in elevation is 69 feet.

77. $600 - (-52) = 600 + 52 = 652$
The difference in elevation is 652 feet.

79. $144 - 0 = 144$
The difference in elevation is 144 feet.

81. $867 - (-330) = 867 + 330 = 1197$
The difference in temperatures is 1197°F.

83. $895 - 1677 = 895 + (-1677) = -782$
The trade balance was -$782 billion.

85. The sum of -5 and a number is $-5 + x$.

87. Subtract a number from -20 is $-20 - x$.

89. $\dfrac{100}{20} = 5$

91.
$$\begin{array}{r} 23 \\ \times\ 46 \\ \hline 138 \\ 920 \\ \hline 1058 \end{array}$$

93. answers may vary

95. $9 - (-7) = 9 + 7 = 16$

97. $10 - 30 = 10 + (-30) = -20$

99. $|-3| - |-7| = 3 - 7 = 3 + (-7) = -4$

101. $|-5| - |5| = 5 - 5 = 0$

103. $|-15| - |-29| = 15 - 29 = 15 + (-29) = -14$

105. $|-8 - 3| = |-8 + (-3)| = |-11| = 11$
$8 - 3 = 8 + (-3) = 5$
Since $11 \neq 5$, the statement is false.

107. answers may vary

Section 2.4

Practice Problems

1. $-3 \cdot 7 = -21$

2. $-5(-2) = 10$

3. $0 \cdot (-20) = 0$

4. $10(-5) = -50$

5. $8(-6)(-2) = -48(-2) = 96$

6. $(-9)(-2)(-1) = 18(-1) = -18$

7. $(-3)(-4)(-5)(-1) = 12(-5)(-1)$
$= -60(-1)$
$= 60$

8. $(-2)^4 = (-2)(-2)(-2)(-2)$
$= 4(-2)(-2)$
$= -8(-2)$
$= 16$

9. $-8^2 = -(8 \cdot 8) = -64$

10. $\dfrac{42}{-7} = -6$

11. $-16 \div (-2) = 8$

12. $\dfrac{-80}{10} = -8$

13. $\dfrac{-6}{0}$ is undefined.

14. $\dfrac{0}{-7} = 0$

15. $xy = 5 \cdot (-8) = -40$

16. $\dfrac{x}{y} = \dfrac{-12}{-3} = 4$

17. total score $= 4 \cdot (-13) = -52$
The card player's total score was -52.

Vocabulary and Readiness Check

1. The product of a negative number and a positive number is a <u>negative</u> number.

3. The quotient of two negative numbers is a <u>positive</u> number.

5. The product of a negative number and zero is <u>0</u>.

7. The quotient of a negative number and 0 is <u>undefined</u>.

Exercise Set 2.4

1. $-6(-2) = 12$

3. $-4(9) = -36$

5. $9(-9) = -81$

7. $0(-11) = 0$

9. $6(-2)(-4) = -12(-4) = 48$

11. $-1(-3)(-4) = 3(-4) = -12$

13. $-4(4)(-5) = -16(-5) = 80$

15. $10(-5)(0)(-7) = 0$

17. $-5(3)(-1)(-1) = -15(-1)(-1)$
$$= 15(-1)$$
$$= -15$$

19. $-3^2 = -(3 \cdot 3) = -9$

21. $(-3)^3 = (-3)(-3)(-3) = 9(-3) = -27$

23. $-6^2 = -(6 \cdot 6) = -36$

25. $(-4)^3 = (-4)(-4)(-4) = 16(-4) = -64$

27. $-24 \div 3 = -8$

29. $\dfrac{-30}{6} = -5$

31. $\dfrac{-77}{-11} = 7$

33. $\dfrac{0}{-21} = 0$

35. $\dfrac{-10}{0}$ is undefined.

37. $\dfrac{56}{-4} = -14$

39. $-14(0) = 0$

41. $-5(3) = -15$

43. $-9 \cdot 7 = -63$

45. $-7(-6) = 42$

47. $-3(-4)(-2) = 12(-2) = -24$

49. $(-7)^2 = (-7)(-7) = 49$

51. $-\dfrac{25}{5} = -5$

53. $-\dfrac{72}{8} = -9$

55. $-18 \div 3 = -6$

57. $4(-10)(-3) = -40(-3) = 120$

59. $-30(6)(-2)(-3) = -180(-2)(-3)$
$$= 360(-3)$$
$$= -1080$$

61. $3 \cdot (-8) \cdot 0 = -24 \cdot 0 = 0$

63. $\dfrac{120}{-20} = -6$

65. $280 \div (-40) = \dfrac{280}{-40} = -7$

67. $\dfrac{-12}{-4} = 3$

69. $-1^4 = -(1 \cdot 1 \cdot 1 \cdot 1) = -1$

71. $(-2)^5 = (-2)(-2)(-2)(-2)(-2)$
$= 4(-2)(-2)(-2)$
$= -8(-2)(-2)$
$= 16(-2)$
$= -32$

73. $-2(3)(5)(-6) = -6(5)(-6) = -30(-6) = 180$

75. $(-1)^{32} = 1$, since there are an even number of factors.

77. $-2(-3)(-5) = 6(-5) = -30$

79.
$$\begin{array}{r} 48 \\ \times\ 23 \\ \hline 144 \\ 960 \\ \hline 1104 \end{array}$$
$-48 \cdot 23 = -1104$

81.
$$\begin{array}{r} 35 \\ \times\ 82 \\ \hline 70 \\ 2800 \\ \hline 2870 \end{array}$$
$35 \cdot (-82) = -2870$

83. $ab = -8 \cdot 7 = -56$

85. $ab = 9(-2) = -18$

87. $ab = (-7)(-5) = 35$

89. $\dfrac{x}{y} = \dfrac{5}{-5} = -1$

91. $\dfrac{x}{y} = \dfrac{-15}{0}$ is undefined.

93. $\dfrac{x}{y} = \dfrac{-36}{-6} = 6$

95. $xy = -8 \cdot (-2) = 16$
$\dfrac{x}{y} = \dfrac{-8}{-2} = 4$

97. $xy = 0(-8) = 0$
$\dfrac{x}{y} = \dfrac{0}{-8} = 0$

99. $\dfrac{-54}{9} = -6$
The quotient of -54 and 9 is -6.

101.
$$\begin{array}{r} 42 \\ \times\ 6 \\ \hline 252 \end{array}$$
$-42(-6) = 252$
The product of -42 and -6 is 252.

103. The product of -71 and a number is $-71 \cdot x$ or $-71x$.

105. Subtract a number from -16 is $-16 - x$.

107. -29 increased by a number is $-29 + x$.

109. Divide a number by -33 is $\dfrac{x}{-33}$ or $x \div (-33)$.

111. A loss of 4 yards is represented by -4.
$3 \cdot (-4) = -12$
The team had a total loss of 12 yards.

113. Each move of 20 feet down is represented by -20.
$5 \cdot (-20) = -100$
The diver is at a depth of 100 feet.

115. $3 \cdot (-70) = -210$
The melting point of nitrogen is $-210°C$.

117. $-3 \cdot 63 = -189$
The melting point of argon is $-189°C$.

119. $4 \cdot (-457) = -1828$
The net income will be −$1828 million after four quarters.

121. a. $626 - 50 = 576$
576 million fewer VHS cassettes were shipped to retailers in 2005 than in 2001. This is a change of −576 million.

    b. This is a period of 4 years.
$$\frac{-576,000,000}{4} = -144,000,000$$
The average change was −144,000,000 VHS cassettes per year.

123. $90 + 12^2 - 5^3 = 90 + 144 - 125$
$$= 90 + 144 + (-125)$$
$$= 234 + (-125)$$
$$= 109$$

125. $12 \div 4 - 2 + 7 = 3 - 2 + 7 = 1 + 7 = 8$

127. $-57 \div 3 = -19$

129. $-8 - 20 = -8 + (-20) = -28$

131. $-4 - 15 - (-11) = -4 + (-15) + 11$
$$= -19 + 11$$
$$= -8$$

133. The product of an odd number of negative numbers is negative, so the product of seven negative numbers is negative.

135. $(-2)^{12}$ and $(-5)^{12}$ are positive since there are an even number of factors. Note that $(-5)^{12}$ will be the larger of the two since $|-5| > |-2|$ and the exponent on each is the same.
$(-2)^{17}$ and $(-5)^{17}$ are negative since there are an odd number of factors. Note that $\left|(-5)^{17}\right| > \left|(-2)^{17}\right|$ since $|-5| > |-2|$ and the exponent on each is the same. The numbers

from least to greatest are $(-5)^{17}$, $(-2)^{17}$, $(-2)^{12}$, $(-5)^{12}$.

137. answers may vary

The Bigger Picture

1. $-9 + 14 = 5$

2. $-5(-11) = 55$

3. $5 - 11 = 5 + (-11) = -6$

4. $58 - |-70| = 58 - 70 = 58 + (-70) = -12$

5. $18 + (-30) = -12$

6. $(-9)^2 = (-9)(-9) = 81$

7. $-9^2 = -(9 \cdot 9) = -81$

8. $-10 + (-24) = -34$

9. $1 - (-9) = 1 + 9 = 10$

10. $-15 - 15 = -15 + (-15) = -30$

11. $-3(2)(-5) = -6(-5) = 30$

12. $\dfrac{|-88|}{-|-8|} = \dfrac{88}{-8} = -11$

13. $2 + 4(7 - 9)^3 = 2 + 4(-2)^3$
$$= 2 + 4(-8)$$
$$= 2 + (-32)$$
$$= -30$$

14. $-100 - (-20) = -100 + 20 = -80$

15. $1 + 2(7) = 1 + 14 = 15$

16. $30 \div 2 \cdot 3 = 15 \cdot 3 = 45$

Integrated Review

1. Let 0 represent 0°F. Then 50 degrees below zero is represented by −50 and 122 degrees above zero is represented by +122 or 122.

2.

3. $0 > -10$ since 0 is to the right of −10 on a number line.

4. $-15 < -5$ since −15 is to the left of −5 on a number line.

5. $-4 < 4$ since −4 is to the left of 4 on a number line.

6. $-2 > -7$ since −2 is to the right of −7 on a number line.

7. $|-3| = 3$ because −3 is 3 units from 0.

8. $-|-4| = -4$

9. $|-9| = 9$ because −9 is 9 units from 0.

10. $-(-5) = 5$

11. The opposite of 11 is −11.

12. The opposite of −3 is $-(-3) = 3$.

13. The opposite of 64 is −64.

14. The opposite of 0 is $-0 = 0$.

15. $-3 + 15 = 12$

16. $-9 + (-11) = -20$

17. $-8(-6)(-1) = 48(-1) = -48$

18. $-18 \div 2 = -9$

19. $65 + (-55) = 10$

20. $1000 - 1002 = 1000 + (-1002) = -2$

21. $53 - (-53) = 53 + 53 = 106$

22. $-2 - 1 = -2 + (-1) = -3$

23. $\dfrac{0}{-47} = 0$

24. $\dfrac{-36}{-9} = 4$

25. $-17 - (-59) = -17 + 59 = 42$

26. $-8 + (-6) + 20 = -14 + 20 = 6$

27. $\dfrac{-95}{-5} = 19$

28. $-9(100) = -900$

29. $-12 - 6 - (-6) = -12 + (-6) + 6$
$$= -18 + 6$$
$$= -12$$

30. $-4 + (-8) - 16 - (-9) = -4 + (-8) + (-16) + 9$
$$= -12 + (-16) + 9$$
$$= -28 + 9$$
$$= -19$$

31. $\dfrac{-105}{0}$ is undefined.

32. $7(-16)(0)(-3) = 0$ (since one factor is 0)

33. Subtract −8 from −12 is
$-12 - (-8) = -12 + 8 = -4$.

34. The sum of −17 and −27 is
$-17 + (-27) = -44$.

35. The product of −5 and −25 is
$-5(-25) = 125$.

36. The quotient of −100 and −5 is $\dfrac{-100}{-5} = 20$.

37. Divide a number by −17 is $\dfrac{x}{-17}$ or
$x \div (-17)$.

38. The sum of -3 and a number is $-3 + x$.

39. A number decreased by -18 is $x - (-18)$.

40. The product of -7 and a number is $-7 \cdot x$ or $-7x$.

41. $x + y = -3 + 12 = 9$

42. $x - y = -3 - 12 = -3 + (-12) = -15$

43. $\begin{aligned} 2y - x &= 2(12) - (-3) \\ &= 24 - (-3) \\ &= 24 + 3 \\ &= 27 \end{aligned}$

44. $3y + x = 3(12) + (-3) = 36 + (-3) = 33$

45. $5x = 5(-3) = -15$

46. $\dfrac{y}{x} = \dfrac{12}{-3} = -4$

Section 2.5

Practice Problems

1. $(-2)^4 = (-2)(-2)(-2)(-2) = 16$

2. $-2^4 = -(2)(2)(2)(2) = -16$

3. $3 \cdot 6^2 = 3 \cdot (6 \cdot 6) = 3 \cdot 36 = 108$

4. $\dfrac{-25}{5(-1)} = \dfrac{-25}{-5} = 5$

5. $\dfrac{-18 + 6}{-3 - 1} = \dfrac{-12}{-4} = 3$

6. $\begin{aligned} 30 + 50 + (-4)^3 &= 30 + 50 + (-64) \\ &= 80 + (-64) \\ &= 16 \end{aligned}$

7. $-2^3 + (-4)^2 + 1^5 = -8 + 16 + 1 = 8 + 1 = 9$

8. $2(2-9)+(-12)-3 = 2(-7)+(-12)-3$
$$= -14+(-12)-3$$
$$= -26-3$$
$$= -29$$

9. $(-5)\cdot\left|-8\right|+(-3)+2^3 = (-5)\cdot 8+(-3)+2^3$
$$= (-5)\cdot 8+(-3)+8$$
$$= -40+(-3)+8$$
$$= -43+8$$
$$= -35$$

10. $-4[-6+5(-3+5)]-7 = -4[-6+5(2)]-7$
$$= -4[-6+10]-7$$
$$= -4(4)-7$$
$$= -16-7$$
$$= -23$$

11. $x^2 = (-15)^2 = (-15)(-15) = 225$

   $-x^2 = -(-15)^2 = -(-15)(-15) = -225$

12. $5y^2 = 5(4)^2 = 5(16) = 80$

   $5y^2 = 5(-4)^2 = 5(16) = 80$

13. $x^2+y = (-6)^2+(-3) = 36+(-3) = 33$

14. $4-x^2 = 4-(-8)^2 = 4-64 = -60$

15. $\text{average} = \dfrac{\text{sum of numbers}}{\text{number of numbers}}$
$$= \frac{15+(-1)+(-11)+(-14)+(-16)+(-14)+(-1)}{7}$$
$$= \frac{-42}{7}$$
$$= -6$$

The average of the temperatures is $-6°F$.

Calculator Explorations

1. $\dfrac{-120-360}{-10} = 48$

3. $\dfrac{-316+(-458)}{28+(-25)} = -258$

60

Vocabulary and Readiness Check

1. To simplify $-2 \div 2 \cdot (3)$ which operation should be performed first? <u>division</u>

3. The <u>average</u> of a list of numbers is
$$\frac{\text{sum of numbers}}{\text{number of numbers}}.$$

5. To simplify $-2 + 3(10 - 12) \cdot (-8)$, which operation should be performed first? <u>subtraction</u>

Exercise Set 2.5

1. $(-5)^3 = (-5)(-5)(-5) = -125$

3. $-4^3 = -(4)(4)(4) = -64$

5. $8 \cdot 2^2 = 8 \cdot 4 = 32$

7. $8 - 12 - 4 = -4 - 4 = -8$

9. $7 + 3(-6) = 7 + (-18) = -11$

11. $5(-9) + 2 = -45 + 2 = -43$

13. $(-10) + 4 \div 2 = (-10) + 2 = -8$

15. $6 + 7 \cdot 3 - 10 = 6 + 21 - 10$
$$= 27 - 10$$
$$= 17$$

17. $\dfrac{16 - 13}{-3} = \dfrac{3}{-3} = -1$

19. $\dfrac{24}{10 + (-4)} = \dfrac{24}{6} = 4$

21. $5(-3) - (-12) = -15 + 12 = -3$

23. $[8 + (-4)]^2 = [4]^2 = 16$

25. $8 \cdot 6 - 3 \cdot 5 + (-20) = 48 - 3 \cdot 5 + (-20)$
$$= 48 - 15 + (-20)$$
$$= 33 + (-20)$$
$$= 13$$

27. $4 - (-3)^4 = 4 - 81 = -77$

29. $|7 + 3| \cdot 2^3 = |10| \cdot 2^3 = 10 \cdot 2^3 = 10 \cdot 8 = 80$

31. $7 \cdot 6^2 + 4 = 7 \cdot 36 + 4 = 252 + 4 = 256$

33. $7^2 - (4 - 2^3) = 7^2 - (4 - 8)$
$$= 7^2 - (-4)$$
$$= 49 + 4$$
$$= 53$$

35. $|3 - 15| \div 3 = |-12| \div 3 = 12 \div 3 = 4$

37. $-(-2)^6 = -64$

39. $(5 - 9)^2 \div (4 - 2)^2 = (-4)^2 \div (2)^2$
$$= 16 \div 4$$
$$= 4$$

41. $|8 - 24| \cdot (-2) \div (-2) = |-16| \cdot (-2) \div (-2)$
$$= 16 \cdot (-2) \div (-2)$$
$$= -32 \div (-2)$$
$$= 16$$

43. $(-12 - 20) \div 16 - 25 = (-32) \div 16 - 25$
$$= -2 - 25$$
$$= -27$$

45. $5(5 - 2) + (-5)^2 - 6 = 5(3) + (-5)^2 - 6$
$$= 5(3) + 25 - 6$$
$$= 15 + 25 - 6$$
$$= 40 - 6$$
$$= 34$$

47. $(2 - 7) \cdot (6 - 19) = (-5) \cdot (-13)$
$$= 65$$

49. $(-36 \div 6) - (4 \div 4) = -6 - 1 = -7$

51. $(10 - 4^2)^2 = (10 - 16)^2 = (-6)^2 = 36$

53. $2(8 - 10)^2 - 5(1 - 6)^2 = 2(-2)^2 - 5(-5)^2$
$$= 2(4) - 5(25)$$
$$= 8 - 125$$
$$= -117$$

55. $3(-10) \div [5(-3) - 7(-2)]$
$$= 3(-10) \div [-15 + 14]$$
$$= 3(-10) \div (-1)$$
$$= -30 \div (-1)$$
$$= 30$$

57. $\dfrac{(-7)(-3) - (4)(3)}{3[7 \div (3 - 10)]} = \dfrac{21 - 12}{3[7 \div (-7)]}$
$$= \dfrac{9}{3(-1)}$$
$$= \dfrac{9}{-3}$$
$$= -3$$

59. $-3[5 + 2(-4 + 9)] + 15 = -3[5 + 2(5)] + 15$
$$= -3[5 + 10] + 15$$
$$= -3(15) + 15$$
$$= -45 + 15$$
$$= -30$$

61. $x + y + z = -2 + 4 + (-1) = 2 + (-1) = 1$

63. $2x - 3y - 4z = 2(-2) - 3(4) - 4(-1)$
$$= -4 - 12 + 4$$
$$= -16 + 4$$
$$= -12$$

65. $x^2 - y = (-2)^2 - 4 = 4 - 4 = 0$

67. $\dfrac{5y}{z} = \dfrac{5(4)}{-1} = \dfrac{20}{-1} = -20$

69. $x^2 = (-3)^2 = 9$

71. $-z^2 = -(-4)^2 = -16$

73. $2z^3 = 2(-4)^3 = 2(-64) = -128$

75. $10 - x^2 = 10 - (-3)^2 = 10 - 9 = 1$

77. $2x^3 - z = 2(-3)^3 - (-4)$
$$= 2(-27) + 4$$
$$= -54 + 4$$
$$= -50$$

79. average
$$= \dfrac{-10 + 8 + (-4) + 2 + 7 + (-5) + (-12)}{7}$$
$$= \dfrac{-14}{7}$$
$$= -2$$

81. average $= \dfrac{-17 + (-26) + (-20) + (-13)}{4}$
$$= \dfrac{-76}{4}$$
$$= -19$$

83. The lowest score is -13 and the highest is 15.
$15 - (-13) = 15 + 13 = 28$
The difference between the lowest score and the highest score is 28.

85. average $= \dfrac{-13 + (-2) + 0 + 5 + 7 + 15}{6}$
$$= \dfrac{12}{6}$$
$$= 2$$
The average of the scores is 2.

87. no; answers may vary

89.
$$\begin{array}{r} 45 \\ \times\ 90 \\ \hline 4050 \end{array}$$

91.
$$\begin{array}{r} 90 \\ -\ 45 \\ \hline 45 \end{array}$$

93. $8 + 8 + 8 + 8 = 32$
The perimeter is 32 inches.

95. $9 + 6 + 9 + 6 = 30$
The perimeter is 30 feet.

97. $2 \cdot (7 - 5) \cdot 3 = 2 \cdot 2 \cdot 3 = 4 \cdot 3 = 12$

99. $-6 \cdot (10 - 4) = -6 \cdot 6 = -36$

101. answers may vary

103. answers may vary

105. $(-12)^4 = (-12)(-12)(-12)(-12) = 20,736$

107. $x^3 - y^2 = (21)^3 - (-19)^2$
$ = 9261 - 361$
$ = 8900$

109. $(xy + z)^x = [2(-5) + 7]^2$
$ = [-10 + 7]^2$
$ = [-3]^2$
$ = 9$

Section 2.6

Practice Problems

1. $-4x - 3 = 5$
$-4(-2) - 3 \stackrel{?}{=} 5$
$8 - 3 \stackrel{?}{=} 5$
$5 \stackrel{?}{=} 5 \quad$ True

Since $5 = 5$ is true, -2 is a solution of the equation.

2. $y - 6 = -2$
$y - 6 + 6 = -2 + 6$
$y = 4$
Check: $y - 6 = -2$
$4 - 6 \stackrel{?}{=} -2$
$-2 = -2 \quad$ True
The solution is 4.

3. $-2 = z + 8$
$-2 - 8 = z + 8 - 8$
$-10 = z$
Check: $-2 = z + 8$
$-2 \stackrel{?}{=} -10 + 8$
$-2 = -2 \quad$ True
The solution is -10.

4. $10x = -2 + 9x$
$10x - 9x = -2 + 9x - 9x$
$x = -2$
Check: $10x = -2 + 9x$
$10(-2) \stackrel{?}{=} -2 + 9(-2)$
$-20 \stackrel{?}{=} -2 + (-18)$
$-20 = -20 \quad$ True
The solution is -2.

5. $3y = -18$
$\dfrac{3y}{3} = \dfrac{-18}{3}$
$\dfrac{3}{3} \cdot y = \dfrac{-18}{3}$
$y = -6$
Check: $3y = -18$
$3(-6) \stackrel{?}{=} -18$
$-18 = -18 \quad$ True
The solution is -6.

6. $-32 = 8x$
$\dfrac{-32}{8} = \dfrac{8x}{8}$
$\dfrac{-32}{8} = \dfrac{8}{8} \cdot x$
$-4 = x$
Check: $-32 = 8x$
$-32 \stackrel{?}{=} 8(-4)$
$-32 = -32 \quad$ True
The solution is -4.

7. $-3y = -27$

$$\frac{-3y}{-3} = \frac{-27}{-3}$$

$$\frac{-3}{-3} \cdot y = \frac{-27}{-3}$$

$$y = 9$$

Check: $-3y = -27$

$$-3 \cdot 9 \stackrel{?}{=} -27$$

$$-27 = -27 \quad \text{True}$$

The solution is 9.

8. $\dfrac{x}{-4} = 7$

$$-4 \cdot \frac{x}{-4} = -4 \cdot 7$$

$$\frac{-4}{-4} \cdot x = -4 \cdot 7$$

$$x = -28$$

Check: $\dfrac{x}{-4} = 7$

$$\frac{-28}{-4} \stackrel{?}{=} 7$$

$$7 = 7 \quad \text{True}$$

The solution is -28.

Vocabulary and Readiness Check

1. A combination of operations on variables and numbers is called an <u>expression</u>.

3. An <u>equation</u> contains an equal sign ($=$) while an <u>expression</u> does not.

5. A <u>solution</u> of an equation is a number that when substituted for a variable makes the equation a true statement.

7. By the <u>addition</u> property of equality, the same number may be added to or subtracted from both sides of an equation without changing the solution of the equation.

Exercise Set 2.6

1. $x - 8 = -2$

$$6 - 8 \stackrel{?}{=} -2$$

$$-2 \stackrel{?}{=} -2 \quad \text{True}$$

Since $-2 = -2$ is true, 6 is a solution of the equation.

3. $x + 12 = 17$

$$-5 + 12 \stackrel{?}{=} 17$$

$$7 \stackrel{?}{=} 17 \quad \text{False}$$

Since $7 = 17$ is false, -5 is not a solution of the equation.

5. $-9f = 64 - f$

$$-9(-8) \stackrel{?}{=} 64 - (-8)$$

$$72 \stackrel{?}{=} 64 + 8$$

$$72 \stackrel{?}{=} 72 \quad \text{True}$$

Since $72 = 72$ is true, -8 is a solution of the equation.

7. $5(c - 5) = -10$

$$5(3 - 5) \stackrel{?}{=} -10$$

$$5(-2) \stackrel{?}{=} -10$$

$$-10 \stackrel{?}{=} -10 \quad \text{True}$$

Since $-10 = -10$ is true, 3 is a solution of the equation.

9. $a + 5 = 23$

$$a + 5 - 5 = 23 - 5$$

$$a = 18$$

Check: $a + 5 = 23$

$$18 + 5 \stackrel{?}{=} 23$$

$$23 = 23 \quad \text{True}$$

The solution is 18.

11. $d - 9 = -21$

$$d - 9 + 9 = -21 + 9$$

$$d = -12$$

Check: $d - 9 = -21$

$$-12 - 9 \stackrel{?}{=} -21$$

$$-21 = -21 \quad \text{True}$$

The solution is -12.

13.
$$7 = y - 2$$
$$7 + 2 = y - 2 + 2$$
$$9 = y$$
Check: $7 = y - 2$
$$7 \overset{?}{=} 9 - 2$$
$$7 = 7 \quad \text{True}$$
The solution is 9.

15.
$$11x = 10x - 17$$
$$11x - 10x = 10x - 10x - 17$$
$$1x = -17$$
$$x = -17$$
Check: $11x = 10x - 17$
$$11(-17) \overset{?}{=} 10(-17) - 17$$
$$-187 \overset{?}{=} -170 - 17$$
$$-187 = -187 \quad \text{True}$$
The solution is -17.

17.
$$5x = 20$$
$$\frac{5x}{5} = \frac{20}{5}$$
$$\frac{5}{5} \cdot x = \frac{20}{5}$$
$$x = 4$$
Check: $5x = 20$
$$5(4) \overset{?}{=} 20$$
$$20 = 20 \quad \text{True}$$
The solution is 4.

19.
$$-3z = 12$$
$$\frac{-3z}{-3} = \frac{12}{-3}$$
$$\frac{-3}{-3} \cdot z = \frac{12}{-3}$$
$$z = -4$$
Check: $-3z = 12$
$$-3(-4) \overset{?}{=} 12$$
$$12 = 12 \quad \text{True}$$
The solution is -4.

21.
$$\frac{n}{7} = -2$$
$$7 \cdot \frac{n}{7} = 7 \cdot (-2)$$
$$\frac{7}{7} \cdot n = 7 \cdot (-2)$$
$$n = -14$$
Check: $\dfrac{n}{7} = -2$
$$\frac{-14}{7} \overset{?}{=} -2$$
$$-2 = -2 \quad \text{True}$$
The solution is -14.

23.
$$2z = -34$$
$$\frac{2z}{2} = \frac{-34}{2}$$
$$\frac{2}{2} \cdot z = \frac{-34}{2}$$
$$z = -17$$
Check: $2z = -34$
$$2 \cdot (-17) \overset{?}{=} -34$$
$$-34 = -34 \quad \text{True}$$
The solution is -17.

25.
$$-4y = 0$$
$$\frac{-4y}{-4} = \frac{0}{-4}$$
$$\frac{-4}{-4} \cdot y = \frac{0}{-4}$$
$$y = 0$$
Check: $-4y = 0$
$$-4(0) \overset{?}{=} 0$$
$$0 = 0 \quad \text{True}$$
The solution is 0.

27. $-10x = -10$

$$\frac{-10x}{-10} = \frac{-10}{-10}$$

$$\frac{-10}{-10} \cdot x = \frac{-10}{-10}$$

$$x = 1$$

Check: $-10x = -10$

$$-10 \cdot 1 \overset{?}{=} -10$$

$$-10 = -10 \quad \text{True}$$

The solution is 1.

29. $5x = -35$

$$\frac{5x}{5} = \frac{-35}{5}$$

$$\frac{5}{5} \cdot x = \frac{-35}{5}$$

$$x = -7$$

The solution is −7.

31. $n - 5 = -55$

$$n - 5 + 5 = -55 + 5$$

$$n = -50$$

The solution is −50.

33. $-15 = y + 10$

$$-15 - 10 = y + 10 - 10$$

$$-25 = y$$

The solution is −25.

35. $\dfrac{x}{-6} = -6$

$$-6 \cdot \frac{x}{-6} = -6 \cdot (-6)$$

$$\frac{-6}{-6} \cdot x = -6 \cdot (-6)$$

$$x = 36$$

The solution is 36.

37. $11n = 10n + 21$

$$11n - 10n = 10n - 10n + 21$$

$$1n = 21$$

$$n = 21$$

The solution is 21.

39. $-12y = -144$

$$\frac{-12y}{-12} = \frac{-144}{-12}$$

$$\frac{-12}{-12} \cdot y = \frac{-144}{-12}$$

$$y = 12$$

The solution is 12.

41. $\dfrac{n}{4} = -20$

$$4 \cdot \frac{n}{4} = 4 \cdot (-20)$$

$$\frac{4}{4} \cdot n = 4 \cdot (-20)$$

$$n = -80$$

The solution is −80.

43. $-64 = 32y$

$$\frac{-64}{32} = \frac{32y}{32}$$

$$\frac{-64}{32} = \frac{32}{32} \cdot y$$

$$-2 = y$$

The solution is −2.

45. A number decreased by −2 is $x - (-2)$.

47. The product of −6 and a number is $-6 \cdot x$ or $-6x$.

49. The sum of −15 and a number is $-15 + x$.

51. −8 divided by a number is $-8 \div x$ or $\dfrac{-8}{x}$.

53. $n - 42,860 = -1286$

$$n - 42,860 + 42,860 = -1286 + 42,860$$

$$n = 41,574$$

The solution is 41,574.

55.
$$-38x = 15,542$$
$$\frac{-38x}{-38} = \frac{15,542}{-38}$$
$$\frac{-38}{-38} \cdot x = \frac{15,542}{-38}$$
$$x = -409$$
The solution is −409.

57. answers may vary

59. answers may vary

Chapter 2 Vocabulary Check

1. Two numbers that are the same distance from 0 on the number line but are on opposite sides of 0 are called <u>opposites</u>.

2. The <u>absolute value</u> of a number is that number's distance from 0 on a number line.

3. The <u>integers</u> are ..., −3, −2, −1, 0, 1, 2, 3, ...

4. The <u>negative</u> numbers are numbers less than zero.

5. The <u>positive</u> numbers are numbers greater than zero.

6. The symbols "<" and ">" are called <u>inequality symbols</u>.

7. A <u>solution</u> of an equation is a number that when substituted for a variable makes the equation a true statement.

8. The <u>average</u> of a list of numbers is $\dfrac{\text{sum of numbers}}{\textit{number} \text{ of numbers}}$.

9. A combination of operations on variables and numbers is called and <u>expression</u>.

10. A statement of the form "expression = expression" is called an <u>equation</u>.

11. The sign "<" means <u>is less than</u> and ">" means <u>is greater than</u>.

12. By the <u>addition</u> property of equality, the same number may be added to or subtracted from both sides of an equation without changing the solution of the equation.

13. By the <u>multiplication</u> property of equality, the same nonzero number may be multiplied or divided by both sides of an equation without changing the solution of the equation.

Chapter 2 Review

1. If 0 represents ground level, then 1572 feet below the ground is −1572.

2. If 0 represents sea level, then an elevation of 11,239 feet is +11,239.

3.

4.

5. $|-11| = 11$ since -11 is 11 units from 0 on a number line.

6. $|0| = 0$ since 0 is 0 units from 0 on a number line.

7. $-|8| = -8$

8. $-(-9) = 9$

9. $-|-16| = -16$

10. $-(-2) = 2$

11. $-18 > -20$ since -18 is to the right of -20 on a number line.

12. $-5 < 5$ since -5 is to the left of 5 on a number line.

13. $|-123| = 123$
$-|-198| = -198$
Since $123 > -198$, $|-123| > -|-198|$.

14. $8 - |-12| = 8 - 12 = 8 + (-12) = -4$
$-|-16| = -16$
Since $-4 > -16$, $8 - |-12| > -|-16|$.

15. The opposite of -18 is 18.
$-(-18) = 18$

16. The opposite of 42 is negative 42.
$-(42) = -42$

17. False; consider $a = 1$ and $b = 2$, then $1 < 2$.

18. True

19. True

20. True

21. $|y| = |-2| = 2$

22. $|-x| = |-(-3)| = |3| = 3$

23. $-|-z| = -|-(-5)| = -|5| = -5$

24. $-|-n| = -|-(-10)| = -|10| = -10$

25. The bar that extends the farthest in the negative direction corresponds to Elevator D, so Elevator D extends the farthest below ground.

26. The bar that extends the farthest in the positive direction corresponds to Elevator B, so Elevator B extends the highest above ground.

27. $|5| - |-3| = 5 - 3 = 2$
    $5 > 3$, so the answer is positive.
    $5 + (-3) = 2$

28. $|18| - |-4| = 18 - 4 = 14$
    $18 > 4$, so the answer is positive.
    $18 + (-4) = 14$

29. $|16| - |-12| = 16 - 12 = 4$
    $16 > 12$, so the answer is positive.
    $-12 + 16 = 4$

30. $|40| - |-23| = 40 - 23 = 17$
    $40 > 23$, so the answer is positive.
    $-23 + 40 = 17$

31. $|-8| + |-15| = 8 + 15 = 23$
    The common sign is negative, so
    $-8 + (-15) = -23$.

32. $|-5| + |-17| = 5 + 17 = 22$
    The common sign is negative, so
    $-5 + (-17) = -22$.

33. $|-24| - |3| = 24 - 3 = 21$
    $24 > 3$, so the answer is negative.
    $-24 + 3 = -21$

34. $|-89| - |19| = 89 - 19 = 70$
    $89 > 19$, so the answer is negative.
    $-89 + 19 = -70$

35. $15 + (-15) = 0$

36. $-24 + 24 = 0$

37. $|-43| + |-108| = 43 + 108 = 151$
    The common sign is negative, so
    $-43 + (-108) = -151$.

38. $|-100| + |-506| = 100 + 506 = 606$
    The common sign is negative, so
    $-100 + (-506) = -606$.

39. $-15 + (-5) = -20$
    The temperature at 6 a.m. is $-20°$C.

40. $-127 + (-23) = -150$
    The diver's current depth is -150 feet.

41. $-2 + 0 + (-2) + (-3) = -2 + (-2) + (-3)$
    $\qquad\qquad\qquad\qquad = -4 + (-3)$
    $\qquad\qquad\qquad\qquad = -7$
    His total score was -7.

42. $-7 + 8 = +1$
    The team's score was $+1$.

43. $12 - 4 = 12 + (-4) = 8$

44. $-12 - 4 = -12 + (-4) = -16$

45. $-7 - 17 = -7 + (-17) = -24$

46. $7 - 17 = 7 + (-17) = -10$

47. $7 - (-13) = 7 + 13 = 20$

48. $-6 - (-14) = -6 + 14 = 8$

49. $16 - 16 = 16 + (-16) = 0$

50. $-16 - 16 = -16 + (-16) = -32$

51. $-12 - (-12) = -12 + 12 = 0$

52. $-5 - (-12) = -5 + 12 = 7$

53. $-(-5) - 12 + (-3) = 5 + (-12) + (-3)$
    $\qquad\qquad\qquad\qquad = -7 + (-3)$
    $\qquad\qquad\qquad\qquad = -10$

54. $-8 + (-12) - 10 - (-3)$
    $= -8 + (-12) + (-10) + 3$
    $= -20 + (-10) + 3$
    $= -30 + 3$
    $= -27$

55. $600 - (-92) = 600 + 92 = 692$
The difference in elevations is 692 feet.

56. $142 - 125 + 43 - 85$
$= 142 + (-125) + 43 + (-85)$
$= 17 + 43 + (-85)$
$= 60 + (-85)$
$= -25$
The balance in his account is -25.

57. $85 - 99 = 85 + (-99) = -14$
You are -14 feet or 14 feet below ground at the end of the drop.

58. $66 - (-16) = 66 + 16 = 82$
The total length of the elevator shaft for Elevator C is 82 feet.

59. $|-5| - |-6| = 5 - 6 = 5 + (-6) = -1$
$5 - 6 = 5 + (-6) = -1$
$|-5| - |-6| = 5 - 6$ is true.

60. $|-5 - (-6)| = |-5 + 6| = |1| = 1$
$5 + 6 = 11$
Since $1 \neq 11$, the statement is false.

61. $-3(-7) = 21$

62. $-6(3) = -18$

63. $-4(16) = -64$

64. $-5(-12) = 60$

65. $(-5)^2 = (-5)(-5) = 25$

66. $(-1)^5 = (-1)(-1)(-1)(-1)(-1) = -1$

67. $12(-3)(0) = 0$

68. $-1(6)(2)(-2) = -6(2)(-2) = -12(-2) = 24$

69. $-15 \div 3 = -5$

70. $\dfrac{-24}{-8} = 3$

71. $\dfrac{0}{-3} = 0$

72. $\dfrac{-46}{0}$ is undefined.

73. $\dfrac{100}{-5} = -20$

74. $\dfrac{-72}{8} = -9$

75. $\dfrac{-38}{-1} = 38$

76. $\dfrac{45}{-9} = -5$

77. A loss of 5 yards is represented by -5.
$(-5)(2) = -10$
The total loss is 10 yards.

78. A loss of $50 is represented by -50.
$(-50)(4) = -200$
The total loss is $200.

79. A debt of $1024 is represented by -1024.
$-1024 \div 4 = -256$
Each payment is $256.

80. A drop of 45 degrees is represented by -45.
$\dfrac{-45}{9} = -5$ or $-45 \div 9 = -5$
The average drop each hour is $5°F$.

81. $(-7)^2 = (-7)(-7) = 49$

82. $-7^2 = -(7 \cdot 7) = -49$

83. $5 - 8 + 3 = -3 + 3 = 0$

84. $-3 + 12 + (-7) - 10 = 9 + (-7) - 10$
$= 2 - 10$
$= -8$

85. $-10 + 3 \cdot (-2) = -10 + (-6) = -16$

86. $5 - 10 \cdot (-3) = 5 - (-30) = 5 + 30 = 35$

87. $16 \div (-2) \cdot 4 = -8 \cdot 4 = -32$

88. $-20 \div 5 \cdot 2 = -4 \cdot 2 = -8$

89. $16 + (-3) \cdot 12 \div 4 = 16 + (-36) \div 4$
$$= 16 + (-9)$$
$$= 7$$

90. $(-12) + 10 \div (-5) = -12 + (-2) = -14$

91. $4^3 - (8 - 3)^2 = 4^3 - (5)^2 = 64 - 25 = 39$

92. $(-3)^3 - 90 = -27 - 90 = -117$

93. $\dfrac{(-4)(-3) - (-2)(-1)}{-10 + 5} = \dfrac{12 - 2}{-5} = \dfrac{10}{-5} = -2$

94. $\dfrac{4(12 - 18)}{-10 \div (-2 - 3)} = \dfrac{4(-6)}{-10 \div (-5)} = \dfrac{-24}{2} = -12$

95. average $= \dfrac{-18 + 25 + (-30) + 7 + 0 + (-2)}{6}$
$$= \dfrac{-18}{6}$$
$$= -3$$

96. average $= \dfrac{-45 + (-40) + (-30) + (-25)}{4}$
$$= \dfrac{-140}{4}$$
$$= -35$$

97. $2x - y = 2(-2) - 1 = -4 - 1 = -5$

98. $y^2 + x^2 = 1^2 + (-2)^2 = 1 + 4 = 5$

99. $\dfrac{3x}{6} = \dfrac{3(-2)}{6} = \dfrac{-6}{6} = -1$

100. $\dfrac{5y - x}{-y} = \dfrac{5(1) - (-2)}{-1} = \dfrac{5 + 2}{-1} = \dfrac{7}{-1} = -7$

101. $2n - 6 = 16$
$$2(-5) - 6 \stackrel{?}{=} 16$$
$$-10 - 6 \stackrel{?}{=} 16$$
$$-16 \stackrel{?}{=} 16 \quad \text{False}$$
Since $-16 = 16$ is false, -5 is not a solution of the equation.

102. $2(c - 8) = -20$
$$2(-2 - 8) \stackrel{?}{=} -20$$
$$2(-10) \stackrel{?}{=} -20$$
$$-20 \stackrel{?}{=} -20 \quad \text{True}$$
Since $-20 = -20$ is true, -2 is a solution of the equation.

103. $n - 7 = -20$
$$n - 7 + 7 = -20 + 7$$
$$n = -13$$
The solution is -13.

104. $-5 = n + 15$
$$-5 - 15 = n + 15 - 15$$
$$-20 = n$$
The solution is -20.

105. $10x = -30$
$$\dfrac{10x}{10} = \dfrac{-30}{10}$$
$$\dfrac{10}{10} \cdot x = \dfrac{-30}{10}$$
$$x = -3$$
The solution is -3.

106. $-8x = 72$
$$\dfrac{-8x}{-8} = \dfrac{72}{-8}$$
$$\dfrac{-8}{-8} \cdot x = \dfrac{72}{-8}$$
$$x = -9$$
The solution is -9.

107.
$$9y = 8y - 13$$
$$9y - 8y = 8y - 8y - 13$$
$$1y = -13$$
$$y = -13$$
The solution is -13.

108.
$$6x - 31 = 7x$$
$$6x - 6x - 31 = 7x - 6x$$
$$-31 = 1x$$
$$-31 = x$$
The solution is -31.

109.
$$\frac{n}{-4} = -11$$
$$-4 \cdot \frac{n}{-4} = -4 \cdot (-11)$$
$$\frac{-4}{-4} \cdot n = -4 \cdot (-11)$$
$$n = 44$$
The solution is 44.

110.
$$\frac{x}{-2} = 13$$
$$-2 \cdot \frac{x}{-2} = -2 \cdot 13$$
$$\frac{-2}{-2} \cdot x = -2 \cdot 13$$
$$x = -26$$
The solution is -26.

111.
$$n + 12 = -7$$
$$n + 12 - 12 = -7 - 12$$
$$n = -19$$
The solution is -19.

112.
$$n - 40 = -2$$
$$n - 40 + 40 = -2 + 40$$
$$n = 38$$
The solution is 38.

113.
$$-36 = -6x$$
$$\frac{-36}{-6} = \frac{-6x}{-6}$$
$$\frac{-36}{-6} = \frac{-6}{-6} \cdot x$$
$$6 = x$$
The solution is 6.

114.
$$-40 = 8y$$
$$\frac{-40}{8} = \frac{8y}{8}$$
$$\frac{-40}{8} = \frac{8}{8} \cdot y$$
$$-5 = y$$
The solution is -5.

115. $-6 + (-9) = -15$

116. $-16 - 3 = -16 + (-3) = -19$

117. $-4(-12) = 48$

118. $\dfrac{84}{-4} = -21$

119. $-76 - (-97) = -76 + 97 = 21$

120. $-9 + 4 = -5$

121. $-32 + 23 = -9$
His financial situation can be represented by $-\$9$.

122. $-11 + 17 = 6$
The temperature at noon on Tuesday was $6°C$.

123. $12{,}923 - (-195) = 12{,}923 + 195 = 13{,}118$
The difference in elevations is $13{,}118$ feet.

124. $-18 - 9 = -27$
The temperature on Friday was $-27°C$.

125. $(3-7)^2 \div (6-4)^3 = (-4)^2 \div (2)^3 = 16 \div 8 = 2$

126. $3(4+2)+(-6)-3^2 = 3(6)+(-6)-3^2$
$$= 3(6)+(-6)-9$$
$$= 18+(-6)-9$$
$$= 12-9$$
$$= 3$$

127. $2-4\cdot3+5 = 2-12+5 = -10+5 = -5$

128. $4-6\cdot5+1 = 4-30+1 = -26+1 = -25$

129. $\dfrac{-|-14|-6}{7+2(-3)} = \dfrac{-14-6}{7+(-6)} = \dfrac{-20}{1} = -20$

130. $5(7-6)^3 - 4(2-3)^2 + 2^4$
$$= 5(1)^3 - 4(-1)^2 + 2^4$$
$$= 5(1) - 4(1) + 16$$
$$= 5 - 4 + 16$$
$$= 1 + 16$$
$$= 17$$

131. $n - 9 = -30$
$$n - 9 + 9 = -30 + 9$$
$$n = -21$$
The solution is -21.

132. $n + 18 = 1$
$$n + 18 - 18 = 1 - 18$$
$$n = -17$$
The solution is -17.

133. $-4x = -48$
$$\frac{-4x}{-4} = \frac{-48}{-4}$$
$$\frac{-4}{-4} \cdot x = \frac{-48}{-4}$$
$$x = 12$$
The solution is 12.

134. $9x = -81$
$$\frac{9x}{9} = \frac{-81}{9}$$
$$\frac{9}{9} \cdot x = \frac{-81}{9}$$
$$x = -9$$
The solution is -9.

135. $\dfrac{n}{-2} = 100$
$$-2 \cdot \frac{n}{-2} = -2 \cdot 100$$
$$\frac{-2}{-2} \cdot n = -2 \cdot 100$$
$$n = -200$$
The solution is -200.

136. $\dfrac{y}{-1} = -3$
$$-1 \cdot \frac{y}{-1} = -1(-3)$$
$$\frac{-1}{-1} \cdot y = -1 \cdot (-3)$$
$$y = 3$$
The solution is 3.

Chapter 2 Test

1. $-5 + 8 = 3$

2. $18 - 24 = 18 + (-24) = -6$

3. $5 \cdot (-20) = -100$

4. $(-16) \div (-4) = 4$

5. $-18 + (-12) = -30$

6. $-7 - (-19) = -7 + 19 = 12$

7. $(-5) \cdot (-13) = 65$

8. $\dfrac{-25}{-5} = 5$

9. $|-25| + (-13) = 25 + (-13) = 12$

10. $14 - |-20| = 14 - 20 = 14 + (-20) = -6$

11. $|5| \cdot |-10| = 5 \cdot 10 = 50$

12. $\dfrac{|-10|}{-|-5|} = \dfrac{10}{-5} = -2$

13. $(-8) + 9 \div (-3) = -8 + (-3) = -11$

14. $-7+(-32)-12+5 = -7+(-32)+(-12)+5$
$\qquad\qquad\qquad\qquad = -39+(-12)+5$
$\qquad\qquad\qquad\qquad = -51+5$
$\qquad\qquad\qquad\qquad = -46$

15. $(-5)^3 -24\div(-3) = -125-24\div(-3)$
$\qquad\qquad\qquad\quad = -125-(-8)$
$\qquad\qquad\qquad\quad = -125+8$
$\qquad\qquad\qquad\quad = -117$

16. $(5-9)^2 \cdot (8-2)^3 = (-4)^2 \cdot (6)^3$
$\qquad\qquad\qquad\quad = 16\cdot 216$
$\qquad\qquad\qquad\quad = 3456$

17. $-(-7)^2 \div 7\cdot(-4) = -49\div 7\cdot(-4)$
$\qquad\qquad\qquad\quad = -7\cdot(-4)$
$\qquad\qquad\qquad\quad = 28$

18. $3-(8-2)^3 = 3-6^3$
$\qquad\qquad\quad = 3-216$
$\qquad\qquad\quad = 3+(-216)$
$\qquad\qquad\quad = -213$

19. $\dfrac{4}{2}-\dfrac{8^2}{16} = \dfrac{4}{2}-\dfrac{64}{16} = 2-4 = 2+(-4) = -2$

20. $\dfrac{-3(-2)+12}{-1(-4-5)} = \dfrac{6+12}{-1(-9)} = \dfrac{18}{9} = 2$

21. $\dfrac{|25-30|^2}{2(-6)+7} = \dfrac{|-5|^2}{-12+7} = \dfrac{(5)^2}{-5} = \dfrac{25}{-5} = -5$

22. $5(-8)-[6-(2-4)]+(12-16)^2$
$= 5(-8)-[6-(-2)]+(12-16)^2$
$= 5(-8)-(6+2)+(-4)^2$
$= 5(-8)-8+(-4)^2$
$= 5(-8)-8+16$
$= -40-8+16$
$= -48+16$
$= -32$

23. $7x+3y-4z = 7(0)+3(-3)-4(2)$
$\qquad\qquad\quad = 0+(-9)-8$
$\qquad\qquad\quad = -9-8$
$\qquad\qquad\quad = -17$

24. $10-y^2 = 10-(-3)^2 = 10-9 = 1$

25. $\dfrac{3z}{2y} = \dfrac{3(2)}{2(-3)} = \dfrac{6}{-6} = -1$

26. A descent of 22 feet is represented by –22.
$4(-22) = -88$
Mary is 88 feet below sea level.

27. $129+(-79)+(-40)+35 = 50+(-40)+35$
$\qquad\qquad\qquad\qquad\qquad = 10+35$
$\qquad\qquad\qquad\qquad\qquad = 45$
His new balance can be represented by 45.

28. Subtract the elevation of the Romanche Gap from the elevation of Mt. Washington.
$6288-(-25,354) = 6288+25,354 = 31,642$
The difference in elevations is 31,642 feet.

29. Subtract the depth of the lake from the elevation of the surface.
$1495-5315 = 1495+(-5315) = -3820$
The deepest point of the lake is 3820 feet below sea level.

30. average $= \dfrac{-12+(-13)+0+9}{4} = \dfrac{-16}{4} = -4$

31. a. The product of a number and 17 is $17x$.

    b. Twice a number subtracted from 20 is $20-2x$.

32. $-9n = -45$
$\dfrac{-9n}{-9} = \dfrac{-45}{-9}$
$\dfrac{-9}{-9}\cdot n = \dfrac{-45}{-9}$
$n = 5$
The solution is 5.

33.
$$\frac{n}{-7} = 4$$
$$-7 \cdot \frac{n}{-7} = -7 \cdot 4$$
$$\frac{-7}{-7} \cdot n = -7 \cdot 4$$
$$n = -28$$
The solution is -28.

34.
$$x - 16 = -36$$
$$x - 16 + 16 = -36 + 16$$
$$x = -20$$
The solution is -20.

35.
$$9x = 8x - 4$$
$$9x - 8x = 8x - 8x - 4$$
$$1x = -4$$
$$x = -4$$
The solution is -4.

Cumulative Review Chapters 1–2

1. The place value of 3 in 396,418 is hundred-thousands.

2. The place value of 3 in 4308 is hundreds.

3. The place value of 3 in 93,192 is thousands.

4. The place value of 3 is 693,298 is thousands.

5. The place value of 3 in 534,275,866 is ten-millions.

6. The place value of 3 in 267,301,818 is hundred-thousands.

7. **a.** $-7 < 7$ since -7 is to the left of 7 on a number line.

   b. $0 > -4$ since 0 is to the right of -4 on a number line.

   c. $-9 > -11$ since -9 is to the right of -11 on a number line.

8. **a.** $12 > 4$ since 12 is to the right of 4 on a number line.

   b. $13 < 31$ since 13 is to the left of 31 on a number line.

   c. $82 > 79$ since 82 is to the right of 79 on a number line.

9. $13 + 2 + 7 + 8 + 9 = (13 + 7) + (2 + 8) + 9$
   $$= 20 + 10 + 9$$
   $$= 39$$

10. $11 + 3 + 9 + 16 = (11 + 9) + (3 + 16)$
    $$= 20 + 19$$
    $$= 39$$

11.
$$\begin{array}{r} 7826 \\ -505 \\ \hline 7321 \end{array}$$
Check:
$$\begin{array}{r} 7321 \\ +505 \\ \hline 7826 \end{array}$$

12.
$$\begin{array}{r} 3285 \\ -272 \\ \hline 3013 \end{array}$$
Check:
$$\begin{array}{r} 3013 \\ +272 \\ \hline 3285 \end{array}$$

13. Subtract 7257 from the radius of Jupiter.
$$\begin{array}{r} 43,441 \\ -7\,257 \\ \hline 36,184 \end{array}$$
The radius of Saturn is 36,184 miles.

14. Subtract the cost of the camera from the amount in her account.
$$\begin{array}{r} 762 \\ -237 \\ \hline 525 \end{array}$$
She will have $525 left in her account after buying the camera.

15. To round 568 to the nearest ten, observe that the digit in the ones place is 8. Since this digit is at least 5, we add 1 to the digit in the tens place. The number 568 rounded to the nearest ten is 570.

16. To round 568 to the nearest hundred, observe that the digit in the tens place is 6. Since this digit is at least 5, we add 1 to the digit in the hundreds place. The number 568 rounded to the nearest hundred is 600.

17.

4725	rounds to	4700
− 2879	rounds to	− 2900
		1800

18.

8394	rounds to	8000
− 2913	rounds to	− 3000
		5000

19. a. $5(6 + 5) = 5 \cdot 6 + 5 \cdot 5$

    b. $20(4 + 7) = 20 \cdot 4 + 20 \cdot 7$

    c. $2(7 + 9) = 2 \cdot 7 + 2 \cdot 9$

20. a. $5(2 + 12) = 5 \cdot 2 + 5 \cdot 12$

    b. $9(3 + 6) = 9 \cdot 3 + 9 \cdot 6$

    c. $4(8 + 1) = 4 \cdot 8 + 4 \cdot 1$

21.

$$
\begin{array}{r}
631 \\
\times\ 125 \\
\hline
3\ 155 \\
12\ 620 \\
63\ 100 \\
\hline
78,875
\end{array}
$$

22.

$$
\begin{array}{r}
299 \\
\times\ 104 \\
\hline
1\ 196 \\
29\ 900 \\
\hline
31,096
\end{array}
$$

23. a. $42 \div 7 = 6$ because $6 \cdot 7 = 42$.

    b. $\dfrac{64}{8} = 8$ because $8 \cdot 8 = 64$.

    c. $3\overline{)21}^{\,7}$ because $7 \cdot 3 = 21$.

24. a. $\dfrac{35}{5} = 7$ because $7 \cdot 5 = 35$.

    b. $64 \div 8 = 8$ because $8 \cdot 8 = 64$.

    c. $4\overline{)48}^{\,12}$ because $12 \cdot 4 = 48$.

25.

$$
\begin{array}{r}
741 \\
5\overline{)3705} \\
\underline{-5} \\
20 \\
\underline{-20} \\
05 \\
\underline{-5} \\
0
\end{array}
$$

Check:

$$
\begin{array}{r}
741 \\
\times\ 5 \\
\hline
3705
\end{array}
$$

26.

$$
\begin{array}{r}
456 \\
8\overline{)3648} \\
\underline{-32} \\
44 \\
\underline{-40} \\
48 \\
\underline{-48} \\
0
\end{array}
$$

Check:

$$
\begin{array}{r}
456 \\
\times\ 8 \\
\hline
3648
\end{array}
$$

27. $\dfrac{\text{number of cards}}{\text{for each person}} = \dfrac{\text{number of}}{\text{cards}} \div \dfrac{\text{number of}}{\text{friends}}$

$= 238 \div 19$

$$\begin{array}{r} 12 \text{ R } 10 \\ 19\overline{)238} \\ -19 \\ \hline 48 \\ -38 \\ \hline 10 \end{array}$$

Each friend will receive 12 cards. There will be 10 cards left over.

28. $\dfrac{\text{Cost of each}}{\text{ticket}} = \dfrac{\text{total}}{\text{cost}} \div \text{number of tickets}$

$= 324 \div 36$

$$\begin{array}{r} 9 \\ 36\overline{)324} \\ -324 \\ \hline 0 \end{array}$$

Each ticket cost \$9.

29. $9^2 = 9 \cdot 9 = 81$

30. $5^3 = 5 \cdot 5 \cdot 5 = 125$

31. $6^1 = 6$

32. $4^1 = 4$

33. $5 \cdot 6^2 = 5 \cdot 6 \cdot 6 = 180$

34. $2^3 \cdot 7 = 2 \cdot 2 \cdot 2 \cdot 7 = 56$

35. $\dfrac{7 - 2 \cdot 3 + 3^2}{5(2-1)} = \dfrac{7 - 2 \cdot 3 + 9}{5(1)}$

$= \dfrac{7 - 6 + 9}{5}$

$= \dfrac{10}{5}$

$= 2$

36. $\dfrac{6^2 + 4 \cdot 4 + 2^3}{37 - 5^2} = \dfrac{36 + 4 \cdot 4 + 8}{37 - 25}$

$= \dfrac{36 + 16 + 8}{12}$

$= \dfrac{60}{12}$

$= 5$

37. $x + 6 = 8 + 6 = 14$

38. $5 + x = 5 + 9 = 14$

39. a. $|{-9}| = 9$ because -9 is 9 units from 0.

 b. $|3| = 3$ because 3 is 3 units from 0.

 c. $|0| = 0$ because 0 is 0 units from 0.

40. a. $|4| = 4$ because 4 is 4 units from 0.

 b. $|{-7}| = 7$ because -7 is 7 units from 0.

41. $-2 + 5 = 3$

42. $8 + (-3) = 5$

43. $2a - b = 2(8) - (-6)$

$= 16 - (-6)$

$= 16 + 6$

$= 22$

44. $x - y = -2 - (-7) = -2 + 7 = 5$

45. $-7 \cdot 3 = -21$

46. $5(-2) = -10$

47. $0 \cdot (-4) = 0$

48. $-6 \cdot 9 = -54$

49. $3(4 - 7) + (-2) - 5 = 3(-3) + (-2) - 5$

$= -9 + (-2) - 5$

$= -11 - 5$

$= -16$

50.

$$
\begin{aligned}
4 - 8(7 - 3) - (-1) &= 4 - 8(4) - (-1) \\
&= 4 - 32 - (-1) \\
&= 4 - 32 + 1 \\
&= -28 + 1 \\
&= -27
\end{aligned}
$$

Chapter 3

Practice Problems

1. **a.** $8m - 14m = (8 - 14)m = -6m$

   b. $6a + a = 6a + 1a = (6 + 1)a = 7a$

   c. $-y^2 + 3y^2 + 7 = -1y^2 + 3y^2 + 7$
   $= (-1 + 3)y^2 + 7$
   $= 2y^2 + 7$

2. $6z + 5 + z - 4 = 6z + 5 + 1z + (-4)$
   $= 6z + 1z + 5 + (-4)$
   $= (6 + 1)z + 5 + (-4)$
   $= 7z + 1$

3. $6y + 12y - 6 = 18y - 6$

4. $7y - 5 + y + 8 = 7y + (-5) + 1y + 8$
   $= 7y + 1y + (-5) + 8$
   $= 8y + 3$

5. $-7y + 2 - 2y - 9x + 12 - x$
   $= -7y + 2 + (-2y) + (-9x) + 12 + (-1x)$
   $= -7y + (-2y) + (-9x) + (-1x) + 2 + 12$
   $= -9y - 10x + 14$

6. $6(4a) = (6 \cdot 4)a = 24a$

7. $-8(9x) = (-8 \cdot 9)x = -72x$

8. $8(y + 2) = 8 \cdot y + 8 \cdot 2 = 8y + 16$

9. $(7a - 5) = 3 \cdot 7a - 3 \cdot 5 = 21a - 15$

10. $6(5 - y) = 6 \cdot 5 - 6 \cdot y = 30 - 6y$

11. $5(2y - 3) - 8 = 5(2y) - 5(3) - 8$
    $= 10y - 15 - 8$
    $= 10y - 23$

12. $-7(x - 1) + 5(2x + 3)$
    $= -7(x) - (-7)(1) + 5(2x) + 5(3)$
    $= -7x + 7 + 10x + 15$
    $= 3x + 22$

13. $-(y + 1) + 3y - 12 = -1(y + 1) + 3y - 12$
    $= -1 \cdot y + (-1)(1) + 3y - 12$
    $= -y - 1 + 3y - 12$
    $= 2y - 13$

14. $4(2x) = (4 \cdot 2)x = 8x$
    The perimeter is $8x$ centimeters.

15. $3(12y + 9) = 3 \cdot 12y + 3 \cdot 9 = 36y + 27$
    The area of the garden is $(36y + 27)$ square yards.

Vocabulary and Readiness Check

1. $14y^2 + 2x - 23$ is called an <u>expression</u> while $14y^2$, $2x$, and -23 are each called a <u>term</u>.

3. To simplify and expression like $y + 7y$, we <u>combine like terms</u>.

5. The term $5x$ is called a <u>variable</u> term while the term 7 is called a <u>constant</u> term.

7. By the <u>associative</u> properties, the grouping of adding or multiplying numbers can be changed without changing their sum or product.

9. For the term $-3x^2y$, -3 is called the <u>numerical coefficient</u>.

Exercise Set 3.1

1. $3x + 5x = (3 + 5)x = 8x$

3. $2n - 3n = (2 - 3)n = -1n = -n$

5. $4c + c - 7c = (4 + 1 - 7)c = -2c$

7. $4x - 6x + x - 5x = (4 - 6 + 1 - 5)x = -6x$

9. $3a + 2a + 7a - 5 = (3 + 2 + 7)a - 5$
$= 12a - 5$

11. $6(7x) = (6 \cdot 7)x = 42x$

13. $-3(11y) = (-3 \cdot 11)y = -33y$

15. $12(6a) = (12 \cdot 6)a = 72a$

17. $2(y + 3) = 2 \cdot y + 2 \cdot 3 = 2y + 6$

19. $3(a - 6) = 3 \cdot a - 3 \cdot 6 = 3a - 18$

21. $-4(3x + 7) = -4 \cdot 3x + (-4) \cdot 7 = -12x - 28$

23. $2(x + 4) - 7 = 2 \cdot x + 2 \cdot 4 - 7$
$= 2x + 8 - 7$
$= 2x + 1$

25. $8 + 5(3c - 1) = 8 + 5 \cdot 3c + 5 \cdot (-1)$
$= 8 + 15c - 5$
$= 15c + 8 - 5$
$= 15c + 3$

27. $-4(6n - 5) + 3n = -4 \cdot 6n - (-4) \cdot 5 + 3n$
$= -24n + 20 + 3n$
$= -24n + 3n + 20$
$= -21n + 20$

29. $3 + 6(w + 2) + w = 3 + 6 \cdot w + 6 \cdot 2 + w$
$= 3 + 6w + 12 + w$
$= 6w + w + 3 + 12$
$= 7w + 15$

31. $2(3x + 1) + 5(x - 2)$
$= 2(3x) + 2(1) + 5(x) - 5(2)$
$= 6x + 2 + 5x - 10$
$= 6x + 5x + 2 - 10$
$= 11x - 8$

33. $-(2y - 6) + 10 = -1(2y - 6) + 10$
$= -1 \cdot 2y - (-1) \cdot (6) + 10$
$= -2y + 6 + 10$
$= -2y + 16$

35. $18y - 20y = (18 - 20)y = -2y$

37. $z - 8z = (1 - 8)z = -7z$

39. $9d - 3c - d = 9d - d - 3c$
$= (9 - 1)d - 3c$
$= 8d - 3c$

41. $2y - 6 + 4y - 8 = 2y + 4y - 6 - 8$
$= (2 + 4)y - 6 - 8$
$= 6y - 14$

43. $5q + p - 6q - p = 5q - 6q + p - p$
$= (5 - 6)q + (1 - 1)p$
$= -1q + 0p$
$= -q$

45. $2(x + 1) + 20 = 2 \cdot x + 2 \cdot 1 + 20$
$= 2x + 2 + 20$
$= 2x + 22$

47. $5(x - 7) - 8x = 5 \cdot x - 5 \cdot 7 - 8x$
$= 5x - 35 - 8x$
$= 5x - 8x - 35$
$= -3x - 35$

49. $-5(z + 3) + 2z = -5 \cdot z + (-5) \cdot 3 + 2z$
$= -5z - 15 + 2z$
$= -5z + 2z - 15$
$= -3z - 15$

51. $8 - x + 4x - 2 - 9x = -x + 4x - 9x + 8 - 2$
$= (-1 + 4 - 9)x + 8 - 2$
$= -6x + 6$

53. $-7(x + 5) + 5(2x + 1)$
$= -7 \cdot x + (-7) \cdot 5 + 5 \cdot 2x + 5 \cdot 1$
$= -7x - 35 + 10x + 5$
$= -7x + 10x - 35 + 5$
$= 3x - 30$

55. $3r - 5r + 8 + r = 3r - 5r + r + 8$
$= (3 - 5 + 1)r + 8$
$= -1r + 8$
$= -r + 8$

57.
$$-3(n-1)-4n = -3 \cdot n - (-3) \cdot 1 - 4n$$
$$= -3n+3-4n$$
$$= -3n-4n+3$$
$$= -7n+3$$

59.
$$4(z-3)+5z-2 = 4 \cdot z - 4 \cdot 3 + 5z - 2$$
$$= 4z-12+5z-2$$
$$= 4z+5z-12-2$$
$$= 9z-14$$

61.
$$6(2x-1)-12x = 6 \cdot 2x - 6 \cdot 1 - 12x$$
$$= 12x-6-12x$$
$$= 12x-12x-6$$
$$= -6$$

63.
$$-(4x-5)+5 = -1(4x-5)+5$$
$$= -1 \cdot 4x - (-1)(5) + 5$$
$$= -4x+5+5$$
$$= -4x+10$$

65.
$$-(4xy-10)+2(3xy+5)$$
$$= -1(4xy-10)+2(3xy+5)$$
$$= -1 \cdot 4xy - (-1) \cdot 10 + 2 \cdot 3xy + 2 \cdot 5$$
$$= -4xy+10+6xy+10$$
$$= -4xy+6xy+10+10$$
$$= 2xy+20$$

67.
$$3a+4(a+3) = 3a+4 \cdot a + 4 \cdot 3$$
$$= 3a+4a+12$$
$$= 7a+12$$

69.
$$5y-2(y-1)+3 = 5y+(-2)(y)-(-2)(1)+3$$
$$= 5y-2y+2+3$$
$$= 3y+5$$

71.
$$3y+4y+2y+6+5y+16$$
$$= 3y+4y+2y+5y+6+16$$
$$= (3+4+2+5)y+6+16$$
$$= 14y+22$$
The perimeter is $(14y + 22)$ meters.

73.
$$2a+2a+6+5a+6+2a$$
$$= 2a+2a+5a+2a+6+6$$
$$= (2+2+5+2)a+6+6$$
$$= 11a+12$$
The perimeter is $(11a + 12)$ feet.

75.
$$5(-5x+11) = 5 \cdot (-5x) + 5 \cdot 11 = -25x+55$$
The perimeter is $(-25x + 55)$ inches.

77.
$$\text{Area} = (\text{length}) \cdot (\text{width})$$
$$= (4y) \cdot (9)$$
$$= (4 \cdot 9)y$$
$$= 36y$$
The area is $36y$ square inches.

79.
$$\text{Area} = (\text{length}) \cdot (\text{width})$$
$$= (x-2) \cdot (32)$$
$$= x \cdot 32 - 2 \cdot 32$$
$$= 32x-64$$
The area is $(32x - 64)$ square kilometers.

81.
$$\text{Area} = (\text{length}) \cdot (\text{width})$$
$$= (3y+1) \cdot (20)$$
$$= 3y \cdot 20 + 1 \cdot 20$$
$$= (3 \cdot 20)y + 20$$
$$= 60y+20$$
The area is $(60y + 20)$ square miles.

83.
$$\text{Area} = (\text{length}) \cdot (\text{width}) = 50 \cdot 40 = 2000$$
The area is 2000 square feet.

85.
$$\text{Perimeter} = 2 \cdot (\text{length}) + 2 \cdot (\text{width})$$
$$= 2 \cdot (18) + 2 \cdot (14)$$
$$= 36+28$$
$$= 64$$
The perimeter is 64 feet.

87.
$$5+x+2x+1 = x+2x+5+1$$
$$= (1+2)x+5+1$$
$$= 3x+6$$
The perimeter is $(3x + 6)$ feet.

89. $-13 + 10 = -3$

91. $-4 - (-12) = -4 + 12 = 8$

93. $-4 + 4 = 0$

95.
$$5(3x-2) = 5 \cdot 3x - 5 \cdot 2 = 15x - 10$$
The expressions are not equivalent.

97. $7x-(x+2)=7x+(-1)\cdot(x+2)$
$\quad\quad\quad\quad\quad\quad = 7x+(-1)\cdot x+(-1)\cdot 2$
$\quad\quad\quad\quad\quad\quad = 7x-x-2$
The expressions are equivalent.

99. Since multiplication is distributed over subtraction in $6(2x-3)=12x-18$, this is the distributive property.

101. The order of the terms is not changed, but the grouping is. This is the associative property of addition.

103. Add the areas of each rectangle.
$7(2x+1)+3(2x+3)$
$= 7\cdot 2x+7\cdot 1+3\cdot 2x+3\cdot 3$
$= 14x+7+6x+9$
$= 14x+6x+7+9$
$= 20x+16$
The total area is $(20x+16)$ square miles.

105. $9684q-686-4860q+12{,}960$
$= 9684q-4860q-686+12{,}960$
$= 4824q+12{,}274$

107. answers may vary

109. answers may vary

Section 3.2

Practice Problems

1. $\quad x+6=1-3$
$\quad\quad x+6=-2$
$\quad x+6-6=-2-6$
$\quad\quad\quad\quad x=-8$

2. $\quad 10=2m-4m$
$\quad 10=-2m$
$\quad \dfrac{10}{-2}=\dfrac{-2m}{-2}$
$\quad -5=m$

3. $\quad -8+6=\dfrac{a}{3}$
$\quad\quad -2=\dfrac{a}{3}$
$\quad 3\cdot(-2)=3\cdot\dfrac{a}{3}$
$\quad 3\cdot(-2)=\dfrac{3}{3}\cdot a$
$\quad\quad -6=a$

4. $\quad -6y-1+7y=17+2$
$\quad -6y+7y-1=17+2$
$\quad\quad\quad y-1=19$
$\quad\quad y-1+1=19+1$
$\quad\quad\quad\quad y=20$

5. $\quad -4-10=4y-5y$
$\quad\quad\quad -14=-y$
$\quad\quad \dfrac{-14}{-1}=\dfrac{-1y}{-1}$
$\quad\quad\quad 14=y$

6. $\quad\quad 13x=4(3x-1)$
$\quad\quad 13x=4\cdot 3x-4\cdot 1$
$\quad\quad 13x=12x-4$
$\quad 13x-12x=12x-4-12x$
$\quad\quad\quad x=-4$

7. $\quad\quad 5y+2=17$
$\quad 5y+2-2=17-2$
$\quad\quad\quad 5y=15$
$\quad\quad \dfrac{5y}{5}=\dfrac{15}{5}$
$\quad\quad\quad y=3$

8. $\quad -4(x+2)-60=2-10$
$\quad\quad -4x-8-60=2-10$
$\quad\quad\quad -4x-68=-8$
$\quad -4x-68+68=-8+68$
$\quad\quad\quad\quad -4x=60$
$\quad\quad\quad \dfrac{-4x}{-4}=\dfrac{60}{-4}$
$\quad\quad\quad\quad x=-15$

9. a. The sum of -3 and a number is $-3 + x$.

b. -5 decreased by a number is $-5 - x$.

c. Three times a number is $3x$.

d. A number subtracted from 83 is $83 - x$.

e. The quotient of a number and -4 is $\dfrac{x}{-4}$ or $-\dfrac{x}{4}$.

10. a. The product of 5 and a number, decreased by 25, is $5x - 25$.

b. Twice the sum of a number and 3 is $2(x + 3)$.

c. The quotient of 39 and twice a number is $39 \div (2x)$ or $\dfrac{39}{2x}$.

Vocabulary and Readiness Check

1. The equations $-3x = 51$ and $\dfrac{-3x}{-3} = \dfrac{51}{-3}$ are called <u>equivalent</u> equations.

3. The process of writing $-3x + 10x$ as $7x$ is called <u>simplifying</u> the expression.

5. By the <u>addition</u> property of equality, $x = -2$ and $x + 7 = -2 + 7$ are equivalent equations.

Exercise Set 3.2

1.
$$x - 3 = -1 + 4$$
$$x - 3 = 3$$
$$x - 3 + 3 = 3 + 3$$
$$x = 6$$

3.
$$-7 + 10 = m - 5$$
$$3 = m - 5$$
$$3 + 5 = m - 5 + 5$$
$$8 = m$$

5.
$$2w - 12w = 40$$
$$-10w = 40$$
$$\frac{-10w}{-10} = \frac{40}{-10}$$
$$w = -4$$

7.
$$24 = t + 3t$$
$$24 = 4t$$
$$\frac{24}{4} = \frac{4t}{4}$$
$$6 = t$$

9.
$$2z = 12 - 14$$
$$2z = -2$$
$$\frac{2z}{2} = \frac{-2}{2}$$
$$z = -1$$

11.
$$4 - 10 = \frac{z}{-3}$$
$$-6 = \frac{z}{-3}$$
$$-3 \cdot (-6) = -3 \cdot \frac{z}{-3}$$
$$18 = z$$

13.
$$-3x - 3x = 50 - 2$$
$$-6x = 48$$
$$\frac{-6x}{-6} = \frac{48}{-6}$$
$$x = -8$$

15.
$$\frac{x}{5} = -26 + 16$$
$$\frac{x}{5} = -10$$
$$5 \cdot \frac{x}{5} = 5 \cdot (-10)$$
$$x = -50$$

17.
$$7x + 7 - 6x = 10$$
$$7x - 6x + 7 = 10$$
$$x + 7 = 10$$
$$x + 7 - 7 = 10 - 7$$
$$x = 3$$

19. $-8-9 = 3x+5-2x$
       $-17 = x+5$
       $-17-5 = x+5-5$
       $-22 = x$

21. $2(5x-3) = 11x$
       $2 \cdot 5x - 2 \cdot 3 = 11x$
       $10x-6 = 11x$
       $10x-10x-6 = 11x-10x$
       $-6 = x$

23. $3y = 2(y+12)$
       $3y = 2 \cdot y + 2 \cdot 12$
       $3y = 2y+24$
       $3y-2y = 2y-2y+24$
       $y = 24$

25. $21y = 5(4y-6)$
       $21y = 5 \cdot 4y - 5 \cdot 6$
       $21y = 20y-30$
       $21y-20y = 20y-20y-30$
       $y = -30$

27. $-3(-4-2z) = 3z$
       $-3 \cdot (-4) - (-3)(2z) = 3z$
       $12+6z = 3z$
       $12+6z-6z = 3z-6z$
       $12 = -3z$
       $\dfrac{12}{-3} = \dfrac{-3z}{-3}$
       $-4 = z$

29. $2x-8 = 0$
       $2x-8+8 = 0+8$
       $2x = 8$
       $\dfrac{2x}{2} = \dfrac{8}{2}$
       $x = 4$

31. $7y+3 = 24$
       $7y+3-3 = 24-3$
       $7y = 21$
       $\dfrac{7y}{7} = \dfrac{21}{7}$
       $y = 3$

33. $-7 = 2x-1$
       $-7+1 = 2x-1+1$
       $-6 = 2x$
       $\dfrac{-6}{2} = \dfrac{2x}{2}$
       $-3 = x$

35. $6(6y-4) = 12y$
       $6 \cdot 6y - 6 \cdot 4 = 12y$
       $36y-24 = 12y$
       $36y-36y-24 = 12y-36y$
       $-24 = -24y$
       $\dfrac{-24}{-24} = \dfrac{-24y}{-24}$
       $1 = y$

37. $11(x-6) = -4-7$
       $11 \cdot x - 11 \cdot 6 = -11$
       $11x-66 = -11$
       $11x-66+66 = -11+66$
       $11x = 55$
       $\dfrac{11x}{11} = \dfrac{55}{11}$
       $x = 5$

39. $-3(x+1)-10 = 12+8$
       $-3 \cdot x - 3 \cdot 1 - 10 = 20$
       $-3x-3-10 = 20$
       $-3x-13 = 20$
       $-3x-13+13 = 20+13$
       $-3x = 33$
       $\dfrac{-3x}{-3} = \dfrac{33}{-3}$
       $x = -11$

41. $y-20 = 6y$
       $y-y-20 = 6y-y$
       $-20 = 5y$
       $\dfrac{-20}{5} = \dfrac{5y}{5}$
       $-4 = y$

43. $22 - 42 = 4(x-1) - 4$
$-20 = 4 \cdot x - 4 \cdot 1 - 4$
$-20 = 4x - 4 - 4$
$-20 = 4x - 8$
$-20 + 8 = 4x - 8 + 8$
$-12 = 4x$
$\dfrac{-12}{4} = \dfrac{4x}{4}$
$-3 = x$

45. $-2 - 3 = -4 + x$
$-5 = -4 + x$
$-5 + 4 = -4 + 4 + x$
$-1 = x$

47. $y + 1 = -3 + 4$
$y + 1 = 1$
$y + 1 - 1 = 1 - 1$
$y = 0$

49. $3w - 12w = -27$
$-9w = -27$
$\dfrac{-9w}{-9} = \dfrac{-27}{-9}$
$w = 3$

51. $-4x = 20 - (-4)$
$-4x = 20 + 4$
$-4x = 24$
$\dfrac{-4x}{-4} = \dfrac{24}{-4}$
$x = -6$

53. $18 - 11 = \dfrac{x}{-5}$
$7 = \dfrac{x}{-5}$
$-5 \cdot 7 = -5 \cdot \dfrac{x}{-5}$
$-35 = x$

55. $9x - 12 = 78$
$9x - 12 + 12 = 78 + 12$
$9x = 90$
$\dfrac{9x}{9} = \dfrac{90}{9}$
$x = 10$

57. $10 = 7t - 12t$
$10 = -5t$
$\dfrac{10}{-5} = \dfrac{-5t}{-5}$
$-2 = t$

59. $5 - 5 = 3x + 2x$
$0 = 5x$
$\dfrac{0}{5} = \dfrac{5x}{5}$
$0 = x$

61. $50y = 7(7y + 4)$
$50y = 7 \cdot 7y + 7 \cdot 4$
$50y = 49y + 28$
$50y - 49y = 49y - 49y + 28$
$y = 28$

63. $8x = 2(6x + 10)$
$8x = 2 \cdot 6x + 2 \cdot 10$
$8x = 12x + 20$
$8x - 12x = 12x - 12x + 20$
$-4x = 20$
$\dfrac{-4x}{-4} = \dfrac{20}{-4}$
$x = -5$

65. $7x + 14 - 6x = -4 - 10$
$7x - 6x + 14 = -14$
$x + 14 = -14$
$x + 14 - 14 = -14 - 14$
$x = -28$

67.
$$\frac{x}{-4} = -1 - (-8)$$
$$\frac{x}{-4} = -1 + 8$$
$$\frac{x}{-4} = 7$$
$$-4 \cdot \frac{x}{-4} = -4 \cdot 7$$
$$x = -28$$

69.
$$23x + 8 - 25x = 7 - 9$$
$$23x - 25x + 8 = -2$$
$$-2x + 8 = -2$$
$$-2x + 8 - 8 = -2 - 8$$
$$-2x = -10$$
$$\frac{-2x}{-2} = \frac{-10}{-2}$$
$$x = 5$$

71.
$$-3(x + 9) - 41 = 4 - 60$$
$$-3 \cdot x - 3 \cdot 9 - 41 = 4 - 60$$
$$-3x - 27 - 41 = -56$$
$$-3x - 68 = -56$$
$$-3x - 68 + 68 = -56 + 68$$
$$-3x = 12$$
$$\frac{-3x}{-3} = \frac{12}{-3}$$
$$x = -4$$

73. The sum of -7 and a number is $-7 + x$.

75. Eleven subtracted from a number is $x - 11$.

77. The product of -13 and a number is $-13x$.

79. A number divided by -12 is $\frac{x}{-12}$ or $-\frac{x}{12}$.

81. The product of -11 and a number, increased by 5 is $-11x + 5$.

83. Negative ten decreased by 7 times a number is $-10 - 7x$.

85. Seven added to the product of 4 and a number is $4x + 7$.

87. Twice a number, decreased by 17 is $2x - 17$.

89. The product of -6 and the sum of a number and 15 is $-6(x + 15)$.

91. The quotient of 45 and the product of a number and -5 is $\frac{45}{-5x}$.

93. The quotient of seventeen and a number, increased by -15 is $\frac{17}{x} + (-15)$.

95. The longest bar corresponds to the year in which the number of trumpeter swans was the greatest. The year is 2005.

97. From the length of the bar, there were approximately 35,000 trumpeter swans in 2005.

99. answers may vary

101. no; answers may vary

103. answers may vary

105.
$$\frac{y}{72} = -86 - (-1029)$$
$$\frac{y}{72} = -86 + 1029$$
$$\frac{y}{72} = 943$$
$$72 \cdot \frac{y}{72} = 72 \cdot 943$$
$$y = 67,896$$

107.
$$\frac{x}{-2} = 5^2 - |-10| - (-9)$$
$$\frac{x}{-2} = 25 - |-10| - (-9)$$
$$\frac{x}{-2} = 25 - 10 - (-9)$$
$$\frac{x}{-2} = 25 - 10 + 9$$
$$\frac{x}{-2} = 24$$
$$-2 \cdot \frac{x}{-2} = -2 \cdot 24$$
$$x = -48$$

109.
$$|-13| + 3^2 = 100y - |-20| - 99y$$
$$13 + 3^2 = 100y - 20 - 99y$$
$$13 + 9 = 100y - 99y - 20$$
$$22 = y - 20$$
$$22 + 20 = y - 20 + 20$$
$$42 = y$$

Integrated Review

1. $7x - 5y + 14$ is an expression because it does not contain an equal sign.

2. $7x = 35 + 14$ is an equation because it contains an equal sign.

3. $3(x - 2) = 5(x + 1) - 17$ is an equation because it contains an equal sign.

4. $-9(2x + 1) - 4(x - 2) + 14$ is an expression because it does not contain an equal sign.

5. To <u>simplify</u> an expression, we combine any like terms.

6. To <u>solve</u> an equation, we use properties of equality to find any value of the variable that makes the equation a true statement.

7. $7x + x = (7 + 1)x = 8x$

8. $6y - 10y = (6 - 10)y = -4y$

9. $2a + 5a - 9a - 2 = (2 + 5 - 9)a - 2$
$$= -2a - 2$$

10. $6a - 12 - a - 14 = 6a - a - 12 - 14$
$$= (6 - 1)a - 12 - 14$$
$$= 5a - 26$$

11. $-2(4x + 7) = -2 \cdot 4x + (-2) \cdot 7 = -8x - 14$

12. $-3(2x - 10) = -3(2x) - (-3)(10) = -6x + 30$

13. $5(y + 2) - 20 = 5 \cdot y + 5 \cdot 2 - 20$
$$= 5y + 10 - 20$$
$$= 5y - 10$$

14. $12x + 3(x - 6) - 13 = 12x + 3 \cdot x - 3 \cdot 6 - 13$
$$= 12x + 3x - 18 - 13$$
$$= (12 + 3)x - 18 - 13$$
$$= 15x - 31$$

15. Area $= $ (length) $\cdot$ (width)
$$= 3(4x - 2)$$
$$= 3 \cdot 4x - 3 \cdot 2$$
$$= 12x - 6$$
The area is $(12x - 6)$ square meters.

16. Perimeter $= x + x + 2 + 7 = 2x + 9$
The perimeter is $(2x + 9)$ feet.

17.
$$12 = 11x - 14x$$
$$12 = -3x$$
$$\frac{12}{-3} = \frac{-3x}{-3}$$
$$-4 = x$$
Check: $12 = 11x - 14x$
$$12 \overset{?}{=} 11(-4) - 14(-4)$$
$$12 \overset{?}{=} -44 + 56$$
$$12 = 12 \quad \text{True}$$
The solution is -4.

18. $8y + 7y = -45$

$15y = -45$

$\dfrac{15y}{15} = \dfrac{-45}{15}$

$y = -3$

Check: $\quad 8y + 7y = -45$

$8(-3) + 7(-3) \stackrel{?}{=} -45$

$-24 + (-21) \stackrel{?}{=} -45$

$-45 = -45$ True

The solution is -3.

19. $x - 12 = -45 + 23$

$x - 12 = -22$

$x - 12 + 12 = -22 + 12$

$x = -10$

Check: $\quad x - 12 = -45 + 23$

$-10 - 12 \stackrel{?}{=} -45 + 23$

$-22 = -22$ True

The solution is -10.

20. $6 - (-5) = x + 5$

$6 + 5 = x + 5$

$11 = x + 5$

$11 - 5 = x + 5 - 5$

$6 = x$

Check: $6 - (-5) = x + 5$

$6 - (-5) \stackrel{?}{=} 6 + 5$

$6 + 5 \stackrel{?}{=} 6 + 5$

$11 = 11$ True

The solution is 6.

21. $\dfrac{x}{3} = -14 + 9$

$\dfrac{x}{3} = -5$

$3 \cdot \dfrac{x}{3} = 3(-5)$

$x = -15$

Check: $\dfrac{x}{3} = -14 + 9$

$\dfrac{-15}{3} \stackrel{?}{=} -14 + 9$

$-5 = -5$ True

The solution is -15.

22. $\dfrac{z}{4} = -23 - 7$

$\dfrac{z}{4} = -30$

$4 \cdot \dfrac{z}{4} = 4 \cdot (-30)$

$z = -120$

Check: $\quad \dfrac{z}{4} = -23 - 7$

$\dfrac{-120}{4} \stackrel{?}{=} -23 - 7$

$-30 = -30$ True

The solution is -120.

23. $-6 + 2 = 4x + 1 - 3x$

$-4 = x + 1$

$-4 - 1 = x + 1 - 1$

$-5 = x$

Check: $-6 + 2 = 4x + 1 - 3x$

$-6 + 2 \stackrel{?}{=} 4(-5) + 1 - 3(-5)$

$-4 \stackrel{?}{=} -20 + 1 + 15$

$-4 = -4$ True

The solution is -5.

24. $5 - 8 = 5x + 10 - 4x$

$-3 = x + 10$

$-3 - 10 = x + 10 - 10$

$-13 = x$

Check: $5 - 8 = 5x + 10 - 4x$

$5 - 8 \stackrel{?}{=} 5(-13) + 10 - 4(-13)$

$-3 \stackrel{?}{=} -65 + 10 + 52$

$-3 = -3$ True

The solution is -13.

25. $6(3x - 4) = 19x$

$6 \cdot 3x - 6 \cdot 4 = 19x$

$18x - 24 = 19x$

$18x - 18x - 24 = 19x - 18x$

$-24 = x$

Check: $\quad 6(3x - 4) = 19x$

$6[3(-24) - 4] \stackrel{?}{=} 19(-24)$

$6(-72 - 4) \stackrel{?}{=} -456$

$6(-76) \stackrel{?}{=} -456$

$-456 = -456$ True

The solution is -24.

26.
$$25x = 6(4x - 9)$$
$$25x = 6 \cdot 4x - 6 \cdot 9$$
$$25x = 24x - 54$$
$$25x - 24x = 24x - 24x - 54$$
$$x = -54$$
Check: $25x = 6(4x - 9)$
$$25(-54) \stackrel{?}{=} 6[4(-54) - 9]$$
$$-1350 \stackrel{?}{=} 6(-216 - 9)$$
$$-1350 \stackrel{?}{=} 6(-225)$$
$$-1350 = -1350 \quad \text{True}$$
The solution is -54.

27.
$$-36x - 10 + 37x = -12 - (-14)$$
$$x - 10 = -12 + 14$$
$$x - 10 = 2$$
$$x - 10 + 10 = 2 + 10$$
$$x = 12$$
Check: $-36x - 10 + 37x = -12 - (-14)$
$$-36(12) - 10 + 37(12) \stackrel{?}{=} -12 - (-14)$$
$$-432 - 10 + 444 \stackrel{?}{=} -12 + 14$$
$$2 = 2 \quad \text{True}$$
The solution is 12.

28.
$$-8 + (-14) = -80y + 20 + 81y$$
$$-22 = y + 20$$
$$-22 - 20 = y + 20 - 20$$
$$-42 = y$$
Check:
$$-8 + (-14) = -80y + 20 + 81y$$
$$-8 + (-14) \stackrel{?}{=} -80(-42) + 20 + 81(-42)$$
$$-22 \stackrel{?}{=} 3360 + 20 - 3402$$
$$-22 = -22 \quad \text{True}$$
The solution is -42.

29.
$$3x - 16 = -10$$
$$3x - 16 + 16 = -10 + 16$$
$$3x = 6$$
$$\frac{3x}{3} = \frac{6}{3}$$
$$x = 2$$
Check: $3x - 16 = -10$
$$3(2) - 16 \stackrel{?}{=} -10$$
$$6 - 16 \stackrel{?}{=} -10$$
$$-10 = -10 \quad \text{True}$$
The solution is 2.

30.
$$4x - 21 = -13$$
$$4x - 21 + 21 = -13 + 21$$
$$4x = 8$$
$$\frac{4x}{4} = \frac{8}{4}$$
$$x = 2$$
Check: $4x - 21 = -13$
$$4 \cdot 2 - 21 \stackrel{?}{=} -13$$
$$8 - 21 \stackrel{?}{=} -13$$
$$-13 = -13 \quad \text{True}$$
The solution is 2.

31.
$$-8z - 2z = 26 - (-4)$$
$$-10z = 26 + 4$$
$$-10z = 30$$
$$\frac{-10z}{-10} = \frac{30}{-10}$$
$$z = -3$$
Check: $-8z - 2z = 26 - (-4)$
$$-8(-3) - 2(-3) \stackrel{?}{=} 26 - (-4)$$
$$24 + 6 \stackrel{?}{=} 26 + 4$$
$$30 = 30 \quad \text{True}$$
The solution is -3.

32.
$$-12 + (-13) = 5x - 10x$$
$$-25 = -5x$$
$$\frac{-25}{-5} = \frac{-5x}{-5}$$
$$5 = x$$
Check: $-12 + (-13) = 5x - 10x$
$$-12 + (-13) \stackrel{?}{=} 5(5) - 10(5)$$
$$-25 \stackrel{?}{=} 25 - 50$$
$$-25 = -25 \quad \text{True}$$
The solution is 5.

33.
$$-4(x + 8) - 11 = 3 - 26$$
$$-4 \cdot x + (-4) \cdot 8 - 11 = 3 - 26$$
$$-4x - 32 - 11 = -23$$
$$-4x - 43 = -23$$
$$-4x - 43 + 43 = -23 + 43$$
$$-4x = 20$$
$$\frac{-4x}{-4} = \frac{20}{-4}$$
$$x = -5$$

Check: $-4(x+8)-11 = 3-26$
$$-4(-5+8)-11 \overset{?}{=} 3-26$$
$$-4(3)-11 \overset{?}{=} -23$$
$$-12-11 \overset{?}{=} -23$$
$$-23 = -23 \quad \text{True}$$

The solution is -5.

34. $-6(x-2)+10 = -4-10$
$$-6 \cdot x - (-6)(2)+10 = -4-10$$
$$-6x+12+10 = -14$$
$$-6x+22 = -14$$
$$-6x+22-22 = -14-22$$
$$-6x = -36$$
$$\frac{-6x}{-6} = \frac{-36}{-6}$$
$$x = 6$$

Check: $-6(x-2)+10 = -4-10$
$$-6(6-2)+10 \overset{?}{=} -4-10$$
$$-6(4)+10 \overset{?}{=} -14$$
$$-24+10 \overset{?}{=} -14$$
$$-14 = -14 \quad \text{True}$$

The solution is 6.

35. The difference of a number and 10 is $x - 10$.

36. The sum of -20 and a number is $-20 + x$.

37. The product of 10 and a number is $10x$.

38. The quotient of 10 and a number is $\dfrac{10}{x}$.

39. Five added to the product of -2 and a number is $-2x + 5$.

40. The product of -4 and the difference of a number and 1 is $-4(x - 1)$.

Section 3.3

Practice Problems

1. $7x+12 = 3x-4$
$$7x+12-12 = 3x-4-12$$
$$7x = 3x-16$$
$$7x-3x = 3x-16-3x$$
$$4x = -16$$
$$\frac{4x}{4} = \frac{-16}{4}$$
$$x = -4$$

2. $40-5y+5 = -2y-10-4y$
$$45-5y = -6y-10$$
$$45-5y-45 = -6y-10-45$$
$$-5y = -6y-55$$
$$-5y+6y = -6y-55+6y$$
$$y = -55$$

3. $6(a-5) = 4a+4$
$$6a-30 = 4a+4$$
$$6a-30-4a = 4a+4-4a$$
$$2a-30 = 4$$
$$2a-30+30 = 4+30$$
$$2a = 34$$
$$\frac{2a}{2} = \frac{34}{2}$$
$$a = 17$$

4. $4(x+3)+1 = 13$
$$4x+12+1 = 13$$
$$4x+13 = 13$$
$$4x+13-13 = 13-13$$
$$4x = 0$$
$$\frac{4x}{4} = \frac{0}{4}$$
$$x = 0$$

5. a. The difference of 110 and 80 is 30 translates to $110 - 80 = 30$.

   b. The product of 3 and the sum of -9 and 11 amounts to 6 translates to $3(-9 + 11) = 6$.

c. The quotient of 24 and -6 yields -4 translates to $\dfrac{24}{-6} = -4$.

Calculator Explorations

1. Replace x with 12.
$76(12 - 25) = -988$
Yes

3. Replace x with -170.
$-170 + 562 = 392$
$3 \cdot (-170) + 900 = 390$
No

5. Replace x with -21.
$29 \cdot (-21) - 1034 = -1643$
$61 \cdot (-21) - 362 = -1643$
Yes

Vocabulary and Readiness Check

1. An example of an expression is $\underline{3x - 9 + x - 16}$ while an example of an equation is $\underline{5(2x + 6) - 1 = 39}$.

3. To solve $x - 7 = -10$, we use the <u>addition</u> property of equality.

5. To solve $5(x - 1) = 25$, first use the <u>distributive</u> property.

Exercise Set 3.3

1.
$$3x - 7 = 4x + 5$$
$$3x - 3x - 7 = 4x - 3x + 5$$
$$-7 = x + 5$$
$$-7 - 5 = x + 5 - 5$$
$$-12 = x$$

3.
$$10x + 15 = 6x + 3$$
$$10x + 15 - 15 = 6x + 3 - 15$$
$$10x = 6x - 12$$
$$10x - 6x = 6x - 6x - 12$$
$$4x = -12$$
$$\frac{4x}{4} = \frac{-12}{4}$$
$$x = -3$$

5.
$$19 - 3x = 14 + 2x$$
$$19 - 3x + 3x = 14 + 2x + 3x$$
$$19 = 14 + 5x$$
$$19 - 14 = 14 - 14 + 5x$$
$$5 = 5x$$
$$\frac{5}{5} = \frac{5x}{5}$$
$$1 = x$$

7.
$$-14x - 20 = -12x + 70$$
$$-14x + 12x - 20 = -12x + 12x + 70$$
$$-2x - 20 = 70$$
$$-2x - 20 + 20 = 70 + 20$$
$$-2x = 90$$
$$\frac{-2x}{-2} = \frac{90}{-2}$$
$$x = -45$$

9.
$$x + 20 + 2x = -10 - 2x - 15$$
$$x + 2x + 20 = -10 - 15 - 2x$$
$$3x + 20 = -25 - 2x$$
$$3x + 2x + 20 = -25 - 2x + 2x$$
$$5x + 20 = -25$$
$$5x + 20 - 20 = -25 - 20$$
$$5x = -45$$
$$\frac{5x}{5} = \frac{-45}{5}$$
$$x = -9$$

11.
$$40 + 4y - 16 = 13y - 12 - 3y$$
$$40 - 16 + 4y = 13y - 3y - 12$$
$$24 + 4y = 10y - 12$$
$$24 + 4y - 4y = 10y - 4y - 12$$
$$24 = 6y - 12$$
$$24 + 12 = 6y - 12 + 12$$
$$36 = 6y$$
$$\frac{36}{6} = \frac{6y}{6}$$
$$6 = y$$

13.
$$35 - 17 = 3(x - 2)$$
$$18 = 3x - 6$$
$$18 + 6 = 3x - 6 + 6$$
$$24 = 3x$$
$$\frac{24}{3} = \frac{3x}{3}$$
$$8 = x$$

15.
$$3(x - 1) - 12 = 0$$
$$3x - 3 - 12 = 0$$
$$3x - 15 = 0$$
$$3x - 15 + 15 = 0 + 15$$
$$3x = 15$$
$$\frac{3x}{3} = \frac{15}{3}$$
$$x = 5$$

17.
$$2(y - 3) = y - 6$$
$$2y - 6 = y - 6$$
$$2y - y - 6 = y - y - 6$$
$$y - 6 = -6$$
$$y - 6 + 6 = -6 + 6$$
$$y = 0$$

19.
$$-2(y + 4) = 2$$
$$-2y - 8 = 2$$
$$-2y - 8 + 8 = 2 + 8$$
$$-2y = 10$$
$$\frac{-2y}{-2} = \frac{10}{-2}$$
$$y = -5$$

21.
$$2t - 1 = 3(t + 7)$$
$$2t - 1 = 3t + 21$$
$$2t - 2t - 1 = 3t - 2t + 21$$
$$-1 = t + 21$$
$$-1 - 21 = t + 21 - 21$$
$$-22 = t$$

23.
$$3(5c + 1) - 12 = 13c + 3$$
$$15c + 3 - 12 = 13c + 3$$
$$15c - 9 = 13c + 3$$
$$15c - 13c - 9 = 13c - 13c + 3$$
$$2c - 9 = 3$$
$$2c - 9 + 9 = 3 + 9$$
$$2c = 12$$
$$\frac{2c}{2} = \frac{12}{2}$$
$$c = 6$$

25.
$$-4x = 44$$
$$\frac{-4x}{-4} = \frac{44}{-4}$$
$$x = -11$$

27.
$$x + 9 = 2$$
$$x + 9 - 9 = 2 - 9$$
$$x = -7$$

29.
$$8 - b = 13$$
$$8 - 8 - b = 13 - 8$$
$$-b = 5$$
$$-1 \cdot (-b) = -1 \cdot 5$$
$$b = -5$$

31.
$$-20 - (-50) = \frac{x}{9}$$
$$-20 + 50 = \frac{x}{9}$$
$$30 = \frac{x}{9}$$
$$9 \cdot 30 = 9 \cdot \frac{x}{9}$$
$$270 = x$$

33.
$$3r + 4 = 19$$
$$3r + 4 - 4 = 19 - 4$$
$$3r = 15$$
$$\frac{3r}{3} = \frac{15}{3}$$
$$r = 5$$

35.
$$-7c + 1 = -20$$
$$-7c + 1 - 1 = -20 - 1$$
$$-7c = -21$$
$$\frac{-7c}{-7} = \frac{-21}{-7}$$
$$c = 3$$

37.
$$8y - 13y = -20 - 25$$
$$-5y = -45$$
$$\frac{-5y}{-5} = \frac{-45}{-5}$$
$$y = 9$$

39.
$$6(7x - 1) = 43x$$
$$42x - 6 = 43x$$
$$42x - 42x - 6 = 43x - 42x$$
$$-6 = x$$

41.
$$-4 + 12 = 16x - 3 - 15x$$
$$8 = x - 3$$
$$8 + 3 = x - 3 + 3$$
$$11 = x$$

43.
$$-10(x + 3) + 28 = -16 - 16$$
$$-10x - 30 + 28 = -32$$
$$-10x - 2 = -32$$
$$-10x - 2 + 2 = -32 + 2$$
$$-10x = -30$$
$$\frac{-10x}{-10} = \frac{-30}{-10}$$
$$x = 3$$

45.
$$4x + 3 = 2x + 11$$
$$4x - 2x + 3 = 2x - 2x + 11$$
$$2x + 3 = 11$$
$$2x + 3 - 3 = 11 - 3$$
$$2x = 8$$
$$\frac{2x}{2} = \frac{8}{2}$$
$$x = 4$$

47.
$$-2y - 10 = 5y + 18$$
$$-2y - 10 - 5y = 5y + 18 - 5y$$
$$-7y - 10 = 18$$
$$-7y - 10 + 10 = 18 + 10$$
$$-7y = 28$$
$$\frac{-7y}{-7} = \frac{28}{-7}$$
$$y = -4$$

49.
$$-8n + 1 = -6n - 5$$
$$-8n + 8n + 1 = -6n + 8n - 5$$
$$1 = 2n - 5$$
$$1 + 5 = 2n - 5 + 5$$
$$6 = 2n$$
$$\frac{6}{2} = \frac{2n}{2}$$
$$3 = n$$

51.
$$9 - 3x = 14 + 2x$$
$$9 - 3x - 2x = 14 + 2x - 2x$$
$$9 - 5x = 14$$
$$9 - 5x - 9 = 14 - 9$$
$$-5x = 5$$
$$\frac{-5x}{-5} = \frac{5}{-5}$$
$$x = -1$$

53.
$$9a + 29 + 7 = 0$$
$$9a + 36 = 0$$
$$9a + 36 - 36 = 0 - 36$$
$$9a = -36$$
$$\frac{9a}{9} = \frac{-36}{9}$$
$$a = -4$$

55.
$$7(y-2) = 4y - 29$$
$$7y - 14 = 4y - 29$$
$$7y - 4y - 14 = 4y - 4y - 29$$
$$3y - 14 = -29$$
$$3y - 14 + 14 = -29 + 14$$
$$3y = -15$$
$$\frac{3y}{3} = \frac{-15}{3}$$
$$y = -5$$

57.
$$12 + 5t = 6(t + 2)$$
$$12 + 5t = 6t + 12$$
$$12 + 5t - 5t = 6t - 5t + 12$$
$$12 = t + 12$$
$$12 - 12 = t + 12 - 12$$
$$0 = t$$

59.
$$3(5c - 1) - 2 = 13c + 3$$
$$15c - 3 - 2 = 13c + 3$$
$$15c - 5 = 13c + 3$$
$$15c - 13c - 5 = 13c - 13c + 3$$
$$2c - 5 = 3$$
$$2c - 5 + 5 = 3 + 5$$
$$2c = 8$$
$$\frac{2c}{2} = \frac{8}{2}$$
$$c = 4$$

61.
$$10 + 5(z - 2) = 4z + 1$$
$$10 + 5z - 10 = 4z + 1$$
$$5z = 4z + 1$$
$$5z - 4z = 4z - 4z + 1$$
$$z = 1$$

63.
$$7(6 + w) = 6(2 + w)$$
$$42 + 7w = 12 + 6w$$
$$42 + 7w - 6w = 12 + 6w - 6w$$
$$42 + w = 12$$
$$42 - 42 + w = 12 - 42$$
$$w = -30$$

65. The sum of −42 and 16 is −26 translates to
−42 + 16 = −26.

67. The product of −5 and −29 gives 145 translates to −5(−29) = 145.

69. Three times the difference of −14 and 2 amounts to −48 translates to
3(−14 − 2) = −48.

71. The quotient of 100 and twice 50 is equal to 1 translates to $\dfrac{100}{2(50)} = 1$.

73. From the height of the bar, approximately 97 million returns are expected to be filed electronically in 2010.

75. Subtract the number of returns in 2006 from the number in 2009.
90 million − 70 million = 20 million
The increase from 2006 to 2009 is approximately 20 million returns.

77.
$$x^3 - 2xy = 3^3 - 2(3)(-1)$$
$$= 27 - 2(3)(-1)$$
$$= 27 - (-6)$$
$$= 27 + 6$$
$$= 33$$

79.
$$y^5 - 4x^2 = (-1)^5 - 4(3)^2$$
$$= -1 - 4(9)$$
$$= -1 - 36$$
$$= -37$$

81. The first step in solving $2x - 5 = -7$ is to add 5 to both sides, which is choice b.

83. The first step in solving $-3x = -12$ is to divide both sides by −3, which is choice a.

85. The error is in the second line.
$$2(3x - 5) = 5x - 7$$
$$6x - 10 = 5x - 7$$
$$6x - 10 + 10 = 5x - 7 + 10$$
$$6x = 5x + 3$$
$$6x - 5x = 5x + 3 - 5x$$
$$x = 3$$

87. $3^2 \cdot x = (-9)^3$

$9x = -729$

$\dfrac{9x}{9} = \dfrac{-729}{9}$

$x = -81$

89. $x + 45^2 = 54^2$

$x + 2025 = 2916$

$x + 2025 - 2025 = 2916 - 2025$

$x = 891$

91. no; answers may vary

The Bigger Picture

1. $-8x = 40$

$\dfrac{-8x}{-8} = \dfrac{40}{-8}$

$x = -5$

2. $x - 14 = -3$

$x - 14 + 14 = -3 + 14$

$x = 11$

3. $2x = -14$

$\dfrac{2x}{2} = \dfrac{-14}{2}$

$x = -7$

4. $5y + 7 - 4y - 10 = 100$

$y - 3 = 100$

$y - 3 + 3 = 100 + 3$

$y = 103$

5. $-3n + n = -100 - 50$

$-2n = -150$

$\dfrac{-2n}{-2} = \dfrac{-150}{-2}$

$n = 75$

6. $8x + 9 = -79$

$8x + 9 - 9 = -79 - 9$

$8x = -88$

$\dfrac{8x}{8} = \dfrac{-88}{8}$

$x = -11$

7. $4x - 5 = 2x - 17$

$4x - 2x - 5 = 2x - 2x - 17$

$2x - 5 = -17$

$2x - 5 + 5 = -17 + 5$

$2x = -12$

$\dfrac{2x}{2} = \dfrac{-12}{2}$

$x = -6$

8. $30 + 5(2n - 4) = 100$

$30 + 10n - 20 = 100$

$10n + 10 = 100$

$10n + 10 - 10 = 100 - 10$

$10n = 90$

$\dfrac{10n}{10} = \dfrac{90}{10}$

$n = 9$

Section 3.4

Practice Problems

1. a. "Four times a number is 20" is $4x = 20$.

b. "The sum of a number and −5 yields 32" is $x + (-5) = 32$.

c. "Fifteen subtracted from a number amounts to −23" is $x - 15 = -23$.

d. "Five times the difference of a number and 7 is equal to −8" is $5(x - 7) = -8$.

e. "The quotient of triple a number and 5 gives 1" is $\dfrac{3x}{5} = 1$.

2. "The sum of a number and 2 equals 6 added to three times the number" is

$x + 2 = 6 + 3x$

$x - x + 2 = 6 + 3x - x$

$2 = 6 + 2x$

$2 - 6 = 6 - 6 + 2x$

$-4 = 2x$

$\dfrac{-4}{2} = \dfrac{2x}{2}$

$-2 = x$

3. Let x be the distance from Denver to San Francisco. Since the distance from Cincinnati to Denver is 71 miles less than the distance from Denver to San Francisco, the distance from Cincinnati to Denver is $x - 71$. Since the total of the two distances is 2399, the sum of x and $x - 71$ is 2399.

$$x + x - 71 = 2399$$
$$2x - 71 = 2399$$
$$2x - 71 + 71 = 2399 + 71$$
$$2x = 2470$$
$$\frac{2x}{2} = \frac{2470}{2}$$
$$x = 1235$$

The distance from Denver to San Francisco is 1235 miles.

4. Let x be the amount her son receives. Since her husband receives twice as much as her son, her husband receives $2x$. Since the total estate is $57,000, the sum of x and $2x$ is 57,000.

$$x + 2x = 57,000$$
$$3x = 57,000$$
$$\frac{3x}{3} = \frac{57,000}{3}$$
$$x = 19,000$$
$$2x = 2(19,000) = 38,000$$

Her husband will receive $38,000 and her son will receive $19,000.

Exercise Set 3.4

1. "A number added to −5 is −7" is $-5 + x = -7$.

3. "Three times a number yields 27" is $3x = 27$.

5. "A number subtracted from −20 amounts to 104" is $-20 - x = 104$.

7. "Twice a number gives 108" is $2x = 108$.

9. "The product of 5 and the sum of −3 and a number is −20" is $5(-3 + x) = -20$.

11. "Three times a number, added to 9 is 33" is
$$9 + 3x = 33$$
$$9 - 9 + 3x = 33 - 9$$
$$3x = 24$$
$$\frac{3x}{3} = \frac{24}{3}$$
$$x = 8$$

13. "The sum of 3, 4, and a number amounts to 16" is
$$3 + 4 + x = 16$$
$$7 + x = 16$$
$$x + 7 - 7 = 16 - 7$$
$$x = 9$$

15. "The difference of a number and 3 is equal to the quotient of 10 and 5" is
$$x - 3 = \frac{10}{5}$$
$$x - 3 = 2$$
$$x - 3 + 3 = 2 + 3$$
$$x = 5$$

17. "Thirty less a number is equal to the product of 3 and the sum of the number and 6" is
$$30 - x = 3(x + 6)$$
$$30 - x = 3x + 18$$
$$30 - 30 - x = 3x + 18 - 30$$
$$-x = 3x - 12$$
$$-x - 3x = 3x - 3x - 12$$
$$-4x = -12$$
$$\frac{-4x}{-4} = \frac{-12}{-4}$$
$$x = 3$$

19. "40 subtracted from five times a number is 8 more than the number" is
$$5x - 40 = x + 8$$
$$5x - x - 40 = x - x + 8$$
$$4x - 40 = 8$$
$$4x - 40 + 40 = 8 + 40$$
$$4x = 48$$
$$\frac{4x}{4} = \frac{48}{4}$$
$$x = 12$$

21. "Three times the difference of some number and 5 amounts to the quotient of 108 and 12" is

$$3(x-5) = \frac{108}{12}$$
$$3x - 15 = 9$$
$$3x - 15 + 15 = 9 + 15$$
$$3x = 24$$
$$\frac{3x}{3} = \frac{24}{3}$$
$$x = 8$$

23. "The product of 4 and a number is the same as 30 less twice that same number" is
$$4x = 30 - 2x$$
$$4x + 2x = 30 - 2x + 2x$$
$$6x = 30$$
$$\frac{6x}{6} = \frac{30}{6}$$
$$x = 5$$

25. Let x be the number of votes for California. Since Florida has 28 fewer votes than California, Florida had $x - 28$. Since there was a total of 82 for the two states, the sum of x and $x - 28$ is 82.
$$x + x - 28 = 82$$
$$2x - 28 = 82$$
$$2x - 28 + 28 = 82 + 28$$
$$2x = 110$$
$$\frac{2x}{2} = \frac{110}{2}$$
$$x = 55$$
California has 55 votes and Florida has $55 - 28 = 27$ votes.

27. Let x be the fastest speed of the pheasant. Since a falcon's fastest speed is five times as fast as a pheasant, a falcon's fastest speed is $5x$. Since the total speeds for these two birds is 222 miles per hour, the sum of x and $5x$ is 222.
$$x + 5x = 222$$
$$6x = 222$$
$$\frac{6x}{6} = \frac{222}{6}$$
$$x = 37$$

The pheasant's fastest speed is 37 miles per hour and the falcon's fastest speed is $5(37) = 185$ miles per hour.

29. Let x be the number of universities in the United States. Since India has 2649 more universities than the United States, India has $x + 2649$ universities. Since their combined total is 14,165, the sum of x and $x + 2649$ is 14,165.
$$x + x + 2649 = 14,165$$
$$2x + 2649 = 14,165$$
$$2x + 2649 - 2649 = 14,165 - 2649$$
$$2x = 11,516$$
$$\frac{2x}{2} = \frac{11,516}{2}$$
$$x = 5758$$
Thus, the United States has 5758 universities and India has $5758 + 2649 = 8407$ universities.

31. Let x be the cost of the games. Since the cost of the Xbox 360 is 3 times as much as the games, the cost of the Xbox 360 is $3x$. Since the total cost is $560, the sum of x and $3x$ is 560.
$$x + 3x = 560$$
$$4x = 560$$
$$\frac{4x}{4} = \frac{560}{4}$$
$$x = 140$$
The games cost $140 and the Xbox 360 costs $3(140) = \$420$.

33. Let x be the distance from Los Angeles to Tokyo. Since the distance from New York to London is 2001 miles less than the distance from Los Angeles to Tokyo, the distance from New York to London is $x - 2001$. Since the total of the two distances is 8939, the sum of x and $x - 2001$ is 8939.

$$x + x - 2001 = 8939$$
$$2x - 2001 = 8939$$
$$2x - 2001 + 2001 = 8939 + 2001$$
$$2x = 10,940$$
$$\frac{2x}{2} = \frac{10,940}{2}$$
$$x = 5470$$

The distance from Los Angeles to Tokyo is 5470 miles.

35. Let x be the capacity of Neyland Stadium. Since the capacity of Michigan Stadium is 4647 more than that of Neyland Stadium, the capacity of Michigan Stadium is $x + 4647$. Since the combined capacity is 210,355, the sum of x and $x + 4647$ is 210,355.

$$x + x + 4647 = 210,355$$
$$2x + 4647 = 210,355$$
$$2x + 4647 - 4647 = 210,355 - 4647$$
$$2x = 205,708$$
$$\frac{2x}{2} = \frac{205,708}{2}$$
$$x = 102,854$$

The capacity of Neyland Stadium is 102,854 and the capacity of Michigan Stadium is $102,854 + 4647 = 107,501$.

37. Let x be the number of tourists projected to visit Spain in 2020. Since the number of tourists projected to visit China in 2020 is twice the number projected for Spain, the number projected for China is $2x$. Since the total number of tourists projected for the two countries is 210 million, the sum of x and $2x$ is 210 million.

$$x + 2x = 210$$
$$3x = 210$$
$$\frac{3x}{3} = \frac{210}{3}$$
$$x = 70$$
$$2x = 2(70) = 140$$

70 million tourists are projected to visit Spain in 2020; 140 million are projected to visit China in 2020.

39. Let x be the number of cars manufactured in Spain. Since the number of cars manufactured in Germany is twice as many as those manufactured in Spain, the number of cars manufactured in Germany is $2x$. Since the total number of these cars is 19,827, the sum of x and $2x$ is 19,827.

$$x + 2x = 19,827$$
$$3x = 19,827$$
$$\frac{3x}{3} = \frac{19,827}{3}$$
$$x = 6609$$

the number of cars manufactured in Spain is 6609 per day and the number in Germany is $2(6609) = 13,218$ per day.

41. Let x be the amount Anthony received for the accessories. Since he received five times as much for the bike, he received $5x$ for the bike. Since he received a total of $270 for the bike and accessories, the sum of x and $5x$ is 270.

$$x + 5x = 270$$
$$6x = 270$$
$$\frac{6x}{6} = \frac{270}{6}$$
$$x = 45$$

Thus, Anthony received $5 \cdot \$45 = \$225$ for the bike.

43. Let x be the points scored by the Duke Blue Devils. Since the Maryland Terrapins scored 3 more points than the Duke Blue Devils, the Maryland Terrapins scored $x + 3$ points. Since both teams scored a total of 153 points, the sum of x and $x + 3$ is 153.

$$x + x + 3 = 153$$
$$2x + 3 = 153$$
$$2x + 3 - 3 = 153 - 3$$
$$2x = 150$$
$$\frac{2x}{2} = \frac{150}{2}$$
$$x = 75$$

The Maryland Terrapins scored $75 + 3 = 78$ points.

45. 586 rounded to the nearest ten is 590.

47. 1026 rounded to the nearest hundred is 1000.

49. 2986 rounded to the nearest thousand is 3000.

51. answers may vary

53. Use $P = A + C$, where $P = 165,000$ and $A = 156,750$.

$$P = A + C$$
$$165,000 = 156,750 + C$$
$$165,000 - 156,750 = 156,750 - 156,750 + C$$
$$8250 = C$$

The agent will receive $8250.

55. Use $P = C + M$ where $P = 12$ and $C = 7$.

$$P = C + M$$
$$12 = 7 + M$$
$$12 - 7 = 7 - 7 + M$$
$$5 = M$$

The markup is $5.

Chapter 3 Vocabulary Check

1. An algebraic expression is <u>simplified</u> when all like terms have been <u>combined</u>.

2. Terms that are exactly the same, except that they may have different numerical coefficients, are called <u>like</u> terms.

3. A letter used to represent a number is called a <u>variable</u>.

4. A combination of operations on variables and numbers is called an <u>algebraic expression</u>.

5. The addends of an algebraic expression are called the <u>terms</u> of the expression.

6. The number factor of a variable term is called the <u>numerical coefficient</u>.

7. Replacing a variable in an expression by a number and then finding the value of the expression is called <u>evaluating the expression</u> for the variable.

8. A term that is a number only is called a <u>constant</u>.

9. An <u>equation</u> is of the form expression = expression.

10. A <u>solution</u> of an equation is a value for the variable that makes an equation a true statement.

11. To multiply $-3(2x + 1)$, we use the <u>distributive</u> property.

12. By the <u>multiplication</u> property of equality, we may multiply or divide both sides of an equation by any nonzero number without changing the solution of the equation.

13. By the <u>addition</u> property of equality, the same number may be added to or subtracted from both sides of an equation without changing the solution of the equation.

Chapter 3 Review

1. $3y + 7y - 15 = (3 + 7)y - 15 = 10y - 15$

2. $2y - 10 - 8y = 2y - 8y - 10$
$$= (2 - 8)y - 10$$
$$= -6y - 10$$

3. $8a + a - 7 - 15a = 8a + a - 15a - 7$
$$= (8 + 1 - 15)a - 7$$
$$= -6a - 7$$

4. $y + 3 - 9y - 1 = y - 9y + 3 - 1$
$$= (1 - 9)y + 3 - 1$$
$$= -8y + 2$$

5. $2(x + 5) = 2 \cdot x + 2 \cdot 5 = 2x + 10$

6. $-3(y + 8) = -3 \cdot y + (-3) \cdot 8 = -3y - 24$

7. $7x + 3(x - 4) + x = 7x + 3 \cdot x - 3 \cdot 4 + x$
$$= 7x + 3x - 12 + x$$
$$-7x + 3x + x - 12$$
$$= (7 + 3 + 1)x - 12$$
$$= 11x - 12$$

8. $-(3m+2)-m-10$
$= -1(3m+2)-m-10$
$= -1 \cdot 3m + (-1) \cdot 2 - m - 10$
$= -3m - 2 - m - 10$
$= -3m - m - 2 - 10$
$= (-3-1)m - 2 - 10$
$= -4m - 12$

9. $3(5a-2)-20a+10 = 3 \cdot 5a - 3 \cdot 2 - 20a + 10$
$= 15a - 6 - 20a + 10$
$= 15a - 20a - 6 + 10$
$= (15-20)a - 6 + 10$
$= -5a + 4$

10. $6y+3+2(3y-6) = 6y+3+2 \cdot 3y - 2 \cdot 6$
$= 6y + 3 + 6y - 12$
$= 6y + 6y + 3 - 12$
$= (6+6)y + 3 - 12$
$= 12y - 9$

11. $6y-7+11y-y+2 = 6y+11y-y-7+2$
$= (6+11-1)y - 7 + 2$
$= 16y - 5$

12. $10-x+5x-12-3x = -x+5x-3x+10-12$
$= (-1+5-3)x + 10 - 12$
$= 1x - 2$
$= x - 2$

13. Perimeter $= 2(2x+3)$
$= 2 \cdot 2x + 2 \cdot 3$
$= 4x + 6$
The perimeter is $(4x + 6)$ yards.

14. Perimeter $= 4 \cdot 5y = 20y$
The perimeter is $20y$ meters.

15. Area $=$ (length) $\cdot$ (width)
$= 3 \cdot (2x-1)$
$= 3 \cdot 2x - 3 \cdot 1$
$= 6x - 3$
The area is $(6x - 3)$ square yards.

16. Add the areas of the two rectangles.
$10(x-2)+7(5x+4)$
$= 10 \cdot x - 10 \cdot 2 + 7 \cdot 5x + 7 \cdot 4$
$= 10x - 20 + 35x + 28$
$= (10+35)x - 20 + 28$
$= 45x + 8$
The area is $(45x + 8)$ square centimeters.

17. $z - 5 = -7$
$z - 5 + 5 = -7 + 5$
$z = -2$

18. $3x + 10 = 4x$
$3x - 3x + 10 = 4x - 3x$
$10 = x$

19. $3y = -21$
$\dfrac{3y}{3} = \dfrac{-21}{3}$
$y = -7$

20. $-3a = -15$
$\dfrac{-3a}{-3} = \dfrac{-15}{-3}$
$a = 5$

21. $\dfrac{x}{-6} = 2$
$-6 \cdot \dfrac{x}{-6} = -6 \cdot 2$
$x = -12$

22. $\dfrac{y}{-15} = -3$
$-15 \cdot \dfrac{y}{-15} = -15 \cdot (-3)$
$y = 45$

23. $n + 18 = 10 - (-2)$
$n + 18 = 10 + 2$
$n + 18 = 12$
$n + 18 - 18 = 12 - 18$
$n = -6$

24.
$$c - 5 = -13 + 7$$
$$c - 5 = -6$$
$$c - 5 + 5 = -6 + 5$$
$$c = -1$$

25.
$$7x + 5 - 6x = -20$$
$$x + 5 = -20$$
$$x + 5 - 5 = -20 - 5$$
$$x = -25$$

26.
$$17x = 2(8x - 4)$$
$$17x = 2 \cdot 8x - 2 \cdot 4$$
$$17x = 16x - 8$$
$$17x - 16x = 16x - 16x - 8$$
$$x = -8$$

27.
$$5x + 7 = -3$$
$$5x + 7 - 7 = -3 - 7$$
$$5x = -10$$
$$\frac{5x}{5} = \frac{-10}{5}$$
$$x = -2$$

28.
$$-14 = 9y + 4$$
$$-14 - 4 = 9y + 4 - 4$$
$$-18 = 9y$$
$$\frac{-18}{9} = \frac{9y}{9}$$
$$-2 = y$$

29.
$$\frac{z}{4} = -8 - (-6)$$
$$\frac{z}{4} = -8 + 6$$
$$\frac{z}{4} = -2$$
$$4 \cdot \frac{z}{4} = 4 \cdot (-2)$$
$$z = -8$$

30.
$$-1 + (-8) = \frac{x}{5}$$
$$-9 = \frac{x}{5}$$
$$5 \cdot (-9) = 5 \cdot \frac{x}{5}$$
$$-45 = x$$

31.
$$6y - 7y = 100 - 105$$
$$-y = -5$$
$$\frac{-y}{-1} = \frac{-5}{-1}$$
$$y = 5$$

32.
$$19x - 16x = 45 - 60$$
$$3x = -15$$
$$\frac{3x}{3} = \frac{-15}{3}$$
$$x = -5$$

33.
$$9(2x - 7) = 19x$$
$$9 \cdot 2x - 9 \cdot 7 = 19x$$
$$18x - 63 = 19x$$
$$18x - 18x - 63 = 19x - 18x$$
$$-63 = x$$

34.
$$-5(3x + 3) = -14x$$
$$-5 \cdot 3x - 5 \cdot 3 = -14x$$
$$-15x - 15 = -14x$$
$$-15x + 15x - 15 = -14x + 15x$$
$$-15 = x$$

35.
$$3x - 4 = 11$$
$$3x - 4 + 4 = 11 + 4$$
$$3x = 15$$
$$\frac{3x}{3} = \frac{15}{3}$$
$$x = 5$$

36.
$$6y + 1 = 73$$
$$6y + 1 - 1 = 73 - 1$$
$$6y = 72$$
$$\frac{6y}{6} = \frac{72}{6}$$
$$y = 12$$

37.
$$2(x+4)-10=-2(7)$$
$$2\cdot x+2\cdot4-10=-14$$
$$2x+8-10=-14$$
$$2x-2=-14$$
$$2x-2+2=-14+2$$
$$2x=-12$$
$$\frac{2x}{2}=\frac{-12}{2}$$
$$x=-6$$

38.
$$-3(x-6)+13=20-1$$
$$-3\cdot x-(-3)\cdot6+13=19$$
$$-3x+18+13=19$$
$$-3x+31=19$$
$$-3x+31-31=19-31$$
$$-3x=-12$$
$$\frac{-3x}{-3}=\frac{-12}{-3}$$
$$x=4$$

39. The product of -5 and a number is $-5x$.

40. Three subtracted from a number is $x-3$.

41. The sum of -5 and a number is $-5+x$.

42. The quotient of -2 and a number is $\dfrac{-2}{x}$.

43. Eleven added to twice a number is $2x+11$.

44. The product of -5 and a number, decreased by 50 is $-5x-50$.

45. The quotient of 70 and the sum of a number and 6 is $\dfrac{70}{x+6}$.

46. Twice the difference of a number and 13 is $2(x-13)$.

47.
$$2x+5=7x-100$$
$$2x-2x+5=7x-2x-100$$
$$5=5x-100$$
$$5+100=5x-100+100$$
$$105=5x$$
$$\frac{105}{5}=\frac{5x}{5}$$
$$21=x$$

48.
$$-6x-4=x+66$$
$$-6x+6x-4=x+6x+66$$
$$-4=7x+66$$
$$-4-66=7x+66-66$$
$$-70=7x$$
$$\frac{-70}{7}=\frac{7x}{7}$$
$$-10=x$$

49.
$$2x+7=6x-1$$
$$2x-2x+7=6x-2x-1$$
$$7=4x-1$$
$$7+1=4x-1+1$$
$$8=4x$$
$$\frac{8}{4}=\frac{4x}{4}$$
$$2=x$$

50.
$$5x-18=-4x$$
$$5x-5x-18=-4x-5x$$
$$-18=-9x$$
$$\frac{-18}{-9}=\frac{-9x}{-9}$$
$$2=x$$

51.
$$5(n-3)=7+3n$$
$$5n-15=7+3n$$
$$5n-3n-15=7+3n-3n$$
$$2n-15=7$$
$$2n-15+15=7+15$$
$$2n=22$$
$$\frac{2n}{2}=\frac{22}{2}$$
$$n=11$$

52.
$$7(2+x) = 4x - 1$$
$$14 + 7x = 4x - 1$$
$$14 + 7x - 4x = 4x - 4x - 1$$
$$14 + 3x = -1$$
$$14 - 14 + 3x = -1 - 14$$
$$3x = -15$$
$$\frac{3x}{3} = \frac{-15}{3}$$
$$x = -5$$

53.
$$6x + 3 - (-x) = -20 + 5x - 7$$
$$6x + 3 + x = -20 + 5x - 7$$
$$7x + 3 = -27 + 5x$$
$$7x - 5x + 3 = -27 + 5x - 5x$$
$$2x + 3 = -27$$
$$2x + 3 - 3 = -27 - 3$$
$$2x = -30$$
$$\frac{2x}{2} = \frac{-30}{2}$$
$$x = -15$$

54.
$$x - 25 + 2x = -5 + 2x - 10$$
$$-25 + 3x = -15 + 2x$$
$$-25 + 3x - 2x = -15 + 2x - 2x$$
$$-25 + x = -15$$
$$-25 + 25 + x = -15 + 25$$
$$x = 10$$

55.
$$3(x - 4) = 5x - 8$$
$$3x - 12 = 5x - 8$$
$$3x - 3x - 12 = 5x - 3x - 8$$
$$-12 = 2x - 8$$
$$-12 + 8 = 2x - 8 + 8$$
$$-4 = 2x$$
$$\frac{-4}{2} = \frac{2x}{2}$$
$$-2 = x$$

56.
$$4(x - 3) = -2x - 48$$
$$4x - 12 = -2x - 48$$
$$4x + 2x - 12 = -2x + 2x - 48$$
$$6x - 12 = -48$$
$$6x - 12 + 12 = -48 + 12$$
$$6x = -36$$
$$\frac{6x}{6} = \frac{-36}{6}$$
$$x = -6$$

57.
$$6(2n - 1) + 18 = 0$$
$$12n - 6 + 18 = 0$$
$$12n + 12 = 0$$
$$12n + 12 - 12 = 0 - 12$$
$$12n = -12$$
$$\frac{12n}{12} = \frac{-12}{12}$$
$$n = -1$$

58.
$$7(3y - 2) - 7 = 0$$
$$21y - 14 - 7 = 0$$
$$21y - 21 = 0$$
$$21y - 21 + 21 = 0 + 21$$
$$21y = 21$$
$$\frac{21y}{21} = \frac{21}{21}$$
$$y = 1$$

59.
$$95x - 14 = 20x - 10 + 10x - 4$$
$$95x - 14 = 30x - 14$$
$$95x - 14 + 14 = 30x - 14 + 14$$
$$95x = 30x$$
$$95x - 30x = 30x - 30x$$
$$65x = 0$$
$$\frac{65x}{65} = \frac{0}{65}$$
$$x = 0$$

60.
$$32z + 11 - 28z = 50 + 2z - (-1)$$
$$4z + 11 = 50 + 2z + 1$$
$$4z + 11 = 51 + 2z$$
$$4z - 2z + 11 = 51 + 2z - 2z$$
$$2z + 11 = 51$$
$$2z + 11 - 11 = 51 - 11$$
$$2z = 40$$
$$\frac{2z}{2} = \frac{40}{2}$$
$$z = 20$$

61. The difference of 20 and −8 is 28 translates to $20 - (-8) = 28$.

62. Five times the sum of 2 and −6 yields −20 translates to $5[2 + (-6)] = -20$.

63. The quotient of −75 and the sum of 5 and 20 is equal to −3 translates to $\dfrac{-75}{5 + 20} = -3$.

64. Nineteen subtracted from −2 amounts to −21 translates to $-2 - 19 = -21$.

65. "Twice a number minus 8 is 40" is $2x - 8 = 40$.

66. "The product of some number and 6 is equal to the sum of the number and 2" is $6x = x + 2$.

67. "Twelve subtracted from the quotient of a number and 2 is 10" is $\dfrac{x}{2} - 12 = 10$.

68. "The difference of a number and 3 is the quotient of the number and 4" is $x - 3 = \dfrac{x}{4}$.

69. "Five times a number subtracted from 40 is the same as three times the number" is
$$40 - 5x = 3x$$
$$40 - 5x + 5x = 3x + 5x$$
$$40 = 8x$$
$$\frac{40}{8} = \frac{8x}{8}$$
$$5 = x$$
The number is 5.

70. "The product of a number and 3 is twice the difference of that number and 8" is
$$3x = 2(x - 8)$$
$$3x = 2x - 16$$
$$3x - 2x = 2x - 2x - 16$$
$$x = -16$$
The number is −16.

71. Let x be the number of votes for the Independent candidate. Since the Democratic candidate received 272 more votes than the Independent candidate, the Democratic candidate received $x + 272$ votes. The total number of votes is 18,500.
$$x + x + 272 + 14,000 = 18,500$$
$$2x + 14,272 = 18,500$$
$$2x + 14,272 - 14,272 = 18,500 - 14,272$$
$$2x = 4228$$
$$\frac{2x}{2} = \frac{4228}{2}$$
$$x = 2114$$
The Democratic candidate received $2114 + 272 = 2386$ votes.

72. Let x be the number of movies on videotapes. Since the number of movies on DVDs is twice the number on videotapes, the number of movies on DVDs is $2x$. Since the total number of movies is 126, the sum of x and $2x$ is 126.
$$x + 2x = 126$$
$$3x = 126$$
$$\frac{3x}{3} = \frac{126}{3}$$
$$x = 42$$
He has $2(42) = 84$ movies on DVDs.

73. $9x - 20x = (9 - 20)x = -11x$

74. $-5(7x) = (-5 \cdot 7)x = -35x$

75. $12x + 5(2x - 3) - 4 = 12x + 10x - 15 - 4$
$$= 22x - 19$$

76. $-7(x + 6) - 2(x - 5) = -7x - 42 - 2x + 10$
$$= -7x - 2x - 42 + 10$$
$$= -9x - 32$$

77. $c - 5 = -13 + 7$
$$c - 5 = -6$$
$$c - 5 + 5 = -6 + 5$$
$$c = -1$$

78. $7x + 5 - 6x = -20$
$$x + 5 = -20$$
$$x + 5 - 5 = -20 - 5$$
$$x = -25$$

79. $-7x + 3x = -50 - 2$
$$-4x = -52$$
$$\frac{-4x}{-4} = \frac{-52}{-4}$$
$$x = 13$$

80. $-x + 8x = -38 - 4$
$$7x = -42$$
$$\frac{7x}{7} = \frac{-42}{7}$$
$$x = -6$$

81. $9x + 12 - 8x = -6 + (-4)$
$$x + 12 = -10$$
$$x + 12 - 12 = -10 - 12$$
$$x = -22$$

82. $-17x + 14 + 20x - 2x = 5 - (-3)$
$$-17x + 20x - 2x + 14 = 5 + 3$$
$$x + 14 = 8$$
$$x + 14 - 14 = 8 - 14$$
$$x = -6$$

83. $5(2x - 3) = 11x$
$$10x - 15 = 11x$$
$$10x - 10x - 15 = 11x - 10x$$
$$-15 = x$$

84. $\dfrac{y}{-3} = -1 - 5$
$$\dfrac{y}{-3} = -6$$
$$-3 \cdot \dfrac{y}{-3} = -3 \cdot (-6)$$
$$y = 18$$

85. $12y - 10 = -70$
$$12y - 10 + 10 = -70 + 10$$
$$12y = -60$$
$$\frac{12y}{12} = \frac{-60}{12}$$
$$y = -5$$

86. $-6(x - 3) = x + 4$
$$-6x + 18 = x + 4$$
$$-6x + 6x + 18 = x + 6x + 4$$
$$18 = 7x + 4$$
$$18 - 4 = 7x + 4 - 4$$
$$14 = 7x$$
$$\frac{14}{7} = \frac{7x}{7}$$
$$2 = x$$

87. $4n - 8 = 2n + 14$
$$4n - 2n - 8 = 2n - 2n + 14$$
$$2n - 8 = 14$$
$$2n - 8 + 8 = 14 + 8$$
$$2n = 22$$
$$\frac{2n}{2} = \frac{22}{2}$$
$$n = 11$$

88.
$$9(3x-4)+63=0$$
$$27x-36+63=0$$
$$27x+27=0$$
$$27x+27-27=0-27$$
$$27x=-27$$
$$\frac{27x}{27}=\frac{-27}{27}$$
$$x=-1$$

89.
$$-5z+3z-7=8z-1-6$$
$$-2z-7=8z-7$$
$$-2z-8z-7=8z-8z-7$$
$$-10z-7=-7$$
$$-10z-7+7=-7+7$$
$$-10z=0$$
$$\frac{-10z}{-10}=\frac{0}{-10}$$
$$z=0$$

90.
$$4x-3+6x=5x-3-30$$
$$10x-3=5x-33$$
$$10x-5x-3=5x-5x-33$$
$$5x-3=-33$$
$$5x-3+3=-33+3$$
$$5x=-30$$
$$\frac{5x}{5}=\frac{-30}{5}$$
$$x=-6$$

91. "Three times a number added to twelve is 27" is
$$12+3x=27$$
$$12-12+3x=27-12$$
$$3x=15$$
$$\frac{3x}{3}=\frac{15}{3}$$
$$x=5$$
The number is 5.

92. "Twice the sum of a number and four is ten" is
$$2(x+4)=10$$
$$2x+8=10$$
$$2x+8-8=10-8$$
$$2x=2$$
$$\frac{2x}{2}=\frac{2}{2}$$
$$x=1$$
The number is 1.

93. Let x be the number of roadway miles in Hawaii. Since Delaware has 1585 more roadway miles than Hawaii, Delaware has $x + 1585$ roadway miles. Since the total number of roadway miles is 10,203, the sum of x and $x + 1585$ is 10,203.
$$x+x+1585=10,203$$
$$2x+1585=10,203$$
$$2x+1585-1585=10,203-1585$$
$$2x=8618$$
$$\frac{2x}{2}=\frac{8618}{2}$$
$$x=4309$$
Hawaii has 4309 roadway miles and Delaware has 4309 + 1585 = 5894 roadway miles.

94. Let x be the number of roadway miles in South Dakota. Since North Dakota has 3094 more roadway miles than South Dakota, North Dakota has $x + 3094$ roadway miles. Since the total number of roadway miles is 170,470, the sum of x and $x + 3094$ is 170,470.
$$x+x+3094=170,470$$
$$2x+3094=170,470$$
$$2x+3094-3094=170,470-3094$$
$$2x=167,376$$
$$\frac{2x}{2}=\frac{167,376}{2}$$
$$x=83,688$$
South Dakota has 83,688 roadway miles and North Dakota has 83,688 + 3094 = 86,782 roadway miles.

Chapter 3 Test

1. $7x - 5 - 12x + 10 = 7x - 12x - 5 + 10$
$$= (7 - 12)x - 5 + 10$$
$$= -5x + 5$$

2. $-2(3y + 7) = -2 \cdot 3y + (-2) \cdot 7 = -6y - 14$

3. $-(3z + 2) - 5z - 18$
$$= -1(3z + 2) - 5z - 18$$
$$= -1 \cdot 3z + (-1) \cdot 2 - 5z - 18$$
$$= -3z - 2 - 5z - 18$$
$$= -3z - 5z - 2 - 18$$
$$= -8z - 20$$

4. perimeter $= 3(5x + 5)$
$$= 3 \cdot 5x + 3 \cdot 5$$
$$= 15x + 15$$
The perimeter is $(15x + 15)$ inches.

5. Area $= (\text{length}) \cdot (\text{width})$
$$= 4 \cdot (3x - 1)$$
$$= 4 \cdot 3x - 4 \cdot 1$$
$$= 12x - 4$$
The area is $(12x - 4)$ square meters.

6. $12 = y - 3y$
$$12 = -2y$$
$$\frac{12}{-2} = \frac{-2y}{-2}$$
$$-6 = y$$

7. $\dfrac{x}{2} = -5 - (-2)$
$$\frac{x}{2} = -5 + 2$$
$$\frac{x}{2} = -3$$
$$2 \cdot \frac{x}{2} = 2 \cdot (-3)$$
$$x = -6$$

8. $5x + 12 - 4x - 14 = 22$
$$x - 2 = 22$$
$$x - 2 + 2 = 22 + 2$$
$$x = 24$$

9. $-4x + 7 = 15$
$$-4x + 7 - 7 = 15 - 7$$
$$-4x = 8$$
$$\frac{-4x}{-4} = \frac{8}{-4}$$
$$x = -2$$

10. $2(x - 6) = 0$
$$2x - 12 = 0$$
$$2x - 12 + 12 = 0 + 12$$
$$2x = 12$$
$$\frac{2x}{2} = \frac{12}{2}$$
$$x = 6$$

11. $-4(x - 11) - 34 = 10 - 12$
$$-4x + 44 - 34 = 10 - 12$$
$$-4x + 10 = -2$$
$$-4x + 10 - 10 = -2 - 10$$
$$-4x = -12$$
$$\frac{-4x}{-4} = \frac{-12}{-4}$$
$$x = 3$$

12. $5x - 2 = x - 10$
$$5x - x - 2 = x - x - 10$$
$$4x - 2 = -10$$
$$4x - 2 + 2 = -10 + 2$$
$$4x = -8$$
$$\frac{4x}{4} = \frac{-8}{4}$$
$$x = -2$$

13. $4(5x + 3) = 2(7x + 6)$
$$20x + 12 = 14x + 12$$
$$20x + 12 - 14x = 14x + 12 - 14x$$
$$6x + 12 = 12$$
$$6x + 12 - 12 = 12 - 12$$
$$6x = 0$$
$$\frac{6x}{6} = \frac{0}{6}$$
$$x = 0$$

14. $6 + 2(3n - 1) = 28$
$6 + 6n - 2 = 28$
$6n + 4 = 28$
$6n + 4 - 4 = 28 - 4$
$6n = 24$
$\dfrac{6n}{6} = \dfrac{24}{6}$
$n = 4$

15. The sum of -23 and a number translates to $-23 + x$.

16. Three times a number, subtracted from -2 translates to $-2 - 3x$.

17. The sum of twice 5 and -15 is -5 translates to $2(5) + (-15) = -5$.

18. Six added to three times a number equals -30 translates to $3x + 6 = -30$.

19. The difference of three times a number and five times the same number is 4 translates to
$3x - 5x = 4$
$-2x = 4$
$\dfrac{-2x}{-2} = \dfrac{4}{-2}$
$x = -2$
The number is -2.

20. Let x be the number of free throws Maria made. Since Paula made twice as many free throws as Maria, Paula made $2x$ free throws. Since the total number of free throws was 12, the sum of x and $2x$ is 12.
$x + 2x = 12$
$3x = 12$
$\dfrac{3x}{3} = \dfrac{12}{3}$
$x = 4$
Paula made $2(4) = 8$ free throws.

21. Let x be the number of women runners entered in the race. Since the number of men entered in the race is 112 more than the number of women, the number of men is $x + 112$. Since the total number of runners in

the race is 600, the sum of x and $x + 112$ is 600.
$x + x + 112 = 600$
$2x + 112 = 600$
$2x + 112 - 112 = 600 - 112$
$2x = 488$
$\dfrac{2x}{2} = \dfrac{488}{2}$
$x = 244$
244 women entered the race.

Cumulative Review Chapters 1–3

1. 308,063,557 in words is three hundred eight million, sixty-three thousand, five hundred fifty-seven.

2. 276,004 in words is two hundred seventy-six thousand, four.

3. $2 + 3 + 1 + 3 + 4 = 13$
The perimeter is 13 inches.

4. $6 + 3 + 6 + 3 = 18$
The perimeter is 18 inches.

5.
$$\begin{array}{r} 900 \\ -\ 174 \\ \hline 726 \end{array} \qquad Check: \begin{array}{r} 726 \\ +\ 174 \\ \hline 900 \end{array}$$

6.
$$\begin{array}{r} 17{,}801 \\ -\ 8216 \\ \hline 9585 \end{array} \qquad Check: \begin{array}{r} 9585 \\ +\ 8216 \\ \hline 17{,}801 \end{array}$$

7. To round 248,982 to the nearest hundred, observe that the digit in the tens place is 8. Since this digit is at least 5, we add 1 to the digit in the hundreds place. The number 248,982 rounded to the nearest hundred is 249,000.

8. To round 844,497 to the nearest thousand, observe that the digit in the hundreds place is 4. Since this digit is less than 5, we do not add 1 to the digit in the thousands place. The number 844,497 rounded to the nearest thousand is 844,000.

9.
$$
\begin{array}{r}
25 \\
\times\ 8 \\
\hline
200
\end{array}
$$

10.
$$
\begin{array}{r}
395 \\
\times\ 74 \\
\hline
1\ 580 \\
27\ 650 \\
\hline
29{,}230
\end{array}
$$

11.
$$
\begin{array}{r}
208 \\
9\overline{)\ 1872} \\
-18 \\
\hline
07 \\
-0 \\
\hline
72 \\
-72 \\
\hline
0
\end{array}
$$

Check:
$$
\begin{array}{r}
208 \\
\times\ 9 \\
\hline
1872
\end{array}
$$

12.
$$
\begin{array}{r}
86 \\
46\overline{)\ 3956} \\
-368 \\
\hline
276 \\
-276 \\
\hline
0
\end{array}
$$

Check:
$$
\begin{array}{r}
86 \\
\times\ 46 \\
\hline
516 \\
3440 \\
\hline
3956
\end{array}
$$

13. $2 \cdot 4 - 3 \div 3 = 8 - 3 \div 3 = 8 - 1 = 7$

14. $8 \cdot 4 + 9 \div 3 = 32 + 9 \div 3 = 32 + 3 = 35$

15. $x^2 + z - 3 = 5^2 + 4 - 3$
$$
\begin{aligned}
&= 25 + 4 - 3 \\
&= 29 - 3 \\
&= 26
\end{aligned}
$$

16. $2a^2 + 5 - c = 2 \cdot 2^2 + 5 - 3$
$$
\begin{aligned}
&= 2 \cdot 4 + 5 - 3 \\
&= 8 + 5 - 3 \\
&= 13 - 3 \\
&= 10
\end{aligned}
$$

17. $2n - 30 = 10$
Let $n = 26$.
$2 \cdot 26 - 30 \overset{?}{=} 10$
$52 - 30 \overset{?}{=} 10$
$22 = 10$ False
26 is not a solution.
Let $n = 40$.
$2 \cdot 40 - 30 \overset{?}{=} 10$
$80 - 30 \overset{?}{=} 10$
$50 = 10$ False
40 is not a solution.
Let $n = 20$.
$2 \cdot 20 - 30 \overset{?}{=} 10$
$40 - 30 \overset{?}{=} 10$
$10 = 10$ True
20 is a solution.

18. a. $-14 < 0$ because -14 is to the left of 0 on the number line.

     b. $-(-7) = 7$, so $-(-7) > -8$ because 7 is to the right of -8 on the number line.

19. $5 + (-2) = 3$

20. $-3 + (-4) = -7$

21. $-15 + (-10) = -25$

22. $3 + (-7) = -4$

23. $-2 + 5 = 3$

24. $21 + 15 + (-19) = 36 + (-19) = 17$

25. $-4 - 10 = -4 + (-10) = -14$

26. $-2 - 3 = -2 + (-3) = -5$

27. $6 - (-5) = 6 + 5 = 11$

28. $19 - (-10) = 19 + 10 = 29$

29. $-11 - (-7) = -11 + 7 = -4$

30. $-16 - (-13) = -16 + 13 = -3$

31. $\dfrac{-12}{6} = -2$

32. $\dfrac{-30}{-5} = 6$

33. $-20 \div (-4) = 5$

34. $26 \div (-2) = -13$

35. $\dfrac{48}{-3} = -16$

36. $\dfrac{-120}{12} = -10$

37. $(-3)^2 = (-3)(-3) = 9$

38. $-2^5 = -(2 \cdot 2 \cdot 2 \cdot 2 \cdot 2) = -32$

39. $-3^2 = -(3 \cdot 3) = -9$

40. $(-5)^2 = (-5)(-5) = 25$

41. $2y - 6 + 4y + 8 = 2y + 4y - 6 + 8$
$= (2 + 4)y - 6 + 8$
$= 6y + 2$

42. $6x + 2 - 3x + 7 = 6x - 3x + 2 + 7$
$= (6 - 3)x + 2 + 7$
$= 3x + 9$

43. $3y + 1 = 3$
$3(-1) + 1 \stackrel{?}{=} 3$
$-3 + 1 \stackrel{?}{=} 3$
$-2 = 3$ False
Since $-2 = 3$ is false, -1 is not a solution of the equation.

44. $5x - 3 = 7$
$5(2) - 3 \stackrel{?}{=} 7$
$10 - 3 \stackrel{?}{=} 7$
$7 = 7$ True
Since $7 = 7$ is true, 2 is a solution of the equation.

45. $-12x = -36$
$\dfrac{-12x}{-12} = \dfrac{-36}{-12}$
$x = 3$

46. $-3y = 15$
$\dfrac{-3y}{-3} = \dfrac{15}{-3}$
$y = -5$

47. $2x - 6 = 18$
$2x - 6 + 6 = 18 + 6$
$2x = 24$
$\dfrac{2x}{2} = \dfrac{24}{2}$
$x = 12$

48. $3a + 5 = -1$
$3a + 5 - 5 = -1 - 5$
$3a = -6$
$\dfrac{3a}{3} = \dfrac{-6}{3}$
$a = -2$

49. Let x be the price of the software. Since the price of the computer system is four times the price of the software, the price of the computer system is $4x$. Since the combined price is 2100, the sum of x and $4x$ is 2100.
$x + 4x = 2100$
$5x = 2100$
$\dfrac{5x}{5} = \dfrac{2100}{5}$
$x = 420$
The price of the software is $420 and the price of the computer system is $4(\$420) = \$1680$.

50. Let x be the number. "Two times the number plus four is the same amount as three times the number minus seven" translates to

$$2x + 4 = 3x - 7$$
$$2x - 2x + 4 = 3x - 2x - 7$$
$$4 = x - 7$$
$$4 + 7 = x - 7 + 7$$
$$11 = x$$

The number is 11.

Chapter 4

Section 4.1

Practice Problems

1. In the fraction $\dfrac{11}{2}$, the numerator is 11 and the denominator is 2.

2. In the fraction $\dfrac{10y}{17}$, the numerator is 10y and the denominator is 17.

3. 3 out of 8 equal parts are shaded: $\dfrac{3}{8}$

4. 1 out of 6 equal parts are shaded: $\dfrac{1}{6}$

5. 7 out of 10 equal parts are shaded: $\dfrac{7}{10}$

6. 9 out of 16 equal parts are shaded: $\dfrac{9}{16}$

7. answers may vary; for example,

8. answers may vary; for example,

9. $\dfrac{\text{number of planets farther} \to 5}{\text{number of planets in our solar system} \to 8}$

   $\dfrac{5}{8}$ of the planets in our solar system are farther from the Sun than Earth.

10. Each part is $\dfrac{1}{3}$ of a whole and there are 8 parts shaded, or 2 wholes and 2 more parts.

    $\dfrac{8}{3}$; $2\dfrac{2}{3}$

11. Each part is $\dfrac{1}{4}$ of a whole and there are 5 parts shaded, or 1 whole and 1 more part.

    $\dfrac{5}{4}$; $1\dfrac{1}{4}$

12. a.

    b.

    c.

13. a.

    b.

    c.

14. $\dfrac{9}{9} = 1$

15. $\dfrac{-6}{-6} = 1$

16. $\dfrac{0}{-1} = 0$

17. $\dfrac{4}{1} = 4$

18. $\dfrac{-13}{0}$ is undefined.

19. $\dfrac{-13}{1} = -13$

20. a. $5\dfrac{2}{7} = \dfrac{7 \cdot 5 + 2}{7} = \dfrac{35 + 2}{7} = \dfrac{37}{7}$

b. $6\frac{2}{3} = \frac{3 \cdot 6 + 2}{3} = \frac{18 + 2}{3} = \frac{20}{3}$

c. $10\frac{9}{10} = \frac{10 \cdot 10 + 9}{10} = \frac{100 + 9}{10} = \frac{109}{10}$

d. $4\frac{1}{5} = \frac{5 \cdot 4 + 1}{5} = \frac{20 + 1}{5} = \frac{21}{5}$

21. a. $7)\overline{9} \quad \begin{array}{r} 1 \\ \underline{5} \\ 4 \end{array}$

$\frac{9}{5} = 1\frac{4}{5}$

b. $9)\overline{23} \quad \begin{array}{r} 2 \\ \underline{18} \\ 5 \end{array}$

$\frac{23}{9} = 2\frac{5}{9}$

c. $4)\overline{48} \quad \begin{array}{r} 12 \\ \underline{4} \\ 8 \\ \underline{8} \\ 0 \end{array}$

$\frac{48}{4} = 12$

d. $13)\overline{62} \quad \begin{array}{r} 4 \\ \underline{52} \\ 10 \end{array}$

$\frac{62}{13} = 4\frac{10}{13}$

e. $7)\overline{51} \quad \begin{array}{r} 7 \\ \underline{49} \\ 2 \end{array}$

$\frac{51}{7} = 7\frac{2}{7}$

f. $20)\overline{21} \quad \begin{array}{r} 1 \\ \underline{20} \\ 1 \end{array}$

$\frac{21}{20} = 1\frac{1}{20}$

Vocabulary and Readiness Check

1. The number $\frac{17}{31}$ is called a <u>fraction</u>. The number 31 is called its <u>denominator</u> and 17 is called its <u>numerator</u>.

3. The fraction $\frac{8}{3}$ is called an <u>improper</u> fraction, the fraction $\frac{3}{8}$ is called a <u>proper</u> fraction, and $10\frac{3}{8}$ is called a <u>mixed number</u>.

Exercise Set 4.1

1. In the fraction $\frac{1}{2}$, the numerator is 1 and the denominator is 2. Since $1 < 2$, the fraction is proper.

3. In the fraction $\frac{10}{3}$, the numerator is 10 and the denominator is 3. Since $10 > 3$, the fraction is improper.

5. In the fraction $\frac{15}{15}$, the numerator is 15 and the denominator is 15. Since $15 \geq 15$, the fraction is improper.

7. 1 out of 3 equal parts is shaded: $\frac{1}{3}$

9. Each part is $\frac{1}{4}$ of a whole and there are 11 parts shaded, or 2 wholes and 3 more parts.

    a. $\frac{11}{4}$

    b. $2\frac{3}{4}$

11. Each part is $\frac{1}{6}$ of a whole and there are 23 parts shaded, or 3 wholes and 5 more parts.

    a. $\frac{23}{6}$

    b. $3\frac{5}{6}$

13. 7 out of 12 equal parts are shaded: $\frac{7}{12}$

15. 3 out of 7 equal parts are shaded: $\frac{3}{7}$

17. 4 out of 9 equal parts are shaded: $\frac{4}{9}$

19. Each part is $\frac{1}{3}$ of a whole and there are 4 parts shaded, or 1 whole and 1 more part.

    a. $\frac{4}{3}$

    b. $1\frac{1}{3}$

21. Each part is $\frac{1}{2}$ of a whole and there are 11 parts shaded, or 5 wholes and 1 more part.

    a. $\frac{11}{2}$

    b. $5\frac{1}{2}$

23. 1 out of 6 equal parts are shaded: $\frac{1}{6}$

25. 5 of 8 equal parts are shaded: $\frac{5}{8}$

27. answers may vary; for example,

29. answers may vary; for example,

31. answers may vary; for example,

33. freshmen $\rightarrow \dfrac{42}{}$

    students $\rightarrow \overline{131}$

    $\dfrac{42}{131}$ of the students are freshmen.

35. a. $\begin{matrix}\text{number of students} \\ \text{not freshmen}\end{matrix} = 131 - 42$

                    $= 89$

    b. not freshmen $\rightarrow \dfrac{89}{}$

       students $\rightarrow \overline{131}$

       $\dfrac{89}{131}$ of the students are not freshmen.

37. born in Virginia → 8
U.S. presidents → $\overline{43}$

$\dfrac{8}{43}$ of U.S. presidents were born in Virginia.

39. number turned into hurricanes → 15
number of tropical storms → $\overline{28}$

$\dfrac{15}{28}$ of the tropical storms turned into hurricanes.

41. 11 of 31 days of March is $\dfrac{11}{31}$ of the month.

43. number of sophomores → 10
number of students in class → $\overline{31}$

$\dfrac{10}{31}$ of the class is sophomores.

45. There are 50 states total. 33 states contain federal Indian reservations.

a. $\dfrac{33}{50}$ of the states contain federal Indian reservations.

b. $50 - 33 = 17$
17 states do not contain federal Indian reservations.

c. $\dfrac{17}{50}$ of the states do not contain federal Indian reservations.

47. a. blue → 21
total → $\overline{50}$

$\dfrac{21}{50}$ of the marbles are blue.

b. $50 - 21 = 29$
29 of the marbles are red.

c. red → 29
total → $\overline{50}$

$\dfrac{29}{50}$ of the marbles are red.

49.

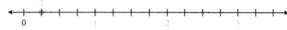

51.

53.

55.

57. $\dfrac{12}{12} = 1$

59. $\dfrac{-5}{1} = -5$

61. $\dfrac{0}{-2} = 0$

63. $\dfrac{-8}{-8} = 1$

65. $\dfrac{-9}{0}$ is undefined

67. $\dfrac{3}{1} = 3$

69. $2\dfrac{1}{3} = \dfrac{3 \cdot 2 + 1}{3} = \dfrac{6 + 1}{3} = \dfrac{7}{3}$

71. $3\dfrac{3}{5} = \dfrac{5 \cdot 3 + 3}{5} = \dfrac{15 + 3}{5} = \dfrac{18}{5}$

73. $6\dfrac{5}{8} = \dfrac{8 \cdot 6 + 5}{8} = \dfrac{48 + 5}{8} = \dfrac{53}{8}$

75. $11\dfrac{6}{7} = \dfrac{7 \cdot 11 + 6}{7} = \dfrac{77 + 6}{7} = \dfrac{83}{7}$

77. $9\dfrac{7}{20} = \dfrac{20 \cdot 9 + 7}{20} = \dfrac{180 + 7}{20} = \dfrac{187}{20}$

79. $166\dfrac{2}{3} = \dfrac{3 \cdot 166 + 2}{3} = \dfrac{498 + 2}{3} = \dfrac{500}{3}$

81.
$$5\overline{)17} \quad \begin{array}{r} 3 \\ \hline 17 \\ 15 \\ \hline 2 \end{array}$$

$\dfrac{17}{5} = 3\dfrac{2}{5}$

83.
$$8\overline{)37} \quad \begin{array}{r} 4 \\ \hline 37 \\ 32 \\ \hline 5 \end{array}$$

$\dfrac{37}{8} = 4\dfrac{5}{8}$

85.
$$15\overline{)47} \quad \begin{array}{r} 3 \\ \hline 47 \\ 45 \\ \hline 2 \end{array}$$

$\dfrac{47}{15} = 3\dfrac{2}{15}$

87.
$$15\overline{)225} \quad \begin{array}{r} 15 \\ \hline 225 \\ 15 \\ \hline 75 \\ 75 \\ \hline 0 \end{array}$$

$\dfrac{225}{15} = 15$

89.
$$175\overline{)182} \quad \begin{array}{r} 1 \\ \hline 182 \\ 175 \\ \hline 7 \end{array}$$

$\dfrac{182}{175} = 1\dfrac{7}{175}$

91.
$$112\overline{)737} \quad \begin{array}{r} 6 \\ \hline 737 \\ 672 \\ \hline 65 \end{array}$$

$\dfrac{737}{112} = 6\dfrac{65}{112}$

93. $3^2 = 3 \cdot 3 = 9$

95. $5^3 = 5 \cdot 5 \cdot 5 = 125$

97. $-\dfrac{11}{2} = \dfrac{-11}{2} = \dfrac{11}{-2}$

99. $\dfrac{-13}{15} = \dfrac{13}{-15} = -\dfrac{13}{15}$

101. answers may vary

103. △△ △△ △△

105. 11 is close to 12, so $\dfrac{11}{12}$ is close to 1. $5\dfrac{11}{12}$ rounded to the nearest whole number is 6.

107. $87 + 57 + 20 + 7 = 171$
57 of the 171 licensees are universities or colleges: $\dfrac{57}{171}$

109. $5 + 3 + 2 + 1 + 1 + 1 = 13$
Five of the 13 USMC training centers are in California: $\dfrac{5}{13}$.

Section 4.2

Practice Problems

1. a. $30 = 2 \cdot 15$
$\qquad\quad \downarrow \ \ \downarrow \searrow$
$\qquad\quad 2 \cdot 3 \ \cdot 5$

b. $56 = 2 \cdot 28$

$2 \cdot 2 \cdot 14$

$2 \cdot 2 \cdot 2 \cdot 7 = 2^3 \cdot 7$

c. $72 = 2 \cdot 36$

$2 \cdot 2 \cdot 18$

$2 \cdot 2 \cdot 2 \cdot 9$

$2 \cdot 2 \cdot 2 \cdot 3 \cdot 3 = 2^3 \cdot 3^2$

2. $60 = 2 \cdot 30$

$2 \cdot 2 \cdot 15$

$2 \cdot 2 \cdot 3 \cdot 5 = 2^2 \cdot 3 \cdot 5$

3.
$$3 \overline{)33} = 11$$
$$3 \overline{)99}$$
$$3 \overline{)297}$$

The prime factorization of 297 is $3^3 \cdot 11$.

4. $\dfrac{30}{45} = \dfrac{15 \cdot 2}{15 \cdot 3} = \dfrac{2}{3}$

5. $\dfrac{39x}{51} = \dfrac{3 \cdot 13 \cdot x}{3 \cdot 17} = \dfrac{13 \cdot x}{17} = \dfrac{13x}{17}$

6. $-\dfrac{9}{50} = -\dfrac{3 \cdot 3}{2 \cdot 5 \cdot 5}$

Since 9 and 50 have no common factors, $-\dfrac{9}{50}$ is already in simplest form.

7. $\dfrac{49}{112} = \dfrac{7 \cdot 7}{7 \cdot 16} = \dfrac{7}{16}$

8. $-\dfrac{64}{20} = -\dfrac{4 \cdot 16}{4 \cdot 5} = -\dfrac{16}{5}$

9. $\dfrac{7a^3}{56a^2} = \dfrac{7 \cdot a \cdot a \cdot a}{7 \cdot 8 \cdot a \cdot a} = \dfrac{a}{8}$

10. $\dfrac{7}{9}$ is in simplest form.

$\dfrac{21}{27} = \dfrac{3 \cdot 7}{3 \cdot 9} = \dfrac{7}{9}$

Since $\dfrac{7}{9}$ and $\dfrac{21}{27}$ both simplify to $\dfrac{7}{9}$, they are equivalent.

11. Not equivalent, since the cross products are not equal: $13 \cdot 5 = 65$ and $4 \cdot 18 = 72$

12. $\dfrac{6 \text{ parks in Washington}}{58 \text{ national parks}} = \dfrac{2 \cdot 3}{2 \cdot 29} = \dfrac{3}{29}$

$\dfrac{3}{29}$ of the national parks are in Washington state.

Calculator Explorations

1. $\dfrac{128}{224} = \dfrac{4}{7}$

3. $\dfrac{340}{459} = \dfrac{20}{27}$

5. $\dfrac{432}{810} = \dfrac{8}{15}$

7. $\dfrac{54}{243} = \dfrac{2}{9}$

Vocabulary and Readiness Check

1. The number 40 equals $2 \cdot 2 \cdot 2 \cdot 5$. Since each factor is prime, we call $2 \cdot 2 \cdot 2 \cdot 5$ the prime factorization of 40.

3. A natural number that has exactly two different factors, 1 and itself, is called a <u>prime</u> number.

5. Fractions that represent the same portion of a whole are called <u>equivalent</u> fractions.

Exercise Set 4.2

1.
$$20 = 2 \cdot 10$$
$$\downarrow \quad \swarrow\searrow$$
$$2 \cdot 2 \quad \cdot \quad 5 = 2^2 \cdot 5$$

3.
$$48 = 2 \cdot 24$$
$$\downarrow \quad \downarrow\searrow$$
$$2 \cdot 2 \cdot 12$$
$$\downarrow \quad \downarrow \quad \downarrow\searrow$$
$$2 \cdot 2 \cdot 2 \cdot 6$$
$$\downarrow \quad \downarrow \quad \downarrow \quad \downarrow\searrow$$
$$2 \cdot 2 \cdot 2 \cdot 2 \cdot 3 = 2^4 \cdot 3$$

5.
$$81 = 9 \quad \cdot \quad 9$$
$$\swarrow\searrow \quad \swarrow\searrow$$
$$3 \cdot 3 \cdot 3 \cdot 3 = 3^4$$

7.
$$162 = 2 \quad \cdot \quad 81$$
$$\downarrow \quad \swarrow \quad \searrow$$
$$2 \cdot 9 \cdot 9$$
$$\downarrow \quad \swarrow\searrow \quad \swarrow\searrow$$
$$2 \cdot 3 \cdot 3 \cdot 3 \cdot 3 = 2 \cdot 3^4$$

9.
$$110 = 2 \cdot 55$$
$$\downarrow \quad \downarrow\searrow$$
$$2 \cdot 5 \cdot 11 = 2 \cdot 5 \cdot 11$$

11. $85 = 5 \cdot 17$

13.
$$240 = 2 \cdot 120$$
$$\downarrow \quad \downarrow\searrow$$
$$2 \cdot 2 \cdot 60$$
$$\downarrow \quad \downarrow \quad \downarrow\searrow$$
$$2 \cdot 2 \cdot 2 \cdot 30$$
$$\downarrow \quad \downarrow \quad \downarrow \quad \downarrow\searrow$$
$$2 \cdot 2 \cdot 2 \cdot 2 \cdot 15$$
$$\downarrow \quad \downarrow \quad \downarrow \quad \downarrow \quad \downarrow\searrow$$
$$2 \cdot 2 \cdot 2 \cdot 2 \cdot 3 \cdot 5 = 2^4 \cdot 3 \cdot 5$$

15.
$$828 = 2 \cdot 414$$
$$\downarrow \quad \downarrow\searrow$$
$$2 \cdot 2 \cdot 207$$
$$\downarrow \quad \downarrow \quad \downarrow\searrow$$
$$2 \cdot 2 \cdot 3 \cdot 69$$
$$\downarrow \quad \downarrow \quad \downarrow \quad \downarrow\searrow$$
$$2 \cdot 2 \cdot 3 \cdot 3 \cdot 23 = 2^2 \cdot 3^2 \cdot 23$$

17. $\dfrac{3}{12} = \dfrac{3 \cdot 1}{3 \cdot 4} = \dfrac{1}{4}$

19. $\dfrac{4x}{42} = \dfrac{2 \cdot 2 \cdot x}{2 \cdot 3 \cdot 7} = \dfrac{2 \cdot x}{3 \cdot 7} = \dfrac{2x}{21}$

21. $\dfrac{14}{16} = \dfrac{2 \cdot 7}{2 \cdot 8} = \dfrac{7}{8}$

23. $\dfrac{20}{30} = \dfrac{2 \cdot 10}{3 \cdot 10} = \dfrac{2}{3}$

25. $\dfrac{35a}{50a} = \dfrac{5 \cdot 7 \cdot a}{5 \cdot 10 \cdot a} = \dfrac{7}{10}$

27. $-\dfrac{63}{81} = -\dfrac{9 \cdot 7}{9 \cdot 9} = -\dfrac{7}{9}$

29. $\dfrac{30x^2}{36x} = \dfrac{5 \cdot 6 \cdot x \cdot x}{6 \cdot 6 \cdot x} = \dfrac{5 \cdot x}{6} = \dfrac{5x}{6}$

31. $\dfrac{27}{64} = \dfrac{3 \cdot 3 \cdot 3}{4 \cdot 4 \cdot 4}$

Since 27 and 64 have no common factors, $\dfrac{27}{64}$ is already in simplest form.

33. $\dfrac{25xy}{40y} = \dfrac{5 \cdot 5 \cdot x \cdot y}{5 \cdot 8 \cdot y} = \dfrac{5 \cdot x}{8} = \dfrac{5x}{8}$

35. $-\dfrac{40}{64} = -\dfrac{8 \cdot 5}{8 \cdot 8} = -\dfrac{5}{8}$

37. $\dfrac{36x^3y^2}{24xy} = \dfrac{3 \cdot 12 \cdot x \cdot x \cdot x \cdot y \cdot y}{2 \cdot 12 \cdot x \cdot y}$

$= \dfrac{3 \cdot x \cdot x \cdot y}{2}$

$= \dfrac{3x^2y}{2}$

39. $\dfrac{90}{120} = \dfrac{30 \cdot 3}{30 \cdot 4} = \dfrac{3}{4}$

41. $\dfrac{40xy}{64xyz} = \dfrac{5 \cdot 8 \cdot x \cdot y}{8 \cdot 8 \cdot x \cdot y \cdot z} = \dfrac{5}{8 \cdot z} = \dfrac{5}{8z}$

43. $\dfrac{66}{308} = \dfrac{22 \cdot 3}{22 \cdot 14} = \dfrac{3}{14}$

45. $-\dfrac{55}{85y} = -\dfrac{5 \cdot 11}{5 \cdot 17 \cdot y} = -\dfrac{11}{17 \cdot y} = -\dfrac{11}{17y}$

47. $\dfrac{189z}{216z} = \dfrac{7 \cdot 27 \cdot z}{8 \cdot 27 \cdot z} = \dfrac{7}{8}$

49. $\dfrac{224a^3b^4c^2}{16ab^4c^2} = \dfrac{14 \cdot 16 \cdot a \cdot a \cdot a \cdot b \cdot b \cdot b \cdot b \cdot c \cdot c}{1 \cdot 16 \cdot a \cdot b \cdot b \cdot b \cdot b \cdot c \cdot c}$

$= \dfrac{14 \cdot a \cdot a}{1}$

$= 14a^2$

51. Equivalent, since the cross products are equal: $2 \cdot 12 = 24$ and $6 \cdot 4 = 24$

53. Not equivalent, since the cross products are not equal: $7 \cdot 8 = 56$ and $5 \cdot 11 = 55$

55. Equivalent, since the cross products are equal: $10 \cdot 9 = 90$ and $15 \cdot 6 = 90$

57. Equivalent, since the cross products are equal: $3 \cdot 18 = 54$ and $9 \cdot 6 = 54$

59. Not equivalent, since the cross products are not equal: $10 \cdot 15 = 150$ and $13 \cdot 13 = 169$

61. Not equivalent, since the cross products are not equal: $8 \cdot 24 = 192$ and $12 \cdot 18 = 216$

63. $\dfrac{2 \text{ hours}}{8 \text{ hours}} = \dfrac{1 \cdot 2}{4 \cdot 2} = \dfrac{1}{4}$

2 hours represents $\dfrac{1}{4}$ of a work shift.

65. $\dfrac{2640 \text{ feet}}{5280 \text{ feet}} = \dfrac{2640 \cdot 1}{2640 \cdot 2} = \dfrac{1}{2}$

2640 feet represents $\dfrac{1}{2}$ of a mile.

67. a. $\dfrac{15}{50} = \dfrac{3 \cdot 5}{10 \cdot 5} = \dfrac{3}{10}$

$\dfrac{3}{10}$ of the states can claim at least one Ritz-Carlton hotel.

b. $50 - 15 = 35$

35 states do not have a Ritz-Carlton hotel.

c. $\dfrac{35}{50} = \dfrac{7 \cdot 5}{10 \cdot 5} = \dfrac{7}{10}$

$\dfrac{7}{10}$ of the states do not have a Ritz-Carlton hotel.

69. $\dfrac{10 \text{ inches}}{24 \text{ inches}} = \dfrac{2 \cdot 5}{2 \cdot 12} = \dfrac{5}{12}$

$\dfrac{5}{12}$ of the wall is concrete.

71. a. $50 - 22 = 28$
28 states do not have this type of Web site.

b. $\dfrac{28}{50} = \dfrac{2 \cdot 14}{2 \cdot 25} = \dfrac{14}{25}$

$\dfrac{14}{25}$ of the states do not have this type of Web site.

73. $\dfrac{291 \text{ individuals}}{449 \text{ individuals}} = \dfrac{291}{449}$

$\dfrac{291}{449}$ of the individuals who had flown in space were American.

75. $\dfrac{x^3}{9} = \dfrac{(-3)^3}{9} = \dfrac{-27}{9} = -3$

77. $2y = 2(-7) = -14$

79. answers may vary

81. $\dfrac{3975}{6625} = \dfrac{3 \cdot 1325}{5 \cdot 1325} = \dfrac{3}{5}$

83. 36 blood donors have blood type A Rh-positive.

$\dfrac{36 \text{ donors}}{100 \text{ donors}} = \dfrac{4 \cdot 9}{4 \cdot 25} = \dfrac{9}{25}$

$\dfrac{9}{25}$ of blood donors have type A Rh-positive blood type.

85. $3 + 1 = 4$ blood donors have an AB blood type.

$\dfrac{4 \text{ donors}}{100 \text{ donors}} = \dfrac{4 \cdot 1}{4 \cdot 25} = \dfrac{1}{25}$

$\dfrac{1}{25}$ of blood donors have an AB blood type.

87. $34{,}020 = 2 \cdot 17{,}010$
$\downarrow\ \downarrow\searrow$
$2 \cdot 2 \cdot 8505$
$\downarrow\ \downarrow\ \downarrow\searrow$
$2 \cdot 2 \cdot 3 \cdot 2835$
$\downarrow\ \downarrow\ \downarrow\ \downarrow\searrow$
$2 \cdot 2 \cdot 3 \cdot 3 \cdot 945$
$\downarrow\ \downarrow\ \downarrow\ \downarrow\ \downarrow\searrow$
$2 \cdot 2 \cdot 3 \cdot 3 \cdot 3 \cdot 315$
$\downarrow\ \downarrow\ \downarrow\ \downarrow\ \downarrow\ \downarrow\searrow$
$2 \cdot 2 \cdot 3 \cdot 3 \cdot 3 \cdot 3 \cdot 105$
$\downarrow\ \downarrow\ \downarrow\ \downarrow\ \downarrow\ \downarrow\ \downarrow\searrow$
$2 \cdot 2 \cdot 3 \cdot 3 \cdot 3 \cdot 3 \cdot 3 \cdot 35$
$\downarrow\ \downarrow\ \downarrow\ \downarrow\ \downarrow\ \downarrow\ \downarrow\ \downarrow\searrow$
$2 \cdot 2 \cdot 3 \cdot 3 \cdot 3 \cdot 3 \cdot 3 \cdot 5 \cdot 7$

$34{,}020 = 2^2 \cdot 3^5 \cdot 5 \cdot 7$

89. answers may vary

91. no; answers may vary

93. The piece representing biological sciences is labeled $\dfrac{2}{25}$, so $\dfrac{2}{25}$ of entering college freshmen plan to major in the biological sciences.

95. answers may vary

97. 1235, 2235, 85, 105, 900, and 1470 are divisible by 5 because each number ends with a 0 or 5.
8691, 786, 2235, 105, 222, 900, and 1470 are divisible by 3 because the sum of each number's digits is divisible by 3.
2235, 105, 900, and 1470 are divisible by both 3 and 5.

99. 15; answers may vary

Section 4.3

Practice Problems

1. $\dfrac{3}{7} \cdot \dfrac{5}{11} = \dfrac{3 \cdot 5}{7 \cdot 11} = \dfrac{15}{77}$

2. $\dfrac{1}{3} \cdot \dfrac{1}{9} = \dfrac{1 \cdot 1}{3 \cdot 9} = \dfrac{1}{27}$

3. $\dfrac{6}{77} \cdot \dfrac{7}{8} = \dfrac{6 \cdot 7}{77 \cdot 8} = \dfrac{2 \cdot 3 \cdot 7}{7 \cdot 11 \cdot 2 \cdot 4} = \dfrac{3}{11 \cdot 4} = \dfrac{3}{44}$

4. $\dfrac{4}{27} \cdot \dfrac{3}{8} = \dfrac{4 \cdot 3}{27 \cdot 8} = \dfrac{1 \cdot 4 \cdot 3}{3 \cdot 9 \cdot 4 \cdot 2} = \dfrac{1}{9 \cdot 2} = \dfrac{1}{18}$

5. $\dfrac{1}{2} \cdot \left(-\dfrac{11}{28} \right) = -\dfrac{1 \cdot 11}{2 \cdot 28} = -\dfrac{11}{56}$

6. $\left(-\dfrac{4}{11} \right)\left(-\dfrac{33}{16} \right) = \dfrac{4 \cdot 33}{11 \cdot 16} = \dfrac{4 \cdot 3 \cdot 11}{11 \cdot 4 \cdot 4} = \dfrac{3}{4}$

7. $\dfrac{1}{6} \cdot \dfrac{3}{10} \cdot \dfrac{25}{16} = \dfrac{1 \cdot 3 \cdot 25}{6 \cdot 10 \cdot 16}$

$\qquad = \dfrac{1 \cdot 3 \cdot 5 \cdot 5}{2 \cdot 3 \cdot 2 \cdot 5 \cdot 16}$

$\qquad = \dfrac{1 \cdot 5}{2 \cdot 2 \cdot 16}$

$\qquad = \dfrac{5}{64}$

8. $\dfrac{2}{3} \cdot \dfrac{3y}{2} = \dfrac{2 \cdot 3 \cdot y}{3 \cdot 2 \cdot 1} = \dfrac{y}{1} = y$

9. $\dfrac{a^3}{b^2} \cdot \dfrac{b}{a^2} = \dfrac{a^3 \cdot b}{b^2 \cdot a^2} = \dfrac{a \cdot a \cdot a \cdot b}{b \cdot b \cdot a \cdot a} = \dfrac{a}{b}$

10. a. $\left(\dfrac{3}{4} \right)^3 = \dfrac{3}{4} \cdot \dfrac{3}{4} \cdot \dfrac{3}{4} = \dfrac{3 \cdot 3 \cdot 3}{4 \cdot 4 \cdot 4} = \dfrac{27}{64}$

b. $\left(-\dfrac{4}{5} \right)^2 = \left(-\dfrac{4}{5} \right) \cdot \left(-\dfrac{4}{5} \right) = \dfrac{4 \cdot 4}{5 \cdot 5} = \dfrac{16}{25}$

11. $\dfrac{8}{7} \div \dfrac{2}{9} = \dfrac{8}{7} \cdot \dfrac{9}{2} = \dfrac{8 \cdot 9}{7 \cdot 2} = \dfrac{2 \cdot 4 \cdot 9}{7 \cdot 2} = \dfrac{4 \cdot 9}{7} = \dfrac{36}{7}$

12. $\dfrac{4}{9} \div \dfrac{1}{2} = \dfrac{4}{9} \cdot \dfrac{2}{1} = \dfrac{4 \cdot 2}{9 \cdot 1} = \dfrac{8}{9}$

13. $-\dfrac{10}{4} \div \dfrac{2}{9} = -\dfrac{10}{4} \cdot \dfrac{9}{2}$

$\qquad = -\dfrac{10 \cdot 9}{4 \cdot 2}$

$\qquad = -\dfrac{2 \cdot 5 \cdot 9}{4 \cdot 2}$

$\qquad = -\dfrac{5 \cdot 9}{4}$

$\qquad = -\dfrac{45}{4}$

14. $\dfrac{3y}{4} \div 5y^3 = \dfrac{3y}{4} \div \dfrac{5y^3}{1}$

$\qquad = \dfrac{3y}{4} \cdot \dfrac{1}{5y^3}$

$\qquad = \dfrac{3y \cdot 1}{4 \cdot 5y^3}$

$\qquad = \dfrac{3 \cdot y \cdot 1}{4 \cdot 5 \cdot y \cdot y \cdot y}$

$\qquad = \dfrac{3 \cdot 1}{4 \cdot 5 \cdot y \cdot y}$

$\qquad = \dfrac{3}{20y^2}$

15. $\left(-\dfrac{2}{3} \cdot \dfrac{9}{14} \right) \div \dfrac{7}{15} = \left(-\dfrac{2 \cdot 9}{3 \cdot 14} \right) \div \dfrac{7}{15}$

$\qquad = \left(-\dfrac{2 \cdot 3 \cdot 3}{3 \cdot 7 \cdot 2} \right) \div \dfrac{7}{15}$

$\qquad = \left(-\dfrac{3}{7} \right) \div \dfrac{7}{15}$

$\qquad = \left(-\dfrac{3}{7} \right) \cdot \dfrac{15}{7}$

$\qquad = -\dfrac{3 \cdot 15}{7 \cdot 7}$

$\qquad = -\dfrac{45}{49}$

16. a. $xy = -\dfrac{3}{4} \cdot \dfrac{9}{2} = -\dfrac{3 \cdot 9}{4 \cdot 2} = -\dfrac{27}{8}$

b. $x \div y = -\dfrac{3}{4} \div \dfrac{9}{2}$

$\qquad\qquad = -\dfrac{3}{4} \cdot \dfrac{2}{9}$

$\qquad\qquad = -\dfrac{3 \cdot 2}{4 \cdot 9}$

$\qquad\qquad = -\dfrac{1 \cdot 3 \cdot 2}{2 \cdot 2 \cdot 3 \cdot 3}$

$\qquad\qquad = -\dfrac{1}{2 \cdot 3}$

$\qquad\qquad = -\dfrac{1}{6}$

17. $\qquad 2x = -\dfrac{9}{4}$

$\qquad 2\left(-\dfrac{9}{8}\right) \overset{?}{=} -\dfrac{9}{4}$

$\qquad \dfrac{2}{1} \cdot \left(-\dfrac{9}{8}\right) \overset{?}{=} -\dfrac{9}{4}$

$\qquad -\dfrac{2 \cdot 9}{1 \cdot 8} \overset{?}{=} -\dfrac{9}{4}$

$\qquad -\dfrac{2 \cdot 9}{1 \cdot 2 \cdot 4} \overset{?}{=} -\dfrac{9}{4}$

$\qquad -\dfrac{9}{4} = -\dfrac{9}{4}$ True

Yes, $-\dfrac{9}{8}$ is a solution of the equation.

18. $\dfrac{1}{6} \cdot 60 = \dfrac{1}{6} \cdot \dfrac{60}{1} = \dfrac{1 \cdot 60}{6 \cdot 1} = \dfrac{1 \cdot 6 \cdot 10}{6 \cdot 1} = 10$

Thus, there are 10 roller coasters in Hershey Park.

Vocabulary and Readiness Check

1. To multiply two fractions, we write

$\dfrac{a}{b} \cdot \dfrac{c}{d} = \dfrac{a \cdot c}{b \cdot d}$.

3. The expression $\dfrac{2^3}{7} = \dfrac{2 \cdot 2 \cdot 2}{7}$ while

$\left(\dfrac{2}{7}\right)^3 = \dfrac{2}{7} \cdot \dfrac{2}{7} \cdot \dfrac{2}{7}$.

5. To divide two fractions, we write

$\dfrac{a}{b} \div \dfrac{c}{d} = \dfrac{a \cdot d}{b \cdot c}$.

Exercise Set 4.3

1. $\dfrac{6}{11} \cdot \dfrac{3}{7} = \dfrac{6 \cdot 3}{11 \cdot 7} = \dfrac{18}{77}$

3. $-\dfrac{2}{7} \cdot \dfrac{5}{8} = -\dfrac{2 \cdot 5}{7 \cdot 8} = -\dfrac{2 \cdot 5}{7 \cdot 2 \cdot 4} = -\dfrac{5}{7 \cdot 4} = -\dfrac{5}{28}$

5. $-\dfrac{1}{2} \cdot \dfrac{2}{15} = \dfrac{1 \cdot 2}{2 \cdot 15} = \dfrac{1}{15}$

7. $\dfrac{18x}{20} \cdot \dfrac{36}{99} = \dfrac{18x \cdot 36}{20 \cdot 99}$

$\qquad = \dfrac{2 \cdot 9 \cdot x \cdot 2 \cdot 18}{2 \cdot 2 \cdot 5 \cdot 9 \cdot 11}$

$\qquad = \dfrac{18 \cdot x}{5 \cdot 11}$

$\qquad = \dfrac{18x}{55}$

9. $3a^2 \cdot \dfrac{1}{4} = \dfrac{3a^2}{1} \cdot \dfrac{1}{4} = \dfrac{3a^2 \cdot 1}{1 \cdot 4} = \dfrac{3a^2}{4}$

11. $\dfrac{x^3}{y^3} \cdot \dfrac{y^2}{x} = \dfrac{x^3 \cdot y^2}{y^3 \cdot x}$

$\qquad = \dfrac{x \cdot x \cdot x \cdot y \cdot y}{y \cdot y \cdot y \cdot x}$

$\qquad = \dfrac{x \cdot x}{y}$

$\qquad = \dfrac{x^2}{y}$

13. $0 \cdot \dfrac{8}{9} = 0$

15. $-\dfrac{17y}{20} \cdot \dfrac{4}{5y} = -\dfrac{17y \cdot 4}{20 \cdot 5y}$

$= -\dfrac{17 \cdot y \cdot 4}{5 \cdot 4 \cdot 5 \cdot y}$

$= -\dfrac{17}{5 \cdot 5}$

$= -\dfrac{17}{25}$

17. $\dfrac{11}{20} \cdot \dfrac{1}{7} \cdot \dfrac{5}{22} = \dfrac{11 \cdot 1 \cdot 5}{20 \cdot 7 \cdot 22}$

$= \dfrac{11 \cdot 1 \cdot 5}{5 \cdot 2 \cdot 2 \cdot 7 \cdot 11 \cdot 2}$

$= \dfrac{1}{2 \cdot 2 \cdot 7 \cdot 2}$

$= \dfrac{1}{56}$

19. $\left(\dfrac{1}{5}\right)^3 = \left(\dfrac{1}{5}\right)\left(\dfrac{1}{5}\right)\left(\dfrac{1}{5}\right) = \dfrac{1 \cdot 1 \cdot 1}{5 \cdot 5 \cdot 5} = \dfrac{1}{125}$

21. $\left(-\dfrac{2}{3}\right)^2 = -\dfrac{2}{3} \cdot -\dfrac{2}{3} = \dfrac{2 \cdot 2}{3 \cdot 3} = \dfrac{4}{9}$

23. $\left(-\dfrac{2}{3}\right)^3 \cdot \dfrac{1}{2} = \left(-\dfrac{2}{3}\right)\left(-\dfrac{2}{3}\right)\left(-\dfrac{2}{3}\right) \cdot \dfrac{1}{2}$

$= -\dfrac{2 \cdot 2 \cdot 2 \cdot 1}{3 \cdot 3 \cdot 3 \cdot 2}$

$= -\dfrac{2 \cdot 2 \cdot 1}{3 \cdot 3 \cdot 3}$

$= -\dfrac{4}{27}$

25. $\dfrac{2}{3} \div \dfrac{5}{6} = \dfrac{2}{3} \cdot \dfrac{6}{5} = \dfrac{2 \cdot 6}{3 \cdot 5} = \dfrac{2 \cdot 2 \cdot 3}{3 \cdot 5} = \dfrac{2 \cdot 2}{5} = \dfrac{4}{5}$

27. $-\dfrac{6}{15} \div \dfrac{12}{5} = -\dfrac{6}{15} \cdot \dfrac{5}{12}$

$= -\dfrac{6 \cdot 5}{15 \cdot 12}$

$= -\dfrac{6 \cdot 5 \cdot 1}{5 \cdot 3 \cdot 6 \cdot 2}$

$= -\dfrac{1}{3 \cdot 2}$

$= -\dfrac{1}{6}$

29. $-\dfrac{8}{9} \div \dfrac{x}{2} = -\dfrac{8}{9} \cdot \dfrac{2}{x} = -\dfrac{8 \cdot 2}{9 \cdot x} = -\dfrac{16}{9x}$

31. $\dfrac{11y}{20} \div \dfrac{3}{11} = \dfrac{11y}{20} \cdot \dfrac{11}{3} = \dfrac{11y \cdot 11}{20 \cdot 3} = \dfrac{121y}{60}$

33. $-\dfrac{2}{3} \div 4 = -\dfrac{2}{3} \div \dfrac{4}{1}$

$= -\dfrac{2}{3} \cdot \dfrac{1}{4}$

$= -\dfrac{2 \cdot 1}{3 \cdot 4}$

$= -\dfrac{2 \cdot 1}{3 \cdot 2 \cdot 2}$

$= -\dfrac{1}{3 \cdot 2}$

$= -\dfrac{1}{6}$

35. $\dfrac{1}{5x} \div \dfrac{5}{x^2} = \dfrac{1}{5x} \cdot \dfrac{x^2}{5}$

$= \dfrac{1 \cdot x^2}{5x \cdot 5}$

$= \dfrac{1 \cdot x \cdot x}{5 \cdot x \cdot 5}$

$= \dfrac{1 \cdot x}{5 \cdot 5}$

$= \dfrac{x}{25}$

37. $\dfrac{2}{3} \cdot \dfrac{5}{9} = \dfrac{2 \cdot 5}{3 \cdot 9} = \dfrac{10}{27}$

39. $\dfrac{3x}{7} \div \dfrac{5}{6x} = \dfrac{3x}{7} \cdot \dfrac{6x}{5} = \dfrac{3x \cdot 6x}{7 \cdot 5} = \dfrac{18x^2}{35}$

41. $\dfrac{16}{27y} \div \dfrac{8}{15y} = \dfrac{16}{27y} \cdot \dfrac{15y}{8}$

$= \dfrac{16 \cdot 15y}{27y \cdot 8}$

$= \dfrac{8 \cdot 2 \cdot 3 \cdot 5 \cdot y}{3 \cdot 9 \cdot y \cdot 8}$

$= \dfrac{2 \cdot 5}{9}$

$= \dfrac{10}{9}$

43. $-\dfrac{5}{28} \cdot \dfrac{35}{25} = -\dfrac{5 \cdot 35}{28 \cdot 25} = -\dfrac{5 \cdot 7 \cdot 5 \cdot 1}{7 \cdot 4 \cdot 5 \cdot 5} = -\dfrac{1}{4}$

45. $\left(-\dfrac{3}{4}\right)^2 = -\dfrac{3}{4} \cdot -\dfrac{3}{4} = \dfrac{3 \cdot 3}{4 \cdot 4} = \dfrac{9}{16}$

47. $\dfrac{x^2}{y} \cdot \dfrac{y^3}{x} = \dfrac{x^2 \cdot y^3}{y \cdot x}$

$= \dfrac{x \cdot x \cdot y \cdot y \cdot y}{y \cdot x \cdot 1}$

$= \dfrac{x \cdot y \cdot y}{1}$

$= xy^2$

49. $7 \div \dfrac{2}{11} = \dfrac{7}{1} \div \dfrac{2}{11} = \dfrac{7}{1} \cdot \dfrac{11}{2} = \dfrac{7 \cdot 11}{1 \cdot 2} = \dfrac{77}{2}$

51. $-3x \div \dfrac{x^2}{12} = -\dfrac{3x}{1} \div \dfrac{x^2}{12}$

$= -\dfrac{3x}{1} \cdot \dfrac{12}{x^2}$

$= -\dfrac{3x \cdot 12}{1 \cdot x^2}$

$= -\dfrac{3 \cdot x \cdot 12}{1 \cdot x \cdot x}$

$= -\dfrac{3 \cdot 12}{1 \cdot x}$

$= -\dfrac{36}{x}$

53. $\left(\dfrac{2}{7} \div \dfrac{7}{2}\right) \dfrac{3}{4} = \left(\dfrac{2}{7} \cdot \dfrac{2}{7}\right) \dfrac{3}{4}$

$= \dfrac{2 \cdot 2 \cdot 3}{7 \cdot 7 \cdot 4}$

$= \dfrac{2 \cdot 2 \cdot 3}{7 \cdot 7 \cdot 2 \cdot 2}$

$= \dfrac{3}{7 \cdot 7}$

$= \dfrac{3}{49}$

55. $-\dfrac{19}{63y} \cdot 9y^2 = -\dfrac{19}{63y} \cdot \dfrac{9y^2}{1}$

$= -\dfrac{19 \cdot 9y^2}{63y \cdot 1}$

$= -\dfrac{19 \cdot 9 \cdot y \cdot y}{9 \cdot 7 \cdot y \cdot 1}$

$= -\dfrac{19 \cdot y}{7 \cdot 1}$

$= -\dfrac{19y}{7}$

57. $-\dfrac{2}{3} \cdot -\dfrac{6}{11} = \dfrac{2 \cdot 6}{3 \cdot 11} = \dfrac{2 \cdot 2 \cdot 3}{3 \cdot 11} = \dfrac{2 \cdot 2}{11} = \dfrac{4}{11}$

59. $\dfrac{4}{8} \div \dfrac{3}{16} = \dfrac{4}{8} \cdot \dfrac{16}{3} = \dfrac{4 \cdot 16}{8 \cdot 3} = \dfrac{4 \cdot 8 \cdot 2}{8 \cdot 3} = \dfrac{4 \cdot 2}{3} = \dfrac{8}{3}$

61. $\dfrac{21x^2}{10y} \div \dfrac{14x}{25y} = \dfrac{21x^2}{10y} \cdot \dfrac{25y}{14x}$

$= \dfrac{21x^2 \cdot 25y}{10y \cdot 14x}$

$= \dfrac{3 \cdot 7 \cdot x \cdot x \cdot 5 \cdot 5 \cdot y}{2 \cdot 5 \cdot y \cdot 2 \cdot 7 \cdot x}$

$= \dfrac{3 \cdot x \cdot 5}{2 \cdot 2}$

$= \dfrac{15x}{4}$

63. $\left(1 \div \dfrac{3}{4}\right) \cdot \dfrac{2}{3} = \left(\dfrac{1}{1} \div \dfrac{3}{4}\right) \cdot \dfrac{2}{3}$

$= \left(\dfrac{1}{1} \cdot \dfrac{4}{3}\right) \cdot \dfrac{2}{3}$

$= \dfrac{1 \cdot 4 \cdot 2}{1 \cdot 3 \cdot 3}$

$= \dfrac{8}{9}$

65. $\dfrac{a^3}{2} \div 30a^3 = \dfrac{a^3}{2} \div \dfrac{30a^3}{1}$

$= \dfrac{a^3}{2} \cdot \dfrac{1}{30a^3}$

$= \dfrac{a^3 \cdot 1}{2 \cdot 30a^3}$

$= \dfrac{a \cdot a \cdot a \cdot 1}{2 \cdot 30 \cdot a \cdot a \cdot a}$

$= \dfrac{1}{2 \cdot 30}$

$= \dfrac{1}{60}$

67. $\dfrac{ab^2}{c} \cdot \dfrac{c}{ab} = \dfrac{ab^2 \cdot c}{c \cdot ab} = \dfrac{a \cdot b \cdot b \cdot c}{c \cdot a \cdot b} = b$

69. $\left(\dfrac{1}{2} \cdot \dfrac{2}{3}\right) \div \dfrac{5}{6} = \left(\dfrac{1 \cdot 2}{2 \cdot 3}\right) \div \dfrac{5}{6}$

$= \dfrac{1}{3} \div \dfrac{5}{6}$

$= \dfrac{1}{3} \cdot \dfrac{6}{5}$

$= \dfrac{1 \cdot 6}{3 \cdot 5}$

$= \dfrac{1 \cdot 2 \cdot 3}{3 \cdot 5}$

$= \dfrac{1 \cdot 2}{5}$

$= \dfrac{2}{5}$

71. $-\dfrac{4}{7} \div \left(\dfrac{4}{5} \cdot \dfrac{3}{7}\right) = -\dfrac{4}{7} \div \left(\dfrac{4 \cdot 3}{5 \cdot 7}\right)$

$= -\dfrac{4}{7} \div \dfrac{12}{35}$

$= -\dfrac{4}{7} \cdot \dfrac{35}{12}$

$= -\dfrac{4 \cdot 35}{7 \cdot 12}$

$= -\dfrac{4 \cdot 7 \cdot 5}{7 \cdot 4 \cdot 3}$

$= -\dfrac{5}{3}$

73. a. $xy = \dfrac{2}{5} \cdot \dfrac{5}{6} = \dfrac{2 \cdot 5}{5 \cdot 6} = \dfrac{2 \cdot 5}{5 \cdot 2 \cdot 3} = \dfrac{1}{3}$

b. $x \div y = \dfrac{2}{5} \div \dfrac{5}{6} = \dfrac{2}{5} \cdot \dfrac{6}{5} = \dfrac{2 \cdot 6}{5 \cdot 5} = \dfrac{12}{25}$

75. a. $xy = -\dfrac{4}{5} \cdot \dfrac{9}{11} = -\dfrac{4 \cdot 9}{5 \cdot 11} = -\dfrac{36}{55}$

b. $x \div y = -\dfrac{4}{5} \div \dfrac{9}{11}$

$= -\dfrac{4}{5} \cdot \dfrac{11}{9}$

$= -\dfrac{4 \cdot 11}{5 \cdot 9}$

$= -\dfrac{44}{45}$

77. $3x = -\dfrac{5}{6}$

$3\left(-\dfrac{5}{18}\right) \overset{?}{=} -\dfrac{5}{6}$

$\dfrac{3}{1} \cdot -\dfrac{5}{18} \overset{?}{=} -\dfrac{5}{6}$

$-\dfrac{3 \cdot 5}{1 \cdot 18} \overset{?}{=} -\dfrac{5}{6}$

$-\dfrac{3 \cdot 5}{3 \cdot 6} \overset{?}{=} -\dfrac{5}{6}$

$-\dfrac{5}{6} = -\dfrac{5}{6}$ True

Yes, $-\dfrac{5}{18}$ is a solution to the equation.

79. $-\dfrac{1}{2}z = \dfrac{1}{10}$

$-\dfrac{1}{2} \cdot \dfrac{2}{5} \overset{?}{=} \dfrac{1}{10}$

$-\dfrac{1 \cdot 2}{2 \cdot 5} \overset{?}{=} \dfrac{1}{10}$

$-\dfrac{1}{5} = \dfrac{1}{10}$ False

No, $\dfrac{2}{5}$ is not a solution of the equation.

81. $\dfrac{1}{4}$ of $200 = \dfrac{1}{4} \cdot 200$

$= \dfrac{1}{4} \cdot \dfrac{200}{1}$

$= \dfrac{1 \cdot 200}{4 \cdot 1}$

$= \dfrac{1 \cdot 4 \cdot 50}{4 \cdot 1}$

$= \dfrac{50}{1}$

$= 50$

83. $\dfrac{5}{6}$ of $24 = \dfrac{5}{6} \cdot 24$

$= \dfrac{5}{6} \cdot \dfrac{24}{1}$

$= \dfrac{5 \cdot 24}{6 \cdot 1}$

$= \dfrac{5 \cdot 6 \cdot 4}{6 \cdot 1}$

$= \dfrac{5 \cdot 4}{1}$

$= \dfrac{20}{1}$

$= 20$

85. $\dfrac{4}{25}$ of $800 = \dfrac{4}{25} \cdot 800$

$= \dfrac{4}{25} \cdot \dfrac{800}{1}$

$= \dfrac{4 \cdot 800}{25 \cdot 1}$

$= \dfrac{4 \cdot 25 \cdot 32}{25 \cdot 1}$

$= \dfrac{4 \cdot 32}{1}$

$= 128$

128 of the students would be expected to major in business.

87. $\frac{2}{5}$ of $2170 = \frac{2}{5} \cdot 2170$

$$= \frac{2}{5} \cdot \frac{2170}{1}$$

$$= \frac{2 \cdot 2170}{5 \cdot 1}$$

$$= \frac{2 \cdot 5 \cdot 434}{5 \cdot 1}$$

$$= \frac{2 \cdot 434}{1}$$

$$= 868$$

He hiked 868 miles.

89. $\frac{1}{2}$ of $\frac{3}{8} = \frac{1}{2} \cdot \frac{3}{8} = \frac{1 \cdot 3}{2 \cdot 8} = \frac{3}{16}$

The radius of the circle is $\frac{3}{16}$ inch.

91. $\frac{2}{3}$ of $2757 = \frac{2}{3} \cdot 2757$

$$= \frac{2}{3} \cdot \frac{2757}{1}$$

$$= \frac{2 \cdot 2757}{3 \cdot 1}$$

$$= \frac{2 \cdot 3 \cdot 919}{3 \cdot 1}$$

$$= \frac{2 \cdot 919}{1}$$

$$= 1838$$

The sale price is $1838.

93. $\frac{1}{9}$ of $180 = \frac{1}{9} \cdot 180$

$$= \frac{1}{9} \cdot \frac{180}{1}$$

$$= \frac{1 \cdot 180}{9 \cdot 1}$$

$$= \frac{1 \cdot 9 \cdot 20}{9 \cdot 1}$$

$$= \frac{20}{1}$$

$$= 20$$

20 contestants participated more than once.

95. area = length $\cdot$ width $= \frac{5}{14} \cdot \frac{1}{5} = \frac{5 \cdot 1}{14 \cdot 5} = \frac{1}{14}$

The area is $\frac{1}{14}$ square foot.

97. $\frac{8}{25} \cdot 12,000 = \frac{8}{25} \cdot \frac{12,000}{1}$

$$= \frac{8 \cdot 25 \cdot 480}{25 \cdot 1}$$

$$= \frac{8 \cdot 480}{1}$$

$$= 3840$$

The family drove 3840 miles for work.

99. $\frac{1}{5}$ of $12,000 = \frac{1}{5} \cdot 12,000$

$$= \frac{1}{5} \cdot \frac{12,000}{1}$$

$$= \frac{1 \cdot 5 \cdot 2400}{5 \cdot 1}$$

$$= 2400$$

The family drove 2400 miles for family business.

101.
$$\begin{array}{r} 27 \\ 76 \\ + 98 \\ \hline 201 \end{array}$$

103.
$$\begin{array}{r} 968 \\ - 772 \\ \hline 196 \end{array}$$

105. answers may vary

107. $\frac{42}{25} \cdot \frac{125}{36} \div \frac{7}{6} = \frac{42}{25} \cdot \frac{125}{36} \cdot \frac{6}{7}$

$$= \frac{42 \cdot 125 \cdot 6}{25 \cdot 36 \cdot 7}$$

$$= \frac{6 \cdot 7 \cdot 5 \cdot 25 \cdot 6}{25 \cdot 6 \cdot 6 \cdot 7}$$

$$= 5$$

109.
$$\frac{3}{25} \text{ of } 290{,}810{,}000 = \frac{3}{25} \cdot 290{,}810{,}000$$
$$= \frac{3}{25} \cdot \frac{290{,}810{,}000}{1}$$
$$= \frac{3 \cdot 290{,}810{,}000}{25 \cdot 1}$$
$$= \frac{3 \cdot 25 \cdot 11{,}632{,}400}{25 \cdot 1}$$
$$= \frac{3 \cdot 11{,}632{,}400}{1}$$
$$= 34{,}897{,}200$$

The population of California is approximately 34,897,200.

111. $\frac{3}{4}$ of 36 on the first bus and $\frac{1}{3}$ of 30 on the second bus are girls.

$$\frac{3}{4} \cdot 36 + \frac{1}{3} \cdot 30 = \frac{3}{4} \cdot \frac{36}{1} + \frac{1}{3} \cdot \frac{30}{1}$$
$$= \frac{3 \cdot 36}{4 \cdot 1} + \frac{1 \cdot 30}{3 \cdot 1}$$
$$= \frac{3 \cdot 4 \cdot 9}{4 \cdot 1} + \frac{1 \cdot 3 \cdot 10}{3 \cdot 1}$$
$$= \frac{3 \cdot 9}{1} + \frac{10}{1}$$
$$= 27 + 10$$
$$= 37$$

There are 37 girls on the two buses.

The Bigger Picture

1. $\dfrac{2}{3} \cdot \dfrac{8}{9} = \dfrac{2 \cdot 8}{3 \cdot 9} = \dfrac{16}{27}$

2.
$$-\frac{2}{3} \div \frac{8}{9} = -\frac{2}{3} \cdot \frac{9}{8}$$
$$= -\frac{2 \cdot 9}{3 \cdot 8}$$
$$= -\frac{2 \cdot 3 \cdot 3}{3 \cdot 2 \cdot 4}$$
$$= -\frac{3}{4}$$

3. $\dfrac{12}{1} \cdot \dfrac{1}{9} = \dfrac{12 \cdot 1}{1 \cdot 9} = \dfrac{3 \cdot 4 \cdot 1}{1 \cdot 3 \cdot 3} = \dfrac{4}{3}$

4.
$$\frac{9x}{2} \div \frac{15x}{8} = \frac{9x}{2} \cdot \frac{8}{15x}$$
$$= \frac{9x \cdot 8}{2 \cdot 15x}$$
$$= \frac{3 \cdot 3 \cdot x \cdot 2 \cdot 4}{2 \cdot 3 \cdot 5 \cdot x}$$
$$= \frac{3 \cdot 4}{5}$$
$$= \frac{12}{5}$$

5. $80 \div 20 \cdot 2 = 4 \cdot 2 = 8$

6. $3^2 \cdot 2^3 = 3 \cdot 3 \cdot 2 \cdot 2 \cdot 2 = 72$

7.
$$\frac{11}{20} \cdot \frac{5}{8} \cdot \frac{4}{33} = \frac{11 \cdot 5 \cdot 4}{20 \cdot 8 \cdot 33}$$
$$= \frac{11 \cdot 5 \cdot 4 \cdot 1}{4 \cdot 5 \cdot 8 \cdot 3 \cdot 11}$$
$$= \frac{1}{8 \cdot 3}$$
$$= \frac{1}{24}$$

8. $20 \div \dfrac{1}{2} = \dfrac{20}{1} \div \dfrac{1}{2} = \dfrac{20}{1} \cdot \dfrac{2}{1} = \dfrac{20 \cdot 2}{1 \cdot 1} = \dfrac{40}{1} = 40$

9.
$$3 + 4(18 - 16)^3 = 3 + 4(2)^3$$
$$= 3 + 4(8)$$
$$= 3 + 32$$
$$= 35$$

10. $100 - 76 = 24$

11.
$$-12x + 7 = 7$$
$$-12x + 7 - 7 = 7 - 7$$
$$-12x = 0$$
$$\frac{-12x}{-12} = \frac{0}{-12}$$
$$x = 0$$

12.
$$9n - 14 = 6n - 32$$
$$9n - 6n - 14 = 6n - 6n - 32$$
$$3n - 14 = -32$$
$$3n - 14 + 14 = -32 + 14$$
$$3n = -18$$
$$\frac{3n}{3} = \frac{-18}{3}$$
$$n = -6$$

Section 4.4

Practice Problems

1. $\dfrac{6}{13} + \dfrac{2}{13} = \dfrac{6+2}{13} = \dfrac{8}{13}$

2. $\dfrac{5}{8} + \dfrac{1}{8} = \dfrac{5+1}{8} = \dfrac{6}{8} = \dfrac{2 \cdot 3}{2 \cdot 4} = \dfrac{3}{4}$

3. $\dfrac{20}{11} + \dfrac{6}{11} + \dfrac{7}{11} = \dfrac{20+6+7}{11} = \dfrac{33}{11}$ or 3

4. $\dfrac{11}{12} - \dfrac{6}{12} = \dfrac{11-6}{12} = \dfrac{5}{12}$

5. $\dfrac{7}{15} - \dfrac{2}{15} = \dfrac{7-2}{15} = \dfrac{5}{15} = \dfrac{1 \cdot 5}{3 \cdot 5} = \dfrac{1}{3}$

6. $-\dfrac{8}{17} + \dfrac{4}{17} = \dfrac{-8+4}{17} = \dfrac{-4}{17}$ or $-\dfrac{4}{17}$

7. $\dfrac{2}{5} - \dfrac{7y}{5} = \dfrac{2-7y}{5}$

8. $\dfrac{4}{11} - \dfrac{6}{11} - \dfrac{3}{11} = \dfrac{4-6-3}{11} = \dfrac{-5}{11}$ or $-\dfrac{5}{11}$

9. $x + y = -\dfrac{10}{12} + \dfrac{5}{12} = \dfrac{-10+5}{12} = \dfrac{-5}{12}$ or $-\dfrac{5}{12}$

10. perimeter $= \dfrac{3}{20} + \dfrac{3}{20} + \dfrac{3}{20} + \dfrac{3}{20}$
$= \dfrac{3+3+3+3}{20}$
$= \dfrac{12}{20}$
$= \dfrac{3 \cdot 4}{5 \cdot 4}$
$= \dfrac{3}{5}$

The perimeter is $\dfrac{3}{5}$ mile.

11. $\dfrac{13}{4} - \dfrac{11}{4} = \dfrac{13-11}{4} = \dfrac{2}{4} = \dfrac{1 \cdot 2}{2 \cdot 2} = \dfrac{1}{2}$

He ran $\dfrac{1}{2}$ mile farther on Monday.

12. 16 is a multiple of 8, so the LCD of $\dfrac{7}{8}$ and $\dfrac{11}{16}$ is 16.

13.
$30 \cdot 1 = 30$	Not a multiple of 25.
$30 \cdot 2 = 60$	Not a multiple of 25.
$30 \cdot 3 = 90$	Not a multiple of 25.
$30 \cdot 4 = 120$	Not a multiple of 25.
$30 \cdot 5 = 150$	A multiple of 25.

The LCD of $\dfrac{23}{25}$ and $\dfrac{1}{30}$ is 150.

14. $40 = \boxed{2 \cdot 2 \cdot 2 \cdot 5}$
$108 = 2 \cdot 2 \cdot \boxed{3 \cdot 3 \cdot 3}$
LCD $= 2 \cdot 2 \cdot 2 \cdot 3 \cdot 3 \cdot 3 \cdot 5 = 1080$
The LCD of $-\dfrac{3}{40}$ and $\dfrac{11}{108}$ is 1080.

15. $20 = 2 \cdot 2 \cdot \boxed{5}$
$24 = \boxed{2 \cdot 2 \cdot 2} \cdot 3$
$45 = \boxed{3 \cdot 3} \cdot 5$
LCD $= 2 \cdot 2 \cdot 2 \cdot 3 \cdot 3 \cdot 5 = 360$
The LCD of $\dfrac{7}{20}, \dfrac{1}{24},$ and $\dfrac{13}{45}$ is 360.

16. $y = \boxed{y}$
$11 = \boxed{11}$

The LCD of $\dfrac{7}{y}$ and $\dfrac{6}{11}$ is $11y$.

17. $\dfrac{7}{8} = \dfrac{7}{8} \cdot \dfrac{7}{7} = \dfrac{7 \cdot 7}{8 \cdot 7} = \dfrac{49}{56}$

18. $\dfrac{1}{4} = \dfrac{1}{4} \cdot \dfrac{5}{5} = \dfrac{1 \cdot 5}{4 \cdot 5} = \dfrac{5}{20}$

19. $\dfrac{3x}{7} = \dfrac{3x}{7} \cdot \dfrac{6}{6} = \dfrac{3x \cdot 6}{7 \cdot 6} = \dfrac{18x}{42}$

20. $4 = \dfrac{4}{1} \cdot \dfrac{6}{6} = \dfrac{4 \cdot 6}{1 \cdot 6} = \dfrac{24}{6}$

21. $\dfrac{9}{4x} = \dfrac{9}{4x} \cdot \dfrac{9}{9} = \dfrac{9 \cdot 9}{4x \cdot 9} = \dfrac{81}{36x}$

Vocabulary and Readiness Check

1. The fractions $\dfrac{9}{11}$ and $\dfrac{13}{11}$ are called <u>like</u> fractions while $\dfrac{3}{4}$ and $\dfrac{1}{3}$ are called <u>unlike</u> fractions.

3. As long as b is not 0, $\dfrac{-a}{b} = \dfrac{a}{-b} = -\dfrac{a}{b}$.

5. The smallest positive number divisible by all the denominators of a list of fractions is called the <u>least common denominator (LCD)</u>.

Exercise Set 4.4

1. $\dfrac{5}{11} + \dfrac{2}{11} = \dfrac{5+2}{11} = \dfrac{7}{11}$

3. $\dfrac{2}{9} + \dfrac{4}{9} = \dfrac{2+4}{9} = \dfrac{6}{9} = \dfrac{3 \cdot 2}{3 \cdot 3} = \dfrac{2}{3}$

5. $-\dfrac{6}{20} + \dfrac{1}{20} = \dfrac{-6+1}{20} = \dfrac{-5}{20} = -\dfrac{1 \cdot 5}{4 \cdot 5} = -\dfrac{1}{4}$

7. $-\dfrac{3}{14} + \left(-\dfrac{4}{14}\right) = \dfrac{-3+(-4)}{14}$
$= \dfrac{-7}{14}$
$= -\dfrac{1 \cdot 7}{2 \cdot 7}$
$= -\dfrac{1}{2}$

9. $\dfrac{2}{9x} + \dfrac{4}{9x} = \dfrac{2+4}{9x} = \dfrac{6}{9x} = \dfrac{2 \cdot 3}{3 \cdot 3 \cdot x} = \dfrac{2}{3x}$

11. $-\dfrac{7x}{18} + \dfrac{3x}{18} + \dfrac{2x}{18} = \dfrac{-7x+3x+2x}{18}$
$= \dfrac{-2x}{18}$
$= -\dfrac{2 \cdot x}{2 \cdot 9}$
$= -\dfrac{x}{9}$

13. $\dfrac{10}{11} - \dfrac{4}{11} = \dfrac{10-4}{11} = \dfrac{6}{11}$

15. $\dfrac{7}{8} - \dfrac{1}{8} = \dfrac{7-1}{8} = \dfrac{6}{8} = \dfrac{3 \cdot 2}{4 \cdot 2} = \dfrac{3}{4}$

17. $\dfrac{1}{y} - \dfrac{4}{y} = \dfrac{1-4}{y} = \dfrac{-3}{y} = -\dfrac{3}{y}$

19. $-\dfrac{27}{33} - \left(-\dfrac{8}{33}\right) = -\dfrac{27}{33} + \dfrac{8}{33}$
$= \dfrac{-27+8}{33}$
$= \dfrac{-19}{33}$
$= -\dfrac{19}{33}$

21. $\dfrac{20}{21} - \dfrac{10}{21} - \dfrac{17}{21} = \dfrac{20-10-17}{21}$

$\qquad\qquad\qquad = \dfrac{-7}{21}$

$\qquad\qquad\qquad = -\dfrac{1\cdot 7}{3\cdot 7}$

$\qquad\qquad\qquad = -\dfrac{1}{3}$

23. $\dfrac{7a}{4} - \dfrac{3}{4} = \dfrac{7a-3}{4}$

25. $-\dfrac{9}{100} + \dfrac{99}{100} = \dfrac{-9+99}{100} = \dfrac{90}{100} = \dfrac{9\cdot 10}{10\cdot 10} = \dfrac{9}{10}$

27. $-\dfrac{13x}{28} - \dfrac{13x}{28} = \dfrac{-13x-13x}{28}$

$\qquad\qquad\qquad = \dfrac{-26x}{28}$

$\qquad\qquad\qquad = -\dfrac{2\cdot 13\cdot x}{2\cdot 14}$

$\qquad\qquad\qquad = -\dfrac{13x}{14}$

29. $\dfrac{9x}{15} + \dfrac{1}{15} = \dfrac{9x+1}{15}$

31. $\dfrac{7x}{16} - \dfrac{15x}{16} = \dfrac{7x-15x}{16} = \dfrac{-8x}{16} = -\dfrac{8\cdot x}{8\cdot 2} = -\dfrac{x}{2}$

33. $\dfrac{9}{12} - \dfrac{7}{12} - \dfrac{10}{12} = \dfrac{9-7-10}{12}$

$\qquad\qquad\qquad = \dfrac{-8}{12}$

$\qquad\qquad\qquad = -\dfrac{2\cdot 4}{3\cdot 4}$

$\qquad\qquad\qquad = -\dfrac{2}{3}$

35. $\dfrac{x}{4} + \dfrac{3x}{4} - \dfrac{2x}{4} + \dfrac{x}{4} = \dfrac{x+3x-2x+x}{4} = \dfrac{3x}{4}$

37. $x + y = \dfrac{3}{4} + \dfrac{2}{4} = \dfrac{3+2}{4} = \dfrac{5}{4}$

39. $x - y = -\dfrac{1}{5} - \left(-\dfrac{3}{5}\right) = -\dfrac{1}{5} + \dfrac{3}{5} = \dfrac{-1+3}{5} = \dfrac{2}{5}$

41. $\dfrac{4}{20} + \dfrac{7}{20} + \dfrac{9}{20} = \dfrac{4+7+9}{20} = \dfrac{20}{20} = 1$

The perimeter is 1 inch.

43. $\dfrac{5}{12} + \dfrac{7}{12} + \dfrac{5}{12} + \dfrac{7}{12} = \dfrac{5+7+5+7}{12} = \dfrac{24}{12} = 2$

The perimeter is 2 meters.

45. North America makes up $\dfrac{16}{100}$ of the world's land area, while South America makes up $\dfrac{12}{100}$ of the land area.

$\dfrac{16}{100} + \dfrac{12}{100} = \dfrac{16+12}{100} = \dfrac{28}{100} = \dfrac{4\cdot 7}{4\cdot 25} = \dfrac{7}{25}$

$\dfrac{7}{25}$ of the world's land area is within North America and South America.

47. Antarctica makes up $\dfrac{9}{100}$ of the world's land area, while Europe makes up $\dfrac{7}{100}$ of the world's land area.

$\dfrac{9}{100} - \dfrac{7}{100} = \dfrac{9-7}{100} = \dfrac{2}{100} = \dfrac{1\cdot 2}{50\cdot 2} = \dfrac{1}{50}$

Antarctica's land area is $\dfrac{1}{50}$ greater than that of Europe.

49. To find the remaining amount of track to be inspected, subtract the $\dfrac{5}{20}$ mile that has already been inspected from the $\dfrac{19}{20}$ mile total that must be inspected.

$\dfrac{19}{20} - \dfrac{5}{20} = \dfrac{19-5}{20} = \dfrac{14}{20} = \dfrac{2\cdot 7}{2\cdot 10} = \dfrac{7}{10}$

$\dfrac{7}{10}$ of a mile of track remains to be inspected.

51. To find the fraction that had speed limits less than 70 mph, subtract the $\dfrac{17}{50}$ that have 70 mph speed limits from the $\dfrac{37}{50}$ that have speed limits up to and including 70.

$$\frac{37}{50} - \frac{17}{50} = \frac{37-17}{50} = \frac{20}{50} = \frac{2 \cdot 10}{5 \cdot 10} = \frac{2}{5}$$

$\dfrac{2}{5}$ of the states have speed limits that were less than 70 mph.

53. Multiples of 9: 9, 18, 27, 36, $\boxed{45}$, 54
Multiples of 15: 15, 30, $\boxed{45}$, 60
LCD: 45

55. Multiples of 36: 36, $\boxed{72}$
Multiples of 24: 24, 48, $\boxed{72}$
LCD: 72

57. $6 = \boxed{2} \cdot 3$
$15 = \boxed{3} \cdot 5$
$25 = \boxed{5 \cdot 5}$
LCD $= 2 \cdot 3 \cdot 5 \cdot 5 = 150$

59. $24 = \boxed{2 \cdot 2 \cdot 2 \cdot 3}$
$x = \boxed{x}$
LCD $= 2 \cdot 2 \cdot 2 \cdot 3 \cdot x = 24x$

61. $18 = \boxed{2 \cdot 3 \cdot 3}$
$21 = 3 \cdot \boxed{7}$
LCD $= 2 \cdot 3 \cdot 3 \cdot 7 = 126$

63. $3 = \boxed{3}$
$21 = 3 \cdot \boxed{7}$
$56 = \boxed{2 \cdot 2 \cdot 2} \cdot 7$
LCD $= 2 \cdot 2 \cdot 2 \cdot 3 \cdot 7 = 168$

65. $\dfrac{2}{3} = \dfrac{2}{3} \cdot \dfrac{7}{7} = \dfrac{2 \cdot 7}{3 \cdot 7} = \dfrac{14}{21}$

67. $\dfrac{4}{7} = \dfrac{4}{7} \cdot \dfrac{5}{5} = \dfrac{4 \cdot 5}{7 \cdot 5} = \dfrac{20}{35}$

69. $\dfrac{1}{2} = \dfrac{1}{2} \cdot \dfrac{25}{25} = \dfrac{1 \cdot 25}{2 \cdot 25} = \dfrac{25}{50}$

71. $\dfrac{14x}{17} = \dfrac{14x}{17} \cdot \dfrac{4}{4} = \dfrac{14x \cdot 4}{17 \cdot 4} = \dfrac{56x}{68}$

73. $\dfrac{2y}{3} = \dfrac{2y}{3} \cdot \dfrac{4}{4} = \dfrac{2y \cdot 4}{3 \cdot 4} = \dfrac{8y}{12}$

75. $\dfrac{5}{9} = \dfrac{5}{9} \cdot \dfrac{4a}{4a} = \dfrac{5 \cdot 4a}{9 \cdot 4a} = \dfrac{20a}{36a}$

77. Denmark: $\dfrac{89}{100} = \dfrac{89}{100}$

Finland: $\dfrac{9}{10} = \dfrac{9 \cdot 10}{10 \cdot 10} = \dfrac{90}{100}$

Israel: $\dfrac{24}{25} = \dfrac{24 \cdot 4}{25 \cdot 4} = \dfrac{96}{100}$

Spain: $\dfrac{23}{25} = \dfrac{23 \cdot 4}{25 \cdot 4} = \dfrac{92}{100}$

Japan: $\dfrac{17}{25} = \dfrac{17 \cdot 4}{25 \cdot 4} = \dfrac{68}{100}$

Norway: $\dfrac{91}{100} = \dfrac{91}{100}$

Singapore: $\dfrac{4}{5} = \dfrac{4 \cdot 20}{5 \cdot 20} = \dfrac{80}{100}$

South Korea: $\dfrac{69}{100} = \dfrac{69}{100}$

India: $\dfrac{2}{25} = \dfrac{2 \cdot 4}{25 \cdot 4} = \dfrac{8}{100}$

United States: $\dfrac{7}{10} = \dfrac{7 \cdot 10}{10 \cdot 10} = \dfrac{70}{100}$

79. $\dfrac{8}{100}$ is the smallest fraction, so India has the smallest fraction of cell phone users.

81. $3^2 = 3 \cdot 3 = 9$

83. $5^3 = 5 \cdot 5 \cdot 5 = 125$

85. $7^2 = 7 \cdot 7 = 49$

87. $2^3 \cdot 3 = 2 \cdot 2 \cdot 2 \cdot 3 = 24$

89. $\dfrac{2}{7} + \dfrac{9}{7} = \dfrac{2+9}{7} = \dfrac{11}{7}$

91. answers may vary

93. $\dfrac{16}{100} + \dfrac{12}{100} + \dfrac{7}{100} + \dfrac{20}{100} + \dfrac{30}{100} + \dfrac{6}{100} + \dfrac{9}{100}$

$= \dfrac{16+12+7+20+30+6+9}{100}$

$= \dfrac{100}{100}$

$= 1$

answers may vary

95. $\dfrac{37x}{165} = \dfrac{37x}{165} \cdot \dfrac{22}{22} = \dfrac{37x \cdot 22}{165 \cdot 22} = \dfrac{814x}{3630}$

97. answers may vary

99. $\dfrac{2}{3} = \dfrac{2 \cdot 5}{3 \cdot 5} = \dfrac{10}{15}$

$\dfrac{2}{3} = \dfrac{2 \cdot 20}{3 \cdot 20} = \dfrac{40}{60}$

$\dfrac{2}{3} = \dfrac{2 \cdot 100}{3 \cdot 100} = \dfrac{200}{300}$

a, b, and d are equivalent to $\dfrac{2}{3}$.

Section 4.5

Practice Problems

1. The LCD of 7 and 21 is 21.

$\dfrac{2}{7} + \dfrac{8}{21} = \dfrac{2 \cdot 3}{7 \cdot 3} + \dfrac{8}{21}$

$= \dfrac{6}{21} + \dfrac{8}{21}$

$= \dfrac{14}{21}$

$= \dfrac{2 \cdot 7}{3 \cdot 7}$

$= \dfrac{2}{3}$

2. The LCD of 6 and 9 is 18.

$\dfrac{5y}{6} + \dfrac{2y}{9} = \dfrac{5y \cdot 3}{6 \cdot 3} + \dfrac{2y \cdot 2}{9 \cdot 2} = \dfrac{15y}{18} + \dfrac{4y}{18} = \dfrac{19y}{18}$

3. The LCD of 5 and 20 is 20.

$-\dfrac{1}{5} + \dfrac{9}{20} = -\dfrac{1 \cdot 4}{5 \cdot 4} + \dfrac{9}{20}$

$= -\dfrac{4}{20} + \dfrac{9}{20}$

$= \dfrac{5}{20}$

$= \dfrac{1 \cdot 5}{4 \cdot 5}$

$= \dfrac{1}{4}$

4. The LCD of 7 and 10 is 70.

$\dfrac{5}{7} - \dfrac{9}{10} = \dfrac{5 \cdot 10}{7 \cdot 10} - \dfrac{9 \cdot 7}{10 \cdot 7} = \dfrac{50}{70} - \dfrac{63}{70} = -\dfrac{13}{70}$

5. The LCD of 8, 3, and 12 is 24.

$\dfrac{5}{8} - \dfrac{1}{3} - \dfrac{1}{12} = \dfrac{5 \cdot 3}{8 \cdot 3} - \dfrac{1 \cdot 8}{3 \cdot 8} - \dfrac{1 \cdot 2}{12 \cdot 2}$

$= \dfrac{15}{24} - \dfrac{8}{24} - \dfrac{2}{24}$

$= \dfrac{5}{24}$

6. Recall that $5 = \dfrac{5}{1}$. The LCD of 1 and 4 is 4.

$$\frac{5}{1} - \frac{y}{4} = \frac{5 \cdot 4}{1 \cdot 4} - \frac{y}{4} = \frac{20}{4} - \frac{y}{4} = \frac{20 - y}{4}$$

7. The LCD of 8 and 20 is 40. Write each fraction as an equivalent fraction with a denominator of 40.

$$\frac{5}{8} = \frac{5 \cdot 5}{8 \cdot 5} = \frac{25}{40}$$

$$\frac{11}{20} = \frac{11 \cdot 2}{20 \cdot 2} = \frac{22}{40}$$

Since $25 > 22$, $\dfrac{25}{40} > \dfrac{22}{40}$, so $\dfrac{5}{8} > \dfrac{11}{20}$.

8. The LCD of 20 and 5 is 20. Write each fraction as an equivalent fraction with a denominator of 20.

$$-\frac{17}{20}$$

$$-\frac{4}{5} = -\frac{4 \cdot 4}{5 \cdot 4} = -\frac{16}{20}$$

Since $-17 < -16$, $-\dfrac{17}{20} < -\dfrac{16}{20}$, so

$$-\frac{17}{20} < -\frac{4}{5}.$$

9. The LCD of 11 and 9 is 99.

$$\begin{aligned} x - y &= \frac{5}{11} - \frac{4}{9} \\ &= \frac{5 \cdot 9}{11 \cdot 9} - \frac{4 \cdot 11}{9 \cdot 11} \\ &= \frac{45}{99} - \frac{44}{99} \\ &= \frac{1}{99} \end{aligned}$$

10. The LCD of 5, 10, and 15 is 30.

$$\begin{aligned} \frac{3}{5} + \frac{3}{10} + \frac{1}{15} &= \frac{3 \cdot 6}{5 \cdot 6} + \frac{3 \cdot 3}{10 \cdot 3} + \frac{1 \cdot 2}{15 \cdot 2} \\ &= \frac{18}{30} + \frac{9}{30} + \frac{2}{30} \\ &= \frac{29}{30} \end{aligned}$$

The homeowner needs $\dfrac{29}{30}$ cubic yard of cement.

11. The LCD of 4 and 3 is 12.

$$\frac{3}{4} - \frac{2}{3} = \frac{3 \cdot 3}{4 \cdot 3} - \frac{2 \cdot 4}{3 \cdot 4} = \frac{9}{12} - \frac{8}{12} = \frac{1}{12}$$

The difference in length is $\dfrac{1}{12}$ foot.

Calculator Explorations

1. $\dfrac{1}{16} + \dfrac{2}{5} = \dfrac{37}{80}$

3. $\dfrac{4}{9} + \dfrac{7}{8} = \dfrac{95}{72}$

5. $\dfrac{10}{17} + \dfrac{12}{19} = \dfrac{394}{323}$

Vocabulary and Readiness Check

1. To add or subtract unlike fractions, we first write the fractions as <u>equivalent</u> fractions with a common denominator. The common denominator we use is called the <u>least common denominator</u>.

3. $\dfrac{1}{6} + \dfrac{5}{8} = \dfrac{1}{6} \cdot \dfrac{4}{4} + \dfrac{5}{8} \cdot \dfrac{3}{3} = \dfrac{4}{\underline{24}} + \dfrac{15}{\underline{24}} = \dfrac{19}{\underline{24}}.$

5. $x - y$ is an <u>expression</u> while $3x = \dfrac{1}{5}$ is an <u>equation</u>.

Exercise Set 4.5

1. The LCD of 3 and 6 is 6.
$$\frac{2}{3}+\frac{1}{6}=\frac{2\cdot2}{3\cdot2}+\frac{1}{6}=\frac{4}{6}+\frac{1}{6}=\frac{5}{6}$$

3. The LCD of 2 and 3 is 6.
$$\frac{1}{2}-\frac{1}{3}=\frac{1\cdot3}{2\cdot3}-\frac{1\cdot2}{3\cdot2}=\frac{3}{6}-\frac{2}{6}=\frac{1}{6}$$

5. The LCD of 11 and 33 is 33.
$$-\frac{2}{11}+\frac{2}{33}=-\frac{2\cdot3}{11\cdot3}+\frac{2}{33}=-\frac{6}{33}+\frac{2}{33}=-\frac{4}{33}$$

7. The LCD of 14 and 7 is 14.
$$\frac{3}{14}-\frac{3}{7}=\frac{3}{14}-\frac{3\cdot2}{7\cdot2}=\frac{3}{14}-\frac{6}{14}=-\frac{3}{14}$$

9. The LCD of 35 and 7 is 35.
$$\frac{11x}{35}+\frac{2x}{7}=\frac{11x}{35}+\frac{2x\cdot5}{7\cdot5}$$
$$=\frac{11x}{35}+\frac{10x}{35}$$
$$=\frac{21x}{35}$$
$$=\frac{3\cdot7\cdot x}{5\cdot7}$$
$$=\frac{3x}{5}$$

11. The LCD of 1 and 12 is 12.
$$2-\frac{y}{12}=\frac{2}{1}-\frac{y}{12}$$
$$=\frac{2\cdot12}{1\cdot12}-\frac{y}{12}$$
$$=\frac{24}{12}-\frac{y}{12}$$
$$=\frac{24-y}{12}$$

13. The LCD of 12 and 9 is 36.
$$\frac{5}{12}-\frac{1}{9}=\frac{5\cdot3}{12\cdot3}-\frac{1\cdot4}{9\cdot4}=\frac{15}{36}-\frac{4}{36}=\frac{11}{36}$$

15. The LCD of 1 and 7 is 7.
$$-7+\frac{5}{7}=-\frac{7}{1}+\frac{5}{7}$$
$$=-\frac{7\cdot7}{1\cdot7}+\frac{5}{7}$$
$$=-\frac{49}{7}+\frac{5}{7}$$
$$=-\frac{44}{7}$$

17. The LCD of 11 and 9 is 99.
$$\frac{5a}{11}+\frac{4a}{9}=\frac{5a\cdot9}{11\cdot9}+\frac{4a\cdot11}{9\cdot11}$$
$$=\frac{45a}{99}+\frac{44a}{99}$$
$$=\frac{89a}{99}$$

19. The LCD of 3 and 6 is 6.
$$\frac{2y}{3}-\frac{1}{6}=\frac{2y\cdot2}{3\cdot2}-\frac{1}{6}=\frac{4y}{6}-\frac{1}{6}=\frac{4y-1}{6}$$

21. The LCD of 2 and x is $2x$.
$$\frac{1}{2}+\frac{3}{x}=\frac{1\cdot x}{2\cdot x}+\frac{3\cdot2}{x\cdot2}=\frac{x}{2x}+\frac{6}{2x}=\frac{x+6}{2x}$$

23. The LCD of 11 and 33 is 33.
$$-\frac{2}{11}-\frac{2}{33}=-\frac{2\cdot3}{11\cdot3}-\frac{2}{33}=-\frac{6}{33}-\frac{2}{33}=-\frac{8}{33}$$

25. The LCD of 14 and 7 is 14.
$$\frac{9}{14}-\frac{3}{7}=\frac{9}{14}-\frac{3\cdot2}{7\cdot2}=\frac{9}{14}-\frac{6}{14}=\frac{3}{14}$$

27. The LCD of 35 and 7 is 35.
$$\frac{11y}{35}-\frac{2}{7}=\frac{11y}{35}-\frac{2\cdot5}{7\cdot5}=\frac{11y}{35}-\frac{10}{35}=\frac{11y-10}{35}$$

29. The LCD of 9 and 12 is 36.
$$\frac{1}{9}-\frac{5}{12}=\frac{1\cdot4}{9\cdot4}-\frac{5\cdot3}{12\cdot3}=\frac{4}{36}-\frac{15}{36}=-\frac{11}{36}$$

31. The LCD of 15 and 12 is 60.

$$\frac{7}{15} - \frac{5}{12} = \frac{7 \cdot 4}{15 \cdot 4} - \frac{5 \cdot 5}{12 \cdot 5}$$
$$= \frac{28}{60} - \frac{25}{60}$$
$$= \frac{3}{60}$$
$$= \frac{1 \cdot 3}{20 \cdot 3}$$
$$= \frac{1}{20}$$

33. The LCD of 7 and 8 is 56.

$$\frac{5}{7} - \frac{1}{8} = \frac{5 \cdot 8}{7 \cdot 8} - \frac{1 \cdot 7}{8 \cdot 7} = \frac{40}{56} - \frac{7}{56} = \frac{33}{56}$$

35. The LCD of 8 and 16 is 16.

$$\frac{7}{8} + \frac{3}{16} = \frac{7 \cdot 2}{8 \cdot 2} + \frac{3}{16} = \frac{14}{16} + \frac{3}{16} = \frac{17}{16}$$

37. $\dfrac{3}{9} - \dfrac{5}{9} = \dfrac{3-5}{9} = -\dfrac{2}{9}$

39. The LCD of 5, 3, and 10 is 30.

$$-\frac{2}{5} + \frac{1}{3} - \frac{3}{10} = -\frac{2 \cdot 6}{5 \cdot 6} + \frac{1 \cdot 10}{3 \cdot 10} - \frac{3 \cdot 3}{10 \cdot 3}$$
$$= -\frac{12}{30} + \frac{10}{30} - \frac{9}{30}$$
$$= -\frac{11}{30}$$

41. The LCD of 11 and 3 is 33.

$$\frac{5}{11} + \frac{y}{3} = \frac{5 \cdot 3}{11 \cdot 3} + \frac{y \cdot 11}{3 \cdot 11}$$
$$= \frac{15}{33} + \frac{11y}{33}$$
$$= \frac{15 + 11y}{33}$$

43. The LCD of 6 and 7 is 42.

$$-\frac{5}{6} - \frac{3}{7} = -\frac{5 \cdot 7}{6 \cdot 7} - \frac{3 \cdot 6}{7 \cdot 6} = -\frac{35}{42} - \frac{18}{42} = -\frac{53}{42}$$

45. The LCD of 2, 4, and 16 is 16.

$$\frac{x}{2} + \frac{x}{4} + \frac{2x}{16} = \frac{x \cdot 8}{2 \cdot 8} + \frac{x \cdot 4}{4 \cdot 4} + \frac{2x}{16}$$
$$= \frac{8x}{16} + \frac{4x}{16} + \frac{2x}{16}$$
$$= \frac{14x}{16}$$
$$= \frac{2 \cdot 7x}{2 \cdot 8}$$
$$= \frac{7x}{8}$$

47. The LCD of 9 and 6 is 18.

$$\frac{7}{9} - \frac{1}{6} = \frac{7 \cdot 2}{9 \cdot 2} - \frac{1 \cdot 3}{6 \cdot 3} = \frac{14}{18} - \frac{3}{18} = \frac{11}{18}$$

49. The LCD of 3 and 13 is 39.

$$\frac{2a}{3} + \frac{6a}{13} = \frac{2a \cdot 13}{3 \cdot 13} + \frac{6a \cdot 3}{13 \cdot 3}$$
$$= \frac{26a}{39} + \frac{18a}{39}$$
$$= \frac{44a}{39}$$

51. The LCD of 30 and 12 is 60.

$$\frac{7}{30} - \frac{5}{12} = \frac{7 \cdot 2}{30 \cdot 2} - \frac{5 \cdot 5}{12 \cdot 5} = \frac{14}{60} - \frac{25}{60} = -\frac{11}{60}$$

53. The LCD of 9 and y is $9y$.

$$\frac{5}{9} + \frac{1}{y} = \frac{5 \cdot y}{9 \cdot y} + \frac{1 \cdot 9}{y \cdot 9} = \frac{5y}{9y} + \frac{9}{9y} = \frac{5y + 9}{9y}$$

55. The LCD of 5, 4, and 2 is 20.

$$\frac{6}{5} - \frac{3}{4} + \frac{1}{2} = \frac{6 \cdot 4}{5 \cdot 4} - \frac{3 \cdot 5}{4 \cdot 5} + \frac{1 \cdot 10}{2 \cdot 10}$$
$$= \frac{24}{20} - \frac{15}{20} + \frac{10}{20}$$
$$= \frac{19}{20}$$

57. The LCD of 5 and 9 is 45.

$$\frac{4}{5} + \frac{4}{9} = \frac{4 \cdot 9}{5 \cdot 9} + \frac{4 \cdot 5}{9 \cdot 5} = \frac{36}{45} + \frac{20}{45} = \frac{56}{45}$$

59. The LCD of $9x$ and 8 is $72x$.

$$\frac{5}{9x}+\frac{1}{8}=\frac{5\cdot 8}{9x\cdot 8}+\frac{1\cdot 9x}{8\cdot 9x}$$
$$=\frac{40}{72x}+\frac{9x}{72x}$$
$$=\frac{40+9x}{72x}$$

61. The LCD of 12, 24, and 6 is 24.

$$-\frac{9}{12}+\frac{17}{24}-\frac{1}{6}=-\frac{9\cdot 2}{12\cdot 2}+\frac{17}{24}-\frac{1\cdot 4}{6\cdot 4}$$
$$=-\frac{18}{24}+\frac{17}{24}-\frac{4}{24}$$
$$=-\frac{5}{24}$$

63. The LCD of 8, 7, and 14 is 56.

$$\frac{3x}{8}+\frac{2x}{7}-\frac{5}{14}=\frac{3x\cdot 7}{8\cdot 7}+\frac{2x\cdot 8}{7\cdot 8}-\frac{5\cdot 4}{14\cdot 4}$$
$$=\frac{21x}{56}+\frac{16x}{56}-\frac{20}{56}$$
$$=\frac{37x-20}{56}$$

65. The LCD of 7 and 10 is 70. Write each fraction as an equivalent fraction with a denominator of 70.

$$\frac{2}{7}=\frac{2\cdot 10}{7\cdot 10}=\frac{20}{70}$$
$$\frac{3}{10}=\frac{3\cdot 7}{10\cdot 7}=\frac{21}{70}$$

Since $20<21$, $\frac{20}{70}<\frac{21}{70}$, so $\frac{2}{7}<\frac{3}{10}$.

67. A positive fraction is greater than a negative fraction.

$$\frac{5}{6}>-\frac{13}{15}$$

69. The LCD of 4 and 14 is 28. Write each fraction as an equivalent fraction with a denominator of 28.

$$-\frac{3}{4}=-\frac{3\cdot 7}{4\cdot 7}=-\frac{21}{28}$$

$$-\frac{11}{14}=-\frac{11\cdot 2}{14\cdot 2}=-\frac{22}{28}$$

Since $-21>-22$, $-\frac{21}{28}>-\frac{22}{28}$, so

$$-\frac{3}{4}>-\frac{11}{14}.$$

71. The LCD of 3 and 4 is 12.

$$x+y=\frac{1}{3}+\frac{3}{4}=\frac{1\cdot 4}{3\cdot 4}+\frac{3\cdot 3}{4\cdot 3}=\frac{4}{12}+\frac{9}{12}=\frac{13}{12}$$

73. $xy=\frac{1}{3}\cdot\frac{3}{4}=\frac{1\cdot 3}{3\cdot 4}=\frac{1}{4}$

75. $2y+x=2\left(\frac{3}{4}\right)+\frac{1}{3}$
$$=\frac{6}{4}+\frac{1}{3}$$
$$=\frac{2\cdot 3}{2\cdot 2}+\frac{1}{3}$$
$$=\frac{3}{2}+\frac{1}{3}$$
$$=\frac{3\cdot 3}{2\cdot 3}+\frac{1\cdot 2}{3\cdot 2}$$
$$=\frac{9}{6}+\frac{2}{6}$$
$$=\frac{11}{6}$$

77. The LCD of 3 and 5 is 15.

$$\frac{4}{5}+\frac{1}{3}+\frac{4}{5}+\frac{1}{3}=\frac{4}{5}\cdot\frac{3}{3}+\frac{1}{3}\cdot\frac{5}{5}+\frac{4}{5}\cdot\frac{3}{3}+\frac{1}{3}\cdot\frac{5}{5}$$
$$=\frac{12}{15}+\frac{5}{15}+\frac{12}{15}+\frac{5}{15}$$
$$=\frac{34}{15}$$

The perimeter is $\frac{34}{15}$ or $2\frac{4}{15}$ centimeters.

79. The LCD of 4, 5, and 2 is 20.

$$\frac{1}{4}+\frac{1}{5}+\frac{1}{2}+\frac{3}{4}=\frac{1\cdot5}{4\cdot5}+\frac{1\cdot4}{5\cdot4}+\frac{1\cdot10}{2\cdot10}+\frac{3\cdot5}{4\cdot5}$$
$$=\frac{5}{20}+\frac{4}{20}+\frac{10}{20}+\frac{15}{20}$$
$$=\frac{34}{20}$$
$$=\frac{17\cdot2}{10\cdot2}$$
$$=\frac{17}{10}\ or\ 1\frac{7}{10}$$

The perimeter is $\frac{17}{10}$ meters or $1\frac{7}{10}$ meters.

81. The sum of a number and $\frac{1}{2}$ translates as

$x+\frac{1}{2}$.

83. A number subtracted from $-\frac{3}{8}$ translates as

$-\frac{3}{8}-x$.

85. $1-\frac{3}{16}-\frac{3}{16}=\frac{16}{16}-\frac{3}{16}-\frac{3}{16}=\frac{10}{16}=\frac{5}{8}$

The inner diameter is $\frac{5}{8}$ inch.

87. The LCD of 10 and 100 is 100.

$$\frac{17}{100}-\frac{1}{10}=\frac{17}{100}-\frac{1\cdot10}{10\cdot10}=\frac{17}{100}-\frac{10}{100}=\frac{7}{100}$$

The sloth can travel $\frac{7}{100}$ mph faster in trees.

89. The LCD of 2, 16, and 32 is 32.

$$\frac{1}{2}+\frac{11}{16}+\frac{9}{32}=\frac{1}{2}\cdot\frac{16}{16}+\frac{11}{16}\cdot\frac{2}{2}+\frac{9}{32}$$
$$=\frac{16}{32}+\frac{22}{32}+\frac{9}{32}$$
$$=\frac{47}{32}$$

The total length is $\frac{47}{32}$ inches.

91. The LCD of 20 and 25 is 100.

$$\frac{13}{20}-\frac{4}{25}=\frac{13\cdot5}{20\cdot5}-\frac{4\cdot4}{25\cdot4}=\frac{65}{100}-\frac{16}{100}=\frac{49}{100}$$

$\frac{49}{100}$ of the American students ages 10 to 17 name math or science as their favorite subject in school.

93. The LCD of 50 and 2 is 50.

$$\frac{13}{50}+\frac{1}{2}=\frac{13}{50}+\frac{1}{2}\cdot\frac{25}{25}$$
$$=\frac{13}{50}+\frac{25}{50}$$
$$=\frac{38}{50}$$
$$=\frac{2\cdot19}{2\cdot25}$$
$$=\frac{19}{25}$$

The Pacific and Atlantic Oceans account for $\frac{19}{25}$ of the world's water surface.

95. The LCD of 25 and 10 is 50.

$$\frac{3}{25}+\frac{1}{10}=\frac{3\cdot2}{25\cdot2}+\frac{1\cdot5}{10\cdot5}=\frac{6}{50}+\frac{5}{50}=\frac{11}{50}$$

$\frac{11}{50}$ of the beginning freshmen plan to major in arts and humanities or education.

97. $1 - \dfrac{4}{25} = \dfrac{25}{25} - \dfrac{4}{25} = \dfrac{21}{25}$

$\dfrac{21}{25}$ of the beginning freshmen do not plan to major in business.

99. $-50 \div 5 \cdot 2 = -10 \cdot 2 = -20$

101. $(8 - 6) \cdot (4 - 7) = 2 \cdot (-3) = -6$

103. a.

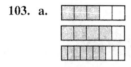

b. There seems to be an error.

c. $\dfrac{3}{5} + \dfrac{4}{5} = \dfrac{3+4}{5} = \dfrac{7}{5} \left(\text{or } \dfrac{14}{10} \text{ not } \dfrac{7}{10} \right)$

105. The LCD of 3, 4, and 540 is 540.

$\dfrac{2}{3} - \dfrac{1}{4} - \dfrac{2}{540} = \dfrac{2}{3} \cdot \dfrac{180}{180} - \dfrac{1}{4} \cdot \dfrac{135}{135} - \dfrac{2}{540}$

$= \dfrac{360}{540} - \dfrac{135}{540} - \dfrac{2}{540}$

$= \dfrac{225}{540} - \dfrac{2}{540}$

$= \dfrac{223}{540}$

107. The LCD of 55 and 1760 is 1760.

$\dfrac{30}{55} + \dfrac{1000}{1760} = \dfrac{30 \cdot 32}{55 \cdot 32} + \dfrac{1000}{1760}$

$= \dfrac{960}{1760} + \dfrac{1000}{1760}$

$= \dfrac{1960}{1760}$

$= \dfrac{49 \cdot 40}{44 \cdot 40}$

$= \dfrac{49}{44}$

109. answers may vary

111. The LCD of 53 and 106 is 106. Write each fraction as an equivalent fraction with a denominator of 106.

$\dfrac{24}{53} = \dfrac{24 \cdot 2}{53 \cdot 2} = \dfrac{48}{106}$

$\dfrac{51}{106}$

Since $51 > 48$, $\dfrac{51}{106} > \dfrac{48}{106}$, so $\dfrac{51}{106} > \dfrac{24}{53}$.

Standard mail accounted for the greater portion of the mail handled by volume.

Integrated Review

1. 3 out of 7 equal parts are shaded: $\dfrac{3}{7}$

2. Each part is $\dfrac{1}{4}$ of a whole and there are 5 parts shaded, or 1 whole and 1 more part: $\dfrac{5}{4}$ or $1\dfrac{1}{4}$

3. number that get fewer than 8 $\rightarrow 73$
total number of people $\rightarrow 85$

$\dfrac{73}{85}$ of people get fewer than 8 hours of sleep each night.

4.

5. $\dfrac{11}{-11} = -1$

6. $\dfrac{17}{1} = 17$

7. $\dfrac{0}{-3} = 0$

8. $\dfrac{7}{0}$ is undefined

9. $65 = 5 \cdot 13$

10. $70 = 2 \cdot 35$

$$\downarrow \quad \downarrow \searrow$$

$$2 \cdot \ 5 \cdot 7$$

$$70 = 2 \cdot 5 \cdot 7$$

11. $315 = 3 \cdot 105$

$$\downarrow \quad \downarrow \searrow$$

$$3 \cdot 3 \cdot 35$$

$$\downarrow \ \downarrow \ \downarrow \searrow$$

$$3 \cdot 3 \cdot 5 \cdot 7$$

$$315 = 3^2 \cdot 5 \cdot 7$$

12. $441 = 3 \cdot 147$

$$\downarrow \quad \downarrow \searrow$$

$$3 \cdot 3 \cdot 49$$

$$\downarrow \ \downarrow \ \downarrow \searrow$$

$$3 \cdot 3 \cdot 7 \cdot 7$$

$$441 = 3^2 \cdot 7^2$$

13. $\dfrac{2}{14} = \dfrac{2 \cdot 1}{2 \cdot 7} = \dfrac{1}{7}$

14. $\dfrac{24}{20} = \dfrac{6 \cdot 4}{5 \cdot 4} = \dfrac{6}{5}$

15. $-\dfrac{56}{60} = -\dfrac{14 \cdot 4}{15 \cdot 4} = -\dfrac{14}{15}$

16. $-\dfrac{72}{80} = -\dfrac{8 \cdot 9}{8 \cdot 10} = -\dfrac{9}{10}$

17. $\dfrac{54x}{135} = \dfrac{27 \cdot 2 \cdot x}{27 \cdot 5} = \dfrac{2x}{5}$

18. $\dfrac{90}{240y} = \dfrac{30 \cdot 3}{30 \cdot 8 \cdot y} = \dfrac{3}{8y}$

19. $\dfrac{165z^3}{210z} = \dfrac{15 \cdot 11 \cdot z \cdot z \cdot z}{15 \cdot 14 \cdot z} = \dfrac{11 \cdot z \cdot z}{14} = \dfrac{11z^2}{14}$

20. $\dfrac{245ab}{385a^2 b^3} = \dfrac{35 \cdot 7 \cdot a \cdot b}{35 \cdot 11 \cdot a \cdot a \cdot b \cdot b \cdot b}$

$$= \dfrac{7}{11 \cdot a \cdot b \cdot b}$$

$$= \dfrac{7}{11ab^2}$$

21. Not equivalent, since the cross products are not equal: $7 \cdot 10 = 70,\ 8 \cdot 9 = 72$

22. Equivalent, since the cross products are equal: $10 \cdot 18 = 180,\ 12 \cdot 15 = 180$

23. **a.** number not adjacent $\rightarrow 2$
    total number $\rightarrow \overline{50}$

   $\dfrac{2}{50} = \dfrac{1 \cdot 2}{25 \cdot 2} = \dfrac{1}{25}$ of the states are not adjacent to any other states.

  b. $50 - 2 = 48$; 48 states are adjacent to other states.

  c. $\dfrac{48}{50} = \dfrac{2 \cdot 24}{2 \cdot 25} = \dfrac{24}{25}$ of the states are adjacent to other states.

24. **a.** number rated R $\rightarrow 92$
    total number $\rightarrow \overline{460}$

   $\dfrac{92}{460} = \dfrac{92 \cdot 1}{92 \cdot 5} = \dfrac{1}{5}$ of the new films were rated R.

  b. $460 - 92 = 368$; 368 of the new films were rated other than R.

  c. $\dfrac{368}{460} = \dfrac{4 \cdot 92}{5 \cdot 92} = \dfrac{4}{5}$ of the new films were rated other than R.

25. $5 = \boxed{5}$
  $6 = \boxed{2 \cdot 3}$
  LCM $= 2 \cdot 3 \cdot 5 = 30$

26. $2 = \boxed{2}$
$14 = 2 \cdot \boxed{7}$
$\text{LCM} = 2 \cdot 7 = 14$

27. $6 = \boxed{2} \cdot 3$
$18 = 2 \cdot \boxed{3 \cdot 3}$
$30 = 2 \cdot 3 \cdot 3 \cdot \boxed{5}$
$\text{LCM} = 2 \cdot 3 \cdot 3 \cdot 5 = 90$

28. $\dfrac{7}{9} = \dfrac{7}{9} \cdot \dfrac{4}{4} = \dfrac{7 \cdot 4}{9 \cdot 4} = \dfrac{28}{36}$

29. $\dfrac{11}{15} = \dfrac{11}{15} \cdot \dfrac{5}{5} = \dfrac{11 \cdot 5}{15 \cdot 5} = \dfrac{55}{75}$

30. $\dfrac{5}{6} = \dfrac{5}{6} \cdot \dfrac{8}{8} = \dfrac{5 \cdot 8}{6 \cdot 8} = \dfrac{40}{48}$

31. $\dfrac{1}{5} + \dfrac{3}{5} = \dfrac{1+3}{5} = \dfrac{4}{5}$

32. $\dfrac{1}{5} - \dfrac{3}{5} = \dfrac{1-3}{5} = \dfrac{-2}{5} = -\dfrac{2}{5}$

33. $\dfrac{1}{5} \cdot \dfrac{3}{5} = \dfrac{1 \cdot 3}{5 \cdot 5} = \dfrac{3}{25}$

34. $\dfrac{1}{5} \div \dfrac{3}{5} = \dfrac{1}{5} \cdot \dfrac{5}{3} = \dfrac{1 \cdot 5}{5 \cdot 3} = \dfrac{1}{3}$

35. $\dfrac{2}{3} \div \dfrac{5}{6} = \dfrac{2}{3} \cdot \dfrac{6}{5} = \dfrac{2 \cdot 6}{3 \cdot 5} = \dfrac{2 \cdot 2 \cdot 3}{3 \cdot 5} = \dfrac{2 \cdot 2}{5} = \dfrac{4}{5}$

36. $\dfrac{2a}{3} \cdot \dfrac{5}{6a} = \dfrac{2a \cdot 5}{3 \cdot 6a} = \dfrac{2 \cdot a \cdot 5}{3 \cdot 2 \cdot 3 \cdot a} = \dfrac{5}{3 \cdot 3} = \dfrac{5}{9}$

37. The LCD of $3y$ and $6y$ is $6y$.
$$\dfrac{2}{3y} - \dfrac{5}{6y} = \dfrac{2}{3y} \cdot \dfrac{2}{2} - \dfrac{5}{6y}$$
$$= \dfrac{2 \cdot 2}{3y \cdot 2} - \dfrac{5}{6y}$$
$$= \dfrac{4}{6y} - \dfrac{5}{6y}$$
$$= -\dfrac{1}{6y}$$

38. The LCD of 3 and 6 is 6.
$$\dfrac{2x}{3} + \dfrac{5x}{6} = \dfrac{2x}{3} \cdot \dfrac{2}{2} + \dfrac{5x}{6}$$
$$= \dfrac{2x \cdot 2}{3 \cdot 2} + \dfrac{5x}{6}$$
$$= \dfrac{4x}{6} + \dfrac{5x}{6}$$
$$= \dfrac{9x}{6}$$
$$= \dfrac{3 \cdot 3 \cdot x}{3 \cdot 2}$$
$$= \dfrac{3x}{2}$$

39. $-\dfrac{1}{7} \cdot -\dfrac{7}{18} = \dfrac{1 \cdot 7}{7 \cdot 18} = \dfrac{1}{18}$

40. $-\dfrac{4}{9} \cdot -\dfrac{3}{7} = \dfrac{4 \cdot 3}{9 \cdot 7} = \dfrac{4 \cdot 3}{3 \cdot 3 \cdot 7} = \dfrac{4}{3 \cdot 7} = \dfrac{4}{21}$

41. $-\dfrac{7z}{8} \div 6z^2 = -\dfrac{7z}{8} \div \dfrac{6z^2}{1}$
$$= -\dfrac{7z}{8} \cdot \dfrac{1}{6z^2}$$
$$= -\dfrac{7z \cdot 1}{8 \cdot 6z^2}$$
$$= -\dfrac{7 \cdot z \cdot 1}{8 \cdot 6 \cdot z \cdot z}$$
$$= -\dfrac{7}{48z}$$

42.
$$-\frac{9}{10} \div 5 = -\frac{9}{10} \div \frac{5}{1}$$
$$= -\frac{9}{10} \cdot \frac{1}{5}$$
$$= -\frac{9 \cdot 1}{10 \cdot 5}$$
$$= -\frac{9}{50}$$

43. The LCD of 8 and 20 is 40.
$$\frac{7}{8} + \frac{1}{20} = \frac{7 \cdot 5}{8 \cdot 5} + \frac{1 \cdot 2}{20 \cdot 2} = \frac{35}{40} + \frac{2}{40} = \frac{37}{40}$$

44. The LCD of 12 and 9 is 36.
$$\frac{5}{12} - \frac{1}{9} = \frac{5 \cdot 3}{12 \cdot 3} - \frac{1 \cdot 4}{9 \cdot 4} = \frac{15}{36} - \frac{4}{36} = \frac{11}{36}$$

45. The LCD of 9, 18, and 3 is 18.
$$\frac{2}{9} + \frac{1}{18} + \frac{1}{3} = \frac{2 \cdot 2}{9 \cdot 2} + \frac{1}{18} + \frac{1 \cdot 6}{3 \cdot 6}$$
$$= \frac{4}{18} + \frac{1}{18} + \frac{6}{18}$$
$$= \frac{11}{18}$$

46. The LCD of 10, 5, and 25 is 50.
$$\frac{3y}{10} + \frac{y}{5} + \frac{6}{25} = \frac{3y \cdot 5}{10 \cdot 5} + \frac{y \cdot 10}{5 \cdot 10} + \frac{6 \cdot 2}{25 \cdot 2}$$
$$= \frac{15y}{50} + \frac{10y}{50} + \frac{12}{50}$$
$$= \frac{25y}{50} + \frac{12}{50}$$
$$= \frac{25y + 12}{50}$$

47. $\frac{2}{3}$ of a number translates as $\frac{2}{3} \cdot x$ or $\frac{2}{3}x$.

48. The quotient of a number and $-\frac{1}{5}$ translates as $x \div \left(-\frac{1}{5} \right)$.

49. A number subtracted from $-\frac{8}{9}$ translates as
$$-\frac{8}{9} - x.$$

50. $\frac{6}{11}$ increased by a number translates as
$$\frac{6}{11} + x.$$

51.
$$\frac{2}{3} \cdot 1530 = \frac{2}{3} \cdot \frac{1530}{1}$$
$$= \frac{2 \cdot 1530}{3 \cdot 1}$$
$$= \frac{2 \cdot 3 \cdot 510}{3 \cdot 1}$$
$$= \frac{2 \cdot 510}{1}$$
$$= 1020$$
$\frac{2}{3}$ of 1530 is 1020.

52.
$$18 \div \frac{3}{4} = \frac{18}{1} \div \frac{3}{4}$$
$$= \frac{18}{1} \cdot \frac{4}{3}$$
$$= \frac{18 \cdot 4}{1 \cdot 3}$$
$$= \frac{3 \cdot 6 \cdot 4}{1 \cdot 3}$$
$$= \frac{6 \cdot 4}{1}$$
$$= 24$$
He can sell 24 lots.

53. The LCD of 8 and 16 is 16.

$$\frac{7}{8} - \frac{1}{16} - \frac{1}{16} = \frac{7 \cdot 2}{8 \cdot 2} - \frac{1}{16} - \frac{1}{16}$$

$$= \frac{14}{16} - \frac{1}{16} - \frac{1}{16}$$

$$= \frac{12}{16}$$

$$= \frac{4 \cdot 3}{4 \cdot 4}$$

$$= \frac{3}{4}$$

The inner diameter is $\frac{3}{4}$ foot.

Section 4.6

Practice Problems

1. $\dfrac{\frac{7y}{10}}{\frac{1}{5}} = \dfrac{7y}{10} \div \dfrac{1}{5} = \dfrac{7y}{10} \cdot \dfrac{5}{1} = \dfrac{7y \cdot 5}{10 \cdot 1} = \dfrac{7 \cdot y \cdot 5}{2 \cdot 5 \cdot 1} = \dfrac{7y}{2}$

2. $\dfrac{\frac{1}{2} + \frac{1}{6}}{\frac{3}{4} - \frac{2}{3}} = \dfrac{\frac{1 \cdot 3}{2 \cdot 3} + \frac{1}{6}}{\frac{3 \cdot 3}{4 \cdot 3} - \frac{2 \cdot 4}{3 \cdot 4}}$

$$= \dfrac{\frac{3}{6} + \frac{1}{6}}{\frac{9}{12} - \frac{8}{12}}$$

$$= \dfrac{\frac{4}{6}}{\frac{1}{12}}$$

$$= \frac{4}{6} \div \frac{1}{12}$$

$$= \frac{4}{6} \cdot \frac{12}{1}$$

$$= \frac{4 \cdot 2 \cdot 6}{6 \cdot 1}$$

$$= \frac{8}{1} \text{ or } 8$$

3. The LCD of 2, 6, 4, and 3 is 12.

$$\dfrac{\frac{1}{2} + \frac{1}{6}}{\frac{3}{4} - \frac{2}{3}} = \dfrac{12\left(\frac{1}{2} + \frac{1}{6}\right)}{12\left(\frac{3}{4} - \frac{2}{3}\right)}$$

$$= \dfrac{12 \cdot \frac{1}{2} + 12 \cdot \frac{1}{6}}{12 \cdot \frac{3}{4} - 12 \cdot \frac{2}{3}}$$

$$= \frac{6 + 2}{9 - 8}$$

$$= \frac{8}{1} \text{ or } 8$$

4. The LCD of 4, 5, and 1 is 20.

$$\dfrac{\frac{3}{4}}{\frac{x}{5} - 1} = \dfrac{20\left(\frac{3}{4}\right)}{20\left(\frac{x}{5} - 1\right)} = \dfrac{20 \cdot \frac{3}{4}}{20 \cdot \frac{x}{5} - 20 \cdot 1} = \dfrac{15}{4x - 20}$$

5. $\left(\dfrac{2}{3}\right)^3 - 2 = \dfrac{8}{27} - 2 = \dfrac{8}{27} - \dfrac{54}{27} = -\dfrac{46}{27}$

6. $\left(-\dfrac{1}{2} + \dfrac{1}{5}\right)\left(\dfrac{7}{8} + \dfrac{1}{8}\right) = \left(-\dfrac{1 \cdot 5}{2 \cdot 5} + \dfrac{1 \cdot 2}{5 \cdot 2}\right)\left(\dfrac{7}{8} + \dfrac{1}{8}\right)$

$$= \left(-\dfrac{5}{10} + \dfrac{2}{10}\right)\left(\dfrac{7}{8} + \dfrac{1}{8}\right)$$

$$= \left(-\dfrac{3}{10}\right)\left(\dfrac{8}{8}\right)$$

$$= \left(-\dfrac{3}{10}\right)(1)$$

$$= -\dfrac{3}{10}$$

7. $-\dfrac{3}{5} - xy = -\dfrac{3}{5} - \dfrac{3}{10} \cdot \dfrac{2}{3}$

$$= -\dfrac{3}{5} - \dfrac{3 \cdot 2}{5 \cdot 2 \cdot 3}$$

$$= -\dfrac{3}{5} - \dfrac{1}{5}$$

$$= -\dfrac{4}{5}$$

Vocabulary and Readiness Check

1. A fraction whose numerator or denominator or both numerator and denominator contain fractions is called a <u>complex</u> fraction.

3. To simplify $-\dfrac{1}{2} \div \dfrac{2}{3} \cdot \dfrac{7}{8}$, which operation do we perform first? <u>division</u>

5. To simplify $\dfrac{1}{3} \div \dfrac{1}{4} \cdot \left(\dfrac{9}{11} + \dfrac{3}{8}\right)^2$, which operation do we perform first? <u>addition</u>

Exercise Set 4.6

1. $\dfrac{\frac{1}{8}}{\frac{3}{4}} = \dfrac{1}{8} \div \dfrac{3}{4} = \dfrac{1}{8} \cdot \dfrac{4}{3} = \dfrac{1 \cdot 4}{8 \cdot 3} = \dfrac{1 \cdot 4}{2 \cdot 4 \cdot 3} = \dfrac{1}{6}$

3. $\dfrac{\frac{2}{3}}{\frac{2}{7}} = \dfrac{2}{3} \div \dfrac{2}{7} = \dfrac{2}{3} \cdot \dfrac{7}{2} = \dfrac{2 \cdot 7}{3 \cdot 2} = \dfrac{7}{3}$

5. $\dfrac{\frac{2x}{27}}{\frac{4}{9}} = \dfrac{2x}{27} \div \dfrac{4}{9}$

   $= \dfrac{2x}{27} \cdot \dfrac{9}{4}$

   $= \dfrac{2x \cdot 9}{27 \cdot 4}$

   $= \dfrac{2 \cdot x \cdot 9}{3 \cdot 9 \cdot 2 \cdot 2}$

   $= \dfrac{x}{6}$

7. The LCD of 4, 5, and 2 is 20.

   $\dfrac{\frac{3}{4} + \frac{2}{5}}{\frac{1}{2} + \frac{3}{5}} = \dfrac{20 \cdot \left(\frac{3}{4} + \frac{2}{5}\right)}{20 \cdot \left(\frac{1}{2} + \frac{3}{5}\right)}$

   $= \dfrac{20 \cdot \frac{3}{4} + 20 \cdot \frac{2}{5}}{20 \cdot \frac{1}{2} + 20 \cdot \frac{3}{5}}$

   $= \dfrac{15 + 8}{10 + 12}$

   $= \dfrac{23}{22}$

9. The LCD of 4, 1, and 8 is 8.

   $\dfrac{\frac{3x}{4}}{5 - \frac{1}{8}} = \dfrac{8 \cdot \left(\frac{3x}{4}\right)}{8 \cdot \left(5 - \frac{1}{8}\right)}$

   $= \dfrac{6x}{8 \cdot 5 - 8 \cdot \frac{1}{8}}$

   $= \dfrac{6x}{40 - 1}$

   $= \dfrac{6x}{39}$

   $= \dfrac{2 \cdot 3 \cdot x}{3 \cdot 13}$

   $= \dfrac{2x}{13}$

11. $\dfrac{1}{5} + \dfrac{1}{3} \cdot \dfrac{1}{4} = \dfrac{1}{5} + \dfrac{1 \cdot 1}{3 \cdot 4}$

    $= \dfrac{1}{5} + \dfrac{1}{12}$

    $= \dfrac{1 \cdot 12}{5 \cdot 12} + \dfrac{1 \cdot 5}{12 \cdot 5}$

    $= \dfrac{12}{60} + \dfrac{5}{60}$

    $= \dfrac{17}{60}$

13. $\dfrac{5}{6} \div \dfrac{1}{3} \cdot \dfrac{1}{4} = \dfrac{5}{6} \cdot \dfrac{3}{1} \cdot \dfrac{1}{4}$

$\qquad = \dfrac{5 \cdot 3 \cdot 1}{2 \cdot 3 \cdot 1 \cdot 2 \cdot 2}$

$\qquad = \dfrac{5}{2 \cdot 2 \cdot 2}$

$\qquad = \dfrac{5}{8}$

15. $2^2 - \left(\dfrac{1}{3}\right)^2 = 4 - \dfrac{1}{9}$

$\qquad = \dfrac{4}{1} - \dfrac{1}{9}$

$\qquad = \dfrac{4 \cdot 9}{1 \cdot 9} - \dfrac{1}{9}$

$\qquad = \dfrac{36}{9} - \dfrac{1}{9}$

$\qquad = \dfrac{35}{9}$

17. $\left(\dfrac{2}{9} + \dfrac{4}{9}\right)\left(\dfrac{1}{3} - \dfrac{9}{10}\right) = \left(\dfrac{2+4}{9}\right)\left(\dfrac{1}{3} \cdot \dfrac{10}{10} - \dfrac{9}{10} \cdot \dfrac{3}{3}\right)$

$\qquad = \left(\dfrac{6}{9}\right)\left(\dfrac{10}{30} - \dfrac{27}{30}\right)$

$\qquad = \left(\dfrac{6}{9}\right)\left(\dfrac{-17}{30}\right)$

$\qquad = -\dfrac{6 \cdot 17}{9 \cdot 30}$

$\qquad = -\dfrac{6 \cdot 17}{3 \cdot 3 \cdot 5 \cdot 6}$

$\qquad = -\dfrac{17}{3 \cdot 3 \cdot 5}$

$\qquad = -\dfrac{17}{45}$

19. $\left(\dfrac{7}{8} - \dfrac{1}{2}\right) \div \dfrac{3}{11} = \left(\dfrac{7}{8} - \dfrac{1}{2} \cdot \dfrac{4}{4}\right) \div \dfrac{3}{11}$

$\qquad = \left(\dfrac{7}{8} - \dfrac{4}{8}\right) \div \dfrac{3}{11}$

$\qquad = \dfrac{3}{8} \div \dfrac{3}{11}$

$\qquad = \dfrac{3}{8} \cdot \dfrac{11}{3}$

$\qquad = \dfrac{3 \cdot 11}{8 \cdot 3}$

$\qquad = \dfrac{11}{8}$

21. $2 \cdot \left(\dfrac{1}{4} + \dfrac{1}{5}\right) + 2 = 2 \cdot \left(\dfrac{1 \cdot 5}{4 \cdot 5} + \dfrac{1 \cdot 4}{5 \cdot 4}\right) + 2$

$\qquad = 2\left(\dfrac{5}{20} + \dfrac{4}{20}\right) + 2$

$\qquad = 2\left(\dfrac{9}{20}\right) + 2$

$\qquad = \dfrac{2}{1} \cdot \dfrac{9}{20} + 2$

$\qquad = \dfrac{2 \cdot 9}{1 \cdot 2 \cdot 10} + 2$

$\qquad = \dfrac{9}{10} + \dfrac{2 \cdot 10}{1 \cdot 10}$

$\qquad = \dfrac{9}{10} + \dfrac{20}{10}$

$\qquad = \dfrac{29}{10}$

23. $\left(\dfrac{3}{4}\right)^2 \div \left(\dfrac{3}{4} - \dfrac{1}{12}\right) = \left(\dfrac{3}{4}\right)^2 \div \left(\dfrac{3\cdot 3}{4\cdot 3} - \dfrac{1}{12}\right)$

$= \left(\dfrac{3}{4}\right)^2 \div \left(\dfrac{9}{12} - \dfrac{1}{12}\right)$

$= \left(\dfrac{3}{4}\right)^2 \div \dfrac{8}{12}$

$= \dfrac{9}{16} \div \dfrac{8}{12}$

$= \dfrac{9}{16} \cdot \dfrac{12}{8}$

$= \dfrac{9\cdot 3\cdot 4}{4\cdot 4\cdot 8}$

$= \dfrac{27}{32}$

25. $\left(\dfrac{2}{5} - \dfrac{3}{10}\right)^2 = \left(\dfrac{2}{5}\cdot\dfrac{2}{2} - \dfrac{3}{10}\right)^2$

$= \left(\dfrac{4}{10} - \dfrac{3}{10}\right)^2$

$= \left(\dfrac{1}{10}\right)^2$

$= \dfrac{1}{10}\cdot\dfrac{1}{10}$

$= \dfrac{1}{100}$

27. $\left(\dfrac{3}{4} + \dfrac{1}{8}\right)^2 - \left(\dfrac{1}{2} + \dfrac{1}{8}\right)$

$= \left(\dfrac{3\cdot 2}{4\cdot 2} + \dfrac{1}{8}\right)^2 - \left(\dfrac{1\cdot 4}{2\cdot 4} + \dfrac{1}{8}\right)$

$= \left(\dfrac{6}{8} + \dfrac{1}{8}\right)^2 - \left(\dfrac{4}{8} + \dfrac{1}{8}\right)$

$= \left(\dfrac{7}{8}\right)^2 - \dfrac{5}{8}$

$= \dfrac{49}{64} - \dfrac{5}{8}$

$= \dfrac{49}{64} - \dfrac{5\cdot 8}{8\cdot 8}$

$= \dfrac{49}{64} - \dfrac{40}{64}$

$= \dfrac{9}{64}$

29. $5y - z = 5\left(\dfrac{2}{5}\right) - \dfrac{5}{6}$

$= 2 - \dfrac{5}{6}$

$= \dfrac{2}{1}\cdot\dfrac{6}{6} - \dfrac{5}{6}$

$= \dfrac{12}{6} - \dfrac{5}{6}$

$= \dfrac{7}{6}$

31. $\dfrac{x}{z} = \dfrac{-\frac{1}{3}}{\frac{5}{6}} = -\dfrac{1}{3} \div \dfrac{5}{6} = -\dfrac{1}{3}\cdot\dfrac{6}{5} = -\dfrac{1\cdot 3\cdot 2}{3\cdot 5} = -\dfrac{2}{5}$

33. $x^2 - yz = \left(-\dfrac{1}{3}\right)^2 - \left(\dfrac{2}{5}\right)\left(\dfrac{5}{6}\right)$

$\qquad = \left(-\dfrac{1}{3}\right)\left(-\dfrac{1}{3}\right) - \dfrac{2\cdot 5}{5\cdot 2\cdot 3}$

$\qquad = \dfrac{1}{9} - \dfrac{1}{3}$

$\qquad = \dfrac{1}{9} - \dfrac{1}{3}\cdot\dfrac{3}{3}$

$\qquad = \dfrac{1}{9} - \dfrac{3}{9}$

$\qquad = \dfrac{-2}{9}$

$\qquad = -\dfrac{2}{9}$

35. $\dfrac{\frac{5a}{24}}{\frac{1}{12}} = \dfrac{5a}{24} \div \dfrac{1}{12} = \dfrac{5a}{24}\cdot\dfrac{12}{1} = \dfrac{5\cdot a\cdot 12}{12\cdot 2\cdot 1} = \dfrac{5a}{2}$

37. $\left(\dfrac{3}{2}\right)^3 + \left(\dfrac{1}{2}\right)^3 = \dfrac{27}{8} + \dfrac{1}{8} = \dfrac{28}{8} = \dfrac{4\cdot 7}{4\cdot 2} = \dfrac{7}{2}$

39. $\left(-\dfrac{1}{2}\right)^2 + \dfrac{1}{5} = \dfrac{1}{4} + \dfrac{1}{5}$

$\qquad = \dfrac{1\cdot 5}{4\cdot 5} + \dfrac{1\cdot 4}{5\cdot 4}$

$\qquad = \dfrac{5}{20} + \dfrac{4}{20}$

$\qquad = \dfrac{9}{20}$

41. $\dfrac{2+\frac{1}{6}}{1-\frac{4}{3}} = \dfrac{6\left(2+\frac{1}{6}\right)}{6\left(1-\frac{4}{3}\right)}$

$\qquad = \dfrac{6\cdot 2 + 6\cdot\frac{1}{6}}{6\cdot 1 - 6\cdot\frac{4}{3}}$

$\qquad = \dfrac{12+1}{6-8}$

$\qquad = \dfrac{13}{-2}$

$\qquad = -\dfrac{13}{2}$

43. $\left(1-\dfrac{2}{5}\right)^2 = \left(\dfrac{5}{5}-\dfrac{2}{5}\right)^2 = \left(\dfrac{3}{5}\right)^2 = \dfrac{3}{5}\cdot\dfrac{3}{5} = \dfrac{9}{25}$

45. $\left(\dfrac{3}{4}-1\right)\left(\dfrac{1}{8}+\dfrac{1}{2}\right) = \left(\dfrac{3}{4}-\dfrac{4}{4}\right)\left(\dfrac{1}{8}+\dfrac{4}{8}\right)$

$\qquad = \left(-\dfrac{1}{4}\right)\left(\dfrac{5}{8}\right)$

$\qquad = -\dfrac{1\cdot 5}{4\cdot 8}$

$\qquad = -\dfrac{5}{32}$

47. $\left(-\dfrac{2}{9}-\dfrac{7}{9}\right)^4 = \left(-\dfrac{9}{9}\right)^4$

$\qquad = (-1)^4$

$\qquad = (-1)(-1)(-1)(-1)$

$\qquad = 1$

49. $\dfrac{\frac{1}{3}-\frac{5}{6}}{\frac{3}{4}+\frac{1}{2}} = \dfrac{12\left(\frac{1}{3}-\frac{5}{6}\right)}{12\left(\frac{3}{4}+\frac{1}{2}\right)}$

$\qquad = \dfrac{12\cdot\frac{1}{3} - 12\cdot\frac{5}{6}}{12\cdot\frac{3}{4} + 12\cdot\frac{1}{2}}$

$\qquad = \dfrac{4-10}{9+6}$

$\qquad = \dfrac{-6}{15}$

$\qquad = -\dfrac{2\cdot 3}{3\cdot 5}$

$\qquad = -\dfrac{2}{5}$

51. $\left(\dfrac{3}{4} \div \dfrac{6}{5}\right) - \left(\dfrac{3}{4} \cdot \dfrac{6}{5}\right) = \left(\dfrac{3}{4} \cdot \dfrac{5}{6}\right) - \left(\dfrac{3}{4} \cdot \dfrac{6}{5}\right)$

$$= \frac{3 \cdot 5}{4 \cdot 6} - \frac{3 \cdot 6}{4 \cdot 5}$$

$$= \frac{3 \cdot 5}{4 \cdot 2 \cdot 3} - \frac{3 \cdot 2 \cdot 3}{2 \cdot 2 \cdot 5}$$

$$= \frac{5}{8} - \frac{9}{10}$$

$$= \frac{5 \cdot 5}{8 \cdot 5} - \frac{9 \cdot 4}{10 \cdot 4}$$

$$= \frac{25}{40} - \frac{36}{40}$$

$$= -\frac{11}{40}$$

53. $\dfrac{\frac{x}{3}+2}{5+\frac{1}{3}} = \dfrac{3\left(\frac{x}{3}+2\right)}{3\left(5+\frac{1}{3}\right)}$

$$= \frac{3 \cdot \frac{x}{3} + 3 \cdot 2}{3 \cdot 5 + 3 \cdot \frac{1}{3}}$$

$$= \frac{x+6}{15+1}$$

$$= \frac{x+6}{16}$$

55. $3 + \dfrac{1}{2} = \dfrac{3}{1} \cdot \dfrac{2}{2} + \dfrac{1}{2} = \dfrac{6}{2} + \dfrac{1}{2} = \dfrac{7}{2}$ or $3\dfrac{1}{2}$

57. $9 - \dfrac{5}{6} = \dfrac{9}{1} \cdot \dfrac{6}{6} - \dfrac{5}{6} = \dfrac{54}{6} - \dfrac{5}{6} = \dfrac{49}{6}$ or $8\dfrac{1}{6}$

59. no; answers may vary

61. $\dfrac{\frac{1}{2}+\frac{3}{4}}{2} = \dfrac{4\left(\frac{1}{2}+\frac{3}{4}\right)}{4(2)} = \dfrac{4 \cdot \frac{1}{2} + 4 \cdot \frac{3}{4}}{4 \cdot 2} = \dfrac{2+3}{8} = \dfrac{5}{8}$

63. $\dfrac{\frac{1}{4}+\frac{2}{14}}{2} + \dfrac{28\left(\frac{1}{4}+\frac{2}{14}\right)}{28(2)} = \dfrac{28 \cdot \frac{1}{4} + 28 \cdot \frac{2}{14}}{28 \cdot 2}$

$$= \frac{7+4}{56}$$

$$= \frac{11}{56}$$

65. The average of a and b should be halfway between a and b.

67. False; the average cannot be less than the smallest number.

69. False, if the absolute value of the negative fraction is greater than the absolute value of the positive fraction, the sum is negative.

71. True

73. No; answers may vary.

75. Addition, multiplication, subtraction, division

77. Subtraction, multiplication, division, addition

79. $4x + y = 4 \cdot \dfrac{3}{4} + \left(-\dfrac{4}{7}\right)$

$$= \frac{4}{1} \cdot \frac{3}{4} + \left(-\frac{4}{7}\right)$$

$$= \frac{12}{4} - \frac{4}{7}$$

$$= \frac{12 \cdot 7}{4 \cdot 7} - \frac{4 \cdot 4}{7 \cdot 4}$$

$$= \frac{84}{28} - \frac{16}{28}$$

$$= \frac{68}{28}$$

$$= \frac{4 \cdot 17}{4 \cdot 7}$$

$$= \frac{17}{7}$$

81. $\dfrac{\frac{9}{14}}{x+y} = \dfrac{\frac{9}{14}}{\frac{3}{4}+\left(-\frac{4}{7}\right)}$

$\qquad = \dfrac{28\left(\frac{9}{14}\right)}{28\left(\frac{3}{4}-\frac{4}{7}\right)}$

$\qquad = \dfrac{18}{28\cdot\frac{3}{4}-28\cdot\frac{4}{7}}$

$\qquad = \dfrac{18}{21-16}$

$\qquad = \dfrac{18}{5}$

Section 4.7

Practice Problems

1.

2. $1\dfrac{2}{3}\cdot\dfrac{11}{15} = \dfrac{5}{3}\cdot\dfrac{11}{15} = \dfrac{5\cdot11}{3\cdot5\cdot3} = \dfrac{11}{9}$ or $1\dfrac{2}{9}$

3. $\dfrac{5}{6}\cdot18 = \dfrac{5}{6}\cdot\dfrac{18}{1} = \dfrac{5\cdot18}{6\cdot1} = \dfrac{5\cdot6\cdot3}{6\cdot1} = \dfrac{15}{1}$ or 15

4. $3\dfrac{1}{5}\cdot2\dfrac{3}{4} = \dfrac{16}{5}\cdot\dfrac{11}{4}$

$\qquad = \dfrac{16\cdot11}{5\cdot4}$

$\qquad = \dfrac{4\cdot4\cdot11}{5\cdot4}$

$\qquad = \dfrac{44}{5}$ or $8\dfrac{4}{5}$

Estimate: $3\dfrac{1}{5}$ rounds to 3, $2\dfrac{3}{4}$ rounds to 3.

$3 \cdot 3 = 9$, so the answer is reasonable.

5. $3\cdot6\dfrac{7}{15} = \dfrac{3}{1}\cdot\dfrac{97}{15} = \dfrac{3\cdot97}{1\cdot5\cdot3} = \dfrac{97}{5}$ or $19\dfrac{2}{5}$

Estimate: $6\dfrac{7}{15}$ rounds to 6 and $3 \cdot 6 = 18$, so the answer is reasonable.

6. $\dfrac{4}{9}\div7 = \dfrac{4}{9}\div\dfrac{7}{1} = \dfrac{4}{9}\cdot\dfrac{1}{7} = \dfrac{4\cdot1}{9\cdot7} = \dfrac{4}{63}$

7. $\dfrac{8}{15}\div3\dfrac{4}{5} = \dfrac{8}{15}\div\dfrac{19}{5}$

$\qquad = \dfrac{8}{15}\cdot\dfrac{5}{19}$

$\qquad = \dfrac{8\cdot5}{15\cdot19}$

$\qquad = \dfrac{8\cdot5}{5\cdot3\cdot19}$

$\qquad = \dfrac{8}{57}$

8. $3\dfrac{2}{7}\div2\dfrac{3}{14} = \dfrac{23}{7}\div\dfrac{31}{14}$

$\qquad = \dfrac{23}{7}\cdot\dfrac{14}{31}$

$\qquad = \dfrac{23\cdot7\cdot2}{7\cdot31}$

$\qquad = \dfrac{46}{31}$ or $1\dfrac{15}{31}$

9. $\begin{array}{r} 2\dfrac{1}{6} \\ +\ 4\dfrac{2}{5} \\ \hline \end{array}$ $\qquad \begin{array}{r} 2\dfrac{5}{30} \\ +\ 4\dfrac{12}{30} \\ \hline 6\dfrac{17}{30} \end{array}$

Estimate: $2\dfrac{1}{6}$ rounds to 2, $4\dfrac{2}{5}$ rounds to 4, and $2 + 4 = 6$, so the answer is reasonable.

10. $\begin{array}{r} 3\dfrac{5}{14} \\ +\ 2\dfrac{6}{7} \\ \hline \end{array}$ $\qquad \begin{array}{r} 3\dfrac{5}{14} \\ +\ 2\dfrac{12}{14} \\ \hline 5\dfrac{17}{14} = 5+1\dfrac{3}{14} = 6\dfrac{3}{14} \end{array}$

11.

$$\begin{array}{r} 12 \\ 3\dfrac{6}{7} \\ + \ 2\dfrac{1}{5} \\ \hline \end{array}$$

$$\begin{array}{r} 12 \\ 3\dfrac{30}{35} \\ + \ 2\dfrac{7}{35} \\ \hline 17\dfrac{37}{35} = 17 + 1\dfrac{2}{35} = 18\dfrac{2}{35} \end{array}$$

12.

$$\begin{array}{r} 32\dfrac{7}{9} \\ - \ 16\dfrac{5}{18} \\ \hline \end{array}$$

$$\begin{array}{r} 32\dfrac{14}{18} \\ - \ 16\dfrac{5}{18} \\ \hline 16\dfrac{9}{18} = 16\dfrac{1}{2} \end{array}$$

13.

$$\begin{array}{r} 9\dfrac{7}{15} \\ - \ 4\dfrac{3}{5} \\ \hline \end{array}$$

$$\begin{array}{r} 9\dfrac{7}{15} \\ - \ 4\dfrac{9}{15} \\ \hline \end{array}$$

$$\begin{array}{r} 8\dfrac{22}{15} \\ - \ 4\dfrac{9}{15} \\ \hline 4\dfrac{13}{15} \end{array}$$

14.

$$\begin{array}{r} 25 \\ - \ 10\dfrac{2}{9} \\ \hline \end{array}$$

$$\begin{array}{r} 24\dfrac{9}{9} \\ - \ 10\dfrac{2}{9} \\ \hline 14\dfrac{7}{9} \end{array}$$

15.

$$\begin{array}{r} 23\dfrac{1}{4} \\ - \ 19\dfrac{5}{12} \\ \hline \end{array}$$

$$\begin{array}{r} 23\dfrac{3}{12} \\ - \ 19\dfrac{5}{12} \\ \hline \end{array}$$

$$\begin{array}{r} 22\dfrac{15}{12} \\ - \ 19\dfrac{5}{12} \\ \hline 3\dfrac{10}{12} = 3\dfrac{5}{6} \end{array}$$

The girth of the largest known American beech tree is $3\dfrac{5}{6}$ feet larger than the girth of the largest known sugar maple tree.

16. $44 \div 3\dfrac{1}{7} = 44 \div \dfrac{22}{7}$

$$= \dfrac{44}{1} \cdot \dfrac{7}{22}$$

$$= \dfrac{2 \cdot 22 \cdot 7}{1 \cdot 22}$$

$$= \dfrac{14}{1} \text{ or } 14$$

Therefore, 14 dresses can be made from 44 yards.

17. $-9\dfrac{3}{7} = -\dfrac{7 \cdot 9 + 3}{7} = -\dfrac{66}{7}$

18. $-5\dfrac{10}{11} = -\dfrac{11 \cdot 5 + 10}{11} = -\dfrac{65}{11}$

19. $-\dfrac{37}{8} = -4\dfrac{5}{8}$ $8\overline{)37}$ with quotient 4, $\dfrac{32}{5}$

20. $-\dfrac{46}{5} = -9\dfrac{1}{5}$ $5\overline{)46}$ with quotient 9, $\dfrac{45}{1}$

21. $2\dfrac{3}{4} \cdot \left(-3\dfrac{3}{5}\right) = \dfrac{11}{4} \cdot \left(-\dfrac{18}{5}\right)$

$$= -\dfrac{11 \cdot 18}{4 \cdot 5}$$

$$= -\dfrac{11 \cdot 2 \cdot 9}{2 \cdot 2 \cdot 5}$$

$$= -\dfrac{99}{10} \text{ or } -9\dfrac{9}{10}$$

22. $-4\dfrac{2}{7} \div 1\dfrac{1}{4} = -\dfrac{30}{7} \div \dfrac{5}{4}$

$\phantom{-4\dfrac{2}{7} \div 1\dfrac{1}{4}} = -\dfrac{30}{7} \cdot \dfrac{4}{5}$

$\phantom{-4\dfrac{2}{7} \div 1\dfrac{1}{4}} = -\dfrac{30 \cdot 4}{7 \cdot 5}$

$\phantom{-4\dfrac{2}{7} \div 1\dfrac{1}{4}} = -\dfrac{5 \cdot 6 \cdot 4}{7 \cdot 5}$

$\phantom{-4\dfrac{2}{7} \div 1\dfrac{1}{4}} = -\dfrac{24}{7}$ or $-3\dfrac{3}{7}$

23. $\begin{array}{r} 12\dfrac{3}{4} \\ -\ 6\dfrac{2}{3} \\ \hline \end{array}$ $\begin{array}{r} 12\dfrac{9}{12} \\ -\ 6\dfrac{8}{12} \\ \hline 6\dfrac{1}{12} \end{array}$

$6\dfrac{2}{3} - 12\dfrac{3}{4} = -6\dfrac{1}{12}$

24. $\begin{array}{r} 9\dfrac{2}{7} \\ +\ 30\dfrac{11}{14} \\ \hline \end{array}$ $\begin{array}{r} 9\dfrac{4}{14} \\ +\ 30\dfrac{11}{14} \\ \hline 39\dfrac{15}{14} = 40\dfrac{1}{14} \end{array}$

$-9\dfrac{2}{7} - 30\dfrac{11}{14} = -40\dfrac{1}{14}$

Calculator Explorations

1. $25\dfrac{5}{11} = \dfrac{280}{11}$

3. $107\dfrac{31}{35} = \dfrac{3776}{35}$

5. $\dfrac{365}{14} = 26\dfrac{1}{14}$

7. $\dfrac{2769}{30} = 92\dfrac{3}{10}$

Vocabulary and Readiness Check

1. The number $5\dfrac{3}{4}$ is called a <u>mixed number</u>.

3. To estimate operations on mixed numbers, we <u>round</u> mixed numbers to the nearest whole number.

Exercise Set 4.7

1.

3.

5. $2\dfrac{11}{12}$ rounds to 3.

$1\dfrac{1}{4}$ rounds to 1.

$3 \cdot 1 = 3$

The best estimate is b.

7. $12\dfrac{2}{11}$ rounds to 12.

$3\dfrac{9}{10}$ rounds to 4.

$12 \div 4 = 3$

The best estimate is a.

9. $2\dfrac{2}{3} \cdot \dfrac{1}{7} = \dfrac{8}{3} \cdot \dfrac{1}{7} = \dfrac{8}{21}$

11. $7 \div 1\dfrac{3}{5} = \dfrac{7}{1} \div \dfrac{8}{5} = \dfrac{7}{1} \cdot \dfrac{5}{8} = \dfrac{7 \cdot 5}{1 \cdot 8} = \dfrac{35}{8}$ or $4\dfrac{3}{8}$

13. Exact: $2\dfrac{1}{5} \cdot 3\dfrac{1}{2} = \dfrac{11}{5} \cdot \dfrac{7}{2} = \dfrac{77}{10}$ or $7\dfrac{7}{10}$

Estimate: $2\dfrac{1}{5}$ rounds to 2, $3\dfrac{1}{2}$ rounds to 4.

$2 \cdot 4 = 8$ so the answer is reasonable.

15. Exact: $3\frac{4}{5} \cdot 6\frac{2}{7} = \frac{19}{5} \cdot \frac{44}{7} = \frac{836}{35}$ or $23\frac{31}{35}$

Estimate: $3\frac{4}{5}$ rounds to 4, $6\frac{2}{7}$ rounds to 6.

$4 \cdot 6 = 24$

17. $5 \cdot 2\frac{1}{2} = \frac{5}{1} \cdot \frac{5}{2} = \frac{25}{2}$ or $12\frac{1}{2}$

19. $3\frac{2}{3} \cdot 1\frac{1}{2} = \frac{11}{3} \cdot \frac{3}{2} = \frac{11 \cdot 3}{3 \cdot 2} = \frac{11}{2}$ or $5\frac{1}{2}$

21. $2\frac{2}{3} \div \frac{1}{7} = \frac{8}{3} \cdot \frac{7}{1} = \frac{56}{3}$ or $18\frac{2}{3}$

23. $3\frac{7}{8}$ rounds to 4.

$2\frac{1}{5}$ rounds to 2.

$4 + 2 = 6$

The best estimate is a.

25. $8\frac{1}{3}$ rounds to 8.

$1\frac{1}{2}$ rounds to 2.

$8 + 2 = 10$

The best estimate is b.

27. Exact:

$$4\frac{7}{12}$$
$$+\ 2\frac{1}{12}$$
$$\overline{6\frac{8}{12} = 6\frac{2}{3}}$$

Estimate: $4\frac{7}{12}$ rounds to 5, $2\frac{1}{12}$ rounds to 2. $5 + 2 = 7$ so the answer is reasonable.

29. Exact:

$$10\frac{3}{14} \qquad\qquad 10\frac{3}{14}$$
$$+\ 3\frac{4}{7} \qquad\qquad +\ 3\frac{8}{14}$$
$$\overline{} \qquad \overline{13\frac{11}{14}}$$

Estimate: $10\frac{3}{14}$ rounds to 10, $3\frac{4}{7}$ rounds to 4. $10 + 4 = 14$ so the answer is reasonable.

31.
$$9\frac{1}{5} \qquad\qquad 9\frac{5}{25}$$
$$+\ 8\frac{2}{25} \qquad\qquad +\ 8\frac{2}{25}$$
$$\overline{} \qquad \overline{17\frac{7}{25}}$$

33.
$$12\frac{3}{14} \qquad\qquad 12\frac{18}{84}$$
$$10 \qquad\qquad\qquad 10$$
$$+\ 25\frac{5}{12} \qquad\qquad +\ 25\frac{35}{84}$$
$$\overline{} \qquad \overline{47\frac{53}{84}}$$

35.
$$15\frac{4}{7} \qquad\qquad 15\frac{8}{14}$$
$$+\ 9\frac{11}{14} \qquad\qquad +\ 9\frac{11}{14}$$
$$\overline{} \qquad \overline{24\frac{19}{14} = 24 + 1\frac{5}{14} = 25\frac{5}{14}}$$

37.
$$3\frac{5}{8} \qquad\qquad 3\frac{15}{24}$$
$$2\frac{1}{6} \qquad\qquad 2\frac{4}{24}$$
$$+\ 7\frac{3}{4} \qquad\qquad +\ 7\frac{18}{24}$$
$$\overline{} \qquad \overline{12\frac{37}{24} = 12 + 1\frac{13}{24} = 13\frac{13}{24}}$$

39. Exact: $4\dfrac{7}{10}$

$\qquad\quad -\,2\dfrac{1}{10}$

$\qquad\qquad\quad 2\dfrac{6}{10}=2\dfrac{3}{5}$

Estimate: $4\dfrac{7}{10}$ rounds to 5, $2\dfrac{1}{10}$ rounds to

2. $5-2=3$ so the answer is reasonable.

41. Exact: $10\dfrac{13}{14}$ $10\dfrac{13}{14}$

$\qquad\quad -\,3\dfrac{4}{7}$ $-\,3\dfrac{8}{14}$

$\qquad\qquad\qquad\qquad\qquad\quad 7\dfrac{5}{14}$

Estimate: $10\dfrac{13}{14}$ rounds to 11, $3\dfrac{4}{7}$ rounds to

4. $11-4=7$ so the answer is reasonable.

43. $9\dfrac{1}{5}$ $9\dfrac{5}{25}$ $8\dfrac{30}{25}$

$\quad-\,8\dfrac{6}{25}$ $-\,8\dfrac{6}{25}$ $-\,8\dfrac{6}{25}$

$\qquad\qquad\qquad\qquad\qquad\qquad\qquad\quad \dfrac{24}{25}$

45. 6 $5\dfrac{9}{9}$

$\quad-\,2\dfrac{4}{9}$ $-\,2\dfrac{4}{9}$

$\qquad\qquad\qquad\qquad\qquad 3\dfrac{5}{9}$

47. $63\dfrac{1}{6}$ $63\dfrac{2}{12}$ $62\dfrac{14}{12}$

$\quad-\,47\dfrac{5}{12}$ $-\,47\dfrac{5}{12}$ $-\,47\dfrac{5}{12}$

$\qquad\qquad\qquad\qquad\qquad\qquad\quad 15\dfrac{9}{12}=15\dfrac{3}{4}$

49. $2\dfrac{3}{4}$

$\quad+\,1\dfrac{1}{4}$

$\qquad\quad 3\dfrac{4}{4}=3+1=4$

51. $15\dfrac{4}{7}$ $15\dfrac{8}{14}$ $14\dfrac{22}{14}$

$\quad-\,9\dfrac{11}{14}$ $-\,9\dfrac{11}{14}$ $-\,9\dfrac{11}{14}$

$\qquad\qquad\qquad\qquad\qquad\qquad\quad 5\dfrac{11}{14}$

53. $3\dfrac{1}{9}\cdot 2=\dfrac{28}{9}\cdot\dfrac{2}{1}=\dfrac{56}{9}=6\dfrac{2}{9}$

55. $1\dfrac{2}{3}\div 2\dfrac{1}{5}=\dfrac{5}{3}\div\dfrac{11}{5}=\dfrac{5}{3}\cdot\dfrac{5}{11}=\dfrac{25}{33}$

57. $22\dfrac{4}{9}+13\dfrac{5}{18}=22\dfrac{8}{18}+13\dfrac{5}{18}=35\dfrac{13}{18}$

59. $5\dfrac{2}{3}-3\dfrac{1}{6}=5\dfrac{4}{6}-3\dfrac{1}{6}=2\dfrac{3}{6}=2\dfrac{1}{2}$

61. $15\dfrac{1}{5}$ $15\dfrac{6}{30}$

$\quad 20\dfrac{3}{10}$ $20\dfrac{9}{30}$

$\quad+\,37\dfrac{2}{15}$ $+\,37\dfrac{4}{30}$

$\qquad\qquad\qquad\qquad\quad 72\dfrac{19}{30}$

63. $6\dfrac{4}{7}$ $6\dfrac{8}{14}$ $5\dfrac{22}{14}$

$\quad-\,5\dfrac{11}{14}$ $-\,5\dfrac{11}{14}$ $-\,5\dfrac{11}{14}$

$\qquad\qquad\qquad\qquad\qquad\qquad\quad \dfrac{11}{14}$

65. $4\dfrac{2}{7} \cdot 1\dfrac{3}{10} = \dfrac{30}{7} \cdot \dfrac{13}{10}$

$\qquad = \dfrac{30 \cdot 13}{7 \cdot 10}$

$\qquad = \dfrac{3 \cdot 10 \cdot 13}{7 \cdot 10}$

$\qquad = \dfrac{39}{7}$

$\qquad = 5\dfrac{4}{7}$

67.
$$
\begin{array}{r}
6\dfrac{2}{11} \\
3 \\
+\,4\dfrac{10}{33} \\
\hline
\end{array}
\qquad
\begin{array}{r}
6\dfrac{6}{33} \\
3 \\
+\,4\dfrac{10}{33} \\
\hline
13\dfrac{16}{33}
\end{array}
$$

69. $-5\dfrac{2}{7}$ decreased by a number translates as

$-5\dfrac{2}{7} - x.$

71. Multiply $1\dfrac{9}{10}$ by a number translates as

$1\dfrac{9}{10} \cdot x.$

73. Subtract the standard gauge in the U.S. from the standard gauge in Spain.

$$
\begin{array}{r}
65\dfrac{9}{10} \\
-\,56\dfrac{1}{2} \\
\hline
\end{array}
\qquad
\begin{array}{r}
65\dfrac{9}{10} \\
-\,56\dfrac{5}{10} \\
\hline
9\dfrac{4}{10} = 9\dfrac{2}{5}
\end{array}
$$

Spain's standard gauge is $9\dfrac{2}{5}$ inches wider than the U.S. standard gauge.

75.
$$
\begin{array}{r}
11\dfrac{1}{4} \\
-\,3\dfrac{3}{5} \\
\hline
\end{array}
\qquad
\begin{array}{r}
11\dfrac{5}{20} \\
-\,3\dfrac{12}{20} \\
\hline
\end{array}
\qquad
\begin{array}{r}
10\dfrac{25}{20} \\
-\,3\dfrac{12}{20} \\
\hline
7\dfrac{13}{20}
\end{array}
$$

Tucson gets an average of $7\dfrac{13}{20}$ inches more rain than Yuma.

77. $2 \cdot 1\dfrac{3}{4} = \dfrac{2}{1} \cdot \dfrac{7}{4} = \dfrac{2 \cdot 7}{1 \cdot 4} = \dfrac{2 \cdot 7}{1 \cdot 2 \cdot 2} = \dfrac{7}{2}$ or $3\dfrac{1}{2}$

The area is $\dfrac{7}{2}$ or $3\dfrac{1}{2}$ square yards.

79. $\dfrac{3}{4} \cdot 1\dfrac{1}{4} = \dfrac{3}{4} \cdot \dfrac{5}{4} = \dfrac{3 \cdot 5}{4 \cdot 4} = \dfrac{15}{16}$

The area is $\dfrac{15}{16}$ square inch.

81.
$$
\begin{array}{r}
5\dfrac{1}{3} \\
5 \\
7\dfrac{7}{8} \\
+\,3 \\
\hline
\end{array}
\qquad
\begin{array}{r}
5\dfrac{8}{24} \\
5 \\
7\dfrac{21}{24} \\
+\,3 \\
\hline
20\dfrac{29}{24} = 20 + 1\dfrac{5}{24} = 21\dfrac{5}{24}
\end{array}
$$

The perimeter is $21\dfrac{5}{24}$ meters.

83. $15\dfrac{2}{3} - \left(3\dfrac{1}{4} + 2\dfrac{1}{2}\right) = 15\dfrac{2}{3} - \left(3\dfrac{1}{4} + 2\dfrac{2}{4}\right)$

$\qquad\qquad\qquad\qquad = 15\dfrac{2}{3} - 5\dfrac{3}{4}$

$$
\begin{array}{r}
15\dfrac{2}{3} \\
-\,5\dfrac{3}{4} \\
\hline
\end{array}
\qquad
\begin{array}{r}
15\dfrac{8}{12} \\
-\,5\dfrac{9}{12} \\
\hline
\end{array}
\qquad
\begin{array}{r}
14\dfrac{20}{12} \\
-\,5\dfrac{9}{12} \\
\hline
9\dfrac{11}{12}
\end{array}
$$

No; the remaining board is $9\dfrac{11}{12}$ feet which is $\dfrac{1}{12}$ foot short.

85. $12\dfrac{3}{4} \div 4 = \dfrac{51}{4} \div \dfrac{4}{1} = \dfrac{51}{4} \cdot \dfrac{1}{4} = \dfrac{51}{16}$

The patient walked $\dfrac{51}{16} = 3\dfrac{3}{16}$ miles per day.

87. $12 \div 2\dfrac{4}{7} = \dfrac{12}{1} \div \dfrac{18}{7}$

$\qquad = \dfrac{12}{1} \cdot \dfrac{7}{18}$

$\qquad = \dfrac{6 \cdot 2 \cdot 7}{1 \cdot 6 \cdot 3}$

$\qquad = \dfrac{14}{3}$

$\qquad = 4\dfrac{2}{3}$

The length is $4\dfrac{2}{3}$ meters.

89.

$$
\begin{array}{ll}
2\dfrac{9}{20} & 2\dfrac{27}{60} \\[2mm]
2\dfrac{2}{3} & 2\dfrac{40}{60} \\[2mm]
+\,4\dfrac{7}{15} & +\,4\dfrac{28}{60} \\[1mm]
\hline
 & 8\dfrac{95}{60} = 8\dfrac{19}{12} = 8 + 1\dfrac{7}{12} = 9\dfrac{7}{12}
\end{array}
$$

The total duration of the eclipses is

$9\dfrac{7}{12}$ minutes.

91.

$$
\begin{array}{lll}
4\dfrac{7}{15} & 4\dfrac{7}{15} & 3\dfrac{22}{15} \\[2mm]
-\,2\dfrac{2}{3} & -\,2\dfrac{10}{15} & -\,2\dfrac{10}{15} \\[1mm]
\hline
 & & 1\dfrac{12}{15} = 1\dfrac{4}{5}
\end{array}
$$

It will be $1\dfrac{4}{5}$ minutes longer.

93. $-4\dfrac{2}{5} \cdot 2\dfrac{3}{10} = -\dfrac{22}{5} \cdot \dfrac{23}{10}$

$\qquad = -\dfrac{2 \cdot 11 \cdot 23}{5 \cdot 2 \cdot 5}$

$\qquad = -\dfrac{253}{25}$

$\qquad = -10\dfrac{3}{25}$

95. $-5\dfrac{1}{8} - 19\dfrac{3}{4} = -\left(5\dfrac{1}{8} + 19\dfrac{3}{4}\right)$

$\qquad = -\left(5\dfrac{1}{8} + 19\dfrac{6}{8}\right)$

$\qquad = -\left(24\dfrac{7}{8}\right)$

$\qquad = -24\dfrac{7}{8}$

97. $-31\dfrac{2}{15} + 17\dfrac{3}{20} = -31\dfrac{8}{60} + 17\dfrac{9}{60}$

$\qquad = -30\dfrac{68}{60} + 17\dfrac{9}{60}$

$\qquad = -13\dfrac{59}{60}$

99. $1\dfrac{3}{4} \div \left(-3\dfrac{1}{2}\right) = \dfrac{7}{4} \div \left(-\dfrac{7}{2}\right)$

$\qquad = \dfrac{7}{4} \cdot \left(-\dfrac{2}{7}\right)$

$\qquad = -\dfrac{7 \cdot 2}{4 \cdot 7}$

$\qquad = -\dfrac{2}{4}$

$\qquad = -\dfrac{1}{2}$

101. $11\frac{7}{8} - 13\frac{5}{6} = 11\frac{21}{24} - 13\frac{20}{24}$

$\qquad = -\left(13\frac{20}{24} - 11\frac{21}{24}\right)$

$\qquad = -\left(12\frac{44}{24} - 11\frac{21}{24}\right)$

$\qquad = -1\frac{23}{24}$

103. $-7\frac{3}{10} \div (-100) = -\frac{73}{10} \div \left(-\frac{100}{1}\right)$

$\qquad = -\frac{73}{10} \cdot \left(-\frac{1}{100}\right)$

$\qquad = \frac{73 \cdot 1}{10 \cdot 100}$

$\qquad = \frac{73}{1000}$

105. $\frac{1}{3}(3x) = \left(\frac{1}{3} \cdot 3\right)x = 1 \cdot x = x$

107. $\frac{2}{3}\left(\frac{3}{2}a\right) = \left(\frac{2}{3} \cdot \frac{3}{2}\right)a = 1 \cdot a = a$

109. a. $9\frac{5}{5} = 9 + 1 = 10$

b. $9\frac{100}{100} = 9 + 1 = 10$

c. $6\frac{44}{11} = 6 + 4 = 10$

d. $8\frac{13}{13} = 8 + 1 = 9$

a, b, and c are equivalent to 10.

111. Incorrect, to divide mixed number, first write each mixed number as an improper fraction.

113. answers may vary

115. Incorrect; $3\frac{2}{3} \cdot 1\frac{1}{7} = \frac{11}{3} \cdot \frac{8}{7} = \frac{88}{21}$ or $4\frac{4}{21}$

117. answers may vary

The Bigger Picture

1. $\frac{3}{17} + \frac{2}{17} = \frac{3+2}{17} = \frac{5}{17}$

2. $\frac{9}{10} - \frac{1}{10} = \frac{9-1}{10} = \frac{8}{10} = \frac{4}{5}$

3. $\frac{2}{3} + \frac{3}{10} = \frac{2 \cdot 10}{3 \cdot 10} + \frac{3 \cdot 3}{10 \cdot 3} = \frac{20}{30} + \frac{9}{30} = \frac{29}{30}$

4. $\frac{23}{24} - \frac{11}{12} = \frac{23}{24} - \frac{11 \cdot 2}{12 \cdot 2} = \frac{23}{24} - \frac{22}{24} = \frac{1}{24}$

5. $\frac{7}{8} + \frac{19}{20} = \frac{7 \cdot 5}{8 \cdot 5} + \frac{19 \cdot 2}{20 \cdot 2} = \frac{35}{40} + \frac{38}{40} = \frac{73}{40} = 1\frac{33}{40}$

6. $\frac{3^3}{4^3} = \frac{3 \cdot 3 \cdot 3}{4 \cdot 4 \cdot 4} = \frac{27}{64}$

7. $\begin{array}{r} 16 \\ -\ 3\frac{4}{7} \\ \hline \end{array}$ $\begin{array}{r} 15\frac{7}{7} \\ -\ 3\frac{4}{7} \\ \hline 12\frac{3}{7} \end{array}$

8. $\begin{array}{r} 2\frac{5}{8} \\ 1\frac{1}{6} \\ +\ 5\frac{3}{4} \\ \hline \end{array}$ $\begin{array}{r} 2\frac{15}{24} \\ 1\frac{4}{24} \\ +\ 5\frac{18}{24} \\ \hline 8\frac{37}{24} = 8 + 1\frac{13}{24} = 9\frac{13}{24} \end{array}$

9. $\frac{6}{11} \cdot \frac{8}{9} = \frac{6 \cdot 8}{11 \cdot 9} = \frac{2 \cdot 3 \cdot 8}{11 \cdot 3 \cdot 3} = \frac{2 \cdot 8}{11 \cdot 3} = \frac{16}{33}$

10. $2\dfrac{4}{15} \div 1\dfrac{4}{5} = \dfrac{34}{15} \div \dfrac{9}{5}$

$\qquad\qquad = \dfrac{34}{15} \cdot \dfrac{5}{9}$

$\qquad\qquad = \dfrac{34 \cdot 5}{3 \cdot 5 \cdot 9}$

$\qquad\qquad = \dfrac{34}{27}$

$\qquad\qquad = 1\dfrac{7}{27}$

Section 4.8

Practice Problems

1. $\quad y - \dfrac{2}{3} = \dfrac{5}{12}$

$\quad y - \dfrac{2}{3} + \dfrac{2}{3} = \dfrac{5}{12} + \dfrac{2}{3}$

$\qquad\quad y = \dfrac{5}{12} + \dfrac{2 \cdot 4}{3 \cdot 4}$

$\qquad\quad y = \dfrac{5}{12} + \dfrac{8}{12}$

$\qquad\quad y = \dfrac{13}{12}$

Check: $\quad y - \dfrac{2}{3} = \dfrac{5}{12}$

$\qquad\quad \dfrac{13}{12} - \dfrac{2}{3} \overset{?}{=} \dfrac{5}{12}$

$\qquad\quad \dfrac{13}{12} - \dfrac{2 \cdot 4}{3 \cdot 4} \overset{?}{=} \dfrac{5}{12}$

$\qquad\quad \dfrac{13}{12} - \dfrac{8}{12} \overset{?}{=} \dfrac{5}{12}$

$\qquad\qquad\quad \dfrac{5}{12} = \dfrac{5}{12}$ True

The solution is $\dfrac{13}{12}$.

2. $\quad \dfrac{1}{5}y = 2$

$\quad 5 \cdot \dfrac{1}{5}y = 5 \cdot 2$

$\qquad\quad y = 10$

Check: $\quad \dfrac{1}{5}y = 2$

$\qquad\quad \dfrac{1}{5} \cdot 10 \overset{?}{=} 2$

$\qquad\qquad\quad 2 = 2$ True

The solution is 10.

3. $\quad \dfrac{5}{7}b = 25$

$\quad \dfrac{7}{5} \cdot \dfrac{5}{7}b = \dfrac{7}{5} \cdot 25$

$\qquad\quad 1b = \dfrac{7 \cdot 25}{5}$

$\qquad\quad b = 35$

Check: $\quad \dfrac{5}{7}b = 25$

$\qquad\quad \dfrac{5}{7} \cdot 35 \overset{?}{=} 25$

$\qquad\qquad\quad 25 = 25$ True

The solution is 35.

4. $\quad -\dfrac{7}{10}x = \dfrac{2}{5}$

$\quad -\dfrac{10}{7} \cdot \dfrac{7}{10}x = -\dfrac{10}{7} \cdot \dfrac{2}{5}$

$\qquad\qquad x = -\dfrac{10 \cdot 2}{7 \cdot 5}$

$\qquad\qquad x = -\dfrac{4}{7}$

Check: $\quad -\dfrac{7}{10}x = \dfrac{2}{5}$

$\qquad\quad -\dfrac{7}{10} \cdot -\dfrac{4}{7} \overset{?}{=} \dfrac{2}{5}$

$\qquad\qquad\quad \dfrac{28}{70} \overset{?}{=} \dfrac{2}{5}$

$\qquad\qquad\quad \dfrac{2}{5} = \dfrac{2}{5}$ True

The solution is $-\dfrac{4}{7}$.

5.

$$5x = -\frac{3}{4}$$

$$\frac{1}{5} \cdot 5x = \frac{1}{5} \cdot -\frac{3}{4}$$

$$x = -\frac{1 \cdot 3}{5 \cdot 4}$$

$$x = -\frac{3}{20}$$

Check:

$$5x = -\frac{3}{4}$$

$$5 \cdot -\frac{3}{20} \stackrel{?}{=} -\frac{3}{4}$$

$$-\frac{15}{20} \stackrel{?}{=} -\frac{3}{4}$$

$$-\frac{3}{4} = -\frac{3}{4} \quad \text{True}$$

The solution is $-\dfrac{3}{20}$.

6. The LCD of 15 and 5 is 15.

$$\frac{11}{15}x = -\frac{3}{5}$$

$$15 \cdot \frac{11}{15}x = 15 \cdot -\frac{3}{5}$$

$$11x = -9$$

$$\frac{11x}{11} = \frac{-9}{11}$$

$$x = -\frac{9}{11}$$

The solution is $-\dfrac{9}{11}$.

7. The LCD of 8 and 4 is 8.

$$\frac{y}{8} + \frac{3}{4} = 2$$

$$8\left(\frac{y}{8} + \frac{3}{4}\right) = 8(2)$$

$$8 \cdot \frac{y}{8} + 8 \cdot \frac{3}{4} = 8(2)$$

$$y + 6 = 16$$

$$y + 6 - 6 = 16 - 6$$

$$y = 10$$

Check:

$$\frac{y}{8} + \frac{3}{4} = 2$$

$$\frac{10}{8} + \frac{3}{4} \stackrel{?}{=} 2$$

$$\frac{10}{8} + \frac{6}{8} \stackrel{?}{=} 2$$

$$\frac{16}{8} \stackrel{?}{=} 2$$

$$2 = 2 \quad \text{True}$$

The solution is 10.

8.

$$\frac{x}{5} - x = \frac{1}{5}$$

$$5\left(\frac{x}{5} - x\right) = 5\left(\frac{1}{5}\right)$$

$$5\left(\frac{x}{5}\right) - 5(x) = 5\left(\frac{1}{5}\right)$$

$$x - 5x = 1$$

$$-4x = 1$$

$$\frac{-4x}{-4} = \frac{1}{-4}$$

$$x = -\frac{1}{4}$$

To check, replace x with $-\dfrac{1}{4}$ in the original equation to see that a true statement results.

The solution is $-\dfrac{1}{4}$.

9. The LCD of 2 and 5 is 10.

$$\frac{y}{2} = \frac{y}{5} + \frac{3}{2}$$

$$10\left(\frac{y}{2}\right) = 10\left(\frac{y}{5} + \frac{3}{2}\right)$$

$$10\left(\frac{y}{2}\right) = 10\left(\frac{y}{5}\right) + 10\left(\frac{3}{2}\right)$$

$$5y = 2y + 15$$

$$5y - 2y = 2y - 2y + 15$$

$$3y = 15$$

$$\frac{3y}{3} = \frac{15}{3}$$

$$y = 5$$

Check: $\dfrac{y}{2} = \dfrac{y}{5} + \dfrac{3}{2}$

$\dfrac{5}{2} \overset{?}{=} \dfrac{5}{5} + \dfrac{3}{2}$

$2\dfrac{1}{2} \overset{?}{=} 1 + 1\dfrac{1}{2}$

$2\dfrac{1}{2} = 2\dfrac{1}{2}$ True

The solution is 5.

10. $\dfrac{9}{10} - \dfrac{y}{3} = \dfrac{9}{10} \cdot \dfrac{3}{3} - \dfrac{y}{3} \cdot \dfrac{10}{10}$

$= \dfrac{27}{30} - \dfrac{10y}{30}$

$= \dfrac{27 - 10y}{30}$

Exercise Set 4.8

1. $x + \dfrac{1}{3} = -\dfrac{1}{3}$

$x + \dfrac{1}{3} - \dfrac{1}{3} = -\dfrac{1}{3} - \dfrac{1}{3}$

$x = -\dfrac{2}{3}$

Check: $x + \dfrac{1}{3} = -\dfrac{1}{3}$

$-\dfrac{2}{3} + \dfrac{1}{3} \overset{?}{=} -\dfrac{1}{3}$

$\dfrac{-2 + 1}{3} \overset{?}{=} -\dfrac{1}{3}$

$-\dfrac{1}{3} = -\dfrac{1}{3}$ True

The solution is $-\dfrac{2}{3}$.

3. $y - \dfrac{3}{13} = -\dfrac{2}{13}$

$y - \dfrac{3}{13} + \dfrac{3}{13} = -\dfrac{2}{13} + \dfrac{3}{13}$

$y = \dfrac{-2 + 3}{13}$

$y = \dfrac{1}{13}$

Check: $y - \dfrac{3}{13} = -\dfrac{2}{13}$

$\dfrac{1}{13} - \dfrac{3}{13} \overset{?}{=} -\dfrac{2}{13}$

$\dfrac{1 - 3}{13} \overset{?}{=} -\dfrac{2}{13}$

$-\dfrac{2}{13} = -\dfrac{2}{13}$ True

The solution is $\dfrac{1}{13}$.

5. $3x - \dfrac{1}{5} - 2x = \dfrac{1}{5} + \dfrac{2}{5}$

$x - \dfrac{1}{5} = \dfrac{3}{5}$

$x - \dfrac{1}{5} + \dfrac{1}{5} = \dfrac{3}{5} + \dfrac{1}{5}$

$x = \dfrac{4}{5}$

Check: $3x - \dfrac{1}{5} - 2x = \dfrac{1}{5} + \dfrac{2}{5}$

$3 \cdot \dfrac{4}{5} - \dfrac{1}{5} - 2 \cdot \dfrac{4}{5} \overset{?}{=} \dfrac{1}{5} + \dfrac{2}{5}$

$\dfrac{12}{5} - \dfrac{1}{5} - \dfrac{8}{5} \overset{?}{=} \dfrac{1}{5} + \dfrac{2}{5}$

$\dfrac{3}{5} = \dfrac{3}{5}$ True

The solution is $\dfrac{4}{5}$.

7. $x - \dfrac{1}{12} = \dfrac{5}{6}$

$x - \dfrac{1}{12} + \dfrac{1}{12} = \dfrac{5}{6} + \dfrac{1}{12}$

$x = \dfrac{5 \cdot 2}{6 \cdot 2} + \dfrac{1}{12}$

$x = \dfrac{10}{12} + \dfrac{1}{12}$

$x = \dfrac{11}{12}$

Check: $x - \dfrac{1}{12} = \dfrac{5}{6}$

$$\dfrac{11}{12} - \dfrac{1}{12} \stackrel{?}{=} \dfrac{5}{6}$$

$$\dfrac{10}{12} \stackrel{?}{=} \dfrac{5}{6}$$

$$\dfrac{5}{6} = \dfrac{5}{6} \quad \text{True}$$

The solution is $\dfrac{11}{12}$.

9. $\dfrac{2}{5} + y = -\dfrac{3}{10}$

$$\dfrac{2}{5} + y - \dfrac{2}{5} = -\dfrac{3}{10} - \dfrac{2}{5}$$

$$y = -\dfrac{3}{10} - \dfrac{2 \cdot 2}{5 \cdot 2}$$

$$y = -\dfrac{3}{10} - \dfrac{4}{10}$$

$$y = -\dfrac{7}{10}$$

Check: $\dfrac{2}{5} + y = -\dfrac{3}{10}$

$$\dfrac{2}{5} + \left(-\dfrac{7}{10}\right) \stackrel{?}{=} -\dfrac{3}{10}$$

$$\dfrac{2}{5} \cdot \dfrac{2}{2} + \left(-\dfrac{7}{10}\right) \stackrel{?}{=} -\dfrac{3}{10}$$

$$\dfrac{4}{10} + \left(-\dfrac{7}{10}\right) \stackrel{?}{=} -\dfrac{3}{10}$$

$$-\dfrac{3}{10} = -\dfrac{3}{10} \quad \text{True}$$

The solution is $-\dfrac{7}{10}$.

11. $7z + \dfrac{1}{16} - 6z = \dfrac{3}{4}$

$$z + \dfrac{1}{16} = \dfrac{3}{4}$$

$$z + \dfrac{1}{16} - \dfrac{1}{16} = \dfrac{3}{4} - \dfrac{1}{16}$$

$$z = \dfrac{3 \cdot 4}{4 \cdot 4} - \dfrac{1}{16}$$

$$z = \dfrac{12}{16} - \dfrac{1}{16}$$

$$z = \dfrac{11}{16}$$

Check: $7z + \dfrac{1}{16} - 6z = \dfrac{3}{4}$

$$7\left(\dfrac{11}{16}\right) + \dfrac{1}{16} - 6\left(\dfrac{11}{16}\right) \stackrel{?}{=} \dfrac{3}{4}$$

$$\dfrac{77}{16} + \dfrac{1}{16} - \dfrac{66}{16} \stackrel{?}{=} \dfrac{3}{4}$$

$$\dfrac{12}{16} \stackrel{?}{=} \dfrac{3}{4}$$

$$\dfrac{3}{4} = \dfrac{3}{4} \quad \text{True}$$

The solution is $\dfrac{11}{16}$.

13. $-\dfrac{2}{9} = x - \dfrac{5}{6}$

$$-\dfrac{2}{9} + \dfrac{5}{6} = x - \dfrac{5}{6} + \dfrac{5}{6}$$

$$-\dfrac{2}{9} \cdot \dfrac{2}{2} + \dfrac{5}{6} \cdot \dfrac{3}{3} = x$$

$$-\dfrac{4}{18} + \dfrac{15}{18} = x$$

$$\dfrac{11}{18} = x$$

Check: $-\dfrac{2}{9} = x - \dfrac{5}{6}$

$$-\dfrac{2}{9} \overset{?}{=} \dfrac{11}{18} - \dfrac{5}{6}$$

$$-\dfrac{2}{9} \cdot \dfrac{2}{2} \overset{?}{=} \dfrac{11}{18} - \dfrac{5}{6} \cdot \dfrac{3}{3}$$

$$-\dfrac{4}{18} \overset{?}{=} \dfrac{11}{18} - \dfrac{15}{18}$$

$$-\dfrac{4}{18} = -\dfrac{4}{18} \quad \text{True}$$

The solution is $\dfrac{11}{18}$.

15. $7x = 2$

$$\dfrac{1}{7} \cdot 7x = \dfrac{1}{7} \cdot 2$$

$$x = \dfrac{2}{7}$$

17. $\dfrac{1}{4}x = 3$

$$4 \cdot \dfrac{1}{4}x = 4 \cdot 3$$

$$x = 12$$

19. $\dfrac{2}{9}y = -6$

$$\dfrac{9}{2} \cdot \dfrac{2}{9}y = \dfrac{9}{2} \cdot -6$$

$$y = -\dfrac{9 \cdot 6}{2}$$

$$y = -\dfrac{9 \cdot 2 \cdot 3}{2}$$

$$y = -27$$

21. $-\dfrac{4}{9}z = -\dfrac{3}{2}$

$$-\dfrac{9}{4} \cdot \dfrac{4}{9}z = -\dfrac{9}{4} \cdot -\dfrac{3}{2}$$

$$z = \dfrac{27}{8}$$

23. $7a = \dfrac{1}{3}$

$$\dfrac{1}{7} \cdot 7a = \dfrac{1}{7} \cdot \dfrac{1}{3}$$

$$a = \dfrac{1}{21}$$

25. $-3x = -\dfrac{6}{11}$

$$-\dfrac{1}{3} \cdot -3x = -\dfrac{1}{3} \cdot -\dfrac{6}{11}$$

$$x = \dfrac{6}{3 \cdot 11}$$

$$x = \dfrac{2}{11}$$

27. $\dfrac{5}{9}x = -\dfrac{3}{18}$

$$18\left(\dfrac{5}{9}x\right) = 18\left(-\dfrac{3}{18}\right)$$

$$10x = -3$$

$$\dfrac{10x}{10} = \dfrac{-3}{10}$$

$$x = -\dfrac{3}{10}$$

29. $\dfrac{x}{3} + 2 = \dfrac{7}{3}$

$$3\left(\dfrac{x}{3} + 2\right) = 3 \cdot \dfrac{7}{3}$$

$$3 \cdot \dfrac{x}{3} + 3 \cdot 2 = 7$$

$$x + 6 = 7$$

$$x + 6 - 6 = 7 - 6$$

$$x = 1$$

31.
$$\frac{x}{5} - \frac{5}{10} = 1$$
$$10\left(\frac{x}{5} - \frac{5}{10}\right) = 10(1)$$
$$10 \cdot \frac{x}{5} - 10 \cdot \frac{5}{10} = 10$$
$$2x - 5 = 10$$
$$2x - 5 + 5 = 10 + 5$$
$$2x = 15$$
$$\frac{2x}{2} = \frac{15}{2}$$
$$x = \frac{15}{2}$$

33.
$$\frac{1}{2} - \frac{3}{5} = \frac{x}{10}$$
$$10\left(\frac{1}{2} - \frac{3}{5}\right) = 10 \cdot \frac{x}{10}$$
$$10 \cdot \frac{1}{2} - 10 \cdot \frac{3}{5} = x$$
$$5 - 6 = x$$
$$-1 = x$$

35.
$$\frac{x}{3} = \frac{x}{5} - 2$$
$$15\left(\frac{x}{3}\right) = 15\left(\frac{x}{5} - 2\right)$$
$$5x = 15 \cdot \frac{x}{5} - 15 \cdot 2$$
$$5x = 3x - 30$$
$$5x - 3x = 3x - 3x - 30$$
$$2x = -30$$
$$\frac{2x}{2} = \frac{-30}{2}$$
$$x = -15$$

37.
$$\frac{x}{7} - \frac{4}{3} = \frac{x}{7} \cdot \frac{3}{3} - \frac{4}{3} \cdot \frac{7}{7}$$
$$= \frac{3x}{21} - \frac{28}{21}$$
$$= \frac{3x - 28}{21}$$

39. $\dfrac{y}{2} + 5 = \dfrac{y}{2} + \dfrac{5}{1} \cdot \dfrac{2}{2} = \dfrac{y}{2} + \dfrac{10}{2} = \dfrac{y+10}{2}$

41.
$$\frac{3x}{10} + \frac{x}{6} = \frac{3x}{10} \cdot \frac{3}{3} + \frac{x}{6} \cdot \frac{5}{5}$$
$$= \frac{9x}{30} + \frac{5x}{30}$$
$$= \frac{14x}{30}$$
$$= \frac{2 \cdot 7 \cdot x}{2 \cdot 15}$$
$$= \frac{7x}{15}$$

43.
$$\frac{3}{8}x = \frac{1}{2}$$
$$\frac{8}{3} \cdot \frac{3}{8}x = \frac{8}{3} \cdot \frac{1}{2}$$
$$x = \frac{8}{6}$$
$$x = \frac{4}{3}$$

45.
$$\frac{2}{3} - \frac{x}{5} = \frac{4}{15}$$
$$15\left(\frac{2}{3} - \frac{x}{5}\right) = 15 \cdot \frac{4}{15}$$
$$15 \cdot \frac{2}{3} - 15 \cdot \frac{x}{5} = 4$$
$$10 - 3x = 4$$
$$10 - 3x - 10 = 4 - 10$$
$$-3x = -6$$
$$\frac{-3x}{-3} = \frac{-6}{-3}$$
$$x = 2$$

47.
$$\frac{9}{14}z = \frac{27}{20}$$
$$\frac{14}{9} \cdot \frac{9}{14}z = \frac{14}{9} \cdot \frac{27}{20}$$
$$z = \frac{2 \cdot 7 \cdot 9 \cdot 3}{9 \cdot 2 \cdot 10}$$
$$z = \frac{21}{10}$$

49. $-3m - 5m = \dfrac{4}{7}$

$\qquad -8m = \dfrac{4}{7}$

$\qquad -\dfrac{1}{8} \cdot -8m = -\dfrac{1}{8} \cdot \dfrac{4}{7}$

$\qquad\qquad m = -\dfrac{1 \cdot 4}{2 \cdot 4 \cdot 7}$

$\qquad\qquad m = -\dfrac{1}{14}$

51. $\dfrac{x}{4} + 1 = \dfrac{1}{4}$

$\qquad 4\left(\dfrac{x}{4} + 1\right) = 4\left(\dfrac{1}{4}\right)$

$\qquad 4 \cdot \dfrac{x}{4} + 4 \cdot 1 = 1$

$\qquad\qquad x + 4 = 1$

$\qquad x + 4 - 4 = 1 - 4$

$\qquad\qquad x = -3$

53. $\dfrac{5}{9} - \dfrac{2}{3} = \dfrac{5}{9} - \dfrac{2}{3} \cdot \dfrac{3}{3} = \dfrac{5}{9} - \dfrac{6}{9} = -\dfrac{1}{9}$

55. $\dfrac{1}{5}y = 10$

$\qquad 5 \cdot \dfrac{1}{5}y = 5 \cdot 10$

$\qquad\qquad y = 50$

57. $\dfrac{5}{7}y = -\dfrac{15}{49}$

$\qquad \dfrac{7}{5} \cdot \dfrac{5}{7}y = \dfrac{7}{5} \cdot -\dfrac{15}{49}$

$\qquad\qquad y = -\dfrac{7 \cdot 15}{5 \cdot 49}$

$\qquad\qquad y = -\dfrac{7 \cdot 3 \cdot 5}{5 \cdot 7 \cdot 7}$

$\qquad\qquad y = -\dfrac{3}{7}$

59. $\dfrac{x}{2} - x = -2$

$\qquad 2\left(\dfrac{x}{2} - x\right) = 2(-2)$

$\qquad 2 \cdot \dfrac{x}{2} - 2 \cdot x = -4$

$\qquad\qquad x - 2x = -4$

$\qquad\qquad -x = -4$

$\qquad\qquad \dfrac{-x}{-1} = \dfrac{-4}{-1}$

$\qquad\qquad x = 4$

61. $-\dfrac{5}{8}y = \dfrac{3}{16} - \dfrac{9}{16}$

$\qquad -\dfrac{5}{8}y = -\dfrac{6}{16}$

$\qquad -\dfrac{5}{8}y = -\dfrac{3}{8}$

$\qquad -\dfrac{8}{5} \cdot -\dfrac{5}{8}y = -\dfrac{8}{5} \cdot -\dfrac{3}{8}$

$\qquad\qquad y = \dfrac{8 \cdot 3}{5 \cdot 8}$

$\qquad\qquad y = \dfrac{3}{5}$

63. $17x - 25x = \dfrac{1}{3}$

$\qquad -8x = \dfrac{1}{3}$

$\qquad -\dfrac{1}{8} \cdot -8x = -\dfrac{1}{8} \cdot \dfrac{1}{3}$

$\qquad\qquad x = -\dfrac{1}{24}$

65.
$$\frac{7}{6}x = \frac{1}{4} - \frac{2}{3}$$
$$12 \cdot \frac{7}{6}x = 12\left(\frac{1}{4} - \frac{2}{3}\right)$$
$$14x = 12 \cdot \frac{1}{4} - 12 \cdot \frac{2}{3}$$
$$14x = 3 - 8$$
$$14x = -5$$
$$\frac{14x}{14} = \frac{-5}{14}$$
$$x = -\frac{5}{14}$$

67.
$$\frac{b}{4} = \frac{b}{12} + \frac{2}{3}$$
$$12 \cdot \frac{b}{4} = 12\left(\frac{b}{12} + \frac{2}{3}\right)$$
$$3b = 12 \cdot \frac{b}{12} + 12 \cdot \frac{2}{3}$$
$$3b = b + 8$$
$$3b - b = b + 8 - b$$
$$2b = 8$$
$$\frac{2b}{2} = \frac{8}{2}$$
$$b = 4$$

69.
$$\frac{x}{3} + 2 = \frac{x}{2} + 8$$
$$6\left(\frac{x}{3} + 2\right) = 6\left(\frac{x}{2} + 8\right)$$
$$6 \cdot \frac{x}{3} + 6 \cdot 2 = 6 \cdot \frac{x}{2} + 6 \cdot 8$$
$$2x + 12 = 3x + 48$$
$$2x + 12 - 2x = 3x + 48 - 2x$$
$$12 = x + 48$$
$$12 - 48 = x + 48 - 48$$
$$-36 = x$$

71. 57,236 rounded to the nearest hundred is 57,200.

73. 327 rounded to the nearest ten is 330.

75. answers may vary

77.
$$\frac{19}{53} = \frac{353x}{1431} + \frac{23}{27}$$
$$1431 \cdot \frac{19}{53} = 1431\left(\frac{353x}{1431} + \frac{23}{27}\right)$$
$$27 \cdot 19 = 1431 \cdot \frac{353x}{1431} + 1431 \cdot \frac{23}{27}$$
$$513 = 353x + 1219$$
$$513 - 1219 = 353x + 1219 - 1219$$
$$-706 = 353x$$
$$\frac{-706}{353} = \frac{353x}{353}$$
$$-2 = x$$

Chapter 4 Vocabulary Check

1. Two numbers are <u>reciprocals</u> of each other if their product is 1.

2. A <u>composite number</u> is a natural number greater than 1 that is not prime.

3. Fractions that represent the same portion of a whole are called <u>equivalent</u> fractions.

4. An <u>improper fraction</u> is a fraction whose numerator is greater than or equal to its denominator.

5. A <u>prime number</u> is a natural number greater than 1 whose only factors are 1 and itself.

6. A fraction is in <u>simplest form</u> when the numerator and the denominator have no factors in common other than 1.

7. A <u>proper fraction</u> is one whose numerator is less than its denominator.

8. A <u>mixed number</u> contains a whole number part and a fraction part.

9. In the fraction $\frac{7}{9}$, the 7 is called the <u>numerator</u> and the 9 is called the <u>denominator</u>.

10. The <u>prime factorization</u> of a number is the factorization in which all the factors are prime numbers.

11. The fraction $\dfrac{3}{0}$ is <u>undefined</u>.

12. The fraction $\dfrac{0}{5} = \underline{0}$.

13. Fractions that have the same denominator are called <u>like</u> fractions.

14. The LCM of the denominators in a list of fractions is called the <u>least common denominator</u>.

15. A fraction whose numerator or denominator or both numerator and denominator contain fractions is called a <u>complex fraction</u>.

16. In $\dfrac{a}{b} = \dfrac{c}{d}$, $a \cdot d$ and $b \cdot c$ are called <u>cross products</u>.

Chapter 4 Review

1. 2 out of 6 equal parts are shaded: $\dfrac{2}{6}$

2. 4 out of 7 equal parts are shaded: $\dfrac{4}{7}$

3. Each part is $\dfrac{1}{3}$ of a whole and there are 7 parts shaded, or 2 wholes and 1 more part: $\dfrac{7}{3}$ or $2\dfrac{1}{3}$

4. Each part is $\dfrac{1}{4}$ of a whole and there are 13 parts shaded, or 3 wholes and 1 more part: $\dfrac{13}{4}$ or $3\dfrac{1}{4}$

5. $\begin{array}{l}\text{successful free throws} \rightarrow 11 \\ \text{total free throws} \qquad\rightarrow \overline{12}\end{array}$

   $\dfrac{11}{12}$ of the free throws were made.

6. **a.** $131 - 23 = 108$
   108 cars are not blue.

   b. $\begin{array}{l}\text{not blue} \rightarrow 108 \\ \text{total} \qquad \rightarrow \overline{131}\end{array}$

   $\dfrac{108}{131}$ of the cars on the lot are not blue.

7. $\dfrac{3}{-3} = -1$

8. $\dfrac{-20}{-20} = 1$

9. $\dfrac{0}{-1} = 0$

10. $\dfrac{4}{0}$ is undefined

11.

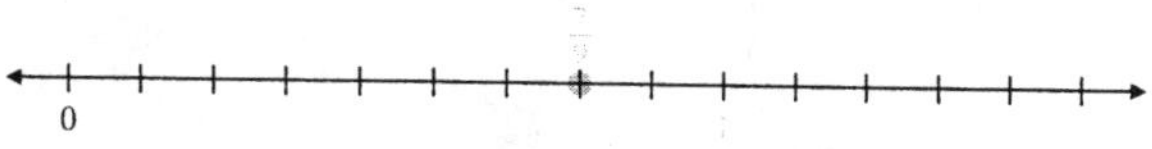

12.

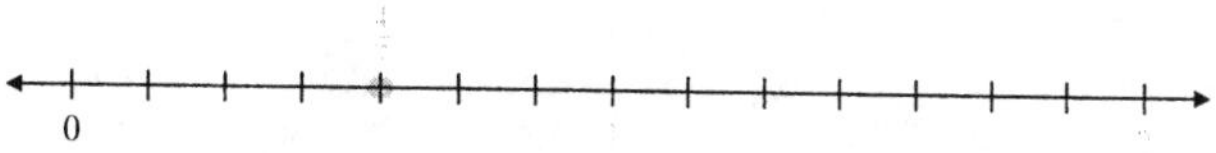

13.

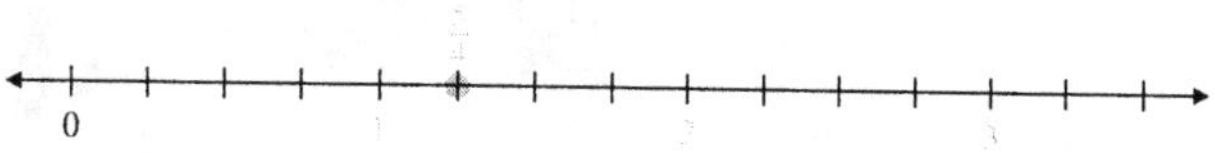

14.

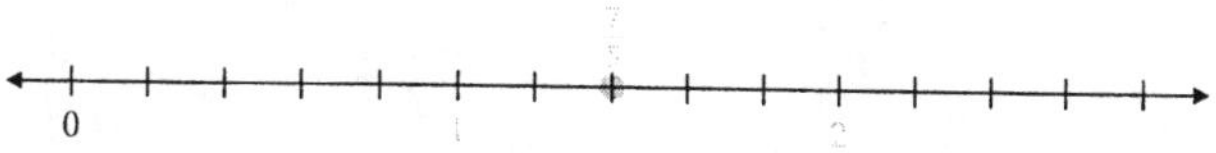

15. $\begin{array}{r} 3 \\ 4\overline{)15} \\ \underline{12} \\ 3 \end{array}$

$\dfrac{15}{4} = 3\dfrac{3}{4}$

16. $\begin{array}{r} 3 \\ 13\overline{)39} \\ \underline{39} \\ 0 \end{array}$

$\dfrac{39}{13} = 3$

17. $2\dfrac{1}{5} = \dfrac{5 \cdot 2 + 1}{5} = \dfrac{10 + 1}{5} = \dfrac{11}{5}$

18. $3\dfrac{8}{9} = \dfrac{9 \cdot 3 + 8}{9} = \dfrac{27 + 8}{9} = \dfrac{35}{9}$

19. $\dfrac{12}{28} = \dfrac{4 \cdot 3}{7 \cdot 3} = \dfrac{4}{7}$

20. $\dfrac{15}{27} = \dfrac{3 \cdot 5}{3 \cdot 9} = \dfrac{5}{9}$

21. $-\dfrac{25x}{75x^2} = -\dfrac{1 \cdot 25 \cdot x}{3 \cdot 25 \cdot x \cdot x} = -\dfrac{1}{3x}$

22. $-\dfrac{36y^3}{72y} = -\dfrac{1 \cdot 36 \cdot y \cdot y \cdot y}{2 \cdot 36 \cdot y} = -\dfrac{y^2}{2}$

23. $\dfrac{29ab}{32abc} = \dfrac{29 \cdot a \cdot b}{32 \cdot a \cdot b \cdot c} = \dfrac{29}{32c}$

24. $\dfrac{18xyz}{23xy} = \dfrac{18 \cdot x \cdot y \cdot z}{23 \cdot x \cdot y} = \dfrac{18z}{23}$

25. $\dfrac{45x^2y}{27xy^3} = \dfrac{9 \cdot 5 \cdot x \cdot x \cdot y}{9 \cdot 3 \cdot x \cdot y \cdot y \cdot y} = \dfrac{5x}{3y^2}$

26. $\dfrac{42ab^2c}{30abc^3} = \dfrac{6 \cdot 7 \cdot a \cdot b \cdot b \cdot c}{6 \cdot 5 \cdot a \cdot b \cdot c \cdot c \cdot c} = \dfrac{7b}{5c^2}$

27. $\dfrac{8 \text{ inches}}{12 \text{ inches}} = \dfrac{8}{12} = \dfrac{4 \cdot 2}{4 \cdot 3} = \dfrac{2}{3}$

8 inches represents $\dfrac{2}{3}$ of a foot.

28. $15 - 6 = 9$ cars are not white.

$\dfrac{9 \text{ non-white cars}}{15 \text{ total cars}} = \dfrac{9}{15} = \dfrac{3 \cdot 3}{3 \cdot 5} = \dfrac{3}{5}$

$\dfrac{3}{5}$ of the cars are not white.

29. Not equivalent, since the cross products are not equal: $34 \cdot 4 = 136$ and $10 \cdot 14 = 140$

30. Equivalent, since the cross products are equal: $30 \cdot 15 = 450$ and $50 \cdot 9 = 450$

31. $\dfrac{3}{5} \cdot \dfrac{1}{2} = \dfrac{3 \cdot 1}{5 \cdot 2} = \dfrac{3}{10}$

32. $-\dfrac{6}{7} \cdot \dfrac{5}{12} = -\dfrac{6 \cdot 5}{7 \cdot 12} = -\dfrac{6 \cdot 5}{7 \cdot 6 \cdot 2} = -\dfrac{5}{7 \cdot 2} = -\dfrac{5}{14}$

33. $-\dfrac{24x}{5} \cdot \left(-\dfrac{15}{8x^3}\right) = \dfrac{24x \cdot 15}{5 \cdot 8x^3}$

$= \dfrac{8 \cdot 3 \cdot x \cdot 5 \cdot 3}{5 \cdot 8 \cdot x \cdot x \cdot x}$

$= \dfrac{3 \cdot 3}{x \cdot x}$

$= \dfrac{9}{x^2}$

34. $\dfrac{27y^3}{21} \cdot \dfrac{7}{18y^2} = \dfrac{27y^3 \cdot 7}{21 \cdot 18y^2}$

$= \dfrac{3 \cdot 9 \cdot y \cdot y \cdot y \cdot 7}{3 \cdot 7 \cdot 9 \cdot 2 \cdot y \cdot y}$

$= \dfrac{y}{2}$

35. $\left(-\dfrac{1}{3}\right)^3 = \left(-\dfrac{1}{3}\right)\left(-\dfrac{1}{3}\right)\left(-\dfrac{1}{3}\right)$

$= -\dfrac{1 \cdot 1 \cdot 1}{3 \cdot 3 \cdot 3}$

$= -\dfrac{1}{27}$

36. $\left(-\dfrac{5}{12}\right)^2 = \left(-\dfrac{5}{12}\right)\left(-\dfrac{5}{12}\right) = \dfrac{5 \cdot 5}{12 \cdot 12} = \dfrac{25}{144}$

37. $-\dfrac{3}{4} \div \dfrac{3}{8} = -\dfrac{3}{4} \cdot \dfrac{8}{3}$

$= -\dfrac{3 \cdot 8}{4 \cdot 3}$

$= -\dfrac{3 \cdot 4 \cdot 2}{1 \cdot 4 \cdot 3}$

$= -\dfrac{2}{1}$

$= -2$

38. $\dfrac{21a}{4} \div \dfrac{7a}{5} = \dfrac{21a}{4} \cdot \dfrac{5}{7a}$

$= \dfrac{21a \cdot 5}{4 \cdot 7a}$

$= \dfrac{7 \cdot 3 \cdot a \cdot 5}{4 \cdot 7 \cdot a}$

$= \dfrac{3 \cdot 5}{4}$

$= \dfrac{15}{4}$

39. $-\dfrac{9}{2} \div -\dfrac{1}{3} = -\dfrac{9}{2} \cdot \left(-\dfrac{3}{1}\right) = \dfrac{9 \cdot 3}{2 \cdot 1} = \dfrac{27}{2}$

40. $-\dfrac{5}{3} \div 2y = -\dfrac{5}{3} \div \dfrac{2y}{1}$

$= -\dfrac{5}{3} \cdot \dfrac{1}{2y}$

$= -\dfrac{5 \cdot 1}{3 \cdot 2y}$

$= -\dfrac{5}{6y}$

41. $x \div y = \dfrac{9}{7} \div \dfrac{3}{4}$

$= \dfrac{9}{7} \cdot \dfrac{4}{3}$

$= \dfrac{9 \cdot 4}{7 \cdot 3}$

$= \dfrac{3 \cdot 3 \cdot 4}{7 \cdot 3}$

$= \dfrac{3 \cdot 4}{7}$

$= \dfrac{12}{7}$

42. $ab = -7 \cdot \dfrac{9}{10} = -\dfrac{7}{1} \cdot \dfrac{9}{10} = -\dfrac{7 \cdot 9}{1 \cdot 10} = -\dfrac{63}{10}$

43. area $=$ length $\cdot$ width $= \dfrac{11}{6} \cdot \dfrac{7}{8} = \dfrac{11 \cdot 7}{6 \cdot 8} = \dfrac{77}{48}$

The area is $\dfrac{77}{48}$ square feet.

44. area $=$ side $\cdot$ side $= \dfrac{2}{3} \cdot \dfrac{2}{3} = \dfrac{2 \cdot 2}{3 \cdot 3} = \dfrac{4}{9}$

The area is $\dfrac{4}{9}$ square meter.

45. $\dfrac{7}{11} + \dfrac{3}{11} = \dfrac{7+3}{11} = \dfrac{10}{11}$

46. $\dfrac{4}{9} + \dfrac{2}{9} = \dfrac{4+2}{9} = \dfrac{6}{9} = \dfrac{2 \cdot 3}{3 \cdot 3} = \dfrac{2}{3}$

47. $\dfrac{1}{12} - \dfrac{5}{12} = \dfrac{1-5}{12} = \dfrac{-4}{12} = -\dfrac{1 \cdot 4}{3 \cdot 4} = -\dfrac{1}{3}$

48. $\dfrac{11x}{15} + \dfrac{x}{15} = \dfrac{11x+x}{15} = \dfrac{12x}{15} = \dfrac{3 \cdot 4 \cdot x}{3 \cdot 5} = \dfrac{4x}{5}$

49. $\dfrac{4y}{21} - \dfrac{3}{21} = \dfrac{4y-3}{21}$

50. $\dfrac{4}{15} - \dfrac{3}{15} - \dfrac{2}{15} = \dfrac{4-3-2}{15} = \dfrac{-1}{15} = -\dfrac{1}{15}$

51. $3 = 3$

$x = x$

LCD $= 3 \cdot x = 3x$

52. $4 = 2 \cdot 2$

$8 = \boxed{2 \cdot 2 \cdot 2}$

$12 = 2 \cdot 2 \cdot \boxed{3}$

LCD $= 2 \cdot 2 \cdot 2 \cdot 3 = 24$

53. $\dfrac{2}{3} = \dfrac{2}{3} \cdot \dfrac{10}{10} = \dfrac{2 \cdot 10}{3 \cdot 10} = \dfrac{20}{30}$

54. $\dfrac{5}{8} = \dfrac{5}{8} \cdot \dfrac{7}{7} = \dfrac{5 \cdot 7}{8 \cdot 7} = \dfrac{35}{56}$

55. $\dfrac{7a}{6} = \dfrac{7a}{6} \cdot \dfrac{7}{7} = \dfrac{7a \cdot 7}{6 \cdot 7} = \dfrac{49a}{42}$

56. $\dfrac{9b}{4} = \dfrac{9b}{4} \cdot \dfrac{5}{5} = \dfrac{9b \cdot 5}{4 \cdot 5} = \dfrac{45b}{20}$

57. $\dfrac{4}{5x} = \dfrac{4}{5x} \cdot \dfrac{10}{10} = \dfrac{4 \cdot 10}{5x \cdot 10} = \dfrac{40}{50x}$

58. $\dfrac{5}{9y} = \dfrac{5}{9y} \cdot \dfrac{2}{2} = \dfrac{5 \cdot 2}{9y \cdot 2} = \dfrac{10}{18y}$

59. $\dfrac{3}{8} + \dfrac{2}{8} + \dfrac{1}{8} = \dfrac{3+2+1}{8} = \dfrac{6}{8} = \dfrac{2 \cdot 3}{2 \cdot 4} = \dfrac{3}{4}$

He did $\dfrac{3}{4}$ of his homework that evening.

60. $\dfrac{9}{16} + \dfrac{3}{16} + \dfrac{9}{16} + \dfrac{3}{16} = \dfrac{9+3+9+3}{16}$

$= \dfrac{24}{16}$

$= \dfrac{3 \cdot 8}{2 \cdot 8}$

$= \dfrac{3}{2}$

The perimeter is $\dfrac{3}{2}$ miles.

61. The LCD of 18 and 9 is 18.

$\dfrac{7}{18} + \dfrac{2}{9} = \dfrac{7}{18} + \dfrac{2 \cdot 2}{9 \cdot 2} = \dfrac{7}{18} + \dfrac{4}{18} = \dfrac{11}{18}$

62. The LCD of 13 and 26 is 26.

$\dfrac{4}{13} - \dfrac{1}{26} = \dfrac{4 \cdot 2}{13 \cdot 2} - \dfrac{1}{26} = \dfrac{8}{26} - \dfrac{1}{26} = \dfrac{7}{26}$

63. The LCD of 3 and 4 is 12.

$-\dfrac{1}{3} + \dfrac{1}{4} = -\dfrac{1 \cdot 4}{3 \cdot 4} + \dfrac{1 \cdot 3}{4 \cdot 3} = -\dfrac{4}{12} + \dfrac{3}{12} = -\dfrac{1}{12}$

64. The LCD of 3 and 4 is 12.

$-\dfrac{2}{3} + \dfrac{1}{4} = -\dfrac{2 \cdot 4}{3 \cdot 4} + \dfrac{1 \cdot 3}{4 \cdot 3} = -\dfrac{8}{12} + \dfrac{3}{12} = -\dfrac{5}{12}$

65. The LCD of 11 and 55 is 55.

$\dfrac{5x}{11} + \dfrac{2}{55} = \dfrac{5x \cdot 5}{11 \cdot 5} + \dfrac{2}{55} = \dfrac{25x}{55} + \dfrac{2}{55} = \dfrac{25x+2}{55}$

66. The LCD of 15 and 5 is 15.

$\dfrac{4}{15} + \dfrac{b}{5} = \dfrac{4}{15} + \dfrac{b \cdot 3}{5 \cdot 3} = \dfrac{4}{15} + \dfrac{3b}{15} = \dfrac{4+3b}{15}$

67. The LCD of 12 and 9 is 36.

$\dfrac{5y}{12} - \dfrac{2y}{9} = \dfrac{5y \cdot 3}{12 \cdot 3} - \dfrac{2y \cdot 4}{9 \cdot 4} = \dfrac{15y}{36} - \dfrac{8y}{36} = \dfrac{7y}{36}$

68. The LCD of 18 and 9 is 18.

$\dfrac{7x}{18} + \dfrac{2x}{9} = \dfrac{7x}{18} + \dfrac{2x \cdot 2}{9 \cdot 2} = \dfrac{7x}{18} + \dfrac{4x}{18} = \dfrac{11x}{18}$

69. The LCD of 9 and y is $9y$.

$\dfrac{4}{9} + \dfrac{5}{y} = \dfrac{4 \cdot y}{9 \cdot y} + \dfrac{5 \cdot 9}{y \cdot 9} = \dfrac{4y}{9y} + \dfrac{45}{9y} = \dfrac{4y+45}{9y}$

70. The LCD of 14 and 7 is 14.

$-\dfrac{9}{14} - \dfrac{3}{7} = -\dfrac{9}{14} - \dfrac{3 \cdot 2}{7 \cdot 2} = -\dfrac{9}{14} - \dfrac{6}{14} = -\dfrac{15}{14}$

71. The LCD of 25, 75, and 50 is 150.

$\dfrac{4}{25} + \dfrac{23}{75} + \dfrac{7}{50} = \dfrac{4 \cdot 6}{25 \cdot 6} + \dfrac{23 \cdot 2}{75 \cdot 2} + \dfrac{7 \cdot 3}{50 \cdot 3}$

$= \dfrac{24}{150} + \dfrac{46}{150} + \dfrac{21}{150}$

$= \dfrac{91}{150}$

72. The LCD of 3, 9, and 6 is 18.

$\dfrac{2}{3} - \dfrac{2}{9} - \dfrac{1}{6} = \dfrac{2 \cdot 6}{3 \cdot 6} - \dfrac{2 \cdot 2}{9 \cdot 2} - \dfrac{1 \cdot 3}{6 \cdot 3}$

$= \dfrac{12}{18} - \dfrac{4}{18} - \dfrac{3}{18}$

$= \dfrac{5}{18}$

73. The LCD of 9 and 6 is 18.

$$\frac{2}{9}+\frac{5}{6}+\frac{2}{9}+\frac{5}{6}=\frac{2\cdot2}{9\cdot2}+\frac{5\cdot3}{6\cdot3}+\frac{2\cdot2}{9\cdot2}+\frac{5\cdot3}{6\cdot3}$$

$$=\frac{4}{18}+\frac{15}{18}+\frac{4}{18}+\frac{15}{18}$$

$$=\frac{38}{18}$$

$$=\frac{2\cdot19}{2\cdot9}$$

$$=\frac{19}{9}$$

The perimeter is $\frac{19}{9}$ meters.

74. The LCD of 5 and 10 is 10.

$$\frac{1}{5}+\frac{3}{5}+\frac{7}{10}=\frac{1\cdot2}{5\cdot2}+\frac{3\cdot2}{5\cdot2}+\frac{7}{10}$$

$$=\frac{2}{10}+\frac{6}{10}+\frac{7}{10}$$

$$=\frac{15}{10}$$

$$=\frac{3\cdot5}{2\cdot5}$$

$$=\frac{3}{2}$$

The perimeter is $\frac{3}{2}$ feet.

75. The LCD of 25 and 50 is 50.

$$\frac{9}{25}+\frac{3}{50}=\frac{9\cdot2}{25\cdot2}+\frac{3}{50}=\frac{18}{50}+\frac{3}{50}=\frac{21}{50}$$

$\frac{21}{50}$ of the donors have type A blood.

76. The LCD of 12 and 3 is 12.

$$\frac{2}{3}-\frac{5}{12}=\frac{2\cdot4}{3\cdot4}-\frac{5}{12}=\frac{8}{12}-\frac{5}{12}=\frac{3}{12}=\frac{1\cdot3}{4\cdot3}=\frac{1}{4}$$

The difference in length is $\frac{1}{4}$ yard.

77. $\dfrac{\frac{2x}{5}}{\frac{7}{10}}=\dfrac{2x}{5}\div\dfrac{7}{10}$

$$=\frac{2x}{5}\cdot\frac{10}{7}$$

$$=\frac{2x\cdot10}{5\cdot7}$$

$$=\frac{2\cdot x\cdot5\cdot2}{5\cdot7}$$

$$=\frac{4x}{7}$$

78. $\dfrac{\frac{3y}{7}}{\frac{11}{7}}=\dfrac{3y}{7}\div\dfrac{11}{7}=\dfrac{3y}{7}\cdot\dfrac{7}{11}=\dfrac{3y\cdot7}{7\cdot11}=\dfrac{3y}{11}$

79. $\dfrac{\frac{2}{5}-\frac{1}{2}}{\frac{3}{4}-\frac{7}{10}}=\dfrac{20\left(\frac{2}{5}-\frac{1}{2}\right)}{20\left(\frac{3}{4}-\frac{7}{10}\right)}$

$$=\frac{20\cdot\frac{2}{5}-20\cdot\frac{1}{2}}{20\cdot\frac{3}{4}-20\cdot\frac{7}{10}}$$

$$=\frac{8-10}{15-14}$$

$$=\frac{-2}{1}$$

$$=-2$$

80. $\dfrac{\frac{5}{6}-\frac{1}{4}}{\frac{-1}{12y}}=\dfrac{12y\left(\frac{5}{6}-\frac{1}{4}\right)}{12y\left(\frac{-1}{12y}\right)}$

$$=\frac{12y\cdot\frac{5}{6}-12y\cdot\frac{1}{4}}{-1}$$

$$=\frac{10y-3y}{-1}$$

$$=\frac{7y}{-1}$$

$$=-7y$$

81. $\dfrac{x}{y+z} = \dfrac{\frac{1}{2}}{-\frac{2}{3}+\frac{4}{5}}$

$= \dfrac{30 \cdot \frac{1}{2}}{30\left(-\frac{2}{3}+\frac{4}{5}\right)}$

$= \dfrac{15}{30\left(-\frac{2}{3}\right)+30\cdot\frac{4}{5}}$

$= \dfrac{15}{-20+24}$

$= \dfrac{15}{4}$

82. $\dfrac{x+y}{z} = \dfrac{\frac{1}{2}+\left(-\frac{2}{3}\right)}{\frac{4}{5}}$

$= \dfrac{30\left(\frac{1}{2}-\frac{2}{3}\right)}{30\cdot\frac{4}{5}}$

$= \dfrac{30\cdot\frac{1}{2}-30\cdot\frac{2}{3}}{6\cdot4}$

$= \dfrac{15-20}{24}$

$= -\dfrac{5}{24}$

83. $\dfrac{5}{13}\div\dfrac{1}{2}\cdot\dfrac{4}{5} = \dfrac{5}{13}\cdot\dfrac{2}{1}\cdot\dfrac{4}{5} = \dfrac{5\cdot2\cdot4}{13\cdot1\cdot5} = \dfrac{8}{13}$

84. $\dfrac{2}{27}-\left(\dfrac{1}{3}\right)^2 = \dfrac{2}{27}-\dfrac{1}{9}$

$= \dfrac{2}{27}-\dfrac{1\cdot3}{9\cdot3}$

$= \dfrac{2}{27}-\dfrac{3}{27}$

$= -\dfrac{1}{27}$

85. $\dfrac{9}{10}\cdot\dfrac{1}{3}-\dfrac{2}{5}\cdot\dfrac{1}{11} = \dfrac{9\cdot1}{10\cdot3}-\dfrac{2\cdot1}{5\cdot11}$

$= \dfrac{3}{10}-\dfrac{2}{55}$

$= \dfrac{3\cdot11}{10\cdot11}-\dfrac{2\cdot2}{55\cdot2}$

$= \dfrac{33}{110}-\dfrac{4}{110}$

$= \dfrac{29}{110}$

86. $-\dfrac{2}{7}\cdot\left(\dfrac{1}{5}+\dfrac{3}{10}\right) = -\dfrac{2}{7}\cdot\left(\dfrac{1\cdot2}{5\cdot2}+\dfrac{3}{10}\right)$

$= -\dfrac{2}{7}\cdot\left(\dfrac{2}{10}+\dfrac{3}{10}\right)$

$= -\dfrac{2}{7}\cdot\dfrac{5}{10}$

$= -\dfrac{2}{7}\cdot\dfrac{1}{2}$

$= -\dfrac{1}{7}$

87. $7\dfrac{3}{8}$ $7\dfrac{9}{24}$

$9\dfrac{5}{6}$ $9\dfrac{20}{24}$

$+\,3\dfrac{1}{12}$ $+\,3\dfrac{2}{24}$

$19\dfrac{31}{24} = 19+1\dfrac{7}{24} = 20\dfrac{7}{24}$

88. $8\dfrac{1}{5}$ rounds to 8.

$5\dfrac{3}{11}$ rounds to 5.

An estimate is 8 + 5 = 13.

$8\dfrac{1}{5}$ $8\dfrac{11}{55}$ $7\dfrac{66}{55}$

$-\,5\dfrac{3}{11}$ $-\,5\dfrac{15}{55}$ $-\,5\dfrac{15}{55}$

$2\dfrac{51}{55}$

89. $1\dfrac{5}{8}$ rounds to 2.

$3\dfrac{1}{5}$ rounds to 3.

An estimate is $2 \cdot 3 = 6$.

$$1\dfrac{5}{8} \cdot 3\dfrac{1}{5} = \dfrac{13}{8} \cdot \dfrac{16}{5}$$
$$= \dfrac{13 \cdot 16}{8 \cdot 5}$$
$$= \dfrac{13 \cdot 8 \cdot 2}{8 \cdot 5}$$
$$= \dfrac{26}{5}$$
$$= 5\dfrac{1}{5}$$

90. $6\dfrac{3}{4} \div 1\dfrac{2}{7} = \dfrac{27}{4} \div \dfrac{9}{7}$
$$= \dfrac{27}{4} \cdot \dfrac{7}{9}$$
$$= \dfrac{9 \cdot 3 \cdot 7}{4 \cdot 9}$$
$$= \dfrac{21}{4}$$
$$= 5\dfrac{1}{4}$$

91. $341 \div 15\dfrac{1}{2} = \dfrac{341}{1} \div \dfrac{31}{2}$
$$= \dfrac{341}{1} \cdot \dfrac{2}{31}$$
$$= \dfrac{11 \cdot 31 \cdot 2}{1 \cdot 31}$$
$$= 22$$

We would expect 22 miles on one gallon.

92. $7\dfrac{1}{3} \cdot 5 = \dfrac{22}{3} \cdot \dfrac{5}{1} = \dfrac{22 \cdot 5}{3 \cdot 1} = \dfrac{110}{3} = 36\dfrac{2}{3}$

There are $\dfrac{110}{3}$ or $36\dfrac{2}{3}$ grams of fat in a 5-ounce hamburger patty.

93. $18\dfrac{7}{8} - 10\dfrac{3}{8} = 8\dfrac{4}{8} = 8\dfrac{1}{2}$

$$8\dfrac{1}{2} \div 2 = \dfrac{17}{2} \div \dfrac{2}{1} = \dfrac{17}{2} \cdot \dfrac{1}{2} = \dfrac{17}{4} = 4\dfrac{1}{4}$$

Each measurement is $4\dfrac{1}{4}$ inches.

94. $1\dfrac{3}{10} - \dfrac{3}{5} = \dfrac{13}{10} - \dfrac{3}{5} = \dfrac{13}{10} - \dfrac{3 \cdot 2}{5 \cdot 2} = \dfrac{13}{10} - \dfrac{6}{10} = \dfrac{7}{10}$

The unknown measurement is $\dfrac{7}{10}$ yard.

95. $-12\dfrac{1}{7} + \left(-15\dfrac{3}{14}\right) = -12\dfrac{2}{14} + \left(-15\dfrac{3}{14}\right)$
$$= -27\dfrac{5}{14}$$

96. $-3\dfrac{1}{5} \div \left(-2\dfrac{7}{10}\right) = -\dfrac{16}{5} \div \left(-\dfrac{27}{10}\right)$
$$= -\dfrac{16}{5} \cdot \left(-\dfrac{10}{27}\right)$$
$$= \dfrac{16 \cdot 5 \cdot 2}{5 \cdot 27}$$
$$= \dfrac{32}{27}$$
$$= 1\dfrac{5}{27}$$

97. $-2\dfrac{1}{4} \cdot 1\dfrac{3}{4} = -\dfrac{9}{4} \cdot \dfrac{7}{4} = -\dfrac{63}{16} = -3\dfrac{15}{16}$

98. $23\dfrac{7}{8} - 24\dfrac{7}{10} = -\left(24\dfrac{7}{10} - 23\dfrac{7}{8}\right)$

$$
\begin{array}{ccc}
24\dfrac{7}{10} & 24\dfrac{28}{40} & 23\dfrac{68}{40} \\[2mm]
-23\dfrac{7}{8} & -23\dfrac{35}{40} & -23\dfrac{35}{40} \\ \hline
 & & \dfrac{33}{40}
\end{array}
$$

$$23\dfrac{7}{8} - 24\dfrac{7}{10} = -\dfrac{33}{40}$$

99.
$$a - \frac{2}{3} = \frac{1}{6}$$
$$a - \frac{2}{3} + \frac{2}{3} = \frac{1}{6} + \frac{2}{3}$$
$$a = \frac{1}{6} + \frac{2 \cdot 2}{3 \cdot 2}$$
$$a = \frac{1}{6} + \frac{4}{6}$$
$$a = \frac{5}{6}$$

100.
$$9x + \frac{1}{5} - 8x = -\frac{7}{10}$$
$$x + \frac{1}{5} = -\frac{7}{10}$$
$$x + \frac{1}{5} - \frac{1}{5} = -\frac{7}{10} - \frac{1}{5}$$
$$x = -\frac{7}{10} - \frac{2}{10}$$
$$x = -\frac{9}{10}$$

101.
$$-\frac{3}{5}x = 6$$
$$-\frac{5}{3} \cdot -\frac{3}{5}x = -\frac{5}{3} \cdot 6$$
$$x = -10$$

102.
$$\frac{2}{9}y = -\frac{4}{3}$$
$$\frac{9}{2} \cdot \frac{2}{9}y = \frac{9}{2} \cdot -\frac{4}{3}$$
$$y = -\frac{9 \cdot 4}{2 \cdot 3}$$
$$y = -\frac{3 \cdot 3 \cdot 2 \cdot 2}{2 \cdot 3}$$
$$y = -\frac{6}{1}$$
$$y = -6$$

103.
$$\frac{x}{7} - 3 = -\frac{6}{7}$$
$$7\left(\frac{x}{7} - 3\right) = 7\left(-\frac{6}{7}\right)$$
$$7 \cdot \frac{x}{7} - 7 \cdot 3 = -6$$
$$x - 21 = -6$$
$$x - 21 + 21 = -6 + 21$$
$$x = 15$$

104.
$$\frac{y}{5} + 2 = \frac{11}{5}$$
$$5\left(\frac{y}{5} + 2\right) = 5\left(\frac{11}{5}\right)$$
$$5 \cdot \frac{y}{5} + 5 \cdot 2 = 11$$
$$y + 10 = 11$$
$$y + 10 - 10 = 11 - 10$$
$$y = 1$$

105.
$$\frac{1}{6} + \frac{x}{4} = \frac{17}{12}$$
$$12\left(\frac{1}{6} + \frac{x}{4}\right) = 12 \cdot \frac{17}{12}$$
$$12 \cdot \frac{1}{6} + 12 \cdot \frac{x}{4} = 17$$
$$2 + 3x = 17$$
$$2 + 3x - 2 = 17 - 2$$
$$3x = 15$$
$$\frac{3x}{3} = \frac{15}{3}$$
$$x = 5$$

106.

$$\frac{x}{5} - \frac{5}{4} = \frac{x}{2} - \frac{1}{20}$$

$$20\left(\frac{x}{5} - \frac{5}{4}\right) = 20\left(\frac{x}{2} - \frac{1}{20}\right)$$

$$20\cdot\frac{x}{5} - 20\cdot\frac{5}{4} = 20\cdot\frac{x}{2} - 20\cdot\frac{1}{20}$$

$$4x - 25 = 10x - 1$$

$$4x - 25 - 10x = 10x - 1 - 10x$$

$$-25 - 6x = -1$$

$$-25 - 6x + 25 = -1 + 25$$

$$-6x = 24$$

$$\frac{-6x}{-6} = \frac{24}{-6}$$

$$x = -4$$

107. $\dfrac{6}{15}\cdot\dfrac{5}{8} = \dfrac{6\cdot 5}{15\cdot 8} = \dfrac{3\cdot 2\cdot 5}{5\cdot 3\cdot 2\cdot 4} = \dfrac{1}{4}$

108.

$$\frac{5x^2}{y} \div \frac{10x^3}{y^3} = \frac{5x^2}{y}\cdot\frac{y^3}{10x^3}$$

$$= \frac{5x^2\cdot y^3}{y\cdot 10x^3}$$

$$= \frac{5\cdot x\cdot x\cdot y\cdot y\cdot y}{y\cdot 5\cdot 2\cdot x\cdot x\cdot x}$$

$$= \frac{y^2}{2x}$$

109. $\dfrac{3}{10} - \dfrac{1}{10} = \dfrac{3-1}{10} = \dfrac{2}{10} = \dfrac{1\cdot 2}{5\cdot 2} = \dfrac{1}{5}$

110. $\dfrac{7}{8x}\cdot\left(-\dfrac{2}{3}\right) = -\dfrac{7\cdot 2}{8x\cdot 3} = -\dfrac{7\cdot 2}{2\cdot 4\cdot x\cdot 3} = -\dfrac{7}{12x}$

111.

$$\frac{2x}{3} + \frac{x}{4} = \frac{2x}{3}\cdot\frac{4}{4} + \frac{x}{4}\cdot\frac{3}{3}$$

$$= \frac{8x}{12} + \frac{3x}{12}$$

$$= \frac{8x + 3x}{12}$$

$$= \frac{11x}{12}$$

112. $-\dfrac{5}{11} + \dfrac{2}{55} = -\dfrac{5}{11}\cdot\dfrac{5}{5} + \dfrac{2}{55} = -\dfrac{25}{55} + \dfrac{2}{55} = -\dfrac{23}{55}$

113.

$$-1\frac{3}{5} \div \frac{1}{4} = -\frac{8}{5} \div \frac{1}{4}$$

$$= -\frac{8}{5}\cdot\frac{4}{1}$$

$$= -\frac{8\cdot 4}{5\cdot 1}$$

$$= -\frac{32}{5} \text{ or } -6\frac{2}{5}$$

114. $2\dfrac{7}{8}$ rounds to 3.

$9\dfrac{1}{2}$ rounds to 10.

An estimate is $3 + 10 = 13$.

$$\begin{array}{r} 2\frac{7}{8} \\ + 9\frac{1}{2} \\ \hline \end{array} \qquad \begin{array}{r} 2\frac{7}{8} \\ + 9\frac{4}{8} \\ \hline 11\frac{11}{8} = 11 + 1\frac{3}{8} = 12\frac{3}{8} \end{array}$$

115. $12\dfrac{1}{7}$ rounds to 12.

$9\dfrac{3}{5}$ rounds to 10.

An estimate is $12 - 10 = 2$.

$$\begin{array}{r} 12\frac{1}{7} \\ - 9\frac{3}{5} \\ \hline \end{array} \qquad \begin{array}{r} 12\frac{5}{35} \\ - 9\frac{21}{35} \\ \hline \end{array} \qquad \begin{array}{r} 11\frac{40}{35} \\ - 9\frac{21}{35} \\ \hline 2\frac{19}{35} \end{array}$$

116.
$$\frac{2+\frac{3}{4}}{1-\frac{1}{8}} = \frac{8\left(2+\frac{3}{4}\right)}{8\left(1-\frac{1}{8}\right)}$$
$$= \frac{8\cdot 2 + 8\cdot\frac{3}{4}}{8\cdot 1 - 8\cdot\frac{1}{8}}$$
$$= \frac{16+6}{8-1}$$
$$= \frac{22}{7} \text{ or } 3\frac{1}{7}$$

117.
$$-\frac{3}{8}\cdot\left(\frac{2}{3}-\frac{4}{9}\right) = -\frac{3}{8}\left(\frac{2\cdot 3}{3\cdot 3}-\frac{4}{9}\right)$$
$$= -\frac{3}{8}\left(\frac{6}{9}-\frac{4}{9}\right)$$
$$= -\frac{3}{8}\left(\frac{2}{9}\right)$$
$$= -\frac{3\cdot 2}{8\cdot 9}$$
$$= -\frac{3\cdot 2}{2\cdot 4\cdot 3\cdot 3}$$
$$= -\frac{1}{12}$$

118.
$$11x - \frac{2}{7} - 10x = -\frac{13}{14}$$
$$x - \frac{2}{7} = -\frac{13}{14}$$
$$x - \frac{2}{7} + \frac{2}{7} = -\frac{13}{14} + \frac{2}{7}$$
$$x = -\frac{13}{14} + \frac{2\cdot 2}{7\cdot 2}$$
$$x = -\frac{13}{14} + \frac{4}{14}$$
$$x = -\frac{9}{14}$$

119.
$$-\frac{3}{5}x = \frac{4}{15}$$
$$-\frac{5}{3}\cdot-\frac{3}{5}x = -\frac{5}{3}\cdot\frac{4}{15}$$
$$x = -\frac{5\cdot 4}{3\cdot 15}$$
$$x = -\frac{5\cdot 4}{3\cdot 5\cdot 3}$$
$$x = -\frac{4}{9}$$

120.
$$\frac{x}{12}+\frac{5}{6} = -\frac{3}{4}$$
$$12\left(\frac{x}{12}+\frac{5}{6}\right) = 12\left(-\frac{3}{4}\right)$$
$$12\cdot\frac{x}{12}+12\cdot\frac{5}{6} = -9$$
$$x+10 = -9$$
$$x+10-10 = -9-10$$
$$x = -19$$

121.
$$50-5\frac{1}{2} = \frac{50}{1}-\frac{11}{2}$$
$$= \frac{50\cdot 2}{1\cdot 2}-\frac{11}{2}$$
$$= \frac{100}{2}-\frac{11}{2}$$
$$= \frac{89}{2}$$
$$= 44\frac{1}{2}$$

The length of the remaining piece is $44\frac{1}{2}$ yards.

122. $5\frac{1}{2}\cdot 7\frac{4}{11} = \frac{11}{2}\cdot\frac{81}{11} = \frac{11\cdot 81}{2\cdot 11} = \frac{81}{2} \text{ or } 40\frac{1}{2}$

The area is $\frac{81}{2}$ or $40\frac{1}{2}$ square feet.

Chapter 4 Test

1. 7 of the 16 equal parts are shaded: $\dfrac{7}{16}$.

2. $7\dfrac{2}{3} = \dfrac{3\cdot 7 + 2}{3} = \dfrac{21+2}{3} = \dfrac{23}{3}$

3. $\begin{array}{r} 18 \\ 4\overline{)75} \\ \underline{4} \\ 35 \\ \underline{32} \\ 3 \end{array}$

   $\dfrac{75}{4} = 18\dfrac{3}{4}$

4. $\dfrac{24}{210} = \dfrac{6\cdot 4}{6\cdot 7\cdot 5} = \dfrac{4}{35}$

5. $-\dfrac{42x}{70} = -\dfrac{3\cdot 14\cdot x}{5\cdot 14} = -\dfrac{3x}{5}$

6. $5\cdot 11 = 55$
   $7\cdot 8 = 56$
   Since $55 \neq 56$, the fractions are not equivalent.

7. $6\cdot 63 = 378$
   $27\cdot 14 = 378$
   Since $378 = 378$, the fractions are equivalent.

8. $84 = 2\cdot 42 = 2\cdot 2\cdot 21 = 2\cdot 2\cdot 3\cdot 7 = 2^2\cdot 3\cdot 7$

9. $495 = 3\cdot 165 = 3\cdot 3\cdot 55 = 3\cdot 3\cdot 5\cdot 11 = 3^2\cdot 5\cdot 11$

10. $\dfrac{4}{4} \div \dfrac{3}{4} = \dfrac{4}{4}\cdot \dfrac{4}{3} = \dfrac{4\cdot 4}{4\cdot 3} = \dfrac{4}{3}$

11. $-\dfrac{4}{3}\cdot \dfrac{4}{4} = -\dfrac{4\cdot 4}{3\cdot 4} = -\dfrac{4}{3}$

12. $\dfrac{7x}{9} + \dfrac{x}{9} = \dfrac{7x+x}{9} = \dfrac{8x}{9}$

13. The LCD of 7 and x is $7x$.
    $\dfrac{1}{7} - \dfrac{3}{x} = \dfrac{1}{7}\cdot \dfrac{x}{x} - \dfrac{3}{x}\cdot \dfrac{7}{7} = \dfrac{x}{7x} - \dfrac{21}{7x} = \dfrac{x-21}{7x}$

14. $\dfrac{xy^3}{z}\cdot \dfrac{z}{xy} = \dfrac{xy^3\cdot z}{z\cdot xy}$
    $= \dfrac{x\cdot y\cdot y\cdot y\cdot z}{x\cdot y\cdot z}$
    $= \dfrac{y\cdot y}{1}$
    $= y^2$

15. $-\dfrac{2}{3}\cdot -\dfrac{8}{15} = \dfrac{2\cdot 8}{3\cdot 15} = \dfrac{16}{45}$

16. $\dfrac{9a}{10} + \dfrac{2}{5} = \dfrac{9a}{10} + \dfrac{2\cdot 2}{5\cdot 2} = \dfrac{9a}{10} + \dfrac{4}{10} = \dfrac{9a+4}{10}$

17. $-\dfrac{8}{15y} - \dfrac{2}{15y} = \dfrac{-8-2}{15y}$
    $= \dfrac{-10}{15y}$
    $= -\dfrac{2\cdot 5}{3\cdot 5\cdot y}$
    $= -\dfrac{2}{3y}$

18. $\dfrac{3a}{8}\cdot \dfrac{16}{6a^3} = \dfrac{3a\cdot 16}{8\cdot 6a^3}$
    $= \dfrac{3\cdot a\cdot 8\cdot 2}{8\cdot 2\cdot 3\cdot a\cdot a\cdot a}$
    $= \dfrac{1}{a\cdot a}$
    $= \dfrac{1}{a^2}$

19. $\dfrac{11}{12} - \dfrac{3}{8} + \dfrac{5}{24} = \dfrac{11 \cdot 2}{12 \cdot 2} - \dfrac{3 \cdot 3}{8 \cdot 3} + \dfrac{5}{24}$

$\qquad\qquad\qquad = \dfrac{22}{24} - \dfrac{9}{24} + \dfrac{5}{24}$

$\qquad\qquad\qquad = \dfrac{22 - 9 + 5}{24}$

$\qquad\qquad\qquad = \dfrac{18}{24}$

$\qquad\qquad\qquad = \dfrac{3 \cdot 6}{4 \cdot 6}$

$\qquad\qquad\qquad = \dfrac{3}{4}$

20.

$$3\tfrac{7}{8} \qquad\qquad 3\tfrac{35}{40}$$
$$7\tfrac{2}{5} \qquad\qquad 7\tfrac{16}{40}$$
$$+\,2\tfrac{3}{4} \qquad\quad +\,2\tfrac{30}{40}$$
$$\overline{\qquad\qquad} \qquad \overline{12\tfrac{81}{40} = 12 + 2\tfrac{1}{40} = 14\tfrac{1}{40}}$$

21.

$$19 \qquad\qquad\quad 18\tfrac{11}{11}$$
$$-\,2\tfrac{3}{11} \qquad\quad -\,2\tfrac{3}{11}$$
$$\overline{\qquad\qquad} \qquad \overline{16\tfrac{8}{11}}$$

22. $-\dfrac{16}{3} \div -\dfrac{3}{12} = -\dfrac{16}{3} \cdot -\dfrac{12}{3}$

$\qquad\qquad\qquad = \dfrac{16 \cdot 12}{3 \cdot 3}$

$\qquad\qquad\qquad = \dfrac{16 \cdot 3 \cdot 4}{3 \cdot 3}$

$\qquad\qquad\qquad = \dfrac{64}{3}$ or $21\tfrac{1}{3}$

23. $3\dfrac{1}{3} \cdot 6\dfrac{3}{4} = \dfrac{10}{3} \cdot \dfrac{27}{4}$

$\qquad\qquad = \dfrac{10 \cdot 27}{3 \cdot 4}$

$\qquad\qquad = \dfrac{2 \cdot 5 \cdot 3 \cdot 9}{3 \cdot 2 \cdot 2}$

$\qquad\qquad = \dfrac{5 \cdot 9}{2}$

$\qquad\qquad = \dfrac{45}{2}$ or $22\tfrac{1}{2}$

24. $-\dfrac{2}{7} \cdot \left(6 - \dfrac{1}{6}\right) = -\dfrac{2}{7} \cdot \left(\dfrac{6}{1} - \dfrac{1}{6}\right)$

$\qquad\qquad\qquad\quad = -\dfrac{2}{7} \cdot \left(\dfrac{6 \cdot 6}{1 \cdot 6} - \dfrac{1}{6}\right)$

$\qquad\qquad\qquad\quad = -\dfrac{2}{7} \cdot \left(\dfrac{36}{6} - \dfrac{1}{6}\right)$

$\qquad\qquad\qquad\quad = -\dfrac{2}{7} \cdot \dfrac{35}{6}$

$\qquad\qquad\qquad\quad = -\dfrac{2 \cdot 35}{7 \cdot 6}$

$\qquad\qquad\qquad\quad = -\dfrac{2 \cdot 7 \cdot 5}{7 \cdot 2 \cdot 3}$

$\qquad\qquad\qquad\quad = -\dfrac{5}{3}$ or $-1\tfrac{2}{3}$

25. $\dfrac{1}{2} \div \dfrac{2}{3} \cdot \dfrac{3}{4} = \dfrac{1}{2} \cdot \dfrac{3}{2} \cdot \dfrac{3}{4} = \dfrac{1 \cdot 3 \cdot 3}{2 \cdot 2 \cdot 4} = \dfrac{9}{16}$

26. $\left(-\dfrac{3}{4}\right)^2 \div \left(\dfrac{2}{3} + \dfrac{5}{6}\right) = \left(-\dfrac{3}{4}\right)^2 \div \left(\dfrac{2}{3} \cdot \dfrac{2}{2} + \dfrac{5}{6}\right)$

$\qquad\qquad\qquad\qquad\quad = \left(-\dfrac{3}{4}\right)^2 \div \left(\dfrac{4}{6} + \dfrac{5}{6}\right)$

$\qquad\qquad\qquad\qquad\quad = \left(-\dfrac{3}{4}\right)^2 \div \dfrac{9}{6}$

$\qquad\qquad\qquad\qquad\quad = \dfrac{9}{16} \div \dfrac{9}{6}$

$\qquad\qquad\qquad\qquad\quad = \dfrac{9}{16} \cdot \dfrac{6}{9}$

$\qquad\qquad\qquad\qquad\quad = \dfrac{9 \cdot 2 \cdot 3}{2 \cdot 8 \cdot 9}$

$\qquad\qquad\qquad\qquad\quad = \dfrac{3}{8}$

27.
$$\left(\frac{5}{6}+\frac{4}{3}+\frac{7}{12}\right)\div 3=\left(\frac{5\cdot 2}{6\cdot 2}+\frac{4\cdot 4}{3\cdot 4}+\frac{7}{12}\right)\div 3$$
$$=\left(\frac{10}{12}+\frac{16}{12}+\frac{7}{12}\right)\div 3$$
$$=\frac{33}{12}\div\frac{3}{1}$$
$$=\frac{33}{12}\cdot\frac{1}{3}$$
$$=\frac{33\cdot 1}{12\cdot 3}$$
$$=\frac{11\cdot 3\cdot 1}{12\cdot 3}$$
$$=\frac{11}{12}$$

28.
$$\frac{\frac{5x}{7}}{\frac{20x^2}{21}}=\frac{5x}{7}\div\frac{20x^2}{21}$$
$$=\frac{5x}{7}\cdot\frac{21}{20x^2}$$
$$=\frac{5\cdot x\cdot 3\cdot 7}{7\cdot 4\cdot 5\cdot x\cdot x}$$
$$=\frac{3}{4x}$$

29.

$$\frac{5+\frac{3}{7}}{2-\frac{1}{2}}=\frac{14\left(5+\frac{3}{7}\right)}{14\left(2-\frac{1}{2}\right)}=\frac{14\cdot 5+14\cdot\frac{3}{7}}{14\cdot 2-14\cdot\frac{1}{2}}=\frac{70+6}{28-7}=\frac{76}{21}$$

30.
$$-\frac{3}{8}x=\frac{3}{4}$$
$$-\frac{8}{3}\cdot-\frac{3}{8}x=-\frac{8}{3}\cdot\frac{3}{4}$$
$$x=-\frac{8\cdot 3}{3\cdot 4}$$
$$x=-\frac{4\cdot 2\cdot 3}{3\cdot 4}$$
$$x=-2$$

31.
$$\frac{x}{5}+x=-\frac{24}{5}$$
$$5\left(\frac{x}{5}+x\right)=5\left(-\frac{24}{5}\right)$$
$$5\cdot\frac{x}{5}+5\cdot x=-24$$
$$x+5x=-24$$
$$6x=-24$$
$$\frac{6x}{6}=\frac{-24}{6}$$
$$x=-4$$

32.
$$\frac{2}{3}+\frac{x}{4}=\frac{5}{12}+\frac{x}{2}$$
$$12\left(\frac{2}{3}+\frac{x}{4}\right)=12\left(\frac{5}{12}+\frac{x}{2}\right)$$
$$12\cdot\frac{2}{3}+12\cdot\frac{x}{4}=12\cdot\frac{5}{12}+12\cdot\frac{x}{2}$$
$$8+3x=5+6x$$
$$8+3x-6x=5+6x-6x$$
$$8-3x=5$$
$$8-8-3x=5-8$$
$$-3x=-3$$
$$\frac{-3x}{-3}=\frac{-3}{-3}$$
$$x=1$$

33. $-5x=-5\left(-\frac{1}{2}\right)=\frac{5}{1}\cdot\frac{1}{2}=\frac{5\cdot 1}{1\cdot 2}=\frac{5}{2}$

34. $x\div y=\frac{1}{2}\div 3\frac{7}{8}$
$$=\frac{1}{2}\div\frac{31}{8}$$
$$=\frac{1}{2}\cdot\frac{8}{31}$$
$$=\frac{1\cdot 8}{2\cdot 31}$$
$$=\frac{1\cdot 2\cdot 4}{2\cdot 31}$$
$$=\frac{4}{31}$$

35.

$$6\frac{1}{2} \qquad 6\frac{2}{4} \qquad 5\frac{6}{4}$$
$$-2\frac{3}{4} \qquad -2\frac{3}{4} \qquad -2\frac{3}{4}$$
$$\overline{} \qquad \overline{} \qquad \overline{3\frac{3}{4}}$$

The remaining piece is $3\frac{3}{4}$ feet.

36. Housing: $\dfrac{8}{25}$

Food: $\dfrac{7}{50}$

$$\frac{8}{25} + \frac{7}{50} = \frac{8\cdot2}{25\cdot2} + \frac{7}{50} = \frac{16}{50} + \frac{7}{50} = \frac{16+7}{50} = \frac{23}{50}$$

$\dfrac{23}{50}$ of spending goes for housing and food combined.

37. Education: $\dfrac{1}{50}$

Transportation: $\dfrac{1}{5}$

Clothing: $\dfrac{1}{25}$

$$\frac{1}{50} + \frac{1}{5} + \frac{1}{25} = \frac{1}{50} + \frac{1}{5}\cdot\frac{10}{10} + \frac{1}{25}\cdot\frac{2}{2}$$
$$= \frac{1}{50} + \frac{10}{50} + \frac{2}{50}$$
$$= \frac{13}{50}$$

$\dfrac{13}{50}$ of spending goes for education, transportation, and clothing.

38.

$$\frac{3}{50}\cdot47,000 = \frac{3}{50}\cdot\frac{47,000}{1}$$
$$= \frac{3\cdot50\cdot940}{50\cdot1}$$
$$= 2820$$

Expect to spend \$2820 on health care.

39.

$$\text{perimeter} = 1 + \frac{2}{3} + 1 + \frac{2}{3}$$
$$= \frac{3}{3} + \frac{2}{3} + \frac{3}{3} + \frac{2}{3}$$
$$= \frac{10}{3}$$
$$= 3\frac{1}{3}$$

$$\text{area} = \text{length}\cdot\text{width} = 1\cdot\frac{2}{3} = \frac{2}{3}$$

The perimeter is $3\frac{1}{3}$ feet and the area is

$\dfrac{2}{3}$ square feet.

40.

$$258 \div 10\frac{3}{4} = \frac{258}{1} \div \frac{43}{4}$$
$$= \frac{258}{1}\cdot\frac{4}{43}$$
$$= \frac{43\cdot6\cdot4}{1\cdot43}$$
$$= 24$$

Expect to travel 24 miles on 1 gallon of gas.

Cumulative Review Chapters 1–4

1. 546 in words is five hundred forty-six.

2. 115 in words is one hundred fifteen.

3. 27,034 in words is twenty-seven thousand, thirty-four.

4. 6573 in words is six thousand five hundred seventy-three.

5.
$$\begin{array}{r} 46 \\ + 713 \\ \hline 759 \end{array}$$

6.
$$\begin{array}{r} 587 \\ + 44 \\ \hline 631 \end{array}$$

7.

$$
\begin{array}{r}
543 \\
-\ 29 \\
\hline
514
\end{array}
$$

Check:
$$
\begin{array}{r}
514 \\
+\ 29 \\
\hline
543
\end{array}
$$

8.

$$
\begin{array}{r}
995 \\
-\ 62 \\
\hline
933
\end{array}
$$

Check:
$$
\begin{array}{r}
933 \\
+\ 62 \\
\hline
995
\end{array}
$$

9. 278,362 rounded to the nearest thousand is 278,000.

10. 1436 rounded to the nearest ten is 1440.

11.

$$
\begin{array}{r}
4800 \\
\times\ \ \ \ 12 \\
\hline
9\ 600 \\
48,000 \\
\hline
57,600
\end{array}
$$

Therefore, 12 DVDs can hold 57,600 megabytes of information.

12.

$$
\begin{array}{r}
435 \\
\times\ \ \ 3 \\
\hline
1305
\end{array}
$$

He travels 1305 miles in 3 days.

13.

$$
\begin{array}{r}
7089 \\
8\overline{)56{,}717} \\
\underline{56}\ \ \ \ \ \ \\
0\ 7\ \ \ \ \\
\underline{0}\ \ \ \ \\
71\ \ \\
\underline{64}\ \ \\
77 \\
\underline{72} \\
5
\end{array}
$$

$56,717 \div 8 = 7089$ R 5
Check: $7089 \times 8 + 5 = 56,712 + 5 = 56,717$

14.

$$
\begin{array}{r}
379 \\
12\overline{)4558} \\
\underline{36}\ \ \ \ \\
95\ \ \\
\underline{84}\ \ \\
118 \\
\underline{108} \\
10
\end{array}
$$

$4558 \div 12 = 379$ R 10
Check: $379 \times 12 + 10 = 4548 + 10 = 4558$

15. $7 \cdot 7 \cdot 7 = 7^3$

16. $7 \cdot 7 = 7^2$

17. $3 \cdot 3 \cdot 3 \cdot 3 \cdot 9 \cdot 9 \cdot 9 = 3^4 \cdot 9^3$

18. $9 \cdot 9 \cdot 9 \cdot 9 \cdot 5 \cdot 5 = 9^4 \cdot 5^2$

19. $2(x - y) = 2(6 - 3) = 2(3) = 6$

20. $8a + 3(b - 5) = 8 \cdot 5 + 3(9 - 5)$
$$= 8 \cdot 5 + 3 \cdot 4$$
$$= 40 + 12$$
$$= 52$$

21. Let 0 represent the surface of the earth. Then 6824 feet *below* the surface is represented as -6824.

22. Let 0 represent a temperature of $0°F$. Then $21°F$ *below* zero is represented as -21.

23. $-7 + 3 = 4$

24. $-3 + 8 = 5$

25. $7 - 8 - (-5) - 1 = 7 - 8 + 5 - 1$
$$= 7 + (-8) + 5 + (-1)$$
$$= -1 + 5 + (-1)$$
$$= 4 + (-1)$$
$$= 3$$

26. $6 + (-8) - (-9) + 3 = 6 + (-8) + 9 + 3$
$$= -2 + 9 + 3$$
$$= 7 + 3$$
$$= 10$$

27. $(-5)^2 = (-5)(-5) = 25$

28. $-2^4 = -2 \cdot 2 \cdot 2 \cdot 2 = -16$

29. $3(4-7) + (-2) - 5 = 3(-3) + (-2) - 5$
$$= -9 + (-2) + (-5)$$
$$= -11 + (-5)$$
$$= -16$$

30. $(20 - 5^2)^2 = (20 - 25)^2 = (-5)^2 = 25$

31. $2y - 6 + 4y + 8 = (2y + 4y) + (-6 + 8)$
$$= 6y + 2$$

32. $5x - 1 + x + 10 = (5x + x) + (-1 + 10)$
$$= 6x + 9$$

33. $5x + 2 - 4x = 7 - 19$
$$x + 2 = -12$$
$$x + 2 - 2 = -12 - 2$$
$$x = -14$$

34. $9y + 1 - 8y = 3 - 20$
$$y + 1 = -17$$
$$y + 1 - 1 = -17 - 1$$
$$y = -18$$

35. $17 - 7x + 3 = -3x + 21 - 3x$
$$20 - 7x = 21 - 6x$$
$$20 - 7x + 7x = 21 - 6x + 7x$$
$$20 = 21 + x$$
$$20 - 21 = 21 - 21 + x$$
$$-1 = x \text{ or } x = -1$$

36. $9x - 2 = 7x - 24$
$$9x - 7x - 2 = 7x - 7x - 24$$
$$2x - 2 = -24$$
$$2x - 2 + 2 = -24 + 2$$
$$2x = -22$$
$$\frac{2x}{2} = \frac{-22}{2}$$
$$x = -11$$

37. Two of five equal parts are shaded: $\frac{2}{5}$

38. $156 = 2 \cdot 78$
$$\downarrow \ \downarrow \searrow$$
$$2 \cdot 2 \cdot 39$$
$$\downarrow \ \downarrow \ \downarrow \searrow$$
$$2 \cdot 2 \cdot 3 \cdot 13$$
$$156 = 2^2 \cdot 3 \cdot 13$$

39. a. $4\frac{2}{9} = \frac{9 \cdot 4 + 2}{9} = \frac{36 + 2}{9} = \frac{38}{9}$

b. $1\frac{8}{11} = \frac{11 \cdot 1 + 8}{11} = \frac{11 + 8}{11} = \frac{19}{11}$

40. $7\frac{4}{5} = \frac{5 \cdot 7 + 4}{5} = \frac{35 + 4}{5} = \frac{39}{5}$

41. $\frac{42x}{66} = \frac{6 \cdot 7 \cdot x}{6 \cdot 11} = \frac{7x}{11}$

42. $\frac{70}{105y} = \frac{35 \cdot 2}{35 \cdot 3 \cdot y} = \frac{2}{3y}$

43. $3\frac{1}{3} \cdot \frac{7}{8} = \frac{10}{3} \cdot \frac{7}{8}$
$$= \frac{10 \cdot 7}{3 \cdot 8}$$
$$= \frac{2 \cdot 5 \cdot 7}{3 \cdot 2 \cdot 4}$$
$$= \frac{5 \cdot 7}{3 \cdot 4}$$
$$= \frac{35}{12} \text{ or } 2\frac{11}{12}$$

44. $\dfrac{2}{3} \cdot 4 = \dfrac{2}{3} \cdot \dfrac{4}{1} = \dfrac{2 \cdot 4}{3 \cdot 1} = \dfrac{8}{3}$ or $2\dfrac{2}{3}$

45. $\dfrac{5}{16} \div \dfrac{3}{4} = \dfrac{5}{16} \cdot \dfrac{4}{3} = \dfrac{5 \cdot 4}{16 \cdot 3} = \dfrac{5 \cdot 4}{4 \cdot 4 \cdot 3} = \dfrac{5}{4 \cdot 3} = \dfrac{5}{12}$

46. $1\dfrac{1}{10} \div 5\dfrac{3}{5} = \dfrac{11}{10} \div \dfrac{28}{5} = \dfrac{11}{10} \cdot \dfrac{5}{21} = \dfrac{11 \cdot 5}{5 \cdot 2 \cdot 28} = \dfrac{11}{56}$

Chapter 5

Section 5.1

Practice Problems

1. **a.** 0.06 in words is six hundredths.

    b. −200.073 in words is negative two hundred and seventy-three thousandths.

    c. 0.0829 in words in eight hundred twenty-nine ten-thousandths.

2. 87.31 in words is eighty-seven and thirty-one hundredths.

3. 52.1085 in words is fifty-two and one thousand eighty-five ten thousandths.

4. The check should be paid to "CLECO," for the amount of "207.40," which is written in words as "Two hundred seven and $\frac{40}{100}$."

5. Five hundred and ninety-six hundredths is 500.96.

6. Thirty-nine and forty-two thousandths is 39.042.

7. $0.051 = \dfrac{51}{1000}$

8. $29.97 = 29\dfrac{97}{100}$

9. $0.12 = \dfrac{12}{100} = \dfrac{3}{25}$

10. $64.8 = 64\dfrac{8}{10} = 64\dfrac{4}{5}$

11. $-209.986 = -209\dfrac{986}{1000} = -209\dfrac{493}{500}$

12. 29.208   26.28
        ↑   ↑
        0   <   8
     so 26.208 < 26.28

13. 0.12   0.026
      ↑   ↑
      1   >   0
     so 0.12 > 0.026

14. 0.039   0.0309
       ↑   ↑
       9   >   0
     so 0.039 > 0.0309
     Thus, −0.039 < −0.0309.

15. To round 482.7817 to the nearest thousandth, observe that the digit in the ten-thousandths place is 7. Since this digit is at least 5, we add 1 to the digit in the thousandths place. The number 482.7817 rounded to the nearest thousandth is 482.782.

16. To round −0.032 to the nearest hundredth, observe that the digit in the thousandths place is 2. Since this digit is less than 5, we do not add 1 to the digit in the hundredths place. The number −0.032 rounded to the nearest hundredth is −0.03.

17. To round 3.14159265 to the nearest ten-thousandth, observe that the digit in the hundred-thousandths place is 9. Since this digit is at least 5, we add 1 to the digit in the ten-thousandths place. The number 3.14159265 rounded to the nearest the ten-thousandth is 3.1416, or $\pi \approx 3.1416$.

18. $24.62 rounded to the nearest dollar is $24, since $6 \geq 5$.

Vocabulary and Readiness Check

1. The number "twenty and eight hundredths" is written in <u>words</u> and "20.08" is written in <u>standard form</u>.

3. Like fractions, <u>decimals</u> are used to denote part of a whole.

5. The place value <u>tenths</u> is to the right of the decimal point while <u>tens</u> is to the left of the decimal point.

Exercise Set 5.1

1. 5.62 in words is five and sixty-two hundredths.

3. 16.23 in words is sixteen and twenty-three hundredths.

5. −0.205 in words is negative two hundred five thousandths.

7. 167.009 in words is one hundred sixty-seven and nine thousandths.

9. 3000.04 in words is three thousand and four hundredths.

11. 105.6 in words is one hundred five and six tenths.

13. 2.43 in words is two and forty-three hundredths.

15. The check should be paid to "R.W. Financial," for the amount "321.42," which is written in words as "Three hundred twenty-one and $\dfrac{42}{100}$."

17. The check should be paid to "Bell South," for the amount of "59.68," which is written in words as "Fifty-nine and $\dfrac{68}{100}$."

19. Two and eight tenths is 2.8.

21. Nine and eight hundredths is 9.08.

23. Negative seven hundred five and six hundred twenty-five thousandths is −705.625.

25. Forty-six ten-thousandths is 0.0046.

27. $0.7 = \dfrac{7}{10}$

29. $0.27 = \dfrac{27}{100}$

31. $0.4 = \dfrac{4}{10} = \dfrac{2}{5}$

33. $5.4 = 5\dfrac{4}{10} = 5\dfrac{2}{5}$

35. $-0.058 = -\dfrac{58}{1000} = -\dfrac{29}{500}$

37. $7.008 = 7\dfrac{8}{1000} = 7\dfrac{1}{125}$

39. $15.802 = 15\dfrac{802}{1000} = 15\dfrac{401}{500}$

41. $0.3005 = \dfrac{3005}{10,000} = \dfrac{601}{2000}$

43. Eight tenths is 0.8 and as a fraction is $\dfrac{8}{10} = \dfrac{4}{5}$.

45. In words, 0.077 is seventy-seven thousandths. As a fraction, $0.077 = \dfrac{77}{1000}$.

47. 0.15   0.16
5 < 6
so 0.15 < 0.16

49. 0.57   0.54
7 > 4
so 0.57 > 0.54
Thus −0.57 < −0.54.

51. 0.098   0.1
0 < 1
so 0.098 < 0.1

53. 0.54900   0.549
9 = 9
so 0.54900 = 0.549

55. 167.908   167.980
0 < 8
so 167.908 < 167.980

57. 1.062   1.07
6 < 7
so 1.062 < 1.07
Thus, −1.062 > −1.07.

59. −7.052   7.0052
− < +
so −7.052 < 7.0052

61. 0.023   0.024
3 < 4
so 0.023 < 0.024
Thus, −0.023 > −0.024.

63. To round 0.57 to the nearest tenth, observe that the digit in the hundredths place is 7. Since this digit is at least 5, we add 1 to the digit in the tenths place. The number 0.57 rounded to the nearest tenth is 0.6.

65. To round 98,207.23 to the nearest ten, observe that the digit in the ones place is 7. Since this digit is at least 5, we add 1 to the digit in the tens place. The number 98,207.23 rounded to the nearest ten is 98,210.

67. To round −0.234 to the nearest hundredth, observe that the digit in the thousandths place is 4. Since this digit is less than 5, we do not add 1 to the digit in the hundredths place. The number −0.234 rounded to the nearest hundredth is −0.23.

69. To round 0.5942 to the nearest thousandth, observe that the digit in the ten-thousandths place is 2. Since this digit is less than 5, we do not add 1 to the digit in the thousandths place. The number 0.5942 rounded to the nearest thousandth is 0.594.

71. To round $\pi \approx 3.14159265$ to the nearest tenth, observe that the digit in the hundredths place is 4. Since this digit is less than 5, we do not add 1 to the digit in the tenths place. The number $\pi \approx 3.14159265$ rounded to the nearest tenth is 3.1.

73. To round $\pi \approx 3.14159265$ to the nearest thousandth, observe that the digit in the ten-thousandth place is 5. Since this digit is at least 5, we add 1 to the digit in the thousandths place. The number $\pi \approx 3.14159265$ rounded to the nearest thousandth is 3.142.

75. To round 26.95 to the nearest one, observe that the digit in the tenths place is 9. Since this digit is at least 5, we add 1 to the digit in the ones place. The number 26.95 rounded to the nearest one is 27. The amount is $27.

77. To round 0.1992 to the nearest hundredth, observe that the digit in the thousandths place is 9. Since this digit is at least 5, we add 1 to the digit in the hundredths place. The number 0.1992 rounded to the nearest hundredth is 0.2.The amount is $0.20.

79. To round 10.414 to the nearest tenth, observe that the digit in the hundredths place is 1. Since this digit is less than 5, we do not add 1 to the digit in the tenths place. The number 10.414 rounded to the nearest tenth is 10.4. The length is 10.4 centimeters.

81. To round 1.7283 to the nearest hundredth, observe that the digit in the thousandths place is 8. Since this digit is at least 5, we add 1 to the digit in the hundredths place. The number 1.7283 rounded to the nearest hundredth is 1.73.The time is 1.73 hours.

83. To round 47.89 to the nearest one, observe that the digit in the tenths place is 8. Since this digit is at least 5, we add 1 to the digit in the ones place. The number 47.89 rounded to the nearest one is 48. The price is $48.

85. To round 24.6229 to the nearest thousandth, observe that the digit in the ten-thousandths place is 9. Since this digit is at least 5, we add 1 to the digit in the thousandths place. The number 24.6229 rounded to the nearest thousandth is 24.623. The day length is 24.623 hours.

87.
$$\begin{array}{r} 3452 \\ + \ 2314 \\ \hline 5766 \end{array}$$

89.
$$\begin{array}{r} 82 \\ - \ 47 \\ \hline 35 \end{array}$$

91. To round 2849.1738 to the nearest hundred, observe that the digit in the tens place is 4. Since this digit is less than 5, we do not add 1 to the digit in the hundreds place. The number 2849.1738 rounded to the nearest hundred is 2800, which is choice b.

93. To round 2849.1738 to the nearest hundredth, observe that the digit in the thousandths place is 3. Since this digit is less than 5, we do not add 1 to the digit in the hundredths place. 2849.1738 rounded to the nearest hundredth is 2849.17, which is choice a.

95. answers may vary

97. $7\dfrac{12}{100} = 7.12$

99. $0.00026849576 = \dfrac{26{,}849{,}576}{100{,}000{,}000{,}000}$

101. answers may vary

103. answers may vary

105. 0.0612 and 0.0586 rounded to the nearest hundredth are 0.06. 0.066 rounds to 0.07. 0.0506 rounds to 0.05.

107. From smallest to largest, 0.01, 0.0839, 0.09, 0.1

109. 602.2 rounded to the nearest ten is 600. 33.8 rounded to the nearest ten is 30.

$$\dfrac{600}{30} = 20$$

An individual music video has a value of approximately \$20.

Section 5.2

Practice Problems

1. a.
$$\begin{array}{r} 19.520 \\ +\ 5.371 \\ \hline 24.891 \end{array}$$

b.
$$\begin{array}{r} 40.080 \\ +17.612 \\ \hline 57.692 \end{array}$$

c.
$$\begin{array}{r} 0.125 \\ +422.800 \\ \hline 422.925 \end{array}$$

2. a.
$$\begin{array}{r} 34.5670 \\ 129.4300 \\ +\ 2.8903 \\ \hline 166.8873 \end{array}$$

b.
$$\begin{array}{r} 11.210 \\ 46.013 \\ +362.526 \\ \hline 419.749 \end{array}$$

3.
$$\begin{array}{r} 19.000 \\ +26.072 \\ \hline 45.072 \end{array}$$

4. $7.12 + (-9.92)$
Subtract the absolute values.
$$\begin{array}{r} 9.92 \\ -7.12 \\ \hline 2.80 \end{array}$$
Attach the sign of the larger absolute value.
$7.12 + (-9.92) = -2.8$

5. a.
$$\begin{array}{r} 6.70 \\ -3.92 \\ \hline 2.78 \end{array}$$
Check:
$$\begin{array}{r} 2.78 \\ +3.92 \\ \hline 6.70 \end{array}$$

b.
$$\begin{array}{r} 9.720 \\ -4.068 \\ \hline 5.652 \end{array}$$
Check:
$$\begin{array}{r} 5.652 \\ +4.068 \\ \hline 9.720 \end{array}$$

6. a.
$$\begin{array}{r} 73.00 \\ -29.31 \\ \hline 43.69 \end{array}$$
Check:
$$\begin{array}{r} 43.69 \\ +29.31 \\ \hline 73.00 \end{array}$$

b.
$$\begin{array}{r} 210.00 \\ -\ 68.22 \\ \hline 141.78 \end{array}$$
Check:
$$\begin{array}{r} 141.78 \\ +\ 68.22 \\ \hline 210.00 \end{array}$$

7.
$$\begin{array}{r} 25.91 \\ -19.00 \\ \hline 6.91 \end{array}$$

8. $-5.4 - 9.6 = -5.4 + (-9.6)$
Add the absolute values.

$$
\begin{array}{r}
5.4 \\
+\ 9.6 \\
\hline
15.0
\end{array}
$$

Attach the common sign.
$-5.4 - 9.6 = -15$

9. $-1.05 - (-7.23) = -1.05 + 7.23$
Subtract the absolute values.

$$
\begin{array}{r}
7.23 \\
-\ 1.05 \\
\hline
6.18
\end{array}
$$

Attach the sign of the larger absolute value.
$-1.05 - (-7.23) = 6.18$

10. a.

Exact	Estimate 1	Estimate 2
58.10	60	60
+ 326.97	+ 300	+ 330
385.07	360	390

b.

Exact	Estimate 1	Estimate 2
16.080	16	20
− 0.925	− 1	− 1
15.155	15	19

11. $y - z = 11.6 - 10.8 = 0.8$

12.
$$y - 4.3 = 7.8$$
$$12.1 - 4.3 \overset{?}{=} 7.8$$
$$7.8 = 7.8 \quad \text{True}$$
Yes, 12.1 is a solution.

13. $-4.3y + 7.8 - 20.1y + 14.6$
$= -4.3y - 20.1y + 7.8 + 14.6$
$= (-4.3 - 20.1)y + (7.8 + 14.6)$
$= -24.4y + 22.4$

14.

$$
\begin{array}{r}
563.52 \\
52.68 \\
+\ 127.50 \\
\hline
743.70
\end{array}
$$

The total cost is $743.70.

15.

$$
\begin{array}{r}
72.6 \\
-\ 70.8 \\
\hline
1.8
\end{array}
$$

The average height in the Netherlands is 1.8 inches greater than the average height in Czechoslovakia.

Calculator Explorations

1. $315.782 + 12.96 = 328.742$

3. $6.249 - 1.0076 = 5.2414$

5.

$$
\begin{array}{r}
12.555 \\
224.987 \\
5.2 \\
+\ 622.65 \\
\hline
865.392
\end{array}
$$

Vocabulary and Readiness Check

1. The decimal point in a whole number is positioned after the <u>last</u> digit.

3. To simplify an expression, we combine any <u>like</u> terms.

5. To add or subtract decimals, we line up the decimal points <u>vertically</u>.

Exercise Set 5.2

1.

$$
\begin{array}{r}
5.6 \\
+\ 2.1 \\
\hline
7.7
\end{array}
$$

3.

$$
\begin{array}{r}
8.20 \\
+\ 2.15 \\
\hline
10.35
\end{array}
$$

5.

$$
\begin{array}{r}
24.6000 \\
2.3900 \\
+\ 0.0678 \\
\hline
27.0578
\end{array}
$$

7. $-2.6 + (-5.97)$
Add the absolute values.

$$
\begin{array}{r}
2.60 \\
+\ 5.97 \\
\hline
8.57
\end{array}
$$

Attach the common sign.
$-2.6 + (-5.97) = -8.57$

9. $18.56 + (-8.23)$
Subtract the absolute values.

$$
\begin{array}{r}
18.56 \\
-\ 8.23 \\
\hline
10.33
\end{array}
$$

Attach the sign of the larger absolute value.
$18.56 + (-8.23) = 10.33$

11. Exact:

$$
\begin{array}{r}
234.89 \\
+\ 230.67 \\
\hline
465.56
\end{array}
$$

Estimate:

$$
\begin{array}{r}
230 \\
+\ 230 \\
\hline
460
\end{array}
$$

13. Exact:

$$
\begin{array}{r}
100.009 \\
6.080 \\
+\ 9.034 \\
\hline
115.123
\end{array}
$$

Estimate:

$$
\begin{array}{r}
100 \\
6 \\
+\ 9 \\
\hline
115
\end{array}
$$

15.

$$
\begin{array}{r}
39.000 \\
3.006 \\
+\ 8.403 \\
\hline
50.409
\end{array}
$$

17.

$$
\begin{array}{r}
12.6 \\
-\ 8.2 \\
\hline
4.4
\end{array}
$$

Check:

$$
\begin{array}{r}
4.4 \\
+\ 8.2 \\
\hline
12.6
\end{array}
$$

19.

$$
\begin{array}{r}
18.0 \\
-\ 2.7 \\
\hline
15.3
\end{array}
$$

Check:

$$
\begin{array}{r}
15.3 \\
+\ 2.7 \\
\hline
18.0
\end{array}
$$

21.

$$
\begin{array}{r}
654.90 \\
-\ 56.67 \\
\hline
598.23
\end{array}
$$

Check:

$$
\begin{array}{r}
598.23 \\
+\ 56.67 \\
\hline
654.90
\end{array}
$$

23. Exact: $5.9 - 4.07 = 1.83$
Estimate: $6 - 4 = 2$
Check: $1.83 + 4.07 = 5.90$

25. Exact:

$$
\begin{array}{r}
1000.0 \\
-\ 123.4 \\
\hline
876.6
\end{array}
$$

Check:

$$
\begin{array}{r}
876.6 \\
+\ 123.4 \\
\hline
1000.0
\end{array}
$$

Estimate:

$$
\begin{array}{r}
1000 \\
-\ 100 \\
\hline
900
\end{array}
$$

27.

$$
\begin{array}{r}
200.0 \\
-\ 5.6 \\
\hline
194.4
\end{array}
$$

29. $-1.12 - 5.2 = -1.12 + (-5.2)$
Add the absolute values.

$$
\begin{array}{r}
1.12 \\
+\ 5.20 \\
\hline
6.32
\end{array}
$$

Attach the common sign.
$-1.12 - 5.2 = -6.32$

31. $5.21 - 11.36 = 5.21 + (-11.36)$
Subtract the absolute values.

$$
\begin{array}{r}
11.36 \\
-\ 5.21 \\
\hline
6.15
\end{array}
$$

Attach the sign of the larger absolute value.
$5.21 - 11.36 = -6.15$

33. $-2.6 - (-5.7) = -2.6 + 5.7$
Subtract the absolute values.

$$
\begin{array}{r}
5.7 \\
-\ 2.6 \\
\hline
3.1
\end{array}
$$

Attach the sign of the larger absolute value.
$-2.6 - (-5.7) = 3.1$

35.   3.0000
        − 0.0012

        2.9988

37.   23.0
        − 6.7

        16.3

39.   0.9
        + 2.2

        3.1

41. $-6.06 + 0.44$
Subtract the absolute values.
        6.06
        − 0.44

        5.62
Attach the sign of the larger absolute value.
$-6.06 + 0.44 = -5.62$

43.   500.21
        − 136.85

        363.36

45. $50.2 - 600 = 50.2 + (-600)$
Subtract the absolute values.
        600.0
        − 50.2

        549.8
Attach the sign of the larger absolute value.
$50.2 - 600 = -549.8$

47.   923.5
        − 61.9

        861.6

49.   100.009
        6.080
        + 9.034

        115.123

51. $-0.003 + 0.091$
Subtract the absolute values.
        0.091
        − 0.003

        0.088
Attach the sign of the larger absolute value.
$-0.003 + 0.091 = 0.088$

53. $-102.4 - 78.04 = -102.4 + (-78.04)$
Add the absolute values.
        102.40
        + 78.04

        180.44
Attach the common sign.
$-102.4 - 78.04 = -180.44$

55. $-2.9 - (-1.8) = -2.9 + 1.8$
Subtract the absolute values.
        2.9
        − 1.8

        1.1
Attach the sign of the larger absolute value.
$-2.9 - (-1.8) = -1.1$

57. $x + z = 3.6 + 0.21 = 3.81$

59. $x - z = 3.6 - 0.21 = 3.39$

61. $y - x + z = 5 - 3.6 + 0.21$
$= 5.00 - 3.60 + 0.21$
$= 1.40 + 0.21$
$= 1.61$

63. $x + 2.7 = 9.3$
$7 + 2.7 \stackrel{?}{=} 9.3$
$9.7 = 9.3$ False
No, 7 is not a solution.

65. $27.4 + y = 16$
$27.4 + (-11.4) \stackrel{?}{=} 16$
$16 = 16$ True
Yes, −11.4 is a solution.

67. $2.3 + x = 5.3 - x$
$2.3 + 1 \stackrel{?}{=} 5.3 - 1$
$3.3 = 4.3$ False
No, 1 is not a solution.

69. $30.7x + 17.6 - 23.8x - 10.7$
$= 30.7x - 23.8x + 17.6 - 10.7$
$= (30.7 - 23.8)x + (17.6 - 10.7)$
$= 6.9x + 6.9$

71. $-8.61 + 4.23y - 2.36 - 0.76y$
$= 4.23y - 0.76y - 8.61 - 2.36$
$= (4.23 - 0.76)y + (-8.61 - 2.36)$
$= 3.47y - 10.97$

73. Subtract the opening price from the closing price.

$$\begin{array}{r} 22.07 \\ -\ 21.90 \\ \hline 0.17 \end{array}$$

The price of each share increased by $0.17.

75. Perimeter $= 7.14 + 7.14 + 7.14 + 7.14$
$= 28.56$ meters

77. Perimeter $= 3.6 + 2.0 + 3.6 + 2.0$
$= 11.2$ inches

79.
$$\begin{array}{r} 40.00 \\ -\ 32.48 \\ \hline 7.52 \end{array}$$
The change was $7.52.

81. The phrase "How much faster" indicates that we should subtract the average wind speed from the record speed.

$$\begin{array}{r} 321.0 \\ -\ 35.2 \\ \hline 285.8 \end{array}$$

The highest wind speed is 285.8 miles per hour faster than the average wind speed.

83.
$$\begin{array}{r} 10.4 \\ -\ 7.0 \\ \hline 3.4 \end{array}$$
The decrease in revenue is predicted to be $3.4 billion.

85. To find the total, we add.

$$\begin{array}{r} 600.78 \\ 460.99 \\ +\ 436.72 \\ \hline 1498.49 \end{array}$$

The total ticket sales were $1498.49 million.

87. To find the amount of snow in Blue Canyon, add 111.6 to the amount in Marquette.

$$\begin{array}{r} 129.2 \\ +\ 111.6 \\ \hline 240.8 \end{array}$$

Blue Canyon receives on average 240.8 inches each year.

89. Add the lengths of the sides to get the perimeter.

$$\begin{array}{r} 12.40 \\ 29.34 \\ +\ 25.70 \\ \hline 67.44 \end{array}$$

67.44 feet of border material is needed.

91.
$$\begin{array}{r} 160.927 \\ -\ 142.734 \\ \hline 18.193 \end{array}$$
The difference is 18.193 miles per hour.

93. The tallest bar indicates the greatest chocolate consumption per person, so Switzerland has the greatest chocolate consumption per person.

95.
$$\begin{array}{r} 22.0 \\ -\ 13.9 \\ \hline 8.1 \end{array}$$
The difference in consumption is 8.1 pounds per year.

97.

Country	Pounds of Chocolate per Person
Switzerland	22.0
Norway	16.0
Germany	15.8
United Kingdom	14.5
Belgium	13.9

99. $46 \cdot 3 = 138$

101. $\left(\dfrac{1}{5}\right)^3 = \dfrac{1}{5} \cdot \dfrac{1}{5} \cdot \dfrac{1}{5} = \dfrac{1 \cdot 1 \cdot 1}{5 \cdot 5 \cdot 5} = \dfrac{1}{125}$

103. $10.68 - (2.3 + 2.3) = 10.68 - 4.60 = 6.08$
The unknown length is 6.08 inches.

105. 3 nickels, 3 dimes, and 3 quarters:
$0.05 + 0.05 + 0.05 + 0.10 + 0.10 + 0.10$
$+ 0.25 + 0.25 + 0.25 = 1.2$
The value of the coins shown is \$1.20.

107. 1 nickel, 1 dime, and 2 pennies:
$0.05 + 0.10 + 0.01 + 0.01 = 0.17$
3 nickels and 2 pennies:
$0.05 + 0.05 + 0.05 + 0.01 + 0.01 = 0.17$
1 dime and 7 pennies:
$0.10 + 0.01 + 0.01 + 0.01 + 0.01 + 0.01$
$\qquad + 0.01 + 0.01 = 0.17$

109. answers may vary

111.
$$\begin{aligned} 256,436.012 \\ - \ 256,435.235 \\ \hline 0.777 \end{aligned}$$
The difference in measurements is
0.777 mile.

113. $14.271 - 8.968x + 1.333 - 201.815x +$
$101.239x$
$= -8.968x - 201.815x + 101.239x + 14.271$
$\qquad + 1.333$
$= (-8.968 - 201.815 + 101.239)x$
$\qquad + (14.271 + 1.333)$
$= -109.544x + 15.604$

Section 5.3

Practice Problems

1.
$$\begin{array}{r} 34.8 \\ \times \ 0.62 \\ \hline 696 \\ 20\ 880 \\ \hline 21.576 \end{array}$$
1 decimal place
2 decimal places
$1 + 2 = 3$ decimal places

2.
$$\begin{array}{r} 0.0641 \\ \times \quad 27 \\ \hline 4487 \\ 1\ 2820 \\ \hline 1.7307 \end{array}$$
4 decimal places
0 decimal places
$4 + 0 = 4$ decimal places

3. $(7.3)(-0.9) = -6.57$ (Be sure to include the negative sign.)

4. Exact:
$$\begin{array}{r} 30.26 \\ \times \quad 2.89 \\ \hline 2\ 7234 \\ 24\ 2080 \\ 60\ 5200 \\ \hline 87.4514 \end{array}$$
Estimate:
$$\begin{array}{r} 30 \\ \times \ 3 \\ \hline 90 \end{array}$$

5. $46.8 \times 10 = 468$

6. $203.004 \times 100 = 20,300.4$

7. $(-2.33)(1000) = -2330$

8. $6.94 \times 0.1 = 0.694$

9. $3.9 \times 0.01 = 0.039$

10. $(-7682)(-0.001) = 7.682$

11. $76.2 \text{ million} = 76.2 \times 1 \text{ million}$
$= 76.2 \times 1,000,000$
$= 76,200,000$

12. $7y = 7(-0.028) = -0.196$

13. $-6x = 33$

        $-6(-5.5) \stackrel{?}{=} 33$

               $33 = 33$ True

    Yes, -5.5 is a solution.

14. $C = 2\pi r = 2\pi \cdot 11 = 22\pi \approx 22(3.14) = 69.08$

      The circumference is

      22π meters ≈ 69.08 meters.

15.   60.5

      $\times$ 5.6

      ——————

      36 30

      302 50

      ——————

      338.80

      She needs 338.8 ounces of fertilizer.

Vocabulary and Readiness Check

1. When multiplying decimals, the number of decimal places in the product is equal to the <u>sum</u> of the number of decimal place in the factors.

3. When multiplying a decimal number by powers of 10 such as 10, 100, 1000, and so on, we move the decimal point in the number to the <u>right</u> the same number of places as there are <u>zeros</u> in the power of 10.

5. The distance around a circle is called its <u>circumference</u>.

Exercise Set 5.3

1.   0.17   2 decimal places

    $\times$ 8   0 decimal places

    ————

    1.36 $2 + 0 = 2$ decimal places

3.   1.2   1 decimal place

    $\times$ 0.5   1 decimal place

    ————

    0.60 $1 + 1 = 2$ decimal places

5. The product $(-2.3)(7.65)$ is negative.

      7.65   2 decimal places

    $\times$ 2.3   1 decimal place

    ————

    2 295

    15 300

    ————

   -17.595 $2 + 1 = 3$ decimal places and

                 include the negative sign

7. The product $(-5.73)(-9.6)$ is positive.

      5.73   2 decimal places

    $\times$ 9.6   1 decimal place

    ————

    3 438

    51 570

    ————

    55.008 $2 + 1 = 3$ decimal places

9. Exact: 6.8   Estimate: 7

         $\times$ 4.2 $\times$ 4

         ————   ——

         1 36   28

         27 20

         ————

         28.56

11.   0.347   3 decimal places

    $\times$ 0.3   1 decimal place

    ————

    0.1041 $3 + 1 = 4$ decimal places

13. Exact: 1.0047   Estimate: 1

         $\times$ 8.2 $\times$ 8

         ————   ——

         20094   8

         8 03760

         ————

         8.23854

15.   490.2   1 decimal place

    $\times$ 0.023   3 decimal places

    ————

    1 4706

    9 8040

    ————

    11.2746 $1 + 3 = 4$ decimal places

17. $6.5 \times 10 = 65$

19. $8.3 \times 0.1 = 0.83$

21. $(-7.093)(1000) = -7093$

23. $7.093 \times 100 = 709.3$

25. $(-9.83)(-0.01) = 0.0983$

27. $25.23 \times 0.001 = 0.02523$

29.
$$\begin{array}{r} 0.123 \\ \times\ \ 0.4 \\ \hline 0.0492 \end{array}$$

31. $(147.9)(100) = 14{,}790$

33.
$$\begin{array}{r} 8.6 \\ \times\ 0.15 \\ \hline 430 \\ 860 \\ \hline 1.290 \end{array}$$ or 1.29

35. $(937.62)(-0.01) = -9.3762$

37. $562.3 \times 0.001 = 0.5623$

39.
$$\begin{array}{r} 6.32 \\ \times\ \ 5.7 \\ \hline 4\ 424 \\ 31\ 600 \\ \hline 36.024 \end{array}$$

41. 1.5 billion $= 1.5 \times 1$ billion
$= 1.5 \times 1{,}000{,}000{,}000$
$= 1{,}500{,}000{,}000$
The cost at launch was $1,500,000,000.

43. 49.8 million $= 49.8 \times 1$ million
$= 49.8 \times 1{,}000{,}000$
$= 49{,}800{,}000$
The roller coaster has given more than 49,800,000 rides.

45. $xy = 3(-0.2) = -0.6$

47. $xz - y = 3(5.7) - (-0.2)$
$= 17.1 - (-0.2)$
$= 17.1 + 0.2$
$= 17.3$

49. $0.6x = 4.92$
$0.6(14.2) \overset{?}{=} 4.92$
$8.52 = 4.92$ False
No, 14.2 is not a solution.

51. $3.5y = -14$
$3.5(-4) \overset{?}{=} -14$
$-14 = -14$ True
Yes, -4 is a solution.

53. $C = \pi d$ is $\pi(10\text{ cm}) = 10\pi$ cm
$C \approx 10(3.14)\text{ cm} = 31.4$ cm

55. $C = 2\pi r$ is $2\pi \cdot 9.1$ yards $= 18.2$ yards
$C \approx 18.2(3.14)$ yards $= 57.148$ yards

57. Multiply the number of ounces by the number of grams of saturated fat in 1 ounce.
$$\begin{array}{r} 6.2 \\ \times\ \ 4 \\ \hline 24.8 \end{array}$$
There are 24.8 grams of saturated fat in a 4-ounce serving of cream cheese.

59. Area = length $\cdot$ width
$$\begin{array}{r} 3.6 \\ \times\ 2.0 \\ \hline 7.20 \end{array}$$
The area is 7.2 square inches.

61. Circumference = $\pi \cdot$ diameter
$C = \pi \cdot 250 = 250\pi$
$$\begin{array}{r} 250 \\ \times\ \ 3.14 \\ \hline 10\ 00 \\ 25\ 00 \\ 750\ 00 \\ \hline 785.00 \end{array}$$
The circumference is 250π feet, which is approximately 785 feet.

63. $C = \pi \cdot d$

$C = \pi \cdot 135 = 135\pi$

$$\begin{array}{r} 135 \\ \times\,3.14 \\ \hline 5\,40 \\ 13\,50 \\ 405\,00 \\ \hline 423.90 \end{array}$$

He travels 135π meters or approximately 423.9 meters.

65. Multiply her height in meters by the number of inches in 1 meter.

$$\begin{array}{r} 39.37 \\ \times\ \ 1.65 \\ \hline 1\,9685 \\ 23\,6220 \\ 39\,3700 \\ \hline 64.9605 \end{array}$$

She is approximately 64.9605 inches tall.

67.
$$\begin{array}{r} 17.88 \\ \times\ \ \ 40 \\ \hline 715.20 \end{array}$$

His pay for last week was $715.20.

69. a. Circumference = $2 \cdot \pi \cdot$ radius

Smaller circle:

$C = 2 \cdot \pi \cdot 10 = 20\pi$

$C \approx 20(3.14) = 62.8$

The circumference of the smaller circle is approximately 62.8 meters.

Larger circle:

$C = 2 \cdot \pi \cdot 20 = 40\pi$

$C \approx 40(3.14) = 125.6$

The circumference of the larger circle is approximately 125.6 meters.

b. Yes, the circumference gets doubled when the radius is doubled.

71. $3.40 \times 100 = 340$

The farmer received $340.

73.
$$\begin{array}{r} 115.26 \\ \times\ \ \ \ 675 \\ \hline 576\,30 \\ 8\,068\,20 \\ 69\,156\,00 \\ \hline 77,800.50 \end{array}$$

675 U.S. Dollars is equivalent to 77,800.5 Japanese yen.

75.
$$\begin{array}{r} 0.7951 \\ \times\ \ \ \ 900 \\ \hline 715.5900 \end{array}$$

900 U.S. Dollars will buy 715.59 European Union euros.

77.
$$\begin{array}{r} 8 \\ 365\overline{)\,2920} \\ -2920 \\ \hline 0 \end{array}$$

79. $-\dfrac{24}{7} \div \dfrac{8}{21} = -\dfrac{24}{7} \cdot \dfrac{21}{8}$

$\qquad = -\dfrac{8 \cdot 3 \cdot 7 \cdot 3}{7 \cdot 8}$

$\qquad = -\dfrac{3 \cdot 3}{1}$

$\qquad = -9$

81.
$$\begin{array}{r} 3.60 \\ +\,0.04 \\ \hline 3.64 \end{array}$$

83.
$$\begin{array}{r} 3.60 \\ -\,0.04 \\ \hline 3.56 \end{array}$$

85.
$$\begin{array}{r} -0.221 \\ \times\ \ \ \ 0.5 \\ \hline -0.1105 \end{array}$$

87.
```
     20.6
  ×  1.86
  --------
   1 236
  16 480
  20 600
  --------
  38.316
```
$38.316 \times 100,000 = 3,831,600$
The radio wave travels 3,831,600 miles in 20.6 seconds.

89. answers may vary

91. answers may vary

Section 5.4

Practice Problems

1.
```
          46.3
     8) 370.4          Check:  46.3
       -32                   ×    8
       ----                  -------
        50                    370.4
       -48
       ----
         2 4
        -2 4
        -----
          0
```

2.
```
          0.71
    48) 34.08           Check:  0.71
       -33 6                   ×  48
       ------                  ------
          48                    5 68
         -48                   28 40
         ----                  ------
           0                    34.08
```

3. a.
```
           1.135
     14) 15.890          Check:  1.135
        -14                     ×   14
        ----                    -------
         1 8                     4 540
        -1 4                    11 350
        -----                   -------
          49                    15.890
         -42
         ----
          70
         -70
         ----
           0
```
Thus, $-15.89 \div 14 = -1.135$.

b.
```
            0.027
   104) 2.808            Check:  0.027
      -2 08                     ×  104
      ------                    -------
        728                        108
       -728                      2 700
       -----                     -------
          0                      2.808
```
Thus, $-2.808 \div (-104) = 0.027$

4.
```
                              29.8
   5.6) 166.88  becomes  56) 1668.8
                            -112
                            -----
                             548
                            -504
                            -----
                              44 8
                             -44 8
                             ------
                                0
```

5.
```
                               12.35
   0.16) 1.976  becomes  16) 197.60
                            -16
                            ----
                             37
                            -32
                            ----
                             56
                            -48
                            ----
                             80
                            -80
                            ----
                             0
```

6. $0.57\overline{)23.4}$ becomes

$$
\begin{array}{r}
41.052 \approx 41.05 \\
57\overline{)\,2340.000} \\
\underline{-228} \\
60 \\
\underline{-57} \\
3\,00 \\
\underline{-2\,85} \\
150 \\
\underline{-114} \\
36
\end{array}
$$

7. $91.5\overline{)713.7}$ becomes

$$
\begin{array}{r}
7.8 \\
915\overline{)\,7137.0} \\
\underline{-6405} \\
732\,0 \\
\underline{-732\,0} \\
0
\end{array}
$$

Estimate: $100\overline{)700}$ with quotient 7

8. $\dfrac{362.1}{100} = 0.3621$

9. $-\dfrac{0.49}{10} = -0.049$

10. $x \div y = 0.035 \div 0.02$

$0.02\overline{)0.035}$ becomes

$$
\begin{array}{r}
1.75 \\
2\overline{)\,3.50} \\
\underline{-2} \\
1\,5 \\
\underline{-1\,4} \\
10 \\
\underline{-10} \\
0
\end{array}
$$

11. $\dfrac{x}{100} = 3.9$

$\dfrac{39}{100} \overset{?}{=} 3.9$

$0.39 = 3.9$ False

No, 39 is not a solution.

12.

$$
\begin{array}{r}
11.84 \\
1250\overline{)\,14800.00} \\
\underline{-1250} \\
2300 \\
\underline{-1250} \\
10\,50\,0 \\
\underline{-10\,00\,0} \\
50\,0\,0 \\
\underline{-50\,0\,0} \\
0
\end{array}
$$

He needs 11.84 bags or 12 whole bags.

Calculator Explorations

1. $102.62 \times 41.8 \approx 100 \times 40 = 4000$
Since 4000 is not close to 428.9516, it is not reasonable.

3. $1025.68 - 125.42 \approx 1000 - 100 = 900$
Since 900 is close to 900.26, it is reasonable.

Vocabulary and Readiness Check

1. In $6.5 \div 5 = 1.3$, the number 1.3 is called the quotient, 5 is the divisor, and 6.5 is the dividend.

3. To divide a decimal number by a power of 10 such as 10, 100, 1000, or so on, we move the decimal point in the number to the left the same number of places as there are zeros in the power of 10.

Exercise Set 5.4

1.

$$
\begin{array}{r}
4.6 \\
6\overline{)\,27.6} \\
\underline{-24} \\
3\,6 \\
\underline{-3\,6} \\
0
\end{array}
$$

3.
$$
\begin{array}{r}
0.094 \\
5\overline{)0.470} \\
\underline{-45} \\
20 \\
\underline{-20} \\
0
\end{array}
$$

5. $0.06\overline{)18}$ becomes
$$
\begin{array}{r}
300 \\
6\overline{)1800} \\
\underline{-18} \\
000
\end{array}
$$

7. $0.82\overline{)4.756}$ becomes
$$
\begin{array}{r}
5.8 \\
82\overline{)475.6} \\
\underline{-410} \\
65\ 6 \\
\underline{-65\ 6} \\
0
\end{array}
$$

9. Exact: $5.5\overline{)36.3}$ becomes
$$
\begin{array}{r}
6.6 \\
55\overline{)363.0} \\
\underline{-330} \\
33\ 0 \\
\underline{-33\ 0} \\
0
\end{array}
$$

Estimate:
$$
\begin{array}{r}
6 \\
6\overline{)36}
\end{array}
$$

11.
$$
\begin{array}{r}
0.413 \\
18\overline{)7.434} \\
\underline{-7\ 2} \\
23 \\
\underline{-18} \\
54 \\
\underline{-54} \\
0
\end{array}
$$

13. A positive number divided by a negative number is a negative number.

$0.06\overline{)36}$ becomes
$$
\begin{array}{r}
600 \\
6\overline{)3600} \\
\underline{-36} \\
0
\end{array}
$$

$36 \div (-0.06) = -600$

15. A negative number divided by a negative number is a positive number.

$0.6\overline{)4.2}$ becomes
$$
\begin{array}{r}
7 \\
6\overline{)42} \\
\underline{-42} \\
0
\end{array}
$$

$(-4.2) \div (-0.6) = 7$

17. $0.27\overline{)1.296}$ becomes
$$
\begin{array}{r}
4.8 \\
27\overline{)129.6} \\
\underline{-108} \\
21\ 6 \\
\underline{-21\ 6} \\
0
\end{array}
$$

19. $0.02\overline{)42}$ becomes
$$
\begin{array}{r}
2100 \\
2\overline{)4200} \\
\underline{-4} \\
02 \\
\underline{-2} \\
000
\end{array}
$$

21. $0.82\overline{)4.756}$ becomes
$$
\begin{array}{r}
5.8 \\
82\overline{)475.6} \\
\underline{-410} \\
65\ 6 \\
\underline{-65\ 6} \\
0
\end{array}
$$

23. A negative number divided by a negative number is a positive number.

$$
\begin{array}{r}
5.5 \\
6.6\overline{)36.3} \text{ becomes } 66\overline{)363.0} \\
\underline{-330} \\
33\ 0 \\
\underline{-33\ 0} \\
0
\end{array}
$$

$-36.3 \div -6.6 = 5.5$

25. Exact:
$$
\begin{array}{r}
9.8 \\
7.2\overline{)70.56} \text{ becomes } 72\overline{)705.6} \\
\underline{-648} \\
57\ 6 \\
\underline{-57\ 6} \\
0
\end{array}
$$

Estimate:
$$
\begin{array}{r}
10 \\
7\overline{)70}
\end{array}
$$

27.
$$
\begin{array}{r}
9.6 \\
5.4\overline{)51.84} \text{ becomes } 54\overline{)518.4} \\
\underline{-486} \\
32\ 4 \\
\underline{-32\ 4} \\
0
\end{array}
$$

29.
$$
\begin{array}{r}
45 \\
0.027\overline{)1.215} \text{ becomes } 27\overline{)1215} \\
\underline{-108} \\
135 \\
\underline{-135} \\
0
\end{array}
$$

$\dfrac{1.215}{0.027} = 45$

31.
$$
\begin{array}{r}
54.592 \\
0.25\overline{)13.648} \text{ becomes } 25\overline{)1364.800} \\
\underline{-125} \\
114 \\
\underline{-100} \\
14\ 8 \\
\underline{-12\ 5} \\
2\ 30 \\
\underline{-2\ 25} \\
50 \\
\underline{-50} \\
0
\end{array}
$$

33.
$$
\begin{array}{r}
0.0055 \\
3.78\overline{)0.02079} \text{ becomes } 378\overline{)2.0790} \\
\underline{-1\ 890} \\
1890 \\
\underline{-1890} \\
0
\end{array}
$$

35. $0.023\overline{)0.549}$ becomes

$$
\begin{array}{r}
23.869 \approx 23.87 \\
23\overline{)549.000} \\
\underline{-46} \\
89 \\
\underline{-69} \\
20\ 0 \\
\underline{-18\ 4} \\
1\ 60 \\
\underline{-1\ 38} \\
220 \\
\underline{-207} \\
13
\end{array}
$$

37. $0.6\overline{)68.39}$ becomes $\begin{array}{r} 113.98 \approx 114.0 \\ 6\overline{)683.90} \\ \underline{-6} \\ 08 \\ \underline{-6} \\ 23 \\ \underline{-18} \\ 5\;9 \\ \underline{-5\;4} \\ 50 \\ \underline{-48} \\ 2 \end{array}$

39. $\dfrac{83.397}{100} = 0.83397$

41. $\dfrac{26.87}{10} = 2.687$

43. $12.9 \div (-1000) = -0.0129$

45. $\begin{array}{r} 1.26 \\ 7\overline{)88.2} \\ \underline{-7} \\ 18 \\ \underline{-14} \\ 4\;2 \\ \underline{-4\;2} \\ 0 \end{array}$

47. $\dfrac{13.1}{10} = 1.31$

49. $\dfrac{456.25}{10,000} = 0.045625$

51. $\begin{array}{r} 0.413 \\ 3\overline{)1.239} \\ \underline{-1\;2} \\ 03 \\ \underline{-3} \\ 09 \\ \underline{-9} \\ 0 \end{array}$

53. $0.6\overline{)4.2}$ becomes $\begin{array}{r} 7 \\ 6\overline{)42} \\ \underline{-42} \\ 0 \end{array}$

$4.2 \div (-0.6) = -7$

55. $0.27\overline{)1.296}$ becomes $\begin{array}{r} 4.8 \\ 27\overline{)129.6} \\ \underline{-108} \\ 21\;6 \\ \underline{-21\;6} \\ 0 \end{array}$

$-1.296 \div 0.27 = -4.8$

57. $0.02\overline{)42}$ becomes $\begin{array}{r} 2100 \\ 2\overline{)4200} \\ \underline{-4} \\ 02 \\ \underline{-2} \\ 0 \end{array}$

$42 \div 0.02 = 2100$

59. $0.6\overline{)18}$ becomes $\begin{array}{r} 30 \\ 6\overline{)180} \\ \underline{-18} \\ 00 \end{array}$

$-18 \div (-0.6) = 30$

61. $0.0015\overline{)87}$ becomes

$$
\begin{array}{r}
58{,}000 \\
15\overline{)870{,}000} \\
\underline{-75} \\
120 \\
\underline{-120} \\
0\,000
\end{array}
$$

$87 \div (-0.0015) = -58{,}000$

63. $1.6\overline{)1.104}$ becomes

$$
\begin{array}{r}
0.69 \\
16\overline{)11.04} \\
\underline{-9\,6} \\
1\,44 \\
\underline{-1\,44} \\
0
\end{array}
$$

$-1.104 \div 1.6 = -0.69$

65. $-2.4 \div (-100) = \dfrac{-2.4}{-100} = 0.024$

67. $0.071\overline{)4.615}$ becomes

$$
\begin{array}{r}
65 \\
71\overline{)4615} \\
\underline{-426} \\
355 \\
\underline{-355} \\
0
\end{array}
$$

$\dfrac{4.615}{0.071} = 65$

69. $z \div y = 4.52 \div (-0.8)$

$0.8\overline{)4.52}$ becomes

$$
\begin{array}{r}
5.65 \\
8\overline{)45.20} \\
\underline{-40} \\
5\,2 \\
\underline{-4\,8} \\
40 \\
\underline{-40} \\
0
\end{array}
$$

$z \div y = 4.52 \div (-0.8) = -5.65$

71. $x \div y = 5.65 \div (-0.8)$

$0.8\overline{)5.65}$ becomes

$$
\begin{array}{r}
7.0625 \\
8\overline{)56.5000} \\
\underline{-56} \\
0\,5 \\
\underline{-0} \\
50 \\
\underline{-48} \\
20 \\
\underline{-16} \\
40 \\
\underline{-40} \\
0
\end{array}
$$

$x \div y = 5.65 \div (-0.8) = -7.0625$

73.
$$\frac{x}{4} = 3.04$$
$$\frac{12.16}{4} \stackrel{?}{=} 3.04$$
$$3.04 = 3.04 \quad \text{True}$$
Yes, 12.16 is a solution.

75.
$$\frac{z}{100} = 0.8$$
$$\frac{8}{100} \stackrel{?}{=} 0.8$$
$$0.08 = 0.8 \quad \text{False}$$
No, 8 is not a solution.

77.
$$
\begin{array}{r}
10.5 \approx 11 \\
52\overline{)546.0} \\
\underline{-52} \\
26 \\
\underline{-0} \\
26\,0 \\
\underline{-26\,0} \\
\end{array}
$$

Since he must buy whole quarts, 11 quarts are needed.

79. $39.37\overline{)200}$ becomes

$$
\begin{array}{r}
5.08 \approx 5.1 \\
3937\overline{)20000.00} \\
-19685 \\
\hline
315\ 0 \\
-\quad 0 \\
\hline
31500 \\
-31496 \\
\hline
4
\end{array}
$$

There are approximately 5.1 meters in 200 inches.

81. Divide the number of crayons by 64.

$$
\begin{array}{r}
11.40 \approx 11.4 \\
64\overline{)730.00} \\
-64 \\
\hline
90 \\
-64 \\
\hline
26\ 0 \\
-25\ 6 \\
\hline
40
\end{array}
$$

740 crayons is approximately 11.4 boxes.

83. $6 \times 4 = 24$
There are 24 teaspoons in 4 fluid ounces.

85. From Exercise 83, we know that there are 24 teaspoons in 4 fluid ounces. Thus, there are 48 half teaspoons (0.5 tsp) or doses in 4 fluid ounces. To see how long the medicine will last, if a dose is taken every 4 hours, there are $24 \div 4 = 6$ doses taken per day. 48 (doses) $\div$ 6 (per day) = 8 days. The medicine will last 8 days.

87. There are 52 weeks in 1 year.

$$
\begin{array}{r}
248.07 \\
52\overline{)12,900.00} \\
-10\ 4 \\
\hline
2\ 50 \\
-2\ 08 \\
\hline
420 \\
-416 \\
\hline
4\ 0 \\
-0 \\
\hline
400 \\
-364 \\
\hline
36
\end{array}
$$

Americans aged 18–22 drive, on average, 248.1 miles per week.

89.
$$
\begin{array}{r}
52 \\
\times\ 40 \\
\hline
2080
\end{array}
$$

There are 2080 hours in 52 40-hour weeks.

$$
\begin{array}{r}
5109.615 \approx 5109.62 \\
2080\overline{)10,628,000.000} \\
-10\ 400 \\
\hline
228\ 0 \\
-208\ 0 \\
\hline
20\ 00 \\
-\quad 0 \\
\hline
20\ 000 \\
-18\ 720 \\
\hline
1\ 280\ 0 \\
-1\ 248\ 0 \\
\hline
32\ 00 \\
-20\ 80 \\
\hline
11\ 200 \\
-10\ 400 \\
\hline
800
\end{array}
$$

His hourly wage would have been $5109.62.

91. $\dfrac{3}{5} \cdot \dfrac{7}{10} = \dfrac{3 \cdot 7}{5 \cdot 10} = \dfrac{21}{50}$

93. $\dfrac{3}{5} - \dfrac{7}{10} = \dfrac{3}{5} \cdot \dfrac{2}{2} - \dfrac{7}{10} = \dfrac{6}{10} - \dfrac{7}{10} = -\dfrac{1}{10}$

95. $0.3\overline{)1.278}$ becomes

$$
\begin{array}{r}
4.26 \\
3\overline{)12.78} \\
-12 \\
\hline
0\,7 \\
-6 \\
\hline
18 \\
-18 \\
\hline
0
\end{array}
$$

97.
$$
\begin{array}{r}
1.278 \\
+\ 0.300 \\
\hline
1.578
\end{array}
$$

99.
$$
\begin{array}{rl}
8.6 & \text{1 decimal place} \\
\times\ 3.1 & \text{1 decimal place} \\
\hline
86 & \\
25\,80 & \\
\hline
26.66 & 1+1 = 2 \text{ decimal places}
\end{array}
$$
$(-8.6)(3.1) = -26.66$

101.
$$
\begin{array}{r}
1000.00 \\
-\ \ \ 95.71 \\
\hline
904.29
\end{array}
$$

103. 8.62×41.7 is approximately $9 \times 40 = 360$, which is choice c.

105. $78.6 \div 97$ is approximately
$78.6 \div 100 = 0.786$, which is choice b.

107. $\dfrac{86 + 78 + 91 + 87}{4} = \dfrac{342}{4} = 85.5$

109. Area = (length)(width)

$4.5\overline{)38.7}$ becomes
$$
\begin{array}{r}
8.6 \\
45\overline{)387.0} \\
-360 \\
\hline
27\,0 \\
-27\,0 \\
\hline
0
\end{array}
$$
The length is 8.6 feet.

111. answers may vary

113. $1.15\overline{)75}$ becomes
$$
\begin{array}{r}
65.21 \approx 65.2 \\
115\overline{)7500.00} \\
-690 \\
\hline
600 \\
-575 \\
\hline
250 \\
-230 \\
\hline
200 \\
-115 \\
\hline
85
\end{array}
$$

$1.15\overline{)95}$ becomes
$$
\begin{array}{r}
82.60 \approx 82.6 \\
115\overline{)9500.00} \\
-920 \\
\hline
300 \\
-230 \\
\hline
700 \\
-690 \\
\hline
100 \\
-0 \\
\hline
100
\end{array}
$$

The range of wind speeds is
65.2–82.6 knots.

115. First find the length for one round of wire. Then multiply by 4.

$$
\begin{array}{r}
24.280 \\
15.675 \\
24.280 \\
+\ 15.675 \\
\hline
79.910
\end{array}
\qquad
\begin{array}{r}
79.91 \\
\times\ \ \ \ 4 \\
\hline
319.64
\end{array}
$$

He will need 319.64 meters of wire.

The Bigger Picture

1.
$$
\begin{array}{r}
3.600 \\
8.092 \\
+\ 10.480 \\
\hline
22.172
\end{array}
$$

2.
$$
\begin{array}{r}
7.000 \\
-\ 3.049 \\
\hline
3.951
\end{array}
$$

3. $91.332 \times 100 = 9133.2$

4. $-\dfrac{68}{10} = -6.8$

5.
$$
\begin{array}{r}
5.2 \\
\times\, 0.27 \\
\hline
364 \\
1\,040 \\
\hline
1.404
\end{array}
$$
1 decimal place
2 decimal places
$1 + 2 = 3$ decimal places

6.
$$
\begin{array}{r}
8.66 \\
9\,\overline{)\,77.94} \\
-72 \\
\hline
5\,9 \\
-5\,4 \\
\hline
54 \\
-54 \\
\hline
0
\end{array}
$$

7. $0.35\,\overline{)\,0.01785}$ becomes
$$
\begin{array}{r}
0.051 \\
35\,\overline{)\,1.785} \\
-1\,75 \\
\hline
35 \\
-35 \\
\hline
0
\end{array}
$$

8. $2.3 - (0.4)^2 = 2.30 - 0.16 = 2.14$

9. $\dfrac{8}{15} - \dfrac{2}{5} = \dfrac{8}{15} - \dfrac{2}{5} \cdot \dfrac{3}{3} = \dfrac{8}{15} - \dfrac{6}{15} = \dfrac{8-6}{15} = \dfrac{2}{15}$

10. $-\dfrac{8}{15} \cdot \dfrac{2}{5} = -\dfrac{8 \cdot 2}{15 \cdot 5} = -\dfrac{16}{75}$

Integrated Review

1.
$$
\begin{array}{r}
1.60 \\
+\, 0.97 \\
\hline
2.57
\end{array}
$$

2.
$$
\begin{array}{r}
3.20 \\
+\, 0.85 \\
\hline
4.05
\end{array}
$$

3.
$$
\begin{array}{r}
9.8 \\
-\, 0.9 \\
\hline
8.9
\end{array}
$$

4.
$$
\begin{array}{r}
10.2 \\
-\, 6.7 \\
\hline
3.5
\end{array}
$$

5.
$$
\begin{array}{r}
0.8 \\
\times\, 0.2 \\
\hline
0.16
\end{array}
$$

6.
$$
\begin{array}{r}
0.6 \\
\times\, 0.4 \\
\hline
0.24
\end{array}
$$

7.
$$
\begin{array}{r}
0.27 \\
8\,\overline{)\,2.16} \\
-1\,6 \\
\hline
56 \\
-56 \\
\hline
0
\end{array}
$$

8.
$$
\begin{array}{r}
0.52 \\
6\,\overline{)\,3.12} \\
-3\,0 \\
\hline
12 \\
-12 \\
\hline
0
\end{array}
$$

9.
$$
\begin{array}{r}
9.6 \\
\times\, 0.5 \\
\hline
4.80
\end{array}
$$
$(9.6)(-0.5) = -4.8$

10.
$$
\begin{array}{r}
8.7 \\
\times\, 0.7 \\
\hline
6.09
\end{array}
$$
$(-8.7)(-0.7) = 6.09$

11.
$$
\begin{array}{r}
123.60 \\
-\, 48.04 \\
\hline
75.56
\end{array}
$$

12.   325.20
          − 36.08
        ⎯⎯⎯⎯⎯
          289.12

13. Subtract absolute values.
          25.000
        − 0.026
        ⎯⎯⎯⎯⎯
          24.974
Attach the sign of the larger absolute value.
$-25 + 0.026 = -24.974$

14. Subtract absolute values.
          44.000
        − 0.125
        ⎯⎯⎯⎯⎯
          43.875
Attach the sign of the larger absolute value.
$0.125 + (-44) = -43.875$

15. $3.4\overline{)29.24}$ becomes $34\overline{)292.4}$

$$\begin{array}{r} 8.6 \\ 34\overline{)292.4} \\ \underline{-272} \\ 20\ 4 \\ \underline{-20\ 4} \\ 0 \end{array}$$

$29.24 \div (-3.4) = -8.6$

16. $1.9\overline{)10.26}$ becomes $19\overline{)102.6}$

$$\begin{array}{r} 5.4 \\ 19\overline{)102.6} \\ \underline{-95} \\ 7\ 6 \\ \underline{-7\ 6} \\ 0 \end{array}$$

$-10.26 \div (-1.9) = 5.4$

17. $-2.8 \times 100 = -280$

18. $1.6 \times 1000 = 1600$

19.   96.210
            7.028
        + 121.700
        ⎯⎯⎯⎯⎯⎯
          224.938

20.   0.268
            1.930
        + 142.881
        ⎯⎯⎯⎯⎯⎯
          145.079

21.
$$\begin{array}{r} 0.56 \\ 46\overline{)25.76} \\ \underline{-23\ 0} \\ 2\ 76 \\ \underline{-2\ 76} \\ 0 \end{array}$$

$-25.76 \div -46 = 0.56$

22.
$$\begin{array}{r} 0.63 \\ 43\overline{)27.09} \\ \underline{-25\ 8} \\ 1\ 29 \\ \underline{-1\ 29} \\ 0 \end{array}$$

$-27.09 \div 43 = -0.63$

23.   12.004   3 decimal places
        ×   2.3   1 decimal place
        ⎯⎯⎯⎯⎯
          3 6012
          24 0080
        ⎯⎯⎯⎯⎯
          27.6092 $3 + 1 = 4$ decimal places

24.   28.006   3 decimal places
        ×   5.2   1 decimal place
        ⎯⎯⎯⎯⎯
          5 6012
          140 0300
        ⎯⎯⎯⎯⎯
          145.6312 $3 + 1 = 4$ decimal places

25.   10.0
        − 4.6
        ⎯⎯⎯⎯
          5.4

26. Subtract absolute values.
          18.00
        − 0.26
        ⎯⎯⎯⎯
          17.74
Attach the sign of the greater absolute value.
$0.26 - 18 = -17.74$

27. $-268.19 - 146.25 = -268.19 + (-146.25)$
Add absolute values.

$$
\begin{array}{r}
268.19 \\
+ 146.25 \\
\hline
414.44
\end{array}
$$

Attach the common sign.
$-268.19 - 146.25 = -414.44$

28. $-860.18 - 434.85 = -860.18 + (-434.85)$
Add absolute values.

$$
\begin{array}{r}
860.18 \\
+ \ 434.85 \\
\hline
1295.03
\end{array}
$$

Attach the common sign.
$-860.18 - 434.85 = -1295.03$

29. $0.087\overline{)2.958}$ becomes

$$
\begin{array}{r}
34 \\
87\overline{)2958} \\
-261 \\
\hline
348 \\
-348 \\
\hline
0
\end{array}
$$

$$\frac{2.958}{-0.087} = -34$$

30. $0.061\overline{)1.708}$ becomes

$$
\begin{array}{r}
28 \\
61\overline{)1708} \\
-122 \\
\hline
488 \\
-488 \\
\hline
0
\end{array}
$$

$$\frac{-1.708}{0.061} = -28$$

31.
$$
\begin{array}{r}
160.00 \\
- \ 43.19 \\
\hline
116.81
\end{array}
$$

32.
$$
\begin{array}{r}
120.00 \\
- 101.21 \\
\hline
18.79
\end{array}
$$

33. $15.62 \times 10 = 156.2$

34. $15.62 \div 10 = 1.562$

35.
$$
\begin{array}{r}
15.62 \\
+ 10.00 \\
\hline
25.62
\end{array}
$$

36.
$$
\begin{array}{r}
15.62 \\
- 10.00 \\
\hline
5.62
\end{array}
$$

37.
53.7	rounds to	50
79.2	rounds to	80
+ 71.2	rounds to	+ 70
		200

The estimated distance is 200 miles.

38.
$$
\begin{array}{r}
4.20 \\
- 3.30 \\
\hline
0.90
\end{array}
$$

It costs $0.90 more to send the package as Priority Mail.

39.
$$
\begin{array}{r}
23.324 \\
+ \ 1.556 \\
\hline
24.880
\end{array}
$$

$$
\begin{aligned}
24.88 \text{ billion} &= 24.88 \times 1 \text{ billion} \\
&= 24.88 \times 1,000,000,000 \\
&= 24,880,000,000
\end{aligned}
$$

The total amount spent was $24.88 billion or $24,880,000,000.

Section 5.5

Practice Problems

1. a.
$$
\begin{array}{r}
0.4 \\
5\overline{)2.0} \\
-2.0 \\
\hline
0
\end{array}
$$

$$\frac{2}{5} = 0.4$$

b.

$$
\begin{array}{r}
0.225 \\
40\overline{\smash{)}9.000} \\
\underline{-8\ 0} \\
1\ 00 \\
\underline{-80} \\
200 \\
\underline{-200} \\
0
\end{array}
$$

$\dfrac{9}{40} = 0.225$

4.

$$
\begin{array}{r}
2.1538 \approx 2.154 \\
13\overline{\smash{)}28.0000} \\
\underline{-26} \\
2\ 0 \\
\underline{-1\ 3} \\
70 \\
\underline{-65} \\
50 \\
\underline{-39} \\
110 \\
\underline{-104} \\
6
\end{array}
$$

2.

$$
\begin{array}{r}
0.375 \\
8\overline{\smash{)}3.000} \\
\underline{-2\ 4} \\
60 \\
\underline{-56} \\
40 \\
\underline{-40} \\
0
\end{array}
$$

$-\dfrac{3}{8} = -0.375$

5. $3\dfrac{5}{16} = \dfrac{53}{16}$

$$
\begin{array}{r}
3.3125 \\
16\overline{\smash{)}53.0000} \\
\underline{-48} \\
5\ 0 \\
\underline{-4\ 8} \\
20 \\
\underline{-16} \\
40 \\
\underline{-32} \\
80 \\
\underline{-80} \\
0
\end{array}
$$

3. a.

$$
\begin{array}{r}
0.833... \\
6\overline{\smash{)}5.000} \\
\underline{-4\ 8} \\
20 \\
\underline{-18} \\
20 \\
\underline{-18} \\
2
\end{array}
$$

$\dfrac{5}{6} = 0.8\overline{3}$

b.

$$
\begin{array}{r}
0.22... \\
9\overline{\smash{)}2.00} \\
\underline{-1\ 8} \\
20 \\
\underline{-18} \\
2
\end{array}
$$

$\dfrac{2}{9} = 0.\overline{2}$

Thus, $3\dfrac{5}{16} = 3.3125$.

6. $\dfrac{3}{5} = \dfrac{3}{5} \cdot \dfrac{2}{2} = \dfrac{6}{10} = 0.6$

7. $\dfrac{3}{50} = \dfrac{3}{50} \cdot \dfrac{2}{2} = \dfrac{6}{100} = 0.06$

8.

$$
\begin{array}{r}
0.2 \\
5\overline{\smash{)}1.0} \\
\underline{-1\ 0} \\
0
\end{array}
$$

Since $0.2 < 0.25$, then $\dfrac{1}{5} < 0.25$.

9. a. $\frac{1}{2} = 0.5$ and $0.5 < 0.54$, so $\frac{1}{2} < 0.54$.

b.

$$
\begin{array}{r}
0.44... \\
9\overline{)\,4.00} \\
-3\,6 \\
\hline
40 \\
-36 \\
\hline
4
\end{array}
$$

$0.\overline{4} = 0.44...$, so $0.\overline{4} = \frac{4}{9}$.

c.

$$
\begin{array}{r}
0.714 \\
7\overline{)\,5.000} \\
-4\,9 \\
\hline
10 \\
-7 \\
\hline
30 \\
-28 \\
\hline
2
\end{array}
$$

$0.714 < 0.72$, so $\frac{5}{7} < 0.72$.

10. a. $\frac{1}{3} = 0.333...$

$0.302 = 0.302$

$\frac{3}{8} = 0.375$

$0.302,\ \frac{1}{3},\ \frac{3}{8}$

b. $1.26 = 1.26$

$1\frac{1}{4} = 1.25$

$1\frac{2}{5} = 1.40$

$1\frac{1}{4},\ 1.26,\ 1\frac{2}{5}$

c. $0.4 = 0.40$

$0.41 = 0.41$

$\frac{5}{7} \approx 0.71$

$0.4,\ 0.41,\ \frac{5}{7}$

11. $897.8 \div 100 \times 10 = 8.978 \times 10 = 89.78$

12. $-8.69(3.2 - 1.8) = -8.69(1.4) = -12.166$

13. $(-0.7)^2 + 2.1 = 0.49 + 2.10 = 2.59$

14. $\dfrac{20.06 - (1.2)^2 \div 10}{0.02} = \dfrac{20.06 - 1.44 \div 10}{0.02}$

$$
= \dfrac{20.06 - 0.144}{0.02}
$$

$$
= \dfrac{19.916}{0.02}
$$

$$
= 995.8
$$

15.   Area $= \dfrac{1}{2} \cdot \text{base} \cdot \text{height}$

$$
= \dfrac{1}{2} \cdot 7 \cdot 2.1
$$

$$
= 0.5 \cdot 7 \cdot 2.1
$$

$$
= 7.35
$$

The area of the triangle is 7.35 square meters.

16. $1.7y - 2 = 1.7(2.3) - 2 = 3.91 - 2 = 1.91$

Vocabulary and Readiness Check

1. The number $0.\overline{5}$ means 0.555. <u>false</u>

3. $(-1.2)^2$ means $(-1.2)(-1.2)$ or -1.44. <u>false</u>

Exercise Set 5.5

1.

$$
\begin{array}{r}
0.2 \\
5\overline{)\,1\,0} \\
-1.0 \\
\hline
0
\end{array}
\qquad \frac{1}{5} = 0.2
$$

3. $25\overline{)\,17.00}$ quotient 0.68
$$\underline{-15\ 0}$$
$$\quad\ \ 2\ 00$$
$$\underline{-2\ 00}$$
$$\qquad\ \ 0$$

$\dfrac{17}{25} = 0.68$

5. $4\overline{)\,3.00}$ quotient 0.75
$$\underline{-2\ 8}$$
$$\quad\ 20$$
$$\underline{-20}$$
$$\quad\ \ 0$$

$\dfrac{3}{4} = 0.75$

7. $25\overline{)\,2.00}$ quotient 0.08
$$\underline{-2\ 00}$$
$$\qquad 0$$

$-\dfrac{2}{25} = -0.08$

9. $4\overline{)\,9.00}$ quotient 2.25
$$\underline{-8}$$
$$\quad 1\ 0$$
$$\underline{-\ 8}$$
$$\quad\ \ 20$$
$$\underline{-20}$$
$$\qquad 0$$

$\dfrac{9}{4} = 2.25$

11. $12\overline{)\,11.0000}$ quotient $0.9166...$
$$\underline{-10\ 8}$$
$$\qquad 20$$
$$\underline{-12}$$
$$\qquad 80$$
$$\underline{-72}$$
$$\qquad 80$$
$$\underline{-72}$$
$$\qquad\ 8$$

$\dfrac{11}{12} = 0.91\overline{6}$

13. $40\overline{)\,17.000}$ quotient 0.425
$$\underline{-16\ 0}$$
$$\quad\ 1\ 00$$
$$\underline{-80}$$
$$\quad\ 200$$
$$\underline{-200}$$
$$\qquad\ 0$$

$\dfrac{17}{40} = 0.425$

15. $20\overline{)\,9.00}$ quotient 0.45
$$\underline{-8\ 0}$$
$$\quad 1\ 00$$
$$\underline{-1\ 00}$$
$$\qquad 0$$

$\dfrac{9}{20} = 0.45$

17. $3\overline{)\,1.000}$ quotient $0.333...$
$$\underline{-9}$$
$$\quad 10$$
$$\underline{-9}$$
$$\quad 10$$
$$\underline{-9}$$
$$\quad\ 1$$

$-\dfrac{1}{3} = -0.\overline{3}$

19. $16\overline{)\,7.0000}$ quotient 0.4375
$$\underline{-6\ 4}$$
$$\quad\ 60$$
$$\underline{-48}$$
$$\quad 120$$
$$\underline{-112}$$
$$\qquad 80$$
$$\underline{-80}$$
$$\qquad\ 0$$

$\dfrac{7}{16} = 0.4375$

21.
$$11\overline{)7.000000} \quad\quad \frac{7}{11} = 0.\overline{63}$$

with long division:
- 0.636363...
- $-6\,6$
- 40
- -33
- 70
- -66
- 40
- -33
- 70
- -66
- 40
- -33
- 7

23.
$$20\overline{)17.00} \quad\quad 5\frac{17}{20} = 5.85$$

- 0.85
- $-16\,0$
- $1\,00$
- $-1\,00$
- 0

25.
$$125\overline{)78.000} \quad\quad \frac{78}{125} = 0.624$$

- 0.624
- $-75\,0$
- $3\,00$
- $-2\,50$
- 500
- -500
- 0

27. $-\dfrac{1}{3} = -0.33\overline{3} \approx -0.33$

29. $\dfrac{7}{16} = 0.4375 \approx 0.44$

31. $\dfrac{7}{11} = 0.63\overline{63} \approx 0.6$

33.
$$91\overline{)56.000} \quad\quad 0.615 \approx 0.62$$

- $-54\,6$
- $1\,40$
- -91
- 490
- -455
- 35

35.
$$94\overline{)67.000} \quad\quad 0.712 \approx 0.71$$

- $-65\,8$
- $1\,20$
- -94
- 260
- -188
- 72

37.
$$50\overline{)1.00}$$

- 0.02
- $-1\,00$
- 0

39. 0.562   0.569
$$\uparrow \quad\quad \uparrow$$
$$2 \;<\; 9$$
so, $0.562 < 0.569$

41.
$$200\overline{)43.000}$$

- 0.215
- $-40\,0$
- $3\,00$
- $-2\,00$
- $1\,000$
- $-1\,000$
- 0

$$0.215 = \frac{43}{200}$$

43. 0.0932   0.0923
       ↑   ↑
      3   >   2
so, 0.0932 > 0.0923
Thus, −0.0932 < −0.0923.

45.

$$
\begin{array}{r}
0.833... \\
6\overline{)\,5.000} \\
\underline{-4\ 8} \\
20 \\
\underline{-18} \\
20 \\
\underline{-18} \\
2
\end{array}
$$

$\dfrac{5}{6} = 0.8\overline{3}$ and $0.\overline{6} < 0.8\overline{3}$, so $0.\overline{6} < \dfrac{5}{6}$

47.

$$
\begin{array}{r}
0.5604 \approx 0.560 \\
91\overline{)\,51.0000} \\
\underline{-45\ 5} \\
5\,50 \\
\underline{-546} \\
40 \\
\underline{-0} \\
400 \\
\underline{-364} \\
36
\end{array}
$$

$\dfrac{51}{91} \approx 0.560$ and $0.560 < 0.56\overline{4}$, so

$\dfrac{51}{91} < 0.56\overline{4}.$

49.

$$
\begin{array}{r}
0.571 \approx 0.57 \\
7\overline{)\,4.000} \\
\underline{-3\ 5} \\
50 \\
\underline{-49} \\
10 \\
\underline{-7} \\
3
\end{array}
$$

$\dfrac{4}{7} \approx 0.57$ and $0.57 > 0.14$, so $\dfrac{4}{7} > 0.14.$

51.

$$
\begin{array}{r}
1.3846 \approx 1.385 \\
13\overline{)\,18.0000} \\
\underline{-13} \\
5\ 0 \\
\underline{-3\ 9} \\
1\,10 \\
\underline{-1\,04} \\
60 \\
\underline{-52} \\
80 \\
\underline{-78} \\
2
\end{array}
$$

$\dfrac{18}{13} \approx 1.385$ and $1.38 < 1.385$, so $1.38 < \dfrac{18}{13}.$

53.

$$
\begin{array}{r}
7.125 \\
64\overline{)\,456.000} \\
\underline{-448} \\
8\ 0 \\
\underline{-6\ 4} \\
1\,60 \\
\underline{-1\,28} \\
320 \\
\underline{-320} \\
0
\end{array}
$$

$\dfrac{456}{64} = 7.125$ and $7.123 < 7.125$, so

$7.123 < \dfrac{456}{64}.$

55. 0.32, 0.34, 0.35

57. 0.49 = 0.490
0.49, 0.491, 0.498

59. $\dfrac{42}{8} = 5.25$

     5.23, $\dfrac{42}{8}$, 5.34

61. $\dfrac{5}{8} = 0.625$

     $0.612, \dfrac{5}{8}, 0.649$

63. $(0.3)^2 + 0.5 = 0.09 + 0.5 = 0.59$

65. $\dfrac{1 + 0.8}{-0.6} = \dfrac{1.8}{-0.6} = \dfrac{18}{-6} = -3$

67. $(-2.3)^2(0.3 + 0.7) = (-2.3)^2(1.0)$
                        $= 5.29(1.0)$
                        $= 5.29$

69. $(5.6 - 2.3)(2.4 + 0.4) = (3.3)(2.8) = 9.24$

71. $\dfrac{(4.5)^2}{100} = \dfrac{20.25}{100} = 0.2025$

73. $\dfrac{7 + 0.74}{-6} = \dfrac{7.74}{-6} = -1.29$

75. $\dfrac{1}{5} - 2(7.8) = \dfrac{1}{5} - 15.6 = 0.2 - 15.6 = -15.4$

77. $\dfrac{1}{4}(-9.6 - 5.2) = \dfrac{1}{4}(-14.8)$
                     $= 0.25(-14.8)$
                     $= -3.7$

79. Area $= \dfrac{1}{2} \cdot b \cdot h$
            $= \dfrac{1}{2}(5.7)(9)$
            $= 0.5(5.7)(9)$
            $= 25.65$ square inches

81. Area $= l \cdot w$
            $= (0.62)\left(\dfrac{2}{5}\right)$
            $= (0.62)(0.4)$
            $= 0.248$ square yard

83. $z^2 = (-2.4)^2 = 5.76$

85. $x - y = 6 - 0.3 = 5.7$

87. $4y - z = 4 \cdot 0.3 - (-2.4) = 1.2 + 2.4 = 3.6$

89. $\dfrac{9}{10} + \dfrac{16}{25} = \dfrac{9}{10} \cdot \dfrac{5}{5} + \dfrac{16}{25} \cdot \dfrac{2}{2} = \dfrac{45}{50} + \dfrac{32}{50} = \dfrac{77}{50}$

91. $\left(\dfrac{2}{5}\right)\left(\dfrac{5}{2}\right)^2 = \left(\dfrac{2}{5}\right)\left(\dfrac{5}{2}\right)\left(\dfrac{5}{2}\right) = \dfrac{2 \cdot 5 \cdot 5}{5 \cdot 2 \cdot 2} = \dfrac{5}{2}$

93. $1.0 = 1$

95. $1.00001 > 1$

97. $99 < 100$, so $\dfrac{99}{100} < 1$

99.

$$
\begin{array}{r}
0.1889 \approx 0.189 \\
10{,}661\overline{)\ 2014.0000} \\
\underline{-1066\ 1} \\
947\ 90 \\
\underline{-852\ 88} \\
95\ 020 \\
\underline{-85\ 288} \\
9\ 7320 \\
\underline{-9\ 5949} \\
1371
\end{array}
$$

     $\dfrac{2014}{10{,}661} \approx 0.189$

101.

2019	rounds to	2000
1324	rounds to	1300
773	rounds to	800
684	rounds to	700
703	rounds to	700
502	rounds to	500
		6000

     The total number of stations is estimated to be 6000.

103. answers may vary

Section 5.6

Practice Problems

1. $z + 0.9 = 1.3$
$$z + 0.9 - 0.9 = 1.3 - 0.9$$
$$z = 0.4$$

2. $0.17x = -0.34$
$$\frac{0.17x}{0.17} = \frac{-0.34}{0.17}$$
$$x = -2$$

3. $2.9 = 1.7 + 0.3x$
$$2.9 - 1.7 = 1.7 + 0.3x - 1.7$$
$$1.2 = 0.3x$$
$$\frac{1.2}{0.3} = \frac{0.3x}{0.3}$$
$$4 = x$$

4. $8x + 4.2 = 10x + 11.6$
$$8x + 4.2 - 4.2 = 10x + 11.6 - 4.2$$
$$8x = 10x + 7.4$$
$$8x - 10x = 10x - 10x + 7.4$$
$$-2x = 7.4$$
$$\frac{-2x}{-2} = \frac{7.4}{-2}$$
$$x = -3.7$$

5. $6.3 - 5x = 3(x + 2.9)$
$$6.3 - 5x = 3x + 8.7$$
$$6.3 - 5x - 6.3 = 3x + 8.7 - 6.3$$
$$-5x = 3x + 2.4$$
$$-5x - 3x = 3x + 2.4 - 3x$$
$$-8x = 2.4$$
$$\frac{-8x}{-8} = \frac{2.4}{-8}$$
$$x = -0.3$$

6. $0.2y + 2.6 = 4$
$$10(0.2y + 2.6) = 10(4)$$
$$10(0.2y) + 10(2.6) = 10(4)$$
$$2y + 26 = 40$$
$$2y + 26 - 26 = 40 - 26$$
$$2y = 14$$
$$\frac{2y}{2} = \frac{14}{2}$$
$$y = 7$$

Exercise Set 5.6

1. $x + 1.2 = 7.1$
$$x + 1.2 - 1.2 = 7.1 - 1.2$$
$$x = 5.9$$

3. $-5y = 2.15$
$$\frac{-5y}{-5} = \frac{2.15}{-5}$$
$$y = -0.43$$

5. $6.2 = y - 4$
$$6.2 + 4 = y - 4 + 4$$
$$10.2 = y$$

7. $3.1x = -13.95$
$$\frac{3.1x}{3.1} = \frac{-13.95}{3.1}$$
$$x = -4.5$$

9. $-3.5x + 2.8 = -11.2$
$$-3.5x + 2.8 - 2.8 = -11.2 - 2.8$$
$$-3.5x = -14$$
$$\frac{-3.5x}{-3.5} = \frac{-14}{-3.5}$$
$$x = 4$$

11. $6x + 8.65 = 3x + 10$
$$6x + 8.65 - 8.65 = 3x + 10 - 8.65$$
$$6x = 3x + 1.35$$
$$6x - 3x = 3x - 3x + 1.35$$
$$3x = 1.35$$
$$\frac{3x}{3} = \frac{1.35}{3}$$
$$x = 0.45$$

13.
$$2(x-1.3) = 5.8$$
$$2x - 2.6 = 5.8$$
$$2x - 2.6 + 2.6 = 5.8 + 2.6$$
$$2x = 8.4$$
$$\frac{2x}{2} = \frac{8.4}{2}$$
$$x = 4.2$$

15.
$$0.4x + 0.7 = -0.9$$
$$10(0.4x + 0.7) = 10(-0.9)$$
$$4x + 7 = -9$$
$$4x + 7 - 7 = -9 - 7$$
$$4x = -16$$
$$\frac{4x}{4} = \frac{-16}{4}$$
$$x = -4$$

17.
$$7x - 10.8 = x$$
$$10(7x - 10.8) = 10 \cdot x$$
$$70x - 108 = 10x$$
$$70x - 70x - 108 = 10x - 70x$$
$$-108 = -60x$$
$$\frac{-108}{-60} = \frac{-60x}{-60}$$
$$1.8 = x$$

19.
$$2.1x + 5 - 1.6x = 10$$
$$10(2.1x + 5 - 1.6x) = 10 \cdot 10$$
$$21x + 50 - 16x = 100$$
$$5x + 50 = 100$$
$$5x + 50 - 50 = 100 - 50$$
$$5x = 50$$
$$\frac{5x}{5} = \frac{50}{5}$$
$$x = 10$$

21.
$$y - 3.6 = 4$$
$$y - 3.6 + 3.6 = 4 + 3.6$$
$$y = 7.6$$

23.
$$-0.02x = -1.2$$
$$\frac{-0.02x}{-0.02} = \frac{-1.2}{-0.02}$$
$$x = 60$$

25.
$$6.5 = 10x + 7.2$$
$$6.5 - 7.2 = 10x + 7.2 - 7.2$$
$$-0.7 = 10x$$
$$\frac{-0.7}{10} = \frac{10x}{10}$$
$$-0.07 = x$$

27.
$$2.7x - 25 = 1.2x + 5$$
$$2.7x - 25 + 25 = 1.2x + 5 + 25$$
$$2.7x = 1.2x + 30$$
$$2.7x - 1.2x = 1.2x - 1.2x + 30$$
$$1.5x = 30$$
$$\frac{1.5x}{1.5} = \frac{30}{1.5}$$
$$x = 20$$

29.
$$200x - 0.67 = 100x + 0.81$$
$$200x - 0.67 + 0.67 = 100x + 0.81 + 0.67$$
$$200x = 100x + 1.48$$
$$200x - 100x = 100x - 100x + 1.48$$
$$100x = 1.48$$
$$\frac{100x}{100} = \frac{1.48}{100}$$
$$x = 0.0148$$

31.
$$3(x + 2.71) = 2x$$
$$3x + 8.13 = 2x$$
$$3x - 3x + 8.13 = 2x - 3x$$
$$8.13 = -x$$
$$\frac{8.13}{-1} = \frac{-x}{-1}$$
$$-8.13 = x$$

33.
$$8x - 5 = 10x - 8$$
$$8x - 5 + 8 = 10x - 8 + 8$$
$$8x + 3 = 10x$$
$$8x + 3 - 8x = 10x - 8x$$
$$3 = 2x$$
$$\frac{3}{2} = \frac{2x}{2}$$
$$1.5 = x$$

35.
$$1.2 + 0.3x = 0.9$$
$$1.2 + 0.3x - 1.2 = 0.9 - 1.2$$
$$0.3x = -0.3$$
$$\frac{0.3x}{0.3} = \frac{-0.3}{0.3}$$
$$x = -1$$

37.
$$-0.9x + 2.65 = -0.5x + 5.45$$
$$100(-0.9x + 2.65) = 100(-0.5x + 5.45)$$
$$-90x + 265 = -50x + 545$$
$$-90x + 265 + 90x = -50x + 545 + 90x$$
$$265 = 40x + 545$$
$$265 - 545 = 40x + 545 - 545$$
$$-280 = 40x$$
$$\frac{-280}{40} = \frac{40x}{40}$$
$$-7 = x$$

39.
$$4x + 7.6 = 2(3x - 3.2)$$
$$4x + 7.6 = 6x - 6.4$$
$$10(4x + 7.6) = 10(6x - 6.4)$$
$$40x + 76 = 60x - 64$$
$$40x + 76 + 64 = 60x - 64 + 64$$
$$40x + 140 = 60x$$
$$40x - 40x + 140 = 60x - 40x$$
$$140 = 20x$$
$$\frac{140}{20} = \frac{20x}{20}$$
$$7 = x$$

41.
$$0.7x + 13.8 = x - 2.16$$
$$100(0.7x + 13.8) = 100(x - 2.16)$$
$$70x + 1380 = 100x - 216$$
$$70x + 1380 + 216 = 100x - 216 + 216$$
$$70x + 1596 = 100x$$
$$70x + 1596 - 70x = 100x - 70x$$
$$1596 = 30x$$
$$\frac{1596}{30} = \frac{30x}{30}$$
$$53.2 = x$$

43.
$$2x - 7 + x - 9 = (2x + x) + (-7 - 9)$$
$$= 3x - 16$$

45.
$$\frac{6x}{5} \cdot \frac{1}{2x^2} = \frac{6x \cdot 1}{5 \cdot 2x^2} = \frac{2 \cdot 3 \cdot x}{5 \cdot 2 \cdot x \cdot x} = \frac{3}{5x}$$

47.
$$5\frac{1}{3} \div 9\frac{1}{6} = \frac{16}{3} \div \frac{55}{6} = \frac{16}{3} \cdot \frac{6}{55} = \frac{16 \cdot 2 \cdot 3}{3 \cdot 55} = \frac{32}{55}$$

49.
$$b + 4.6 = 8.3$$
$$b + 4.6 - 4.6 = 8.3 - 4.6$$
$$b = 3.7$$

51.
$$2x - 0.6 + 4x - 0.01 = 2x + 4x - 0.6 - 0.01$$
$$= 6x - 0.61$$

53.
$$5y - 1.2 - 7y + 8 = 5y - 7y - 1.2 + 8$$
$$= -2y + 6.8$$

55.
$$2.8 = z - 6.3$$
$$2.8 + 6.3 = z - 6.3 + 6.3$$
$$9.1 = z$$

57.
$$4.7x + 8.3 = -5.8$$
$$4.7x + 8.3 - 8.3 = -5.8 - 8.3$$
$$4.7x = -14.1$$
$$\frac{4.7x}{4.7} = \frac{-14.1}{4.7}$$
$$x = -3$$

59.
$$7.76 + 8z - 12z + 8.91$$
$$= 8z - 12z + 7.76 + 8.91$$
$$= -4z + 16.67$$

61.
$$5(x - 3.14) = 4x$$
$$5 \cdot x - 5 \cdot 3.14 = 4x$$
$$5x - 15.7 = 4x$$
$$5x - 5x - 15.7 = 4x - 5x$$
$$-15.7 = -x$$
$$15.7 = x$$

63.
$$2.6y + 8.3 = 4.6y - 3.4$$
$$10(2.6y + 8.3) = 10(4.6y - 3.4)$$
$$26y + 83 = 46y - 34$$
$$26y + 83 - 83 = 46y - 34 - 83$$
$$26y = 46y - 117$$
$$26y - 46y = 46y - 46y - 117$$
$$-20y = -117$$
$$\frac{-20y}{-20} = \frac{-117}{-20}$$
$$y = 5.85$$

65.
$$9.6z - 3.2 - 11.7z - 6.9 = 9.6z - 11.7z - 3.2 - 6.9$$
$$= -2.1z - 10.1$$

67. answers may vary

69.
$$-5.25x = -40.33575$$
$$\frac{-5.25x}{-5.25} = \frac{-40.33575}{-5.25}$$
$$x = 7.683$$

71.
$$1.95y + 6.834 = 7.65y - 19.8591$$
$$1.95y + 6.834 - 6.834 = 7.65y - 19.8591 - 6.834$$
$$1.95y = 7.65y - 26.6931$$
$$1.95y - 7.65y = 7.65y - 7.65y - 26.6931$$
$$-5.7y = -26.6931$$
$$\frac{-5.7y}{-5.7} = \frac{-26.6931}{-5.7}$$
$$y = 4.683$$

Section 5.7

Practice Problems

1. $\text{Mean} = \dfrac{87 + 75 + 96 + 91 + 78}{5} = \dfrac{427}{5} = 85.4$

2. $\text{gpa} = \dfrac{4 \cdot 2 + 3 \cdot 4 + 2 \cdot 5 + 1 \cdot 2 + 4 \cdot 2}{2 + 4 + 5 + 2 + 2} = \dfrac{40}{15} \approx 2.67$

3. Because the numbers are in numerical order, the median is the middle number, 24.

4. Write the numbers in numerical order:
   36, 65, 71, 78, 88, 91, 95, 95
   The median is the mean of the two middle numbers.

   $$\text{median} = \frac{78+88}{2} = 83$$

5. Mode: 15 because it occurs most often, 3 times.

6. Median: Write the numbers in order.
   15, 15, 15, 16, 18, 26, 26, 30, 31, 35

   Median is mean of middle two numbers, $\frac{18+26}{2} = 22$.

   Mode: 15 because it occurs most often, 3 times.

Vocabulary and Readiness Check

1. Another word for "mean" is <u>average</u>.

3. The <u>mean (or average)</u> of a set of number items is $\dfrac{\text{sum of items}}{\text{number of items}}$.

5. An example of weighted mean is a calculation of <u>grade point average</u>.

Exercise Set 5.7

1. Mean: $\dfrac{15+23+24+18+25}{5} = \dfrac{105}{5} = 21$

   Median: Write the numbers in order:
   15, 18, 23, 24, 25
   The middle number is 23.
   Mode: There is no mode, since each number occurs once.

3. Mean: $\dfrac{7.6+8.2+8.2+9.6+5.7+9.1}{6} = \dfrac{48.4}{6} \approx 8.1$

   Median: Write the numbers in order:
   5.7, 7.6, 8.2, 8.2, 9.1, 9.6

   Median is mean of middle two: $\dfrac{8.2+8.2}{2} = 8.2$

   Mode: 8.2 since this number appears twice.

5. Mean: $\dfrac{0.5+0.2+0.2+0.6+0.3+1.3+0.8+0.1+0.5}{9}$

   $= \dfrac{4.5}{9}$

   $= 0.5$

   Median: Write the numbers in order:
   0.1, 0.2, 0.2, 0.3, 0.5, 0.5, 0.6, 0.8, 1.3
   The middle number is 0.5.
   Mode: Since 0.2 and 0.5 occur twice, there are two modes, 0.2 and 0.5.

7. Mean: $\dfrac{231+543+601+293+588+109+334+268}{8} = \dfrac{2967}{8}$
≈ 370.9

   Median: Write the numbers in order:
   109, 231, 268, 293, 334, 543, 588, 601
   The mean of the middle two: $\dfrac{293+334}{2} = 313.5$
   Mode: There is no mode, since each number occurs once.

9. Mean: $\dfrac{1679+1483+1483+1450+1380}{5} = \dfrac{7475}{5}$
$= 1495$ feet

11. Because the numbers are in numerical order, the median is mean of the middle two (of the top 8),
$\dfrac{1450+1380}{2} = 1415$ feet.

13. answers may vary

15. $\text{gpa} = \dfrac{3\cdot3+2\cdot3+4\cdot4+2\cdot4}{3+3+4+4} = \dfrac{39}{14} \approx 2.79$

17. $\text{gpa} = \dfrac{4\cdot3+4\cdot3+4\cdot4+3\cdot3+2\cdot1}{3+3+4+3+1}$
$= \dfrac{51}{14}$
≈ 3.64

19. Mean: $\dfrac{7.8+6.9+7.5+4.7+6.9+7.0}{6} = \dfrac{40.8}{6} = 6.8$

21. Mode: 6.9, since this number appears twice.

23. Median: Write the numbers in order.
    79, 85, 88, 89, 91, 93
    The mean of the middle two: $\dfrac{88+89}{2} = 88.5$

25. Mean: $\dfrac{\text{sum of 15 pulse rates}}{15} = \dfrac{1095}{15} = 73$

27. Mode: Since 70 and 71 occur twice, there are two modes, 70 and 71.

29. There were 9 rates lower than the mean. They are 66, 68, 71, 64, 71, 70, 65, 70, and 72.

31. $\dfrac{6}{18} = \dfrac{1 \cdot 6}{3 \cdot 6} = \dfrac{1}{3}$

33. $\dfrac{18}{30y} = \dfrac{3 \cdot 6}{5 \cdot 6 \cdot y} = \dfrac{3}{5y}$

35. $\dfrac{55y^2}{75y^2} = \dfrac{5 \cdot 11 \cdot y \cdot y}{5 \cdot 15 \cdot y \cdot y} = \dfrac{11}{15}$

37. Since the mode is 35, 35 must occur at least twice in the set.
Since there are an odd number of numbers in the set, the median, 37 is in the set.
Let n be the remaining unknown number.

$$\text{Mean: } \frac{35 + 35 + 37 + 40 + n}{5} = 38$$
$$\frac{147 + n}{5} = 38$$
$$5 \cdot \frac{147 + n}{5} = 5 \cdot 38$$
$$147 + n = 190$$
$$147 - 147 + n = 190 - 147$$
$$n = 43$$

The missing numbers are 35, 35, 37, and 43.

39. yes; answers may vary

Chapter 5 Vocabulary Check

1. Like fractional notation, <u>decimal</u> notation is used to denote a part of a whole.

2. To write fractions as decimals, divide the <u>numerator</u> by the <u>denominator</u>.

3. To add or subtract decimals, write the decimals so that the decimal points line up <u>vertically</u>.

4. When writing decimals in words, write "<u>and</u>" for the decimal point.

5. When multiplying decimals, the decimal point in the product is placed so that the number of decimal places in the product is equal to the <u>sum</u> of the number of decimal places in the factors.

6. The <u>mode</u> of a set of numbers is the number that occurs most often.

7. The distance around a circle is called the <u>circumference</u>.

8. The <u>median</u> of a set of numbers in numerical order is the middle number. If there are an even number of numbers, the mode is the <u>mean</u> of the two middle numbers.

9. The <u>mean</u> of a list of number items is $\dfrac{\text{sum of items}}{\text{number of items}}$.

10. When 2 million is written as 2,000,000, we say it is written in <u>standard form</u>.

Chapter 5 Review

1. In 23.45, the 4 is in the tenths place.

2. In 0.000345, the 4 is in the hundred-thousandths place.

3. −23.45 in words is negative twenty-three and forty-five hundredths.

4. 0.00345 in words is three hundred forty-five hundred-thousandths.

5. 109.23 in words is one hundred nine and twenty-three hundredths.

6. 200.000032 in words is two hundred and thirty-two millionths.

7. Eight and six hundredths is 8.06.

8. Negative five hundred three and one hundred two thousandths is −503.102.

9. Sixteen thousand twenty-five and fourteen ten-thousandths is 16,025.0014.

10. Fourteen and eleven thousandths is 14.011.

11. $0.16 = \dfrac{16}{100} = \dfrac{4}{25}$

12. $-12.023 = -12\dfrac{23}{1000}$

13. $\dfrac{231}{100,000} = 0.00231$

14. $25\dfrac{1}{4} = 25\dfrac{25}{100} = 25.25$

15. 0.49   0.43
          ↑   ↑
          9 > 3
      so $0.49 > 0.43$

16. $0.973 = 0.9730$

17. 38.0027   38.00056
           ↑   ↑
           2   >   0
      so $38.0027 > 38.00056$
      Thus, $-38.0027 < -38.00056$.

18. 0.230505   0.23505
            ↑   ↑
            0   <   5
      so $0.23050 < 0.23505$
      Thus, $-0.230505 > -0.23505$.

19. To round 0.623 to the nearest tenth, observe that the digit in the hundredths place is 2. Since this digit is less than 5, we do not add 1 to the digit in the tenths place. The number 0.623 rounded to the nearest tenth is 0.6.

20. To round 0.9384 to the nearest hundredth, observe that the digit in the thousandths place is 8. Since this digit is at least 5, we add 1 to the digit in the hundredths place. The number 0.9384 rounded to the nearest hundredth is 0.94.

21. To round −42.895 to the nearest hundredth, observe that the digit in the thousandths place is 5. Since this digit is at least 5, we add 1 to the digit in the hundredths place. The number −42.895 rounded to the nearest hundredth is −42.90.

22. To round 16.34925 to the nearest thousandth, observe that the digit in the ten-thousandths place is 2. Since this digit is less than 5, we do not add 1 to the digit in the thousandths place. The number 16.34925 rounded to the nearest thousandth is 16.349.

23. $\begin{aligned} 887 \text{ million} &= 887 \times 1 \text{ million} \\ &= 887 \times 1,000,000 \\ &= 887,000,000 \end{aligned}$

24. $\begin{aligned} 600 \text{ thousand} &= 600 \times 1 \text{ thousand} \\ &= 600 \times 1000 \\ &= 600,000 \end{aligned}$

25. $\begin{array}{r} 8.6 \\ +\ 9.5 \\ \hline 18.1 \end{array}$

26. $\begin{array}{r} 3.9 \\ +\ 1.2 \\ \hline 5.1 \end{array}$

27. Add the absolute values.
$\begin{array}{r} 6.40 \\ +\ 0.88 \\ \hline 7.28 \end{array}$
Attach the common sign.
$-6.4 + (-0.88) = -7.28$

28. Subtract the absolute values.
$\begin{array}{r} 19.02 \\ -\ \ 6.98 \\ \hline 12.04 \end{array}$
Attach the sign of the larger absolute value.
$-19.02 + 6.98 = -12.04$

29.
$$\begin{array}{r} 200.490 \\ 16.820 \\ +\ 103.002 \\ \hline 320.312 \end{array}$$

30.
$$\begin{array}{r} 0.00236 \\ 100.45000 \\ +\ 48.29000 \\ \hline 148.74236 \end{array}$$

31.
$$\begin{array}{r} 4.9 \\ -\ 3.2 \\ \hline 1.7 \end{array}$$

32.
$$\begin{array}{r} 5.23 \\ -\ 2.74 \\ \hline 2.49 \end{array}$$

33. $-892.1 - 432.4 = -892.1 + (-432.4)$
Add the absolute values.
$$\begin{array}{r} 892.1 \\ +\ 432.4 \\ \hline 1324.5 \end{array}$$
Attach the common sign.
$-892.1 - 432.4 = -1324.5$

34. $0.064 - 10.2 = 0.064 + (-10.2)$
Subtract the absolute values.
$$\begin{array}{r} 10.200 \\ -\ 0.064 \\ \hline 10.136 \end{array}$$
Attach the sign of the larger absolute value.
$0.064 - 10.2 = -10.136$

35.
$$\begin{array}{r} 100.00 \\ -\ 34.98 \\ \hline 65.02 \end{array}$$

36.
$$\begin{array}{r} 200.00000 \\ -\ 0.00198 \\ \hline 199.99802 \end{array}$$

37.
$$\begin{array}{r} 19.9 \\ 15.1 \\ 10.9 \\ +\ 6.7 \\ \hline 52.6 \end{array}$$
The total distance is 52.6 miles.

38. $x - y = 1.2 - 6.9 = -5.7$

39. Perimeter $= 6.2 + 4.9 + 6.2 + 4.9 = 22.2$
The perimeter is 22.2 inches.

40. Perimeter $= 11.8 + 12.9 + 14.2 = 38.9$
The perimeter is 38.9 feet.

41. $7.2 \times 10 = 72$

42. $9.345 \times 1000 = 9345$

43. A negative number multiplied by a positive number is a negative number.
$$\begin{array}{r} 34.02 \quad \text{2 decimal places} \\ \times\ \ 2.3 \quad \text{1 decimal place} \\ \hline 10\ 206 \\ 68\ 040 \\ \hline 78.246 \quad 2+1=3 \text{ decimal places} \end{array}$$
$-34.02 \times 2.3 = -78.246$

44. A negative number multiplied by a negative number is a positive number.
$$\begin{array}{r} 839.02 \quad \text{2 decimal places} \\ \times\ \ 87.3 \quad \text{1 decimal place} \\ \hline 251\ 706 \\ 5873\ 140 \\ 67121\ 600 \\ \hline 73246.446 \quad 2+1=3 \text{ decimal places} \end{array}$$
$-839.02 \times (-87.3) = 73,246.446$

45. $C = 2\pi r = 2\pi \cdot 7 = 14\pi$ meters
$C \approx 14 \cdot 3.14 = 43.96$ meters

46. $C = \pi d = \pi \cdot 20 = 20\pi$ inches
$C \approx 20 \cdot 3.14 = 62.8$ inches

47.
```
      0.0877
   3)0.2631
    -24
     23
    -21
     21
    -21
      0
```

48.
```
       15.825
   20) 316.500
      -20
      116
     -100
      16 5
     -16 0
        50
       -40
       100
      -100
         0
```

49. A negative number divided by a negative number is a positive number.

$0.3\overline{)21}$ becomes
```
       70
   3) 210
     -21
      00
```

$-21 \div (-0.3) = 70$

50. A negative number divided by a positive number is a negative number.

$0.03\overline{)0.0063}$ becomes
```
      0.21
   3)0.63
    -6
     03
     -3
      0
```

$-0.0063 \div 0.03 = -0.21$

51. $0.34\overline{)2.74}$ becomes
```
            8.0588  ≈ 8.059
   34) 274.0000
      -272
         2 0
        - 0
         2 00
        -1 70
           300
          -272
           280
          -272
             8
```

52. $19.8\overline{)601.92}$ becomes
```
           30.4
   198) 6019.2
       -594
         79
        -0
         79 2
        -79 2
            0
```

53. $\dfrac{23.65}{1000} = 0.02365$

54. $\dfrac{93}{-10} = -9.3$

55. $3.28\overline{)24}$ becomes
```
           7.31  ≈ 7.3
   328) 2400.00
       -2296
         104 0
         -98 4
           5 60
          -3 28
           2 32
```

There are approximately 7.3 meters in 24 feet.

56. $69.71\overline{)3136.95}$ becomes

$$6971\overline{)\begin{array}{r}45\\313695\\-27884\\\hline 34855\\-34855\\\hline 0\end{array}}$$

It will take him 45 months to pay off the loan.

57.

$$5\overline{)\begin{array}{r}0.8\\4.0\\-4\,0\\\hline 0\end{array}}\qquad \frac{4}{5}=0.8$$

58.

$$13\overline{)\begin{array}{r}0.9230\approx 0.923\\12.0000\\-11\,7\\\hline 30\\-26\\\hline 40\\-39\\\hline 10\\-0\\\hline 10\end{array}}$$

$$-\frac{12}{13}\approx -0.923$$

59.

$$3\overline{)\begin{array}{r}0.3333...\\1.0000\\-9\\\hline 10\\-9\\\hline 10\\-9\\\hline 10\\-9\\\hline 1\end{array}}$$

$$2\frac{1}{3}=2.\overline{3}\text{ or }2.333$$

60.

$$60\overline{)\begin{array}{r}0.2166...\\13.0000\\-12\,0\\\hline 1\,00\\-60\\\hline 400\\-360\\\hline 400\\-360\\\hline 40\end{array}}$$

$$\frac{13}{16}=0.21\overline{6}\text{ or }0.217$$

61. $0.392=0.392$

62.

$$7\overline{)\begin{array}{r}0.571\approx 0.57\\4.000\\-3\,5\\\hline 50\\-49\\\hline 10\\-7\\\hline 3\end{array}}$$

$\dfrac{4}{7}\approx 0.57$ and $0.57<0.625$, so $\dfrac{4}{7}<0.625$.

63.

$$17\overline{)\begin{array}{r}0.2941\approx 0.294\\5.0000\\-3\,4\\\hline 1\,60\\-1\,53\\\hline 70\\-68\\\hline 20\\-17\\\hline 3\end{array}}$$

$\dfrac{5}{17}\approx 0.294$ and $0.293<0.294$, so

$$0.293<\frac{5}{17}.$$

64. 0.0231   0.0221
　　　 ↑　　　　↑
　　　 3　>　2
so 0.0231 > 0.0221, thus
−0.0231 < −0.0221.

65. 0.832, 0.837, 0.839

66. $\dfrac{5}{8} = 0.625$

$\dfrac{5}{8}$, 0.626, 0.685

67. $\dfrac{3}{7} \approx 0.428$

0.42, $\dfrac{3}{7}$, 0.43

68. $\dfrac{18}{11} = 1.6\overline{363}$

$1.63 = 1.63$

$\dfrac{19}{12} = 1.58\overline{3}$

$\dfrac{19}{12}$, 1.63, $\dfrac{18}{11}$

69. $-7.6 \times 1.9 + 2.5 = -14.44 + 2.5 = -11.94$

70. $(-2.3)^2 - 1.4 = 5.29 - 1.4 = 3.89$

71. $\dfrac{7 + 0.74}{-0.06} = \dfrac{7.74}{-0.06} = -129$

72. $0.9(6.5 - 5.6) = 0.9(0.9) = 0.81$

73. $\dfrac{(1.5)^2 + 0.5}{0.05} = \dfrac{2.25 + 0.5}{0.05} = \dfrac{2.75}{0.05} = 55$

74. $0.0726 \div 10 \times 1000 = 0.00726 \times 1000$
$= 7.26$

75. Area $= \dfrac{1}{2} \cdot b \cdot h$

$= \dfrac{1}{2}(4.6)(3)$

$= 0.5(4.6)(3)$

$= 6.9$ square feet

76. Area $= \dfrac{1}{2} \cdot b \cdot h$

$= \dfrac{1}{2}(5.2)(2.1)$

$= 0.5(5.2)(2.1)$

$= 5.46$ square inches

77.
$$x + 3.9 = 4.2$$
$$x + 3.9 - 3.9 = 4.2 - 3.9$$
$$x = 0.3$$

78.
$$70 = y - 22.81$$
$$70 + 22.81 = y - 22.81 + 22.81$$
$$92.81 = y$$

79.
$$2x = 17.2$$
$$\dfrac{2x}{2} = \dfrac{17.2}{2}$$
$$x = 8.6$$

80.
$$-1.1y = 88$$
$$\dfrac{-1.1y}{-1.1} = \dfrac{88}{-1.1}$$
$$y = -80$$

81.
$$3x - 0.78 = 1.2 + 2x$$
$$3x - 0.78 + 0.78 = 1.2 + 2x + 0.78$$
$$3x = 1.98 + 2x$$
$$3x - 2x = 1.98 + 2x - 2x$$
$$x = 1.98$$

82.
$$-x + 0.6 - 2x = -4x - 0.9$$
$$-3x + 0.6 = -4x - 0.9$$
$$-3x + 0.6 - 0.6 = -4x - 0.9 - 0.6$$
$$-3x = -4x - 1.5$$
$$-3x + 4x = -4x + 4x - 1.5$$
$$x = -1.5$$

83.
$$-1.3x - 9.4 = -0.4x + 8.6$$
$$10(-1.3x - 9.4) = 10(-0.4x + 8.6)$$
$$-13x - 94 = -4x + 86$$
$$-13x - 94 + 94 = -4x + 86 + 94$$
$$-13x = -4x + 180$$
$$-13x + 4x = -4x + 4x + 180$$
$$-9x = 180$$
$$\frac{-9x}{-9} = \frac{180}{-9}$$
$$x = -20$$

84.
$$3(x - 1.1) = 5x - 5.3$$
$$3x - 3.3 = 5x - 5.3$$
$$3x - 3.3 + 3.3 = 5x - 5.3 + 3.3$$
$$3x = 5x - 2$$
$$3x - 5x = 5x - 5x - 2$$
$$-2x = -2$$
$$\frac{-2x}{-2} = \frac{-2}{-2}$$
$$x = 1$$

85. Mean: $\dfrac{13 + 23 + 33 + 14 + 6}{5} = \dfrac{89}{5} = 17.8$

Median: Write the numbers in order.
6, 13, 14, 23, 33
The middle number is 14.
Mode: There is no mode, since each number occurs once.

86. Mean $= \dfrac{45 + 86 + 21 + 60 + 86 + 64 + 45}{7}$

$= \dfrac{407}{7}$

≈ 58.1

Median: Write the numbers in order.
21, 45, 45, 60, 64, 86, 86
The middle number is 60.
Mode: There are 2 numbers that occur twice, so there are two modes, 45 and 86.

87. Mean $= \dfrac{14,000 + 20,000 + 12,000 + 20,000 + 36,000 + 45,000}{6} = \dfrac{147,000}{6} = 24,500$

Median: Write the numbers in order.
12,000, 14,000, 20,000, 20,000, 36,000, 45,000

The mean of the middle two: $\dfrac{20,000 + 20,000}{2} = 20,000$

Mode: 20,000 is the mode because it occurs twice.

88. Mean $= \dfrac{560+620+123+400+410+300+400+780+430+450}{10} = \dfrac{4473}{10} = 447.3$

Median: Write the numbers in order.

123, 300, 400, 400, 410, 430, 450, 560, 620, 780

The mean of the two middle numbers: $\dfrac{410+430}{2} = 420$

Mode: 400 is the mode because it occurs twice.

89. $\text{gpa} = \dfrac{4 \cdot 3 + 4 \cdot 3 + 2 \cdot 2 + 3 \cdot 3 + 2 \cdot 1}{3 + 3 + 2 + 3 + 1} = \dfrac{39}{12} = 3.25$

90. $\text{gpa} = \dfrac{3 \cdot 3 + 3 \cdot 4 + 2 \cdot 2 + 1 \cdot 2 + 3 \cdot 3}{3 + 4 + 2 + 2 + 3} = \dfrac{36}{14} \approx 2.57$

91. 200.0032 in words is two hundred and thirty-two ten thousandths.

92. Negative sixteen and nine hundredths is −16.09 in standard form.

93. $0.0847 = \dfrac{847}{10,000}$

94. $\dfrac{6}{7} \approx 0.857$

$\dfrac{8}{9} \approx 0.889$

$0.75 = 0.750$

$0.75, \dfrac{6}{7}, \dfrac{8}{9}$

95. $-\dfrac{7}{100} = -0.07$

96.
$$
\begin{array}{r}
0.1125 \\
80\overline{)\,9.0000} \\
\underline{-8\,0} \\
1\,00 \\
\underline{-80} \\
200 \\
\underline{-160} \\
400 \\
\underline{-400} \\
0
\end{array}
$$

$\dfrac{9}{80} = 0.1125$

97.

$$\require{enclose}
\begin{array}{r}
51.0571 \approx 51.057 \\
175 \enclose{longdiv}{8935.0000} \\
\end{array}$$

$$
\begin{array}{r}
-875 \\
\hline
185 \\
-175 \\
\hline
10\,0 \\
-0 \\
\hline
10\,00 \\
-8\,75 \\
\hline
1\,250 \\
-1\,225 \\
\hline
250 \\
-175 \\
\hline
75 \\
\end{array}
$$

$$\frac{8935}{175} \approx 51.057$$

98. 402.000032   402.00032

   ↑   ↑

   0   <   3

so 402.000032 < 402.00032

Thus, −402.000032 > −402.00032.

99. $\dfrac{6}{11} = 0.\overline{54}$

$0.\overline{54} < 0.55$, so $\dfrac{6}{11} < 0.55$

100. To round 86.905 to the nearest hundredth, observe that the digit in the thousandths place is 5. Since this digit is at least 5, we add 1 to the digit in the hundredths place. The number 86.905 rounded to the nearest hundredth is 86.91.

101. To round 3.11526 to the nearest thousandth, observe that the digit in the ten-thousandths place is 2. Since this digit is less than 5, we do not add 1 to the digit in the thousandths place. The number 3.11526 rounded to the nearest thousandth is 3.115.

102. To round 123.46 to the nearest one, observe that the digit in the tenths place is 4. Since this digit is less than 5, we do not add 1 to the digit in the ones place. The number $123.46 rounded to the nearest dollar (or one) is $123.00.

103. To round 3645.52 to the nearest one, observe that the digit in the tenths place is 5. Since this digit is at least 5, we add 1 to the digit in the ones place. The number $3645.52 rounded to the nearest dollar (or one) is $3646.00.

104. Subtract absolute values.

$$
\begin{array}{r}
4.9 \\
-\,3.2 \\
\hline
1.7 \\
\end{array}
$$

Attach the sign of the larger absolute value.

$3.2 - 4.9 = -1.7$

105.

$$
\begin{array}{r}
9.12 \\
-\,3.86 \\
\hline
5.26 \\
\end{array}
$$

106. Subtract absolute values.

$$
\begin{array}{r}
102.06 \\
-\,89.30 \\
\hline
12.76 \\
\end{array}
$$

Attach the sign of the larger absolute value.

$-102 + 89.3 = -12.76$

107.

$$
\begin{array}{r}
-4.021 \\
-10.830 \\
(+)\ -0.056 \\
\hline
-14.907 \\
\end{array}
$$

108.

$$
\begin{array}{rl}
2.54 & \text{2 decimal places} \\
\times\ 3.2 & \text{1 decimal place} \\
\hline
508 & \\
7\,620 & \\
\hline
8.128 & \text{2 + 1 = 3 decimal places} \\
\end{array}
$$

109.

$$
\begin{array}{rl}
-3.45 & \text{2 decimal places} \\
\times\ 2.1 & \text{1 decimal place} \\
\hline
345 & \\
6\,900 & \\
\hline
-7.245 & \text{2 + 1 = 3 decimal places} \\
\end{array}
$$

110. $0.005\overline{)24.5}$ becomes
$$\begin{array}{r} 4900 \\ 5\overline{)24500} \\ \underline{-20} \\ 45 \\ \underline{-45} \\ 000 \end{array}$$

111. $2.3\overline{)54.98}$ becomes

$$\begin{array}{r} 23.9043 \approx 23.904 \\ 23\overline{)549.8000} \\ \underline{-46} \\ 89 \\ \underline{-69} \\ 20\;8 \\ \underline{-20\;7} \\ 10 \\ \underline{-0} \\ 100 \\ \underline{-92} \\ 80 \\ \underline{-69} \\ 11 \end{array}$$

112. length $= 115.9 \approx 120$
width $= 77.3 \approx 80$
Area $=$ length $\cdot$ width
$\quad = 120 \cdot 80$
$\quad = 9600$ square feet

113.
\$1.89	rounds to	\$2
\$1.07	rounds to	\$1
\$0.99	rounds to	$\underline{\$1}$
		\$4

Yes, the items can be purchased with a \$5 bill.

114. $\dfrac{(3.2)^2}{100} = \dfrac{10.24}{100} = 0.1024$

115. $(2.6 + 1.4)(4.5 - 3.6) = (4)(0.9) = 3.6$

116. Mean: $\dfrac{73 + 82 + 95 + 68 + 54}{5} = \dfrac{372}{5} = 74.4$
Median: Write the numbers in order.
54, 68, 73, 82, 95
The median is the middle value: 73.
Mode: There is no mode, since each number occurs once.

117. Mean:
$$\dfrac{952 + 327 + 566 + 814 + 327 + 729}{6} = \dfrac{3715}{6}$$
$$\approx 619.17$$
Median: Write the numbers in order.
327, 327, 566, 729, 814, 952
The mean of the two middle values:
$$\dfrac{566 + 729}{2} = 647.5$$
Mode: 327, since this number appears twice.

Chapter 5 Test

1. 45.092 in words is forty-five and ninety-two thousandths.

2. Three thousand and fifty-nine thousandths in standard form is 3000.059.

3.
$$\begin{array}{r} 2.893 \\ 4.210 \\ \underline{+\;10.492} \\ 17.595 \end{array}$$

4. Add the absolute values.
$$\begin{array}{r} 47.92 \\ \underline{+\;\;3.28} \\ 51.20 \end{array}$$
Attach the common sign.
$-47.92 - 3.28 = -51.20$

5. Subtract the absolute values.
$$\begin{array}{r} 30.25 \\ \underline{-\;\;9.83} \\ 20.42 \end{array}$$
Attach the sign of the larger absolute value.
$9.83 - 30.25 = -20.42$

6. $\begin{array}{r} 10.2 \\ \times\ 4.01 \\ \hline 102 \\ 40\ 800 \\ \hline 40.902 \end{array}$ 1 decimal place
2 decimal places

$1 + 2 = 3$ decimal places

7. $0.23\overline{)0.00843}$ becomes $23\overline{)0.8430}$

$$\begin{array}{r} 0.0366 \approx 0.036 \\ 23\overline{)0.8430} \\ \underline{-69} \\ 153 \\ \underline{-138} \\ 150 \\ \underline{-138} \\ 12 \end{array}$$

$(-0.00843) \div (-0.23) \approx 0.037$

8. 34.8923 rounded to the nearest tenth is 34.9.

9. 0.8623 rounded to the nearest thousandth is 0.862.

10. $\begin{array}{cc} 25.0909 & 25.9090 \\ \uparrow & \uparrow \\ 0 & 9 \end{array}$

$0 < 9$

so $25.0909 < 25.9090$

11. $\begin{array}{r} 0.444... \\ 9\overline{)4.000} \\ \underline{-3\ 6} \\ 40 \\ \underline{-36} \\ 40 \\ \underline{-36} \\ 4 \end{array}$

$0.44\overline{4} < 0.445$, so $\dfrac{4}{9} < 0.445$.

12. $0.345 = \dfrac{345}{1000} = \dfrac{5 \cdot 69}{5 \cdot 200} = \dfrac{69}{200}$

13. $-24.73 = -24\dfrac{73}{100}$

14. $-\dfrac{13}{26} = -\dfrac{1 \cdot 13}{2 \cdot 13} = -\dfrac{1}{2} = -\dfrac{1 \cdot 5}{2 \cdot 5} = -\dfrac{5}{10} = -0.5$

15. $\begin{array}{r} 0.9411 \approx 0.941 \\ 17\overline{)16.0000} \\ \underline{-15\ 3} \\ 70 \\ \underline{-68} \\ 20 \\ \underline{-17} \\ 30 \\ \underline{-17} \\ 13 \end{array}$

$\dfrac{16}{17} \approx 0.941$

16. $(-0.6)^2 + 1.57 = 0.36 + 1.57 = 1.93$

17. $\dfrac{0.23 + 1.63}{-0.3} = \dfrac{1.86}{-0.3} = -6.2$

18. $2.4x - 3.6 - 1.9x - 9.8$
$= (2.4x - 1.9x) + (-3.6 - 9.8)$
$= 0.5x - 13.4$

19. $\begin{aligned} 0.2x + 1.3 &= 0.7 \\ 0.2x + 1.3 - 1.3 &= 0.7 - 1.3 \\ 0.2x &= -0.6 \\ \dfrac{0.2x}{0.2} &= \dfrac{-0.6}{0.2} \\ x &= -3 \end{aligned}$

20. $\begin{aligned} 2(x + 5.7) &= 6x - 3.4 \\ 2x + 11.4 &= 6x - 3.4 \\ 2x + 11.4 - 11.4 &= 6x - 3.4 - 11.4 \\ 2x &= 6x - 14.8 \\ 2x - 6x &= 6x - 6x - 14.8 \\ -4x &= -14.8 \\ \dfrac{-4x}{-4} &= \dfrac{-14.8}{-4} \\ x &= 3.7 \end{aligned}$

21. Mean: $\dfrac{26+32+42+43+49}{5} = \dfrac{192}{5} = 38.4$

Median: The numbers are listed in order.
The middle number is 42.
Mode: There is no mode since each number occurs once.

22. Mean:

$\dfrac{8+10+16+16+14+12+12+13}{8} = \dfrac{101}{8}$
$= 12.625$

Median: List the numbers in order.
8, 10, 12, 12, 13, 14, 16, 16

The mean of the middle two: $\dfrac{12+13}{2} = 12.5$

Mode: 12 and 16 each occur twice, so the modes are 12 and 16.

23. $\text{gpa} = \dfrac{4 \cdot 3 + 3 \cdot 3 + 2 \cdot 3 + 3 \cdot 4 + 4 \cdot 1}{3+3+3+4+1}$

$= \dfrac{43}{14} \approx 3.07$

24. $4,583 \text{ million} = 4583 \times 1 \text{ million}$
$= 4583 \times 1,000,000$
$= 4,583,000,000$

25. $\text{Area} = \dfrac{1}{2}(4.2 \text{ miles})(1.1 \text{ miles})$
$= 0.5(4.2)(1.1) \text{ square miles}$
$= 2.31 \text{ square miles}$

26. $C = 2\pi r = 2\pi \cdot 9 = 18\pi \text{ miles}$
$C \approx 18 \cdot 3.14 = 56.52 \text{ miles}$

27. a. $\text{Area} = \text{length} \cdot \text{width}$
$= (123.8) \times (80)$
$= 9904$
The area is 9904 square feet.

b. $9904 \times 0.02 = 198.08$
Vivian needs to purchase 198.08 ounces.

28.
$$\begin{array}{r} 14.2 \\ 16.1 \\ +\ 23.7 \\ \hline 54.0 \end{array}$$
The total distance is 54 miles.

Cumulative Review Chapters 1–5

1. 72 in words is seventy-two.

2. 107 in words is one hundred seven.

3. 546 in words is five hundred forty-six.

4. 5026 in words is five thousand twenty-six.

5.
$$\begin{array}{r} 46 \\ +\ 713 \\ \hline 759 \end{array}$$

6. $3 + 7 + 9 = 19$
The perimeter is 19 inches.

7.
$$\begin{array}{r} 543 \\ -\ 29 \\ \hline 514 \end{array} \qquad Check: \begin{array}{r} 514 \\ +\ 29 \\ \hline 543 \end{array}$$

8.
$$\begin{array}{r} 121 \text{ R } 1 \\ 27 \overline{)\ 3268} \\ \underline{-27} \\ 56 \\ \underline{-54} \\ 28 \\ \underline{-27} \\ 1 \end{array}$$

9. To round 278,362 to the nearest thousand, observe that the digit in the hundreds place is 3. Since this digit is less than 5, we do not add 1 to the digit in the thousands place. The number 278,362 rounded to the nearest thousand is 278,000.

10. $30 = 1 \cdot 30$
$30 = 2 \cdot 15$
$30 = 3 \cdot 10$
$30 = 5 \cdot 6$
The factors of 30 are 1, 2, 3, 5, 6, 10, 15, 30.

11.
$$\begin{array}{r} 236 \\ \times \quad 86 \\ \hline 1\,416 \\ 18\,880 \\ \hline 20,296 \end{array}$$

12. $236 \times 86 \times 0 = 0$

13. a. $1\overline{)7}$ with 7 on top   *Check*: $7 \times 1 = 7$

   b. $12 \div 1 = 12$ *Check*: $12 \times 1 = 12$

   c. $\dfrac{6}{6} = 1$ *Check*: $1 \times 6 = 6$

   d. $9 \div 9 = 1$ *Check*: $1 \times 9 = 9$

   e. $\dfrac{20}{1} = 20$ *Check*: $20 \times 1 = 20$

   f. $18\overline{)18}$ with 1 on top   *Check*: $1 \times 18 = 18$

14. $\dfrac{25 + 17 + 19 + 39}{4} = \dfrac{100}{4} = 5$
The average is 25.

15. $2 \cdot 4 - 3 \div 3 = 8 - 3 \div 3 = 8 - 1 = 7$

16. $77 \div 11 \cdot 7 = 7 \cdot 7 = 49$

17. $9^2 = 9 \cdot 9 = 81$

18. $5^3 = 5 \cdot 5 \cdot 5 = 125$

19. $3^4 = 3 \cdot 3 \cdot 3 \cdot 3 = 81$

20. $10^3 = 10 \cdot 10 \cdot 10 = 1000$

21. $\dfrac{x - 5y}{y} = \dfrac{35 - 5 \cdot 5}{5} = \dfrac{35 - 25}{5} = \dfrac{10}{5} = 2$

22. $\dfrac{2a + 4}{c} = \dfrac{2 \cdot 7 + 4}{3} = \dfrac{14 + 4}{3} = \dfrac{18}{3} = 6$

23. a. The opposite of 13 is −13.

   b. The opposite of −2 is −(−2) = 2.

   c. The opposite of 0 is 0.

24. a. The opposite of −7 is −(−7) = 7.

   b. The opposite of 4 is −4.

   c. The opposite of −1 is −(−1) = 1.

25. $-2 + (-21) = -23$

26. $-7 + (-15) = -22$

27. $5 \cdot 6^2 = 5 \cdot 36 = 180$

28. $4 \cdot 2^3 = 4 \cdot 8 = 32$

29. $-7^2 = -(7 \cdot 7) = -49$

30. $(-2)^5 = (-2)(-2)(-2)(-2)(-2) = -32$

31. $(-5)^2 = (-5)(-5) = 25$

32. $-3^2 = -(3 \cdot 3) = -9$

33. Each part represents $\dfrac{1}{3}$ of a whole. Four parts are shaded, or 1 whole and 1 part.
$\dfrac{4}{3}$ or $1\dfrac{1}{3}$

34. Each part represents $\dfrac{1}{4}$ of a whole. Eleven parts are shaded or 2 wholes and 3 parts.
$\dfrac{11}{4}$ or $2\dfrac{3}{4}$

35. Each part represents $\dfrac{1}{4}$ of a whole. Eleven parts are shaded or 2 wholes and 3 parts.

$\dfrac{11}{4}$ or $2\dfrac{3}{4}$

36. Each part represents $\dfrac{1}{3}$ of a whole. Fourteen parts are shaded, or 4 wholes and 2 parts.

$\dfrac{14}{3}$ or $4\dfrac{2}{3}$

37. $252 = 2 \cdot 126$

$\quad\quad\quad \downarrow\ \downarrow \searrow$

$\quad\quad\quad 2 \cdot 2 \cdot 63$

$\quad\quad\quad \downarrow\ \downarrow\ \downarrow \searrow$

$\quad\quad\quad 2 \cdot 2 \cdot 3 \cdot 21$

$\quad\quad\quad \downarrow\ \downarrow\ \downarrow\ \downarrow \searrow$

$\quad\quad\quad 2 \cdot 2 \cdot 3 \cdot 3 \cdot\ 7$

$\quad 252 = 2^2 \cdot 3^2 \cdot 7$

38.
$$\begin{array}{r} 87 \\ -\ 25 \\ \hline 62 \end{array}$$

39. $-\dfrac{72}{26} = -\dfrac{36 \cdot 2}{13 \cdot 2} = -\dfrac{36}{13}$

40. $9\dfrac{7}{8} = \dfrac{9 \cdot 8 + 7}{8} = \dfrac{72 + 7}{8} = \dfrac{79}{8}$

41. $\dfrac{16}{40} = \dfrac{2 \cdot 8}{5 \cdot 8} = \dfrac{2}{5}$

$\dfrac{10}{25} = \dfrac{2 \cdot 5}{5 \cdot 5} = \dfrac{2}{5}$

The fractions are equivalent.

42. $\dfrac{4}{7} \approx 0.571$

$\dfrac{5}{9} \approx 0.556$

Since $0.571 > 0.556$, then $\dfrac{4}{7} > \dfrac{5}{9}$.

or

$\dfrac{4}{7} \cdot \dfrac{9}{9} = \dfrac{36}{63}$ and $\dfrac{5}{9} \cdot \dfrac{7}{7} = \dfrac{35}{63}$

Since $36 > 35$, then $\dfrac{36}{63} > \dfrac{35}{63}$ and $\dfrac{4}{7} > \dfrac{5}{9}$.

43. $\dfrac{2}{3} \cdot \dfrac{5}{11} = \dfrac{2 \cdot 5}{3 \cdot 11} = \dfrac{10}{33}$

44. $2\dfrac{5}{8} \cdot \dfrac{4}{7} = \dfrac{21}{8} \cdot \dfrac{4}{7} = \dfrac{21 \cdot 4}{8 \cdot 7} = \dfrac{7 \cdot 3 \cdot 4}{4 \cdot 2 \cdot 7} = \dfrac{3}{2}$ or $1\dfrac{1}{2}$

45. $\dfrac{1}{4} \cdot \dfrac{1}{2} = \dfrac{1 \cdot 1}{4 \cdot 2} = \dfrac{1}{8}$

46. $7 \cdot 5\dfrac{2}{7} = \dfrac{7}{1} \cdot \dfrac{37}{7} = \dfrac{7 \cdot 37}{1 \cdot 7} = \dfrac{37}{1} = 37$

47. $\dfrac{z}{-4} = 11 - 5$

$\quad\quad \dfrac{z}{-4} = 6$

$\quad -4 \cdot \dfrac{z}{-4} = -4 \cdot 6$

$\quad\quad\quad\quad z = -24$

48. $6x - 12 - 5x = -20$

$\quad\quad\quad x - 12 = -20$

$\quad x - 12 + 12 = -20 + 12$

$\quad\quad\quad\quad\quad x = -8$

49.
$$\begin{array}{r} 763.7651 \\ 22.0010 \\ +\ \ 43.8900 \\ \hline 829.6561 \end{array}$$

50.
$$\begin{array}{r} 89.2700 \\ 14.3610 \\ +\ 127.2318 \\ \hline 230.8628 \end{array}$$

51.   23.6   1 decimal place
     ×   0.78   2 decimal places
         1 888
     16 520
     18.408   1 + 2 = 3 decimal places

52.   43.8   1 decimal place
     × 0.645   3 decimal places
       2190
     1 7520
    26 2800
     28.2510   1 + 3 = 4 decimal places

Chapter 6

Practice Problems

1. The ratio of 19 to 30 is $\dfrac{19}{30}$.

2. The ratio of \$16 to \$12 is
$$\frac{\$16}{\$12} = \frac{16}{12} = \frac{4 \cdot 4}{4 \cdot 3} = \frac{4}{3}.$$

3. The ratio of 1.68 to 4.8 is
$$\frac{1.68}{4.8} = \frac{1.68 \cdot 100}{4.8 \cdot 100} = \frac{168}{480} = \frac{7 \cdot 24}{20 \cdot 24} = \frac{7}{20}.$$

4. The ratio of $2\frac{2}{3}$ to $1\frac{13}{15}$ is
$$\frac{2\frac{2}{3}}{1\frac{13}{15}} = 2\frac{2}{3} \div 1\frac{13}{15}$$
$$= \frac{8}{3} \div \frac{28}{15}$$
$$= \frac{8}{3} \cdot \frac{15}{28}$$
$$= \frac{4 \cdot 2 \cdot 3 \cdot 5}{3 \cdot 7 \cdot 4}$$
$$= \frac{10}{7}.$$

5. The ratio of work miles to total miles is
$$\frac{4800 \text{ miles}}{15{,}000 \text{ miles}} = \frac{8 \cdot 600}{25 \cdot 600} = \frac{8}{25}.$$

6. **a.** The ratio of the length of the shortest side to the length of the longest side is
$$\frac{7 \text{ meters}}{9 \text{ meters}} = \frac{7}{9}.$$

   b. The ratio of the length of the longest side to the perimeter is
$$\frac{9 \text{ meters}}{24 \text{ meters}} = \frac{9}{24}.$$

7. $\dfrac{\$1350}{6 \text{ weeks}} = \dfrac{\$225}{1 \text{ week}}$

8. $\dfrac{295 \text{ miles}}{15 \text{ gallons}} = \dfrac{59 \text{ miles}}{3 \text{ gallons}}$

9. $\dfrac{3200 \text{ feet}}{8 \text{ seconds}}$

$$8 \overline{)\ 3200}$$
$$\underline{-32}$$
$$000$$
with 400 on top

The unit rate is $\dfrac{400 \text{ feet}}{1 \text{ second}}$ or 400 feet/second

10. $\dfrac{78 \text{ bushels}}{12 \text{ trees}}$

$$12 \overline{)\ 78.0}$$
$$\underline{-72}$$
$$60$$
$$\underline{-60}$$
$$0$$
with 6.5 on top

The unit rate is $\dfrac{6.5 \text{ bushels}}{1 \text{ tree}}$ or 6.5 bushels/ tree.

11. unit price $= \dfrac{\$170}{5 \text{ days}} = \dfrac{\$34}{1 \text{ day}}$ or \$34 per day

12. unit price $= \dfrac{\$2.32}{11 \text{ ounces}} \approx \dfrac{\$0.21}{1 \text{ ounce}}$

    unit price $= \dfrac{\$3.59}{16 \text{ ounces}} \approx \dfrac{\$0.22}{1 \text{ ounce}}$

    Since the 11-ounce bag has a cheaper price per ounce, it is the better buy.

Vocabulary and Readiness Check

1. A rate with a denominator of 1 is called a <u>unit</u> rate.

3. The word <u>per</u> translates to "<u>division</u>".

5. To write a rate as a unit rate, divide the <u>numerator</u> of the rate by the <u>denominator</u>.

7. True or false: The ratio $\dfrac{7}{5}$ means the same

as the ratio $\dfrac{5}{7}$. <u>false</u>

Exercise Set 6.1

1. The ratio of 16 to 24 is $\dfrac{16}{24} = \dfrac{8 \cdot 2}{8 \cdot 3} = \dfrac{2}{3}$.

3. The ratio of 7.7 to 10 is

$\dfrac{7.7}{10} = \dfrac{7.7 \cdot 10}{10 \cdot 10} = \dfrac{77}{100}$.

5. The ratio of 4.63 to 8.21 is

$\dfrac{4.63}{8.21} = \dfrac{4.63 \cdot 100}{8.21 \cdot 100} = \dfrac{463}{821}$.

7. The ratio of 6 ounces to 16 ounces is

$\dfrac{6 \text{ ounces}}{16 \text{ ounces}} = \dfrac{3 \cdot 2}{8 \cdot 2} = \dfrac{3}{8}$.

9. The ratio of \$32 to \$100 is

$\dfrac{\$32}{\$100} = \dfrac{4 \cdot 8}{4 \cdot 25} = \dfrac{8}{25}$.

11. The ratio of 24 days to 14 days is

$\dfrac{24 \text{ days}}{14 \text{ days}} = \dfrac{12 \cdot 2}{7 \cdot 2} = \dfrac{12}{7}$.

13. The ratio of $3\dfrac{1}{2}$ to $12\dfrac{1}{4}$ is

$\dfrac{3\frac{1}{2}}{12\frac{1}{4}} = 3\dfrac{1}{2} \div 12\dfrac{1}{4}$

$= \dfrac{7}{2} \div \dfrac{49}{4}$

$= \dfrac{7}{2} \cdot \dfrac{4}{49}$

$= \dfrac{7 \cdot 2 \cdot 2}{2 \cdot 7 \cdot 7}$

$= \dfrac{2}{7}$.

15. The ratio of $7\dfrac{3}{5}$ hours to $1\dfrac{9}{10}$ hours is

$\dfrac{7\frac{3}{5} \text{ hours}}{1\frac{9}{10} \text{ hours}} = 7\dfrac{3}{5} \div 1\dfrac{9}{10}$

$= \dfrac{38}{5} \div \dfrac{19}{10}$

$= \dfrac{38}{5} \cdot \dfrac{10}{19}$

$= \dfrac{19 \cdot 2 \cdot 2 \cdot 5}{5 \cdot 19 \cdot 1}$

$= \dfrac{4}{1}$.

17. Perimeter = 125 + 130 + 125 + 130 = 510
The ratio of height to perimeter is

$\dfrac{130 \text{ feet}}{510 \text{ feet}} = \dfrac{10 \cdot 13}{10 \cdot 51} = \dfrac{13}{51}$

19. The ratio of the number of gallons of bottled water to the number of gallons of soft drinks

is $\dfrac{25 \text{ gallons}}{55 \text{ gallons}} = \dfrac{5 \cdot 5}{5 \cdot 11} = \dfrac{5}{11}$.

21. The ratio of women to men is

$\dfrac{125 \text{ women}}{100 \text{ men}} = \dfrac{5 \cdot 25}{4 \cdot 25} = \dfrac{5}{4}$.

23. The ratio of red blood cells to platelet cells

is $\dfrac{600 \text{ cells}}{40 \text{ cells}} = \dfrac{40 \cdot 15}{40 \cdot 1} = \dfrac{15}{1}$.

25. The ratio of the base to the height is

$\dfrac{6 \text{ feet}}{18 \text{ feet}} = \dfrac{1 \cdot 6}{3 \cdot 6} = \dfrac{1}{3}$.

27. The ratio of calories from fat to total

calories is $\dfrac{270 \text{ calories}}{570 \text{ calories}} = \dfrac{30 \cdot 9}{30 \cdot 19} = \dfrac{9}{19}$.

29. The ratio of average amount eaten by New Zealanders to the average amount eaten by

Americans is $\dfrac{56 \text{ pints}}{48 \text{ pints}} = \dfrac{7 \cdot 8}{6 \cdot 8} = \dfrac{7}{6}$.

31. The rate of 5 shrubs every 15 feet is

$$\frac{5 \text{ shrubs}}{15 \text{ feet}} = \frac{1 \text{ shrub}}{3 \text{ feet}}.$$

33. The rate of 15 returns for 100 sales is

$$\frac{15 \text{ returns}}{100 \text{ sales}} = \frac{3 \text{ returns}}{20 \text{ sales}}.$$

35. The rate of 6 laser printers for 28 computers

is $\dfrac{6 \text{ laser printers}}{28 \text{ computers}} = \dfrac{3 \text{ laser printers}}{14 \text{ computers}}.$

37. The rate of 18 gallons of pesticide for

4 acres of crops is $\dfrac{18 \text{ gallons}}{4 \text{ acres}} = \dfrac{9 \text{ gallons}}{2 \text{ acres}}.$

39.
$$
\begin{array}{r}
110 \\
3\overline{)330} \\
\underline{-3} \\
03 \\
\underline{-3} \\
0
\end{array}
$$

330 calories in a 3-ounce serving is

$\dfrac{110 \text{ calories}}{1 \text{ ounce}}$ or 110 calories/ounce.

41.
$$
\begin{array}{r}
75 \\
5\overline{)375} \\
\underline{-35} \\
25 \\
\underline{-25} \\
0
\end{array}
$$

375 riders in 5 subway cars is $\dfrac{75 \text{ riders}}{1 \text{ car}}$ or

75 riders/car.

43.
$$
\begin{array}{r}
90 \\
60\overline{)5400} \\
\underline{-540} \\
00
\end{array}
$$

5400 wingbeats per 60 seconds is

$\dfrac{90 \text{ wingbeats}}{1 \text{ second}}$ or 90 wingbeats/second.

45.
$$
\begin{array}{r}
50,000 \\
20\overline{)1,000,000} \\
\underline{-1\ 00} \\
0
\end{array}
$$

$1,000,000 paid over 20 years is $\dfrac{\$50,000}{1 \text{ year}}$

or $50,000/year.

47.
$$
\begin{array}{r}
1,369,022 \\
2\overline{)2,738,044} \\
\underline{-2} \\
0\ 7 \\
\underline{-6} \\
13 \\
\underline{-12} \\
18 \\
\underline{-18} \\
004 \\
\underline{-4} \\
04 \\
\underline{-4} \\
0
\end{array}
$$

2,738,044 voters for 2 senators is

$\dfrac{1,369,022 \text{ voters}}{1 \text{ senator}}$ or

1,369,022 voters/senator.

49.
$$
\begin{array}{r}
300 \\
40\overline{)12,000} \\
\underline{-12\ 0} \\
0
\end{array}
$$

12,000 good products to 40 defective is

$\dfrac{300 \text{ good}}{1 \text{ defective}}$ or 300 good/defective.

51.

$$
\begin{array}{r}
65,000 \\
20\overline{)\,1,300,000} \\
\underline{-1\,20} \\
100 \\
\underline{-100} \\
0
\end{array}
$$

$1,300,000 to build 20 houses is $\dfrac{\$65,000}{1\ \text{house}}$

or $65,000/house.

53. a.

$$
\begin{array}{r}
31.25 \\
8\overline{)\,250.00} \\
\underline{-24} \\
10 \\
\underline{-8} \\
2\,0 \\
\underline{-1\,6} \\
40 \\
\underline{-40} \\
0
\end{array}
$$

The unit rate for Charlie is
31.25 boards/hour.

b.

$$
\begin{array}{r}
33.5 \\
12\overline{)\,402.0} \\
\underline{-36} \\
42 \\
\underline{-36} \\
6\,0 \\
\underline{-6\,0} \\
0
\end{array}
$$

The unit rate for Suellen is
33.5 boards/hour.

c. Suellen can assemble boards faster,
since 33.5 > 31.25.

55. a. $14.5\overline{)\,400}$ becomes

$$
\begin{array}{r}
27.58 \approx 27.6 \\
145\overline{)\,4000.00} \\
\underline{-290} \\
1100 \\
\underline{-1015} \\
850 \\
\underline{-725} \\
1250 \\
\underline{-1160} \\
90
\end{array}
$$

The unit rate for the car is
≈27.6 miles/gallon.

b. $9.25\overline{)\,270}$ becomes

$$
\begin{array}{r}
29.18 \approx 29.2 \\
925\overline{)\,27000.00} \\
\underline{-1850} \\
8500 \\
\underline{-8325} \\
175\,0 \\
\underline{-92\,5} \\
82\,50 \\
\underline{-74\,00} \\
8\,50
\end{array}
$$

The unit rate for the truck is
≈29.2 miles/gallon.

c. The truck has the better gas mileage,
since 29.2 > 27.6.

57.

$$
\begin{array}{r}
11.50 \\
5\overline{)\,57.50} \\
\underline{-5} \\
07 \\
\underline{-5} \\
2\,5 \\
\underline{-2\,5} \\
0
\end{array}
$$

The unit price is $11.50 per compact disc.

59.
$$\begin{array}{r} 0.17 \\ 7{\overline{)1.19}} \\ \underline{-7} \\ 49 \\ \underline{-49} \\ 0 \end{array}$$

The unit price is $0.17 per banana.

61.
$$\begin{array}{r} 0.1487 \approx 0.149 \\ 8{\overline{)1.1900}} \\ \underline{-8} \\ 39 \\ \underline{-32} \\ 70 \\ \underline{-64} \\ 60 \\ \underline{-56} \\ 4 \end{array}$$

The 8-ounce size costs $0.149 per ounce.

$$\begin{array}{r} 0.1325 \approx 0.133 \\ 12{\overline{)1.5900}} \\ \underline{-1\,2} \\ 39 \\ \underline{-36} \\ 30 \\ \underline{-24} \\ 60 \\ \underline{-60} \\ 0 \end{array}$$

The 12-ounce size costs $0.133 per ounce.
The 12-ounce size is the better buy.

63.
$$\begin{array}{r} 0.1056 \approx 0.106 \\ 16{\overline{)1.6900}} \\ \underline{-1\,6} \\ 090 \\ \underline{-80} \\ 100 \\ \underline{-96} \\ 4 \end{array}$$

The 16-ounce size costs $0.106 per ounce.

$$\begin{array}{r} 0.115 \\ 6{\overline{)0.690}} \\ \underline{-6} \\ 09 \\ \underline{-6} \\ 30 \\ \underline{-30} \\ 0 \end{array}$$

The 6-ounce size costs $0.115 per ounce.
The 16-ounce size is the better buy.

65.
$$\begin{array}{r} 0.1908 \approx 0.191 \\ 12{\overline{)2.2900}} \\ \underline{-1\,2} \\ 1\,09 \\ \underline{-1\,08} \\ 100 \\ \underline{-96} \\ 4 \end{array}$$

The 12-ounce size costs $0.191 per ounce.

$$\begin{array}{r} 0.1862 \approx 0.186 \\ 8{\overline{)1.4900}} \\ \underline{-8} \\ 69 \\ \underline{-64} \\ 50 \\ \underline{-48} \\ 20 \\ \underline{-16} \\ 4 \end{array}$$

The 8-ounce size costs $0.186 per ounce.
The 8-ounce size is the better buy.

67.
$$
\begin{array}{r}
0.0059 \approx 0.006 \\
100\overline{)0.5900} \\
\underline{-500} \\
900 \\
\underline{-900} \\
0
\end{array}
$$

The 100-count size costs $0.006 per napkin.

$$
\begin{array}{r}
0.0051 \approx 0.005 \\
180\overline{)0.9300} \\
\underline{-900} \\
300 \\
\underline{-180} \\
120
\end{array}
$$

The 180-count size costs $0.005 per napkin.
The 180-count size is the better buy.

69.
$$
\begin{array}{r}
2.3 \\
9\overline{)20.7} \\
\underline{-18} \\
2\,7 \\
\underline{-2\,7} \\
0
\end{array}
$$

71. $3.7\overline{)0.555}$ becomes
$$
\begin{array}{r}
0.15 \\
37\overline{)5.55} \\
\underline{-3\,7} \\
1\,85 \\
\underline{-1\,85} \\
0
\end{array}
$$

73. no; answers may vary

75. no; $\dfrac{6 \text{ inches}}{15 \text{ inches}} = \dfrac{2 \cdot 3}{5 \cdot 3} = \dfrac{2}{5}$

77. 10 defective to 200 good is
$$\frac{10 \text{ defective}}{200 \text{ good}} = \frac{1 \text{ defective}}{20 \text{ good}}.$$
Yes, the machine should be repaired.

79. $79{,}543 - 79{,}286 = 257$

$13.4\overline{)257}$ becomes
$$
\begin{array}{r}
19.17 \approx 19.2 \\
134\overline{)2570.00} \\
\underline{-134} \\
1230 \\
\underline{-1206} \\
24\,0 \\
\underline{-13\,4} \\
10\,60 \\
\underline{-9\,38} \\
1\,22
\end{array}
$$

$79{,}895 - 79{,}543 = 352$

$15.8\overline{)352}$ becomes
$$
\begin{array}{r}
22.27 \approx 22.3 \\
158\overline{)3520.00} \\
\underline{-316} \\
360 \\
\underline{-316} \\
440 \\
\underline{-316} \\
1240 \\
\underline{-1106} \\
134
\end{array}
$$

$80{,}242 - 79{,}895 = 347$

$16.1\overline{)347}$ becomes
$$
\begin{array}{r}
21.55 \approx 21.6 \\
161\overline{)3470.00} \\
\underline{-322} \\
250 \\
\underline{-161} \\
89\,0 \\
\underline{-80\,5} \\
8\,50 \\
\underline{-8\,05} \\
45
\end{array}
$$

Beginning Odometer Reading	Ending Odometer Reading	Miles Driven	Gallons of Gas Used	Miles per Gallon (to the nearest tenth)
79,286	79,543	257	13.4	19.2
79,543	79,895	352	15.8	22.3
79,895	80,242	347	16.1	21.6

81.

$$
\begin{array}{r}
543.6 \approx 544 \\
88300\overline{)\ 48000000.0} \\
\underline{-441500} \\
385000 \\
\underline{-353200} \\
318000 \\
264900 \\
\overline{531000} \\
\underline{-529800} \\
1200
\end{array}
$$

There are 544 students per school.

83. answers may vary

85. no; answers may vary

Section 6.2

Practice Problems

1. a. $\begin{array}{l} \text{cups} \rightarrow \\ \text{cups} \rightarrow \end{array} \dfrac{24}{6} = \dfrac{4}{1} \begin{array}{l} \leftarrow \text{cups} \\ \leftarrow \text{cups} \end{array}$

    b. $\begin{array}{l} \text{students} \rightarrow \\ \text{instructors} \rightarrow \end{array} \dfrac{560}{25} = \dfrac{112}{5} \begin{array}{l} \leftarrow \text{students} \\ \leftarrow \text{instructors} \end{array}$

2. $\dfrac{4}{8} \stackrel{?}{=} \dfrac{10}{20}$

      $4 \cdot 20 \stackrel{?}{=} 8 \cdot 10$

        $80 = 80$

The proportion is true.

3. $\dfrac{4.2}{6} \stackrel{?}{=} \dfrac{4.8}{8}$

$4.2 \cdot 8 \stackrel{?}{=} 6 \cdot 4.8$

$33.6 \neq 28.8$

The proportion is false.

4. $\dfrac{3\frac{3}{10}}{1\frac{5}{6}} \stackrel{?}{=} \dfrac{4\frac{1}{5}}{2\frac{1}{3}}$

$3\dfrac{3}{10} \cdot 2\dfrac{1}{3} \stackrel{?}{=} 1\dfrac{5}{6} \cdot 4\dfrac{1}{5}$

$\dfrac{33}{10} \cdot \dfrac{7}{3} \stackrel{?}{=} \dfrac{11}{6} \cdot \dfrac{21}{5}$

$\dfrac{231}{30} = \dfrac{231}{30}$

The proportion is true.

5. $\dfrac{2}{5} = \dfrac{x}{25}$

$2 \cdot 25 = 5 \cdot x$

$50 = 5x$

$\dfrac{50}{5} = \dfrac{5x}{5}$

$10 = x$

6. $\dfrac{15}{2} = \dfrac{60}{x}$

$15 \cdot x = 2 \cdot 60$

$15x = 120$

$\dfrac{15x}{15} = \dfrac{120}{15}$

$x = 8$

7. $\dfrac{\frac{7}{8}}{z} = \dfrac{\frac{2}{3}}{\frac{4}{7}}$

$\dfrac{7}{8} \cdot \dfrac{4}{7} = z \cdot \dfrac{2}{3}$

$\dfrac{1}{2} = z \cdot \dfrac{2}{3}$

$\dfrac{1}{2} \cdot \dfrac{3}{2} = z \cdot \dfrac{2}{3} \cdot \dfrac{3}{2}$

$\dfrac{3}{4} = z$

8. $\dfrac{y}{9} = \dfrac{0.6}{1.2}$

$1.2y = 9 \cdot 0.6$

$1.2y = 5.4$

$\dfrac{1.2y}{1.2} = \dfrac{5.4}{1.2}$

$y = 4.5$

9. $\dfrac{17}{z} = \dfrac{8}{10}$

$17 \cdot 10 = z \cdot 8$

$170 = 8z$

$\dfrac{170}{8} = \dfrac{8z}{8}$

$\dfrac{85}{4} = z$

$21.25 = z$

10. $\dfrac{4.5}{1.8} = \dfrac{y}{3}$

$4.5 \cdot 3 = 1.8y$

$13.5 = 1.8y$

$\dfrac{13.5}{1.8} = \dfrac{1.8y}{1.8}$

$7.5 = y$

Vocabulary and Readiness Check

1. $\dfrac{4.2}{8.4} = \dfrac{1}{2}$ is called a <u>proportion</u> while $\dfrac{7}{8}$ is called a <u>ratio</u>.

3. In a proportion, if cross products are equal, the proportion is <u>true</u>.

Exercise Set 6.2

1. $\dfrac{10 \text{ diamonds}}{6 \text{ opals}} = \dfrac{5 \text{ diamonds}}{3 \text{ opals}}$

3. $\dfrac{20 \text{ students}}{5 \text{ microscopes}} = \dfrac{4 \text{ students}}{1 \text{ microscope}}$

5. $\dfrac{6 \text{ eagles}}{58 \text{ sparrows}} = \dfrac{3 \text{ eagles}}{29 \text{ sparrows}}$

7. $\dfrac{2\frac{1}{4} \text{ cups flour}}{24 \text{ cookies}} = \dfrac{6\frac{3}{4} \text{ cups flour}}{72 \text{ cookies}}$

9. $\dfrac{22 \text{ vanilla wafers}}{1 \text{ cup cookie crumbs}}$
$= \dfrac{55 \text{ vanilla wafers}}{2.5 \text{ cups cookie crumbs}}$

11. $\dfrac{15}{9} \overset{?}{=} \dfrac{5}{3}$
$15 \cdot 3 \overset{?}{=} 9 \cdot 5$
$45 = 45$
The proportion is true.

13. $\dfrac{5}{8} \overset{?}{=} \dfrac{4}{7}$
$5 \cdot 7 \overset{?}{=} 8 \cdot 4$
$35 \neq 32$
The proportion is false.

15. $\dfrac{9}{36} \overset{?}{=} \dfrac{2}{8}$
$9 \cdot 8 \overset{?}{=} 36 \cdot 2$
$72 = 72$
The proportion is true.

17. $\dfrac{5}{8} \overset{?}{=} \dfrac{625}{1000}$
$5 \cdot 1000 \overset{?}{=} 8 \cdot 625$
$5000 = 5000$
The proportion is true.

19. $\dfrac{0.8}{0.3} \overset{?}{=} \dfrac{0.2}{0.6}$
$0.8 \cdot 0.6 \overset{?}{=} 0.3 \cdot 0.2$
$0.48 \neq 0.06$
The proportion is false.

21. $\dfrac{4.2}{8.4} \overset{?}{=} \dfrac{5}{10}$
$4.2 \cdot 10 \overset{?}{=} 8.4 \cdot 5$
$42 = 42$
The proportion is true.

23. $\dfrac{\frac{3}{4}}{\frac{4}{3}} \overset{?}{=} \dfrac{\frac{1}{2}}{\frac{8}{9}}$
$\dfrac{3}{4} \cdot \dfrac{8}{9} \overset{?}{=} \dfrac{4}{3} \cdot \dfrac{1}{2}$
$\dfrac{2}{3} = \dfrac{2}{3}$
The proportion is true.

25. $\dfrac{2\frac{2}{5}}{\frac{2}{3}} \overset{?}{=} \dfrac{\frac{10}{9}}{\frac{1}{4}}$
$2\dfrac{2}{5} \cdot \dfrac{1}{4} \overset{?}{=} \dfrac{2}{3} \cdot \dfrac{10}{9}$
$\dfrac{12}{5} \cdot \dfrac{1}{4} \overset{?}{=} \dfrac{20}{27}$
$\dfrac{12}{20} \neq \dfrac{20}{27}$
The proportion is false.

27. $\dfrac{\frac{4}{5}}{6} \overset{?}{=} \dfrac{\frac{6}{5}}{9}$
$\dfrac{4}{5} \cdot 9 \overset{?}{=} 6 \cdot \dfrac{6}{5}$
$\dfrac{36}{5} = \dfrac{36}{5}$
The proportion is true.

29. $\dfrac{10}{15} \overset{?}{=} \dfrac{4}{6}$
$10 \cdot 6 \overset{?}{=} 15 \cdot 4$
$60 = 60$
The proportion $\dfrac{10}{15} = \dfrac{4}{6}$ is true.

31. $\dfrac{11}{4} \overset{?}{=} \dfrac{5}{2}$
$11 \cdot 2 \overset{?}{=} 4 \cdot 5$
$22 \neq 20$
The proportion $\dfrac{11}{4} = \dfrac{5}{2}$ is false.

33. $\dfrac{0.15}{3} \overset{?}{=} \dfrac{0.35}{7}$

$0.15 \cdot 7 \overset{?}{=} 3 \cdot 0.35$

$1.05 = 1.05$

The proportion $\dfrac{0.15}{3} = \dfrac{0.35}{7}$ is true.

35. $\dfrac{\frac{2}{3}}{\frac{1}{5}} \overset{?}{=} \dfrac{\frac{2}{5}}{\frac{1}{9}}$

$\dfrac{2}{3} \cdot \dfrac{1}{9} \overset{?}{=} \dfrac{1}{5} \cdot \dfrac{2}{5}$

$\dfrac{2}{27} \neq \dfrac{2}{25}$

The proportion $\dfrac{\frac{2}{3}}{\frac{1}{5}} = \dfrac{\frac{2}{5}}{\frac{1}{9}}$ is false.

37. $\dfrac{x}{5} = \dfrac{6}{10}$

$x \cdot 10 = 5 \cdot 6$

$10x = 30$

$\dfrac{10x}{10} = \dfrac{30}{10}$

$x = 3$

39. $\dfrac{-18}{54} = \dfrac{3}{n}$

$-18 \cdot n = 54 \cdot 3$

$-18n = 162$

$\dfrac{-18n}{-18} = \dfrac{162}{-18}$

$n = -9$

41. $\dfrac{30}{10} = \dfrac{15}{y}$

$30 \cdot y = 10 \cdot 15$

$30y = 150$

$\dfrac{30y}{30} = \dfrac{150}{30}$

$y = 5$

43. $\dfrac{8}{15} = \dfrac{z}{6}$

$8 \cdot 6 = 15 \cdot z$

$48 = 15z$

$\dfrac{48}{15} = \dfrac{15z}{15}$

$3.2 = z$

45. $\dfrac{24}{x} = \dfrac{60}{96}$

$24 \cdot 96 = x \cdot 60$

$2304 = 60x$

$\dfrac{2304}{60} = \dfrac{60x}{60}$

$38.4 = x$

47. $\dfrac{-3.5}{12.5} = \dfrac{-7}{n}$

$-3.5 \cdot n = 12.5 \cdot (-7)$

$-3.5n = -87.5$

$\dfrac{-3.5n}{-3.5} = \dfrac{-87.5}{-3.5}$

$n = 25$

49. $\dfrac{n}{0.6} = \dfrac{0.05}{12}$

$n \cdot 12 = 0.6 \cdot 0.05$

$12n = 0.030$

$\dfrac{12n}{12} = \dfrac{0.03}{12}$

$n = 0.0025$

51. $\dfrac{8}{\frac{1}{3}} = \dfrac{24}{n}$

$8 \cdot n = \dfrac{1}{3} \cdot 24$

$8n = 8$

$\dfrac{8n}{8} = \dfrac{8}{8}$

$n = 1$

53.

$$\frac{\frac{1}{3}}{\frac{3}{8}} = \frac{\frac{2}{5}}{n}$$

$$\frac{1}{3} \cdot n = \frac{3}{8} \cdot \frac{2}{5}$$

$$\frac{n}{3} = \frac{3}{20}$$

$$3 \cdot \frac{n}{3} = 3 \cdot \frac{3}{20}$$

$$n = \frac{9}{20}$$

55.

$$\frac{12}{n} = \frac{\frac{2}{3}}{\frac{6}{9}}$$

$$12 \cdot \frac{6}{9} = n \cdot \frac{2}{3}$$

$$8 = n \cdot \frac{2}{3}$$

$$\frac{3}{2} \cdot \frac{8}{1} = n \cdot \frac{2}{3} \cdot \frac{3}{2}$$

$$12 = n$$

57.

$$\frac{n}{1\frac{1}{5}} = \frac{4\frac{1}{6}}{6\frac{2}{3}}$$

$$n \cdot 6\frac{2}{3} = 1\frac{1}{5} \cdot 4\frac{1}{6}$$

$$n \cdot \frac{20}{3} = \frac{6}{5} \cdot \frac{25}{6}$$

$$\frac{20}{3} n = 5$$

$$\frac{3}{20} \cdot \frac{20}{3} n = \frac{3}{20} \cdot 5$$

$$n = \frac{3}{4}$$

59.

$$\frac{25}{n} = \frac{3}{\frac{7}{30}}$$

$$25 \cdot \frac{7}{30} = n \cdot 3$$

$$\frac{35}{6} = 3n$$

$$\frac{1}{3} \cdot \frac{35}{6} = 3n \cdot \frac{1}{3}$$

$$\frac{35}{18} = n$$

61.

$$\frac{3.2}{0.3} = \frac{x}{1.4}$$

$$3.2 \cdot 1.4 = 0.3 \cdot x$$

$$4.48 = 0.3x$$

$$\frac{4.48}{0.3} = \frac{0.3x}{0.3}$$

$$14.9 \approx x$$

63.

$$\frac{z}{5.2} = \frac{0.08}{6}$$

$$z \cdot 6 = 5.2 \cdot 0.08$$

$$6z = 0.416$$

$$\frac{6z}{6} = \frac{0.416}{6}$$

$$z \approx 0.07$$

65.

$$\frac{7}{18} = \frac{x}{5}$$

$$7 \cdot 5 = 18 \cdot x$$

$$35 = 18x$$

$$\frac{35}{18} = \frac{18x}{18}$$

$$1.9 \approx x$$

67.

$$\frac{43}{17} = \frac{8}{z}$$

$$43 \cdot z = 17 \cdot 8$$

$$43z = 136$$

$$\frac{43z}{43} = \frac{136}{43}$$

$$z \approx 3.163$$

69. 8.01   8.1
         ↑   ↑
         0   <   1
        $8.01 < 8.1$

71. $\dfrac{1}{2} > \dfrac{1}{3}$

so $2\dfrac{1}{2} > 2\dfrac{1}{3}$

73. $\dfrac{75}{125} = \dfrac{3 \cdot 25}{5 \cdot 25} = \dfrac{3}{5}$

75. $\dfrac{12x}{42} = \dfrac{2 \cdot 6 \cdot x}{7 \cdot 6} = \dfrac{2x}{7}$

77. $\dfrac{9}{15} = \dfrac{3}{5}$

$\dfrac{9}{3} = \dfrac{15}{5}$

$\dfrac{5}{15} = \dfrac{3}{9}$

$\dfrac{15}{9} = \dfrac{5}{3}$

79. $\dfrac{6}{18} = \dfrac{1}{3}$

$\dfrac{6}{1} = \dfrac{18}{3}$

$\dfrac{3}{18} = \dfrac{1}{6}$

$\dfrac{18}{6} = \dfrac{3}{1}$

81. $\dfrac{a}{b} = \dfrac{c}{d}$

$\dfrac{d}{b} = \dfrac{c}{a}$

$\dfrac{a}{c} = \dfrac{b}{d}$

$\dfrac{b}{a} = \dfrac{d}{c}$

83. answers may vary

85. $\dfrac{x}{7} = \dfrac{0}{8}$

$x \cdot 8 = 7 \cdot 0$

$8x = 0$

$\dfrac{8x}{8} = \dfrac{0}{8}$

$x = 0$

87. $\dfrac{z}{1150} = \dfrac{588}{483}$

$483 \cdot z = 1150 \cdot 588$

$483z = 676,200$

$\dfrac{483z}{483} = \dfrac{676,200}{483}$

$z = 1400$

89. $\dfrac{222}{1515} = \dfrac{37}{y}$

$222 \cdot y = 1515 \cdot 37$

$222y = 56,055$

$\dfrac{222y}{222} = \dfrac{56,055}{222}$

$y = 252.5$

The Bigger Picture

1. $\dfrac{7}{20} - \dfrac{1}{10} = \dfrac{7}{20} - \dfrac{1 \cdot 2}{10 \cdot 2} = \dfrac{7}{20} - \dfrac{2}{20} = \dfrac{5}{20} = \dfrac{1}{4}$

2. $\dfrac{7}{20} \cdot \dfrac{1}{10} = \dfrac{7 \cdot 1}{20 \cdot 10} = \dfrac{7}{200}$

3. $\dfrac{7}{20} \div \dfrac{1}{10} = \dfrac{7}{20} \cdot \dfrac{10}{1} = \dfrac{7 \cdot 10}{2 \cdot 10 \cdot 1} = \dfrac{7}{2}$ or $3\dfrac{1}{2}$

4. $\dfrac{7}{20} + \dfrac{1}{10} = \dfrac{7}{20} + \dfrac{1 \cdot 2}{10 \cdot 2} = \dfrac{7}{20} + \dfrac{2}{20} = \dfrac{9}{20}$

5. $7.6 + 0.02 = 7.62$

6. $\begin{array}{r} 7.6 \\ \times\ 0.02 \\ \hline 0.152 \end{array}$

7.
$$\frac{4}{n} = \frac{50}{100}$$
$$4 \cdot 100 = n \cdot 50$$
$$400 = 50n$$
$$\frac{400}{50} = \frac{50n}{50}$$
$$8 = n$$

8.
$$\frac{60}{10} = \frac{15}{n}$$
$$60 \cdot n = 10 \cdot 15$$
$$60n = 10 \cdot 15$$
$$60n = 150$$
$$\frac{60n}{60} = \frac{150}{60}$$
$$n = \frac{5}{2} \text{ or } 2\frac{1}{2}$$

9.
$$\frac{n}{0.8} = \frac{0.06}{12}$$
$$n \cdot 12 = 0.8 \cdot 0.06$$
$$12n = 0.048$$
$$\frac{12n}{12} = \frac{0.048}{12}$$
$$n = 0.004$$

10.
$$\frac{\frac{7}{8}}{\frac{1}{4}} = \frac{n}{\frac{5}{6}}$$
$$\frac{7}{8} \cdot \frac{5}{6} = \frac{1}{4} \cdot n$$
$$\frac{35}{48} = \frac{1}{4}n$$
$$\frac{4}{1} \cdot \frac{35}{48} = \frac{4}{1} \cdot \frac{1}{4}n$$
$$\frac{35}{12} = n \text{ or } n = 2\frac{11}{12}$$

Integrated Review

1. The ratio of 27 to 30 is $\dfrac{27}{30} = \dfrac{9 \cdot 3}{10 \cdot 3} = \dfrac{9}{10}$.

2. The ratio of 18 to 50 is $\dfrac{18}{50} = \dfrac{9 \cdot 2}{25 \cdot 2} = \dfrac{9}{25}$.

3. The ratio of 9.4 to 10 is
$$\frac{9.4}{10} = \frac{9.4 \times 10}{10 \times 10} = \frac{94}{100} = \frac{47 \cdot 2}{50 \cdot 2} = \frac{47}{50}.$$

4. The ratio of 3.2 to 9.2 is
$$\frac{3.2}{9.2} = \frac{3.2 \times 10}{9.2 \times 10} = \frac{32}{92} = \frac{4 \cdot 8}{4 \cdot 23} = \frac{8}{23}.$$

5. The ratio of 8.65 to 6.95 is
$$\frac{8.65}{6.95} = \frac{8.65 \times 100}{6.95 \times 100} = \frac{865}{695} = \frac{173 \cdot 5}{139 \cdot 5} = \frac{173}{139}.$$

6. The ratio of 3.6 to 4.2 is
$$\frac{3.6}{4.2} = \frac{3.6 \times 10}{4.2 \times 10} = \frac{36}{42} = \frac{6 \cdot 6}{7 \cdot 6} = \frac{6}{7}.$$

7. The ratio of $\dfrac{7}{2}$ to 13 is
$$\frac{\frac{7}{2}}{13} = \frac{7}{2} \div 13 = \frac{7}{2} \cdot \frac{1}{13} = \frac{7 \cdot 1}{2 \cdot 13} = \frac{7}{26}.$$

8. The ratio of $1\dfrac{2}{3}$ to $2\dfrac{3}{4}$ is
$$\frac{1\frac{2}{3}}{2\frac{3}{4}} = 1\frac{2}{3} \div 2\frac{3}{4} = \frac{5}{3} \div \frac{11}{4} = \frac{5}{3} \cdot \frac{4}{11} = \frac{5 \cdot 4}{3 \cdot 11} = \frac{20}{33}.$$

9. The ratio of 16 inches to 24 inches is
$$\frac{16 \text{ inches}}{24 \text{ inches}} = \frac{2 \cdot 8}{3 \cdot 8} = \frac{2}{3}.$$

10. The ratio of 5 hours to 40 hours is
$$\frac{5 \text{ hours}}{40 \text{ hours}} = \frac{5 \cdot 1}{5 \cdot 8} = \frac{1}{8}.$$

11. $\dfrac{12 \text{ inches}}{18 \text{ inches}} = \dfrac{6 \cdot 2}{6 \cdot 3} = \dfrac{2}{3}$

12. a. 10 films were rated PG-13.

    b. $\dfrac{10 \text{ films}}{20 \text{ films}} = \dfrac{10 \cdot 1}{10 \cdot 2} = \dfrac{1}{2}$

13. $\dfrac{4 \text{ professors}}{20 \text{ graduate assistants}}$
$= \dfrac{1 \text{ professor}}{5 \text{ graduate assistants}}$

14. $\dfrac{6 \text{ lights}}{20 \text{ feet}} = \dfrac{3 \text{ lights}}{10 \text{ feet}}$

15. $\dfrac{100 \text{ Senators}}{50 \text{ states}} = \dfrac{2 \text{ Senators}}{1 \text{ state}}$

16. $\dfrac{5 \text{ teachers}}{140 \text{ students}} = \dfrac{1 \text{ teacher}}{28 \text{ students}}$

17. $\dfrac{21 \text{ inches}}{7 \text{ seconds}} = \dfrac{3 \text{ inches}}{1 \text{ second}}$

18. $\dfrac{\$40}{5 \text{ hours}} = \dfrac{\$8}{1 \text{ hour}}$

19. $\dfrac{76 \text{ households with computers}}{100 \text{ households}}$
$= \dfrac{19 \text{ households with computers}}{25 \text{ households}}$

20. $\dfrac{538 \text{ electoral votes}}{50 \text{ states}} = \dfrac{269 \text{ electoral votes}}{25 \text{ states}}$

21. $\dfrac{560 \text{ feet}}{4 \text{ seconds}}$

$$
\begin{array}{r}
140 \\
4\overline{)\,560} \\
\underline{-4} \\
16 \\
\underline{-16} \\
00
\end{array}
$$

The unit rate is $\dfrac{140 \text{ feet}}{1 \text{ second}}$ or 140 feet/second

22. $\dfrac{195 \text{ miles}}{3 \text{ hours}}$

$$
\begin{array}{r}
65 \\
3\overline{)\,195} \\
\underline{-18} \\
15 \\
\underline{-15} \\
0
\end{array}
$$

The unit rate is $\dfrac{65 \text{ miles}}{1 \text{ hour}}$ or 65 miles/hour.

23. $\dfrac{63 \text{ employees}}{3 \text{ fax lines}}$

$$
\begin{array}{r}
21 \\
3\overline{)\,63} \\
\underline{-6} \\
03 \\
\underline{-3} \\
0
\end{array}
$$

The unit rate is $\dfrac{21 \text{ employees}}{1 \text{ fax line}}$ or
21 employees/fax line.

24. $\dfrac{85 \text{ phone calls}}{5 \text{ teenagers}}$

$$
\begin{array}{r}
17 \\
5\overline{)\,85} \\
\underline{-5} \\
35 \\
\underline{-35} \\
0
\end{array}
$$

The unit rate is $\dfrac{17 \text{ phone calls}}{1 \text{ teenager}}$ or
17 phone calls/teenager.

25. $\dfrac{156 \text{ miles}}{6 \text{ gallons}}$

$$
\begin{array}{r}
26 \\
6\overline{)\,156} \\
\underline{12} \\
36 \\
\underline{-36} \\
0
\end{array}
$$

The unit rate is $\dfrac{26 \text{ miles}}{1 \text{ gallon}}$ or
26 miles/gallon.

26. $\dfrac{112 \text{ teachers}}{7 \text{ computers}}$

$$
\begin{array}{r}
16 \\
7\overline{)112} \\
\underline{-7} \\
42 \\
\underline{-42} \\
0
\end{array}
$$

The unit rate is $\dfrac{16 \text{ teachers}}{1 \text{ computer}}$ or

16 teachers/computer.

27. $\dfrac{8125 \text{ books}}{1250 \text{ college students}}$

$$
\begin{array}{r}
6.5 \\
1250\overline{)8125.0} \\
\underline{-7500} \\
625\ 0 \\
\underline{-625\ 0} \\
0
\end{array}
$$

The unit rate is $\dfrac{6.5 \text{ books}}{1 \text{ college student}}$ or

6.5 books/student.

28. $\dfrac{2310 \text{ pounds}}{14 \text{ adults}}$

$$
\begin{array}{r}
165 \\
14\overline{)2310} \\
\underline{-14} \\
91 \\
\underline{-84} \\
70 \\
\underline{-70} \\
0
\end{array}
$$

The unit rate is $\dfrac{165 \text{ pounds}}{1 \text{ adult}}$ or

165 pounds/adult.

29.
$$
\begin{array}{r}
0.27 \\
8\overline{)2.16} \\
\underline{-1\ 6} \\
56 \\
\underline{-56} \\
0
\end{array}
$$

The 8-pound size costs $0.27 per pound.

$$
\begin{array}{r}
0.2772 \approx 0.277 \\
18\overline{)4.9900} \\
\underline{-3\ 6} \\
1\ 39 \\
\underline{-1\ 26} \\
130 \\
\underline{-126} \\
40 \\
\underline{-36} \\
4
\end{array}
$$

The 18-pound size costs $0.277 per pound.
The 8-pound size is the better buy.

30.
$$
\begin{array}{r}
0.0198 \approx 0.020 \\
100\overline{)1.9800} \\
\underline{-1\ 00} \\
980 \\
\underline{-900} \\
800 \\
\underline{-800} \\
0
\end{array}
$$

The 100-plate size costs $0.20 per plate.

$$
\begin{array}{r}
0.0179 \approx 0.018 \\
500\overline{)8.9900} \\
\underline{-5\ 00} \\
3\ 990 \\
\underline{-3\ 500} \\
4900 \\
\underline{-4500} \\
400
\end{array}
$$

The 500-plate size costs $0.018 per plate.
The 500 paper plates is the better buy.

31.

$$
\begin{array}{r}
0.7966\ldots \approx 0.800 \\
3\overline{)2.3900} \\
-2\,1 \\
\hline
29 \\
-27 \\
\hline
20 \\
-18 \\
\hline
20 \\
-18 \\
\hline
2
\end{array}
$$

The 3-pack size costs $0.800 per pack.

$$
\begin{array}{r}
0.7487 \approx 0.750 \\
8\overline{)5.9900} \\
-5\,6 \\
\hline
39 \\
-32 \\
\hline
70 \\
-64 \\
\hline
60 \\
-56 \\
\hline
4
\end{array}
$$

The 8-pack size costs $0.750 per pack.
The 8 pack is the better buy.

32.

$$
\begin{array}{r}
0.9225 \approx 0.923 \\
4\overline{)3.6900} \\
-3\,6 \\
\hline
09 \\
-8 \\
\hline
10 \\
-8 \\
\hline
20 \\
-20 \\
\hline
0
\end{array}
$$

The 4 batteries cost $0.923 each.

$$
\begin{array}{r}
0.989 \\
10\overline{)9.890} \\
-9\,0 \\
\hline
89 \\
-80 \\
\hline
90 \\
-90 \\
\hline
0
\end{array}
$$

The 10 batteries cost $0.989 each.
The 4 batteries are the better buy.

33.
$$\frac{7}{4} \overset{?}{=} \frac{5}{3}$$
$$7 \cdot 3 \overset{?}{=} 4 \cdot 5$$
$$21 \neq 20$$
The proportion is false.

34.
$$\frac{8.2}{2} \overset{?}{=} \frac{16.4}{4}$$
$$8.2 \cdot 4 \overset{?}{=} 2 \cdot 16.4$$
$$32.8 = 32.8$$
The proportion is true.

35.
$$\frac{5}{3} = \frac{40}{x}$$
$$5 \cdot x = 3 \cdot 40$$
$$5x = 120$$
$$\frac{5x}{5} = \frac{120}{5}$$
$$x = 24$$

36.
$$\frac{y}{10} = \frac{13}{4}$$
$$y \cdot 4 = 10 \cdot 13$$
$$4y = 130$$
$$\frac{4y}{4} = \frac{130}{4}$$
$$y = 32.5$$

37. $\dfrac{6}{11} = \dfrac{z}{5}$

$6 \cdot 5 = 11 \cdot z$

$30 = 11z$

$\dfrac{30}{11} = \dfrac{11z}{11}$

$2.\overline{72} = z$ or $z = 2\dfrac{8}{11}$

38. $\dfrac{21}{x} = \dfrac{\frac{7}{2}}{3}$

$21 \cdot 3 = x \cdot \dfrac{7}{2}$

$63 = \dfrac{7}{2} \cdot x$

$\dfrac{2}{7} \cdot \dfrac{63}{1} = \dfrac{2}{7} \cdot \dfrac{7}{2} \cdot x$

$18 = x$

Section 6.3

Practice Problems

1. Let x be the length of the wall.

$\begin{array}{l} \text{feet} \rightarrow \\ \text{inches} \rightarrow \end{array} \dfrac{4}{1} = \dfrac{x}{4\frac{1}{4}} \begin{array}{l} \leftarrow \text{feet} \\ \leftarrow \text{inches} \end{array}$

$4 \cdot 4\dfrac{1}{4} = 1 \cdot x$

$\dfrac{4}{1} \cdot \dfrac{17}{4} = x$

$17 = x$

The wall is 17 feet long.

2. $\begin{array}{l} \text{ounces} \rightarrow \\ \text{gallons} \rightarrow \end{array} \dfrac{5}{16} = \dfrac{8}{x} \begin{array}{l} \leftarrow \text{ounces} \\ \leftarrow \text{gallons} \end{array}$

$5 \cdot x = 16 \cdot 8$

$5x = 128$

$\dfrac{5x}{5} = \dfrac{128}{5}$

$x = 25.6$ or $25\dfrac{3}{5}$

Therefore, 25.6 or $25\dfrac{3}{5}$ gallons of gas can be treated with 8 ounces of alcohol.

3. Let x be the number of gallons.

$270 \times 11 = 2970$ square feet

$\begin{array}{l} \text{gallons} \rightarrow \\ \text{square feet} \rightarrow \end{array} \dfrac{1}{450} = \dfrac{x}{2970} \begin{array}{l} \leftarrow \text{gallons} \\ \leftarrow \text{square feet} \end{array}$

$1 \cdot 2970 = 450 \cdot x$

$\dfrac{2970}{450} = \dfrac{450x}{450}$

$6.6 = x$

$7 \approx x$

Therefore, 7 gallons are needed for the wall.

Exercise Set 6.3

1. Let x be the number of baskets made.

$\begin{array}{l} \text{baskets} \rightarrow \\ \text{attempts} \rightarrow \end{array} \dfrac{45}{100} = \dfrac{x}{800} \begin{array}{l} \leftarrow \text{attempts} \\ \leftarrow \text{baskets} \end{array}$

$45 \cdot 800 = 100 \cdot x$

$36,000 = 100x$

$\dfrac{36,000}{100} = \dfrac{100x}{100}$

$360 = x$

He made 360 baskets.

3. Let x be the number of minutes.

$\begin{array}{l} \text{minutes} \rightarrow \\ \text{pages} \rightarrow \end{array} \dfrac{30}{4} = \dfrac{x}{22} \begin{array}{l} \leftarrow \text{minutes} \\ \leftarrow \text{pages} \end{array}$

$30 \cdot 22 = 4 \cdot x$

$660 = 4x$

$\dfrac{660}{4} = \dfrac{4x}{4}$

$165 = x$

It will take her 165 minutes to word process and spell check 22 pages.

5. Let x be the number of applications received.

$\begin{array}{l} \text{accepted} \rightarrow \\ \text{applied} \rightarrow \end{array} \dfrac{2}{7} = \dfrac{180}{x} \begin{array}{l} \leftarrow \text{accepted} \\ \leftarrow \text{applied} \end{array}$

$2 \cdot x = 7 \cdot 180$

$2x = 1260$

$\dfrac{2x}{2} = \dfrac{1260}{2}$

$x = 630$

The school received 630 applications.

7. Let x be the length of the wall.

$$\text{inches} \rightarrow \frac{1}{8} = \frac{2\frac{7}{8}}{x} \leftarrow \text{inches}$$
$$\text{feet} \rightarrow \frac{1}{8} = \frac{2\frac{7}{8}}{x} \leftarrow \text{feet}$$

$$1 \cdot x = 8 \cdot 2\frac{7}{8}$$
$$x = \frac{8}{1} \cdot \frac{23}{8}$$
$$x = 23$$

The wall is 23 feet long.

9. Let x be the number of square feet required.

$$\text{floor space} \rightarrow \frac{9}{1} = \frac{x}{30} \leftarrow \text{floor space}$$
$$\text{students} \rightarrow \frac{9}{1} = \frac{x}{30} \leftarrow \text{students}$$

$$9 \cdot 30 = 1 \cdot x$$
$$270 = x$$

30 students require 270 square feet of floor space.

11. Let x be the number of gallons.

$$\text{miles} \rightarrow \frac{627}{12.3} = \frac{1250}{x} \leftarrow \text{miles}$$
$$\text{gallons} \rightarrow \frac{627}{12.3} = \frac{1250}{x} \leftarrow \text{gallons}$$

$$627 \cdot x = 12.3 \cdot 1250$$
$$627x = 15,375$$
$$\frac{627x}{627} = \frac{15,375}{627}$$
$$x \approx 25$$

He can expect to burn 25 gallons.

13. Let x be the distance between Milan and Rome.

$$\text{kilometers} \rightarrow \frac{30}{1} = \frac{x}{15} \leftarrow \text{kilometers}$$
$$\text{cm on map} \rightarrow \frac{30}{1} = \frac{x}{15} \leftarrow \text{cm on map}$$

$$30 \cdot 15 = 1 \cdot x$$
$$450 = x$$

Milan and Rome are 450 kilometers apart.

15. Let x be the number of bags.

$$\text{bags} \rightarrow \frac{1}{3000} = \frac{x}{260 \cdot 180} \leftarrow \text{bags}$$
$$\text{square feet} \rightarrow \frac{1}{3000} = \frac{x}{260 \cdot 180} \leftarrow \text{square feet}$$

$$1 \cdot 260 \cdot 180 = 3000 \cdot x$$
$$46,800 = 3000x$$
$$\frac{46,800}{3000} = \frac{3000x}{3000}$$
$$15.6 = x$$
$$16 \approx x$$

16 bags of fertilizer should be purchased.

17. Let x be the number of hits the player is expected to get.

hits $\rightarrow \dfrac{3}{8} = \dfrac{x}{40} \leftarrow$ hits
at bats $\rightarrow 8 \quad 40 \leftarrow$ at bats

$$3 \cdot 40 = 8 \cdot x$$
$$120 = 8x$$
$$\frac{120}{8} = \frac{8x}{8}$$
$$15 = x$$

The player would be expected to get 15 hits.

19. Let x be the number that prefer Coke.

Coke $\rightarrow \dfrac{2}{3} = \dfrac{x}{40} \leftarrow$ Coke
Total $\rightarrow 3 \quad 40 \leftarrow$ Total

$$2 \cdot 40 = 3 \cdot x$$
$$80 = 3x$$
$$\frac{80}{3} = \frac{3x}{3}$$
$$27 \approx x$$

About 27 people are likely to prefer Coke.

21. Let x be the number of applications she should expect.

applications $\rightarrow \dfrac{4}{3} = \dfrac{x}{14} \leftarrow$ applications
ounces $\rightarrow 3 \quad 14 \leftarrow$ ounces

$$4 \cdot 14 = 3 \cdot x$$
$$56 = 3x$$
$$\frac{56}{3} = \frac{3x}{3}$$
$$18\frac{2}{3} = x$$

She should expect 18 applications from the 14-ounce bottle.

23. Let x be the number of weeks.

reams $\rightarrow \dfrac{5}{3} = \dfrac{8}{x} \leftarrow$ reams
weeks $\rightarrow 3 \quad x \leftarrow$ weeks

$$5 \cdot x = 3 \cdot 8$$
$$5x = 24$$
$$\frac{5x}{5} = \frac{24}{5}$$
$$x = 4.8$$
$$x \approx 5$$

The case is likely to last 5 weeks.

25. Let x be the number of servings he can make.

$$\text{milk} \rightarrow \frac{1\frac{1}{2}}{4} = \frac{4}{x} \leftarrow \text{milk} \\ \text{servings} \rightarrow \quad\quad \leftarrow \text{servings}$$

$$1\frac{1}{2} \cdot x = 4 \cdot 4$$

$$\frac{3}{2}x = 16$$

$$\frac{2}{3} \cdot \frac{3}{2}x = \frac{2}{3} \cdot 16$$

$$x = \frac{32}{3} = 10\frac{2}{3}$$

He can make $10\frac{2}{3}$ servings.

27. Let x be the number of calories.

$$\text{ounces} \rightarrow \frac{16}{80} = \frac{24}{x} \leftarrow \text{ounces} \\ \text{calories} \rightarrow \quad\quad \leftarrow \text{calories}$$

$$16 \cdot x = 80 \cdot 24$$

$$16x = 1920$$

$$\frac{16x}{16} = \frac{1920}{16}$$

$$x = 120$$

There are 120 calories in 24 ounces.

29. a. Let x be the number of teaspoons of granules needed.

$$\text{water} \rightarrow \frac{25}{1} = \frac{450}{x} \leftarrow \text{water} \\ \text{granules} \rightarrow \quad\quad \leftarrow \text{granules}$$

$$25 \cdot x = 1 \cdot 450$$

$$25x = 450$$

$$\frac{25x}{25} = \frac{450}{25}$$

$$x = 18$$

18 teaspoons of granules are needed.

b. Let x be the number of tablespoons of granules needed.

$$\text{tsp} \rightarrow \frac{3}{1} = \frac{18}{x} \leftarrow \text{tsp} \\ \text{tbsp} \rightarrow \quad\quad \leftarrow \text{tbsp}$$

$$3 \cdot x = 1 \cdot 18$$

$$3x = 18$$

$$\frac{3x}{3} = \frac{18}{3}$$

$$x = 6$$

6 tablespoons of granules are needed.

31. Let x be the number of people.

$$\text{square feet} \rightarrow \frac{625}{1} = \frac{3750}{x} \leftarrow \text{square feet} \\ \text{people} \rightarrow \quad\quad \leftarrow \text{people}$$

$$625 \cdot x = 1 \cdot 3750$$

$$\frac{625x}{625} = \frac{3750}{625}$$

$$x = 6$$

It will provide enough oxygen for 6 people.

33. Let x be the estimated head-to-toe height of the Statue of Liberty.

$$\text{height} \rightarrow \frac{x}{42} = \frac{5\frac{1}{3}}{2} \leftarrow \text{height} \\ \text{arm length} \rightarrow \quad\quad \leftarrow \text{arm length}$$

$$x \cdot 2 = 42 \cdot 5\frac{1}{3}$$

$$2x = 42 \cdot \frac{16}{3}$$

$$2x = 224$$

$$\frac{2x}{2} = \frac{224}{2}$$

$$x = 112$$

The estimated height is 112 feet.

$$112 - 111\frac{1}{12} = \frac{11}{12}$$

The difference is $\frac{11}{12}$ foot or 11 inches.

35. Let x be the number of milligrams.

$$\text{milligrams} \rightarrow \frac{72}{3.5} = \frac{x}{5} \leftarrow \text{milligrams}$$
$$\text{ounces} \rightarrow \quad\quad \leftarrow \text{ounces}$$

$$72 \cdot 5 = 3.5 \cdot x$$
$$360 = 3.5x$$
$$\frac{360}{3.5} = \frac{3.5x}{3.5}$$
$$102.9 \approx x$$

There are about 102.9 milligrams of cholesterol in 5 ounces of lobster.

37. Let x be the estimated height of the Empire State Building.

$$\text{height} \rightarrow \frac{x}{102} = \frac{881}{72} \leftarrow \text{height}$$
$$\text{stories} \rightarrow \quad\quad \leftarrow \text{stories}$$

$$x \cdot 72 = 102 \cdot 881$$
$$72x = 89,862$$
$$\frac{72x}{72} = \frac{89,862}{72}$$
$$x \approx 1248$$

The height of the Empire State Building is approximately 1248 feet.

39. Let x be the number of visits needing a prescription for medication.

$$\text{medication} \rightarrow \frac{7}{10} = \frac{x}{620} \leftarrow \text{medication}$$
$$\text{total} \rightarrow \quad\quad \leftarrow \text{total}$$

$$7 \cdot 620 = 10 \cdot x$$
$$4340 = 10x$$
$$\frac{4340}{10} = \frac{10x}{10}$$
$$434 = x$$

Expect 434 emergency room visits to need a prescription for medication.

41. Let x be the number expected to have worked in the restaurant industry.

$$\text{restaurant} \rightarrow \frac{x}{84} = \frac{1}{3} \leftarrow \text{restaurant}$$
$$\text{workers} \rightarrow \quad\quad \leftarrow \text{workers}$$

$$x \cdot 3 = 84 \cdot 1$$
$$3x = 84$$
$$\frac{3x}{3} = \frac{84}{3}$$
$$x = 28$$

You would expect 28 of the workers to have worked in the restaurant industry.

43. Let x be the cups of salt.

$$\text{ice} \rightarrow \frac{5}{1} = \frac{12}{x} \leftarrow \text{ice}$$
$$\text{salt} \rightarrow \quad\quad \leftarrow \text{salt}$$

$$5 \cdot x = 1 \cdot 12$$
$$\frac{5x}{5} = \frac{12}{5}$$
$$x = 2.4$$

Mix 2.4 cups of salt with the ice.

45. a. Let x be the number of gallons of oil needed.

$$\text{oil} \rightarrow \frac{x}{5} = \frac{1}{50} \leftarrow \text{oil}$$
$$\text{gas} \rightarrow \quad\quad \leftarrow \text{gas}$$

$$x \cdot 50 = 5 \cdot 1$$
$$50x = 5$$
$$\frac{50x}{50} = \frac{5}{50}$$
$$x = \frac{1}{10} = 0.1$$

0.1 gallon of oil is needed.

b. Let x be the number of fluid ounces.

$$\text{gallons} \rightarrow \frac{1}{128} = \frac{0.1}{x} \leftarrow \text{gallons}$$
$$\text{fluid ounces} \rightarrow \quad\quad \leftarrow \text{fluid ounces}$$

$$1 \cdot x = 128 \cdot 0.1$$
$$x = 12.8$$

0.1 gallon is approximately 13 fluid ounces

47. a. Let x be the milligrams of medicine.

$$\text{milligrams} \rightarrow \frac{x}{275} = \frac{150}{20} \leftarrow \text{milligrams}$$
$$\text{pounds} \rightarrow \quad\quad \leftarrow \text{pounds}$$

$$x \cdot 20 = 275 \cdot 150$$
$$20x = 41,250$$
$$\frac{20x}{20} = \frac{41,250}{20}$$
$$x = 2062.5$$

The daily dose is 2062.5 milligrams.

b. $500 \times \dfrac{24}{8} = 500 \times 3 = 1500$

No, he is not receiving the proper dosage.

49. $200 = 2 \cdot 100$

$= 2 \cdot 4 \cdot 25$

$= 2 \cdot 2 \cdot 2 \cdot 5 \cdot 5$

$= 2^3 \cdot 5^2$

51. $32 = 2 \cdot 16$

$2 \cdot 2 \cdot 8$

$2 \cdot 2 \cdot 2 \cdot 4$

$2 \cdot 2 \cdot 2 \cdot 2 \cdot 2$

$= 2^5$

53. Let x be the number of ml.

$$\begin{array}{c} mg \rightarrow \\ ml \rightarrow \end{array} \frac{15}{1} = \frac{12}{x} \begin{array}{c} \leftarrow mg \\ \leftarrow ml \end{array}$$

$15 \cdot x = 1 \cdot 12$

$15x = 12$

$\dfrac{15x}{15} = \dfrac{12}{15}$

$x = \dfrac{4}{5} = 0.8$

0.8 ml of the medicine should be administered.

55. Let x be the number of ml.

$$\begin{array}{c} mg \rightarrow \\ ml \rightarrow \end{array} \frac{8}{1} = \frac{10}{x} \begin{array}{c} \leftarrow mg \\ \leftarrow ml \end{array}$$

$8 \cdot x = 1 \cdot 10$

$8x = 10$

$\dfrac{8x}{8} = \dfrac{10}{8}$

$x = 1.25$

1.25 ml of the medicine should be administered.

57. 11 muffins are approximately 1 dozen (12) muffins.

$1.5 \cdot 8 = 12$

Approximately 12 cups of milk will be needed.

59. $\begin{array}{c} \text{feet} \rightarrow \\ \text{pounds} \rightarrow \end{array} \dfrac{7}{60} = \dfrac{n}{40} \begin{array}{c} \leftarrow \text{feet} \\ \leftarrow \text{pounds} \end{array}$

$7 \cdot 40 = 60 \cdot n$

$280 = 60n$

$\dfrac{280}{60} = \dfrac{60n}{60}$

$4\dfrac{2}{3} = n$

The distance is $4\dfrac{2}{3}$ feet.

61. answers may vary

Section 6.4

Practice Problems

1. $\sqrt{100} = 10$ because $10^2 = 100$.

2. $\sqrt{64} = 8$ because $8^2 = 64$.

3. $\sqrt{121} = 11$ because $11^2 = 121$.

4. $\sqrt{0} = 0$ because $0^2 = 0$.

5. $\sqrt{\dfrac{1}{4}} = \dfrac{1}{2}$ because $\left(\dfrac{1}{2}\right)^2 = \dfrac{1}{4}$.

6. $\sqrt{\dfrac{9}{16}} = \dfrac{3}{4}$ because $\left(\dfrac{3}{4}\right)^2 = \dfrac{9}{16}$.

7. a. $\sqrt{10} \approx 3.162$

  b. $\sqrt{62} \approx 7.874$

8. Let $a = 12$ and $b = 16$.

$a^2 + b^2 = c^2$

$12^2 + 16^2 = c^2$

$144 + 256 = c^2$

$400 = c^2$

$\sqrt{400} = c$

$20 = c$

The hypotenuse is 20 feet.

9. Let $a = 9$ and $b = 7$.

$$a^2 + b^2 = c^2$$
$$9^2 + 7^2 = c^2$$
$$81 + 49 = c^2$$
$$130 = c^2$$
$$\sqrt{130} = c$$
$$11 \approx c$$

The hypotenuse is approximately 11 kilometers.

10. Let $a = 7$ and $c = 13$.

$$a^2 + b^2 = c^2$$
$$7^2 + b^2 = 13^2$$
$$49 + b^2 = 169$$
$$b^2 = 120$$
$$b = \sqrt{120}$$
$$b \approx 10.95$$

The length of the hypotenuse is exactly $\sqrt{120}$ feet or approximately 10.95 feet.

11.

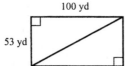

100 yd

53 yd

Let $a = 53$ and $b = 100$.

$$a^2 + b^2 = c^2$$
$$53^2 + 100^2 = c^2$$
$$2809 + 10{,}000 = c^2$$
$$12{,}809 = c^2$$
$$\sqrt{12{,}809} = c$$
$$113 \approx c$$

The diagonal is approximately 113 yards.

Calculator Explorations

1. $\sqrt{1024} = 32$

3. $\sqrt{15} \approx 3.873$

5. $\sqrt{97} \approx 9.849$

Vocabulary and Readiness Check

1. A square root of 100 is <u>10</u> and <u>−10</u> because $10 \cdot 10 = 100$ and $(-10)(-10) = 100$.

3. The <u>radical</u> sign is used to denote the positive square root of a nonnegative number.

5. The numbers 9, 1, and $\dfrac{1}{25}$ are called <u>perfect squares</u>.

7. In the given triangle, $a^2 + \underline{c^2} = \underline{b^2}$.

Exercise Set 6.4

1. $\sqrt{4} = 2$ because $2^2 = 4$.

3. $\sqrt{121} = 11$ because $11^2 = 121$.

5. $\sqrt{\dfrac{1}{81}} = \dfrac{1}{9}$ because $\left(\dfrac{1}{9}\right)^2 = \dfrac{1}{9} \cdot \dfrac{1}{9} = \dfrac{1}{81}$.

7. $\sqrt{\dfrac{16}{64}} = \dfrac{4}{8} = \dfrac{1}{2}$ because $\left(\dfrac{4}{8}\right)^2 = \dfrac{4}{8} \cdot \dfrac{4}{8} = \dfrac{16}{64}$.

9. $\sqrt{3} \approx 1.732$

11. $\sqrt{15} \approx 3.873$

13. $\sqrt{31} \approx 5.568$

15. $\sqrt{26} \approx 5.099$

17. $\sqrt{256} = 16$ because $16^2 = 256$.

19. $\sqrt{92} \approx 9.592$

21. $\sqrt{\dfrac{49}{144}} = \dfrac{7}{12}$ because $\left(\dfrac{7}{12}\right)^2 = \dfrac{7}{12} \cdot \dfrac{7}{12} = \dfrac{49}{144}$.

23. $\sqrt{71} \approx 8.426$

25. Let $a = 5$ and $b = 12$.
$$a^2 + b^2 = c^2$$
$$5^2 + 12^2 = c^2$$
$$25 + 144 = c^2$$
$$169 = c^2$$
$$\sqrt{169} = c$$
$$13 = c$$
The missing length is 13 inches.

27. Let $a = 10$ and $c = 12$.
$$a^2 + b^2 = c^2$$
$$10^2 + b^2 = 12^2$$
$$100 + b^2 = 144$$
$$b^2 = 44$$
$$b = \sqrt{44}$$
$$b \approx 6.633$$
The missing length is approximately 6.633 centimeters.

29. Let $a = 22$ and $b = 48$.
$$c^2 = a^2 + b^2$$
$$c^2 = 22^2 + 48^2$$
$$c^2 = 484 + 2304$$
$$c^2 = 2788$$
$$c = \sqrt{2788}$$
$$c \approx 52.802$$
The missing length is approximately 52.802 meters.

31. Let $a = 108$ and $b = 45$.
$$a^2 + b^2 = c^2$$
$$108^2 + 45^2 = c^2$$
$$11,664 + 2025 = c^2$$
$$13,689 = c^2$$
$$\sqrt{13,689} = c$$
$$117 = c$$
The missing length is 117 millimeters.

33.

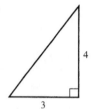

$$\text{hypotenuse} = \sqrt{(\text{leg})^2 + (\text{other leg})^2}$$
$$= \sqrt{(3)^2 + (4)^2}$$
$$= \sqrt{9 + 16}$$
$$= \sqrt{25}$$
$$= 5$$
The hypotenuse has length 5 units.

35.

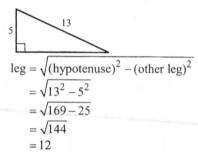

$$\text{leg} = \sqrt{(\text{hypotenuse})^2 - (\text{other leg})^2}$$
$$= \sqrt{13^2 - 5^2}$$
$$= \sqrt{169 - 25}$$
$$= \sqrt{144}$$
$$= 12$$
The leg has length 12 units.

37.

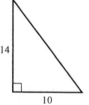

$$\text{hypotenuse} = \sqrt{(\text{leg})^2 + (\text{other leg})^2}$$
$$= \sqrt{(10)^2 + (14)^2}$$
$$= \sqrt{100 + 196}$$
$$= \sqrt{296}$$
$$\approx 17.205$$
The hypotenuse has length of about 17.205 units.

39.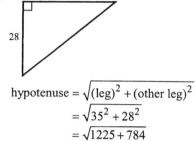

$$\text{hypotenuse} = \sqrt{(\text{leg})^2 + (\text{other leg})^2}$$
$$= \sqrt{35^2 + 28^2}$$
$$= \sqrt{1225 + 784}$$
$$= \sqrt{2009}$$
$$\approx 44.822$$

The hypotenuse has length of about 44.822 units.

41.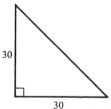

$$\text{hypotenuse} = \sqrt{(\text{leg})^2 + (\text{other leg})^2}$$
$$= \sqrt{(30)^2 + (30)^2}$$
$$= \sqrt{900 + 900}$$
$$= \sqrt{1800}$$
$$\approx 42.426$$

The hypotenuse has length of about 42.426 units.

43.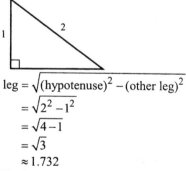

$$\text{leg} = \sqrt{(\text{hypotenuse})^2 - (\text{other leg})^2}$$
$$= \sqrt{2^2 - 1^2}$$
$$= \sqrt{4 - 1}$$
$$= \sqrt{3}$$
$$\approx 1.732$$

The leg has length of about 1.732 units.

45.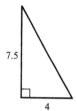

$$\text{hypotenuse} = \sqrt{(\text{leg})^2 + (\text{other leg})^2}$$
$$= \sqrt{(7.5)^2 + (4)^2}$$
$$= \sqrt{56.25 + 16}$$
$$= \sqrt{72.25}$$
$$= 8.5$$

The hypotenuse has length 8.5 units.

47. $\text{hypotenuse} = \sqrt{(\text{leg})^2 + (\text{other leg})^2}$
$$= \sqrt{100^2 + 100^2}$$
$$= \sqrt{10,000 + 10,000}$$
$$= \sqrt{20,000}$$
$$\approx 141.42$$

The length of the diagonal is about 141.42 yards.

49. $\text{leg} = \sqrt{(\text{hypotenuse})^2 - (\text{other leg})^2}$
$$= \sqrt{(32)^2 - (20)^2}$$
$$= \sqrt{1024 - 400}$$
$$= \sqrt{624}$$
$$\approx 25.0$$

The tree is about 25 feet tall.

51. $\text{hypotenuse} = \sqrt{(\text{leg})^2 + (\text{other leg})^2}$
$$= \sqrt{160^2 + 300^2}$$
$$= \sqrt{25,600 + 90,000}$$
$$= \sqrt{115,600}$$
$$= 340$$

The length of the run was 340 feet.

53. $\dfrac{10}{12} = \dfrac{5 \cdot 2}{6 \cdot 2} = \dfrac{5}{6}$

55. $\dfrac{2x}{60} = \dfrac{2 \cdot x}{2 \cdot 30} = \dfrac{x}{30}$

57. $\dfrac{3x}{9} - \dfrac{5}{9} = \dfrac{3x-5}{9}$

59. $\dfrac{9}{8} \cdot \dfrac{x}{8} = \dfrac{9 \cdot x}{8 \cdot 8} = \dfrac{9x}{64}$

61. Since 38 is between $36 = 6 \cdot 6$ and $49 = 7 \cdot 7$, $\sqrt{38}$ is between 6 and 7; $\sqrt{38} \approx 6.16$.

63. Since 101 is between $100 = 10 \cdot 10$ and $121 = 11 \cdot 11$, $\sqrt{101}$ is between 10 and 11; $\sqrt{101} \approx 10.05$.

65. answers may vary

67. $a^2 + b^2 = c^2$
$20^2 + 45^2 \overset{?}{=} 50^2$
$400 + 2025 \overset{?}{=} 2500$
$2425 \neq 2500$
No; the set does not form the lengths of the sides of a right triangle.

The Bigger Picture

1. $\sqrt{81} = 9$ because $9^2 = 81$.

2. $\sqrt{\dfrac{4}{49}} = \dfrac{2}{7}$ because $\left(\dfrac{2}{7}\right)^2 = \dfrac{2}{7} \cdot \dfrac{2}{7} = \dfrac{4}{49}$.

3. $\sqrt{9} \cdot \sqrt{25} = 3 \cdot 5 = 15$

4. $\sqrt{16} \cdot \sqrt{4} = 4 \cdot 2 = 8$

5. $6 \cdot \sqrt{9} + 3 \cdot \sqrt{4} = 6 \cdot 3 + 3 \cdot 2 = 18 + 6 = 24$

6. $3 \cdot \sqrt{25} + 2 \cdot \sqrt{81} = 3 \cdot 5 + 2 \cdot 9 = 15 + 18 = 33$

7. $4 \cdot \sqrt{49} - 0 \div \sqrt{100} = 4 \cdot 7 - 0 \div 10$
$= 28 - 0 \div 10$
$= 28 - 0$
$= 28$

8. $7 \cdot \sqrt{36} - 0 \div \sqrt{64} = 7 \cdot 6 - 0 \div 8$
$= 42 - 0 \div 8$
$= 42 - 0$
$= 42$

9. $\dfrac{\sqrt{4}}{3} + \dfrac{\sqrt{25}}{6} = \dfrac{2}{3} + \dfrac{5}{6}$
$= \dfrac{2 \cdot 2}{3 \cdot 2} + \dfrac{5}{6}$
$= \dfrac{4}{6} + \dfrac{5}{6}$
$= \dfrac{9}{6}$
$= \dfrac{3 \cdot 3}{3 \cdot 2}$
$= \dfrac{3}{2}$ or $1\dfrac{1}{2}$

10. $\dfrac{\sqrt{9}}{16} + \dfrac{\sqrt{49}}{8} = \dfrac{3}{16} + \dfrac{7}{8}$
$= \dfrac{3}{16} + \dfrac{7 \cdot 2}{8 \cdot 2}$
$= \dfrac{3}{16} + \dfrac{14}{16}$
$= \dfrac{17}{16}$ or $1\dfrac{1}{16}$

Section 6.5

Practice Problems

1. a. The triangles are congruent by Side-Angle-Side.

   b. The triangles are not congruent.

2. $\dfrac{9 \text{ mters}}{13 \text{ meters}} = \dfrac{9}{13}$

The ratio of corresponding sides is $\dfrac{9}{13}$.

3. $\dfrac{x}{5} = \dfrac{6}{9}$

$x \cdot 9 = 5 \cdot 6$

$9x = 30$

$\dfrac{9x}{9} = \dfrac{30}{9}$

$x = \dfrac{10}{3}$ or $3\dfrac{1}{3}$

4. $\dfrac{5}{n} = \dfrac{8}{60}$

$5 \cdot 60 = n \cdot 8$

$300 = 8n$

$\dfrac{300}{8} = \dfrac{8n}{8}$

$37.5 = n$

The height of the building is approximately 37.5 feet.

Vocabulary and Readiness Check

1. Two triangles that have the same shape, but not necessarily the same size are congruent. <u>false</u>

3. Congruent triangles are also similar. <u>true</u>

5. For the two similar triangles, the ratio of corresponding sides is $\dfrac{5}{6}$. <u>false</u>

Exercise Set 6.5

1. The triangles are congruent by Side-Side-Side.

3. The triangles are not congruent.

5. The triangles are congruent by Angle-Side-Angle.

7. The triangles are congruent by Side-Angle-Side.

9. $\dfrac{22}{11} = \dfrac{14}{7} = \dfrac{12}{6} = \dfrac{2}{1}$

The ratio of corresponding sides is $\dfrac{2}{1}$.

11. $\dfrac{10.5}{7} = \dfrac{9}{6} = \dfrac{12}{8} = \dfrac{3}{2}$

13. $\dfrac{x}{3} = \dfrac{9}{6}$

$6 \cdot x = 3 \cdot 9$

$6x = 27$

$\dfrac{6x}{6} = \dfrac{27}{6}$

$x = 4.5$

15. $\dfrac{n}{18} = \dfrac{4}{12}$

$n \cdot 12 = 18 \cdot 4$

$12n = 72$

$\dfrac{12n}{12} = \dfrac{72}{12}$

$n = 6$

17. $\dfrac{y}{3.75} = \dfrac{12}{9}$

$9 \cdot y = 12 \cdot 3.75$

$9y = 45$

$\dfrac{9y}{9} = \dfrac{45}{9}$

$y = 5$

19. $\dfrac{z}{18} = \dfrac{30}{40}$

$z \cdot 40 = 18 \cdot 30$

$40z = 540$

$\dfrac{40z}{40} = \dfrac{540}{40}$

$z = 13.5$

21.
$$\frac{x}{3.25} = \frac{17.5}{3.25}$$
$$3.25 \cdot x = 17.5 \cdot 3.25$$
$$3.25x = 56.875$$
$$\frac{3.25x}{3.25} = \frac{56.875}{3.25}$$
$$x = 17.5$$

23.
$$\frac{y}{2} = \frac{18\frac{1}{3}}{3\frac{2}{3}}$$
$$y \cdot 3\frac{2}{3} = 2 \cdot 18\frac{1}{3}$$
$$y \cdot \frac{11}{3} = 2 \cdot \frac{55}{3}$$
$$\frac{11}{3} \cdot y = \frac{110}{3}$$
$$\frac{3}{11} \cdot \frac{11}{3} y = \frac{3}{11} \cdot \frac{110}{3}$$
$$y = 10$$

25.
$$\frac{z}{60} = \frac{15}{32}$$
$$32 \cdot z = 60 \cdot 15$$
$$32z = 900$$
$$\frac{32z}{32} = \frac{900}{32}$$
$$z = 28.125$$

27.
$$\frac{x}{7} = \frac{15}{10\frac{1}{2}}$$
$$x \cdot 10\frac{1}{2} = 7 \cdot 15$$
$$10.5x = 105$$
$$\frac{10.5x}{10.5} = \frac{105}{10.5}$$
$$x = 10$$

29.
$$\frac{x}{25} = \frac{40}{2}$$
$$2 \cdot x = 40 \cdot 25$$
$$2x = 1000$$
$$\frac{2x}{2} = \frac{1000}{2}$$
$$x = 500$$
The building is 500 feet tall.

31.
$$\frac{x}{28} = \frac{100}{5}$$
$$x \cdot 5 = 28 \cdot 100$$
$$5x = 2800$$
$$\frac{5x}{5} = \frac{2800}{5}$$
$$x = 560$$
The fountain is approximately 560 feet high.

33.
$$\frac{x}{18} = \frac{24}{30}$$
$$30 \cdot x = 24 \cdot 18$$
$$30x = 432$$
$$\frac{30x}{30} = \frac{432}{30}$$
$$x = 14.4$$
The shadow of the tree is 14.4 feet long.

35.
$$\frac{x}{55} = \frac{19}{20}$$
$$x \cdot 20 = 55 \cdot 19$$
$$20x = 1045$$
$$\frac{20x}{20} = \frac{1045}{20}$$
$$x = 52.25$$
$$x \approx 52$$
Pete can place 52 neon tetras in the tank.

37. Let $a = 200$ and $c = 430$.
$$a^2 + b^2 = c^2$$
$$200^2 + b^2 = 430^2$$
$$40,000 + b^2 = 184,900$$
$$b^2 = 144,900$$
$$b = \sqrt{144,900}$$
$$b \approx 381$$
The gantry is approximately 381 feet tall.

39.
$$\begin{array}{r} 3.60 \\ + 0.41 \\ \hline 4.01 \end{array}$$

41. $(0.41)(-3) = -1.23$

43. Let x be the new width.
$$\frac{x}{7} = \frac{5}{9}$$
$$x \cdot 9 = 7 \cdot 5$$
$$9x = 35$$
$$\frac{9x}{9} = \frac{35}{9}$$
$$x = 3\frac{8}{9}$$

The new width is $3\frac{8}{9}$ inches. No, it will not

fit on a 3-by-5-inch card because $3\frac{8}{9} > 3$.

45.
$$\frac{n}{58.7} = \frac{11.6}{20.8}$$
$$20.8 \cdot n = 11.6 \cdot 58.7$$
$$20.8n = 680.92$$
$$\frac{20.8n}{20.8} = \frac{680.92}{20.8}$$
$$n \approx 32.7$$

47.
$$\frac{x}{5} = \frac{10}{\frac{1}{4}}$$
$$x \cdot \frac{1}{4} = 5 \cdot 10$$
$$x = 200$$
$$\frac{y}{7\frac{1}{2}} = \frac{10}{\frac{1}{4}}$$
$$y \cdot \frac{1}{4} = 7\frac{1}{2} \cdot 10$$
$$y = 300$$

$$\frac{z}{10\frac{5}{8}} = \frac{10}{\frac{1}{4}}$$
$$z \cdot \frac{1}{4} = 10\frac{5}{8} \cdot 10$$
$$z = 425$$

The actual proposed dimensions are 200 feet by 300 feet by 425 feet.

Chapter 6 Vocabulary Check

1. A <u>ratio</u> is the quotient of two numbers. It can be written as a fraction, using a colon, or using the word *to*.

2. $\dfrac{x}{2} = \dfrac{7}{16}$ is an example of a <u>proportion</u>.

3. A <u>unit rate</u> is a rate with a denominator of 1.

4. A <u>unit price</u> is a "money per item" unit rate.

5. A <u>rate</u> is used to compare different kinds of quantities.

6. In the proportion $\dfrac{x}{2} = \dfrac{7}{16}$, $x \cdot 16$ and $2 \cdot 7$ are called <u>cross products</u>.

7. If cross products are <u>equal</u> the proportion is true.

8. If cross products are <u>not equal</u> the proportion is false.

9. <u>Congruent</u> triangles have the same shape and the same size.

10. <u>Similar</u> triangles have exactly the same shape but not necessarily the same size.

11–13. Label the sides of the right triangle.

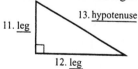

11. <u>leg</u>
12. <u>leg</u>
13. <u>hypotenuse</u>

14. A triangle with one right angle is called a <u>right</u> triangle.

15. In the right triangle,

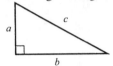

$a^2 + b^2 = c^2$ is called the <u>Pythagorean</u> theorem.

Chapter 6 Review

1. The ratio of 23 to 37 is $\dfrac{23}{37}$.

2. The ratio of \$121 to \$143 is
$$\dfrac{\$121}{\$143} = \dfrac{11 \cdot 11}{11 \cdot 13} = \dfrac{11}{13}.$$

3. The ratio of 4.25 yards to 8.75 yards is
$$\dfrac{4.25 \text{ yards}}{8.75 \text{ yards}} = \dfrac{4.25 \times 100}{8.75 \times 100}$$
$$= \dfrac{425}{875}$$
$$= \dfrac{25 \cdot 17}{25 \cdot 35}$$
$$= \dfrac{17}{35}.$$

4. The ratio of $2\frac{1}{4}$ to $4\frac{3}{8}$ is
$$\dfrac{2\frac{1}{4}}{4\frac{3}{8}} = 2\frac{1}{4} \div 4\frac{3}{8} = \dfrac{9}{4} \div \dfrac{35}{8} = \dfrac{9}{4} \cdot \dfrac{8}{35} = \dfrac{18}{35}.$$

5. $\dfrac{\text{length}}{\text{width}} = \dfrac{4.5 \text{ meters}}{2 \text{ meters}}$
$$= \dfrac{4.5 \times 10}{2 \times 10}$$
$$= \dfrac{45}{20}$$
$$= \dfrac{5 \cdot 9}{5 \cdot 4}$$
$$= \dfrac{9}{4}$$

6. $\dfrac{\text{width}}{\text{perimeter}} = \dfrac{2 \text{ meters}}{(4.5 + 2 + 4.5 + 2) \text{ meters}} = \dfrac{2}{13}$

7. $\dfrac{6000 \text{ people}}{2400 \text{ pets}} = \dfrac{1200 \cdot 5 \text{ people}}{1200 \cdot 2 \text{ pets}} = \dfrac{5 \text{ people}}{2 \text{ pets}}$

8. $\dfrac{15 \text{ pages}}{6 \text{ minutes}} = \dfrac{3 \cdot 5 \text{ pages}}{3 \cdot 2 \text{ minutes}} = \dfrac{5 \text{ pages}}{2 \text{ minutes}}$

9. $\dfrac{468 \text{ miles}}{9 \text{ hours}}$

$$\begin{array}{r} 52 \\ 9{\overline{\smash{\big)}\,468}} \\ \underline{-45} \\ 18 \\ \underline{-18} \\ 0 \end{array}$$

The unit rate is $\dfrac{52 \text{ miles}}{1 \text{ hour}}$ or 52 miles/hour.

10. $\dfrac{180 \text{ feet}}{12 \text{ seconds}}$

$$\begin{array}{r} 15 \\ 12{\overline{\smash{\big)}\,180}} \\ \underline{-12} \\ 60 \\ \underline{-60} \\ 0 \end{array}$$

The unit rate is $\dfrac{15 \text{ feet}}{1 \text{ second}}$ or 15 feet/second.

11. $\dfrac{\$6.96}{4 \text{ diskettes}}$

$$\begin{array}{r} 1.74 \\ 4{\overline{\smash{\big)}\,6.96}} \\ \underline{-4} \\ 29 \\ \underline{-28} \\ 16 \\ \underline{-16} \\ 0 \end{array}$$

The unit rate is $\dfrac{\$1.74}{1 \text{ diskette}}$ or \$1.74/diskette.

12. $\dfrac{104 \text{ bushels}}{8 \text{ trees}}$

$$\begin{array}{r} 13 \\ 8\overline{)104} \\ \underline{-8} \\ 24 \\ \underline{-24} \\ 0 \end{array}$$

The unit rate is $\dfrac{13 \text{ bushels}}{1 \text{ tree}}$ or

13 bushels/tree.

13.
$$\begin{array}{r} 0.1237 \approx 0.124 \\ 8\overline{)0.9900} \\ \underline{-8} \\ 19 \\ \underline{-16} \\ 30 \\ \underline{-24} \\ 60 \\ \underline{-56} \\ 4 \end{array}$$

The 8-ounce size costs $0.124 per ounce.
$$\begin{array}{r} 0.1408 \approx 0.141 \\ 12\overline{)1.6900} \\ \underline{-1\,2} \\ 49 \\ \underline{-48} \\ 100 \\ \underline{-96} \\ 4 \end{array}$$

The 12-ounce size costs $0.141 per ounce.
The 8-ounce size is the better buy.

14.
$$\begin{array}{r} 0.0827 \approx 0.083 \\ 18\overline{)1.4900} \\ \underline{-1\,44} \\ 50 \\ \underline{-36} \\ 140 \\ \underline{-136} \\ 4 \end{array}$$

The 18-ounce size costs $0.083 per ounce.

$$\begin{array}{r} 0.0853 \approx 0.085 \\ 28\overline{)2.3900} \\ \underline{-2\,24} \\ 150 \\ \underline{-140} \\ 100 \\ \underline{-84} \\ 16 \end{array}$$

The 28-ounce size costs $0.085 per ounce.
The 18-ounce size is the better buy.

15. $\dfrac{24 \text{ uniforms}}{8 \text{ players}} = \dfrac{3 \text{ uniforms}}{1 \text{ player}}$

16. $\dfrac{12 \text{ tires}}{3 \text{ cars}} = \dfrac{4 \text{ tires}}{1 \text{ car}}$

17. $\dfrac{19}{8} \overset{?}{=} \dfrac{14}{6}$

$19 \cdot 6 \overset{?}{=} 8 \cdot 14$

$114 \neq 112$

The proportion is false.

18. $\dfrac{3.75}{3} \overset{?}{=} \dfrac{7.5}{6}$

$3.75 \cdot 6 \overset{?}{=} 3 \cdot 7.5$

$22.5 = 22.5$

The proportion is true.

19. $\dfrac{x}{3} = \dfrac{30}{18}$

$x \cdot 18 = 3 \cdot 30$

$18x = 90$

$\dfrac{18x}{18} = \dfrac{90}{18}$

$x = 5$

20. $\dfrac{x}{9} = \dfrac{7}{3}$

$x \cdot 3 = 9 \cdot 7$

$3x = 63$

$\dfrac{3x}{3} = \dfrac{63}{3}$

$x = 21$

21.
$$\frac{-8}{5} = \frac{9}{x}$$
$$-8 \cdot x = 5 \cdot 9$$
$$-8x = 45$$
$$\frac{-8x}{-8} = \frac{45}{-8}$$
$$x = -5.625$$

22.
$$\frac{-27}{\frac{9}{4}} = \frac{x}{-5}$$
$$-27 \cdot (-5) = \frac{9}{4} \cdot x$$
$$135 = \frac{9}{4} \cdot x$$
$$\frac{4}{9} \cdot \frac{135}{1} = \frac{4}{9} \cdot \frac{9}{4} \cdot x$$
$$60 = x$$

23.
$$\frac{0.4}{x} = \frac{2}{4.7}$$
$$0.4 \cdot 4.7 = x \cdot 2$$
$$1.88 = 2x$$
$$\frac{1.88}{2} = \frac{2x}{2}$$
$$0.94 = x$$

24.
$$\frac{x}{4\frac{1}{2}} = \frac{2\frac{1}{10}}{8\frac{2}{5}}$$
$$x \cdot 8\frac{2}{5} = 4\frac{1}{2} \cdot 2\frac{1}{10}$$
$$x \cdot \frac{42}{5} = \frac{9}{2} \cdot \frac{21}{10}$$
$$\frac{42}{5} \cdot x = \frac{189}{20}$$
$$\frac{5}{42} \cdot \frac{42}{5} \cdot x = \frac{5}{42} \cdot \frac{189}{20}$$
$$x = \frac{9}{8} \text{ or } 1\frac{1}{8}$$

25.
$$\frac{x}{0.4} = \frac{4.7}{3}$$
$$x \cdot 3 = 0.4 \cdot 4.7$$
$$3x = 1.88$$
$$\frac{3x}{3} = \frac{1.88}{3}$$
$$x \approx 0.63$$

26.
$$\frac{0.07}{0.3} = \frac{7.2}{n}$$
$$0.07 \cdot n = 0.3 \cdot 7.2$$
$$0.07n = 2.16$$
$$\frac{0.07n}{0.07} = \frac{2.16}{0.07}$$
$$n \approx 30.9$$

27. Let x be the number of completed passes.

completed $\rightarrow \dfrac{3}{7} = \dfrac{x}{32} \leftarrow$ completed
attempted $\rightarrow$ $\leftarrow$ attempted
$$3 \cdot 32 = 7 \cdot x$$
$$96 = 7x$$
$$\frac{96}{7} = \frac{7x}{7}$$
$$14 \approx x$$

He completed 14 passes.

28. Let x be the number of attempted passes.

completed $\rightarrow \dfrac{3}{7} = \dfrac{15}{x} \leftarrow$ completed
attempted $\rightarrow$ $\leftarrow$ attempted
$$3 \cdot x = 7 \cdot 15$$
$$3x = 105$$
$$\frac{3x}{3} = \frac{105}{3}$$
$$x = 35$$

He attempted 35 passes.

29. Let x be the number of bags.

$$\underset{\text{square feet} \rightarrow}{\text{bags} \quad \rightarrow} \frac{1}{4000} = \frac{x}{180 \cdot 175} \underset{\leftarrow \text{ square feet}}{\leftarrow \quad \text{bags}}$$

$$1 \cdot 180 \cdot 175 = 4000 \cdot x$$

$$31,500 = 4000x$$

$$\frac{31,500}{4000} = \frac{4000x}{4000}$$

$$7.875 = x$$

Purchase 8 bags of pesticide.

30. Let x be the number of bags.

$$\underset{\text{square feet} \rightarrow}{\text{bags} \quad \rightarrow} \frac{1}{4000} = \frac{x}{250 \cdot 250} \underset{\leftarrow \text{ square feet}}{\leftarrow \quad \text{bags}}$$

$$1 \cdot 250 \cdot 250 = 4000 \cdot x$$

$$62,500 = 4000x$$

$$\frac{62,500}{4000} = \frac{4000x}{4000}$$

$$15.625 = x$$

Purchase 16 bags of pesticide.

31. Let x be the miles.

$$\underset{\text{inches} \rightarrow}{\text{miles} \rightarrow} \frac{x}{2} = \frac{80}{0.75} \underset{\leftarrow \text{ inches}}{\leftarrow \text{ miles}}$$

$$x \cdot 0.75 = 2 \cdot 80$$

$$0.75x = 160$$

$$\frac{0.75x}{0.75} = \frac{160}{0.75}$$

$$x = \frac{640}{3} \text{ or } 213\frac{1}{3}$$

The distance is $213\frac{1}{3}$ miles.

32. Let x be the inches.

$$\underset{\text{inches} \rightarrow}{\text{miles} \rightarrow} \frac{1025}{x} = \frac{80}{0.75} \underset{\leftarrow \text{ inches}}{\leftarrow \text{ miles}}$$

$$1025 \cdot 0.75 = x \cdot 80$$

$$768.75 = 80x$$

$$\frac{768.75}{80} = \frac{80x}{80}$$

$$9.6 \approx x$$

The distance is about 9.6 inches.

33. $\sqrt{64} = 8$ because $8^2 = 64$.

34. $\sqrt{144} = 12$ because $12^2 = 144$.

35. $\sqrt{12} \approx 3.464$

36. $\sqrt{15} \approx 3.873$

37. $\sqrt{0} = 0$ because $0^2 = 0$.

38. $\sqrt{1} = 1$ because $1^2 = 1$.

39. $\sqrt{50} \approx 7.071$

40. $\sqrt{65} \approx 8.062$

41. $\sqrt{\dfrac{4}{25}} = \dfrac{2}{5}$ because $\left(\dfrac{2}{5}\right)^2 = \dfrac{2}{5} \cdot \dfrac{2}{5} = \dfrac{4}{25}$.

42. $\sqrt{\dfrac{1}{100}} = \dfrac{1}{10}$ because $\left(\dfrac{1}{10}\right)^2 = \dfrac{1}{10} \cdot \dfrac{1}{10} = \dfrac{1}{100}$.

43. hypotenuse $= \sqrt{(\text{leg})^2 + (\text{other leg})^2}$
$= \sqrt{12^2 + 5^2}$
$= \sqrt{144 + 25}$
$= \sqrt{169}$
$= 13$
The leg has length 13 units.

44. hypotenuse $= \sqrt{(\text{leg})^2 + (\text{other leg})^2}$
$= \sqrt{20^2 + 21^2}$
$= \sqrt{400 + 441}$
$= \sqrt{841}$
$= 29$
The leg has length 29 units.

45. leg $= \sqrt{(\text{hypotenuse})^2 - (\text{other leg})^2}$
$= \sqrt{14^2 - 9^2}$
$= \sqrt{196 - 81}$
$= \sqrt{115}$
≈ 10.7
The leg has length of about 10.7 units.

46. hypotenuse $= \sqrt{(\text{leg})^2 + (\text{other leg})^2}$
$= \sqrt{66^2 + 56^2}$
$= \sqrt{4356 + 3136}$
$= \sqrt{7492}$
≈ 86.6
The hypotenuse has length of about 86.6 units.

47. hypotenuse $= \sqrt{(\text{leg})^2 + (\text{other leg})^2}$
$= \sqrt{20^2 + 20^2}$
$= \sqrt{400 + 400}$
$= \sqrt{800}$
≈ 28.28
The diagonal is about 28.28 centimeters.

48. leg $= \sqrt{(\text{hypotenuse})^2 - (\text{other leg})^2}$
$= \sqrt{126^2 - 90^2}$
$= \sqrt{15,876 - 8100}$
$= \sqrt{7776}$
≈ 88.2
The height is about 88.2 feet.

49. The triangles are congruent by Angle-Side-Angle.

50. The triangles are not congruent.

51.
$$\frac{x}{20} = \frac{20}{30}$$
$$x \cdot 30 = 20 \cdot 20$$
$$30x = 400$$
$$\frac{30x}{30} = \frac{400}{30}$$
$$x = \frac{40}{3} \text{ or } 13\frac{1}{3}$$

52.
$$\frac{x}{5.8} = \frac{24}{8}$$
$$x \cdot 8 = 5.8 \cdot 24$$
$$8x = 139.2$$
$$\frac{8x}{8} = \frac{139.2}{8}$$
$$x = 17.4$$

53.
$$\frac{x}{5.5} = \frac{42}{7}$$
$$x \cdot 7 = 5.5 \cdot 42$$
$$7x = 231$$
$$\frac{7x}{7} = \frac{231}{7}$$
$$x = 33$$
The height of the building is approximately 33 feet.

54.
$$\frac{x}{10} = \frac{2}{24} \qquad\qquad \frac{y}{26} = \frac{2}{24}$$
$$x \cdot 24 = 10 \cdot 2 \qquad\quad y \cdot 24 = 26 \cdot 2$$
$$24x = 20 \qquad\qquad 24y = 52$$
$$\frac{24x}{24} = \frac{20}{24} \qquad\qquad \frac{24y}{24} = \frac{52}{24}$$
$$x = \frac{5}{6} \qquad\qquad y = \frac{13}{6} \text{ or } 2\frac{1}{6}$$

The unknown lengths are $x = \dfrac{5}{6}$ inch and $y = 2\dfrac{1}{6}$ inches.

55. The ratio of 15 to 25 is $\dfrac{15}{25} = \dfrac{3}{5}$.

56. The ratio of 3 pints to 81 pints is $\dfrac{3 \text{ pints}}{81 \text{ pints}} = \dfrac{1}{27}$.

57. $\dfrac{2 \text{ teachers}}{18 \text{ students}} = \dfrac{2 \cdot 1 \text{ teachers}}{2 \cdot 9 \text{ students}} = \dfrac{1 \text{ teacher}}{9 \text{ students}}$

58. $\dfrac{6 \text{ nurses}}{24 \text{ patients}} = \dfrac{1 \cdot 6 \text{ nurses}}{4 \cdot 6 \text{ patients}} = \dfrac{1 \text{ nurse}}{4 \text{ patients}}$

59. $\dfrac{136 \text{ miles}}{4 \text{ hours}} = \dfrac{4 \cdot 34 \text{ miles}}{4 \cdot 1 \text{ hour}} = \dfrac{34 \text{ miles}}{1 \text{ hour}}$

The unit rate is $\dfrac{34 \text{ miles}}{1 \text{ hour}}$ or 34 miles/hour.

60. $\dfrac{12 \text{ gallons}}{6 \text{ cows}} = \dfrac{6 \cdot 2 \text{ gallons}}{6 \cdot 1 \text{ cow}} = \dfrac{2 \text{ gallons}}{1 \text{ cow}}$

The unit rate is $\dfrac{2 \text{ gallons}}{1 \text{ cow}}$ or 2 gallons/cow.

61. $\begin{array}{l} \text{Germany} \rightarrow \\ \text{Total} \rightarrow \end{array} \dfrac{29 \text{ medals}}{256 \text{ medals}} = \dfrac{29}{256}$

62. $\dfrac{1576 \text{ steps}}{9.5 \text{ minutes}}$

$9.5\overline{)1576}$ becomes

$$\begin{array}{r} 165.8 \\ 95\overline{)15760.0} \\ -95 \\ \hline 626 \\ -570 \\ \hline 560 \\ -475 \\ \hline 850 \\ -760 \\ \hline 90 \end{array}$$

He ran 166 steps/minute.

63. 4 ounces:

$$\begin{array}{r} 1.235 \\ 4\overline{)4.940} \\ -4 \\ \hline 09 \\ -8 \\ \hline 14 \\ -12 \\ \hline 20 \\ -20 \\ \hline 0 \end{array}$$

The 4-ounce size costs $1.235 per ounce.
8 ounces:

$$\begin{array}{r} 1.2475 \approx 1.248 \\ 8\overline{)\,9.9800} \\ \underline{-8} \\ 1\,9 \\ \underline{-1\,6} \\ 38 \\ \underline{-32} \\ 60 \\ \underline{-56} \\ 40 \\ \underline{-40} \\ 0 \end{array}$$

The 8-ounce size costs $1.248 per ounce.
The 4-ounce size is the better buy.

64. 12 ounces:

$$\begin{array}{r} 0.0541 \approx 0.054 \\ 12\overline{)\,0.6500} \\ \underline{-60} \\ 50 \\ \underline{-48} \\ 20 \\ \underline{-12} \\ 8 \end{array}$$

The 12-ounce size costs $0.054 per ounce.
64 ounces:

$$\begin{array}{r} 0.0465 \approx 0.047 \\ 64\overline{)\,2.9800} \\ \underline{-2\,56} \\ 420 \\ \underline{-384} \\ 360 \\ \underline{-320} \\ 40 \end{array}$$

The 64-ounce size costs $0.047 per ounce.
The 64-ounce size is the better buy.

65. $\dfrac{2 \text{ cups cookie dough}}{30 \text{ cookies}}$

$= \dfrac{4 \text{ cups cookie dough}}{60 \text{ cookies}}$

66. $\dfrac{5 \text{ nickels}}{3 \text{ dollars}} = \dfrac{20 \text{ nickels}}{12 \text{ dollars}}$

67. $\dfrac{3}{x} = \dfrac{15}{8}$

$3 \cdot 8 = x \cdot 15$

$24 = 15x$

$\dfrac{24}{15} = \dfrac{15x}{15}$

$1.6 = x$

68. $\dfrac{5}{4} = \dfrac{x}{20}$

$5 \cdot 20 = 4 \cdot x$

$100 = 4x$

$\dfrac{100}{4} = \dfrac{4x}{4}$

$25 = x$

69. $\dfrac{x}{3} = \dfrac{7.5}{6}$

$x \cdot 6 = 3 \cdot 7.5$

$6x = 22.5$

$\dfrac{6x}{6} = \dfrac{22.5}{6}$

$x = 3.75$

70. $\dfrac{\frac{1}{3}}{25} = \dfrac{x}{30}$

$\dfrac{1}{3} \cdot 30 = 25 \cdot x$

$10 = 25x$

$\dfrac{10}{25} = \dfrac{25x}{25}$

$\dfrac{2}{5} = x$

71. $\sqrt{36} = 6$ because $6^2 = 36$.

72. $\sqrt{\dfrac{16}{81}} = \dfrac{4}{9}$ because $\left(\dfrac{4}{9}\right)^2 = \dfrac{4}{9} \cdot \dfrac{4}{9} = \dfrac{16}{81}$.

73. $\sqrt{105} \approx 10.247$

74. $\sqrt{32} \approx 5.657$

75. hypotenuse $= \sqrt{(\text{leg})^2 + (\text{other leg})^2}$
$= \sqrt{66^2 + 56^2}$
$= \sqrt{4356 + 3136}$
$= \sqrt{7492}$
≈ 86.6

76. leg $= \sqrt{(\text{hypotenuse})^2 - (\text{other leg})^2}$
$= \sqrt{24^2 - 12^2}$
$= \sqrt{576 - 144}$
$= \sqrt{432}$
≈ 20.8

77. $\dfrac{n}{6} = \dfrac{10}{5}$
$n \cdot 5 = 6 \cdot 10$
$5n = 60$
$\dfrac{5n}{5} = \dfrac{60}{5}$
$n = 12$

78. $\dfrac{n}{8\frac{2}{3}} = \dfrac{9\frac{3}{8}}{12\frac{1}{2}}$
$n \cdot 12\frac{1}{2} = 8\frac{2}{3} \cdot 9\frac{3}{8}$
$n \cdot \dfrac{25}{2} = \dfrac{26}{3} \cdot \dfrac{75}{8}$
$\dfrac{25}{2} n = \dfrac{325}{4}$
$\dfrac{2}{25} \cdot \dfrac{25}{2} n = \dfrac{2}{25} \cdot \dfrac{325}{4}$
$n = \dfrac{13}{2}$ or $6\frac{1}{2}$

Chapter 6 Test

1. The ratio of 4500 trees to 6500 trees is
$\dfrac{4500 \text{ trees}}{6500 \text{ trees}} = \dfrac{9 \cdot 500}{13 \cdot 500} = \dfrac{9}{13}$.

2. The ratio of 9 inches of rain in 30 days is
$\dfrac{9 \text{ inches}}{30 \text{ days}} = \dfrac{3 \cdot 3 \text{ inches}}{3 \cdot 10 \text{ days}} = \dfrac{3 \text{ inches}}{10 \text{ days}}$.

3. The ratio of 8.6 to 10 is
$\dfrac{8.6}{10} = \dfrac{8.6 \times 10}{10 \times 10} = \dfrac{86}{100} = \dfrac{43}{50}$.

4. The ratio of $5\frac{7}{8}$ to $9\frac{3}{4}$ is
$\dfrac{5\frac{7}{8}}{9\frac{3}{4}} = 5\frac{7}{8} \div 9\frac{3}{4} = \dfrac{47}{8} \div \dfrac{39}{4} = \dfrac{47}{8} \cdot \dfrac{4}{39} = \dfrac{47}{78}$.

5. The ratio 414 feet to 231 feet is
$\dfrac{414 \text{ feet}}{231 \text{ feet}} = \dfrac{3 \cdot 138}{3 \cdot 77} = \dfrac{138}{77}$.

6. $\dfrac{650 \text{ kilometers}}{8 \text{ hours}}$

$$
\begin{array}{r}
81.25 \\
8\overline{)650.00} \\
\underline{-64} \\
10 \\
\underline{-8} \\
2\,0 \\
\underline{-1\,6} \\
40 \\
\underline{-40} \\
0
\end{array}
$$

The unit rate is $\dfrac{81.25 \text{ kilometers}}{1 \text{ hour}}$ or 81.25 kilometers/hour.

7. $\dfrac{140 \text{ students}}{5 \text{ teachers}}$

$$
\begin{array}{r}
28 \\
5\overline{)140} \\
\underline{-10} \\
40 \\
\underline{-40} \\
0
\end{array}
$$

The unit rate is $\dfrac{28 \text{ students}}{1 \text{ teacher}}$ or 28 students/teacher.

8. $\dfrac{108 \text{ inches}}{12 \text{ seconds}}$

$$12\overline{)108}$$
$$\underline{-108}$$
$$0$$

The unit rate is $\dfrac{9 \text{ inches}}{1 \text{ second}}$ or

9 inches/second.

9.
$$8\overline{)1.1900} \quad 0.1487 \approx 0.149$$
$$\underline{-8}$$
$$39$$
$$\underline{-32}$$
$$70$$
$$\underline{-64}$$
$$60$$
$$\underline{-56}$$
$$4$$

The 8-ounce size costs $0.149/ounce.

$$12\overline{)1.8900} \quad 0.1575 \approx 0.158$$
$$\underline{-1\,2}$$
$$69$$
$$\underline{-60}$$
$$90$$
$$\underline{-84}$$
$$60$$
$$\underline{-60}$$
$$0$$

The 12-ounce size costs $0.158/ounce.
The 8-ounce size is the better buy.

10.
$$16\overline{)1.49000} \quad 0.09312 \approx 0.0931$$
$$\underline{-1\,44}$$
$$50$$
$$\underline{-48}$$
$$20$$
$$\underline{-16}$$
$$40$$
$$\underline{-32}$$
$$8$$

The 16-ounce size costs $0.0931 per ounce.

$$24\overline{)2.39000} \quad 0.09958 \approx 0.0996$$
$$\underline{-2\,16}$$
$$230$$
$$\underline{-216}$$
$$140$$
$$\underline{-120}$$
$$200$$
$$\underline{-192}$$
$$8$$

The 24-ounce size costs $0.0996 per ounce.
The 16-ounce size is the better buy.

11. $\dfrac{28}{16} \overset{?}{=} \dfrac{14}{8}$

$28 \cdot 8 \overset{?}{=} 16 \cdot 14$

$224 = 224$

The proportion is true.

12. $\dfrac{3.6}{2.2} \overset{?}{=} \dfrac{1.9}{1.2}$

$3.6 \cdot 1.2 \overset{?}{=} 2.2 \cdot 1.9$

$4.32 \neq 4.18$

The proportion is false.

13. $\dfrac{n}{3} = \dfrac{15}{9}$

$n \cdot 9 = 3 \cdot 15$

$9n = 45$

$\dfrac{9n}{9} = \dfrac{45}{9}$

$n = 5$

14. $\dfrac{8}{x} = \dfrac{11}{6}$

$8 \cdot 6 = x \cdot 11$

$48 = 11x$

$\dfrac{48}{11} = \dfrac{11x}{11}$

$4\dfrac{4}{11} = x$

15.
$$\frac{4}{\frac{3}{7}} = \frac{y}{\frac{1}{4}}$$

$$4 \cdot \frac{1}{4} = \frac{3}{7} \cdot y$$

$$1 = \frac{3}{7} y$$

$$\frac{7}{3} \cdot 1 = \frac{7}{3} \cdot \frac{3}{7} y$$

$$\frac{7}{3} = y$$

16.
$$\frac{1.5}{5} = \frac{2.4}{n}$$

$$1.5 \cdot n = 5 \cdot 2.4$$

$$1.5n = 12$$

$$\frac{1.5n}{1.5} = \frac{12}{1.5}$$

$$n = 8$$

17. Let x be the length of the home in feet.

feet $\rightarrow x$ $9 \leftarrow$ feet
inches $\rightarrow \overline{11} = \overline{2} \leftarrow$ inches

$$x \cdot 2 = 11 \cdot 9$$

$$2x = 99$$

$$\frac{2x}{2} = \frac{99}{2}$$

$$x = 49\frac{1}{2}$$

The home is $49\frac{1}{2}$ feet long.

18. Let x be the number of hours.

miles $\rightarrow 80$ $100 \leftarrow$ miles
hours $\rightarrow \overline{3} = \overline{x} \leftarrow$ hours

$$80 \cdot x = 3 \cdot 100$$

$$80x = 300$$

$$\frac{80x}{80} = \frac{300}{80}$$

$$x = 3\frac{3}{4}$$

It will take $3\frac{3}{4}$ hours to travel 100 miles.

19. Let x be the number of grams.

grams $\rightarrow 10$ $x \leftarrow$ grams
pounds $\rightarrow \overline{15} = \overline{80} \leftarrow$ pounds

$$10 \cdot 80 = 15 \cdot x$$

$$800 = 15x$$

$$\frac{800}{15} = \frac{15x}{15}$$

$$53\frac{1}{3} = x$$

The standard dose for an 80-pound dog is $53\frac{1}{3}$ grams.

20. $\sqrt{49} = 7$ because $7^2 = 49$.

21. $\sqrt{157} \approx 12.530$

22. $\sqrt{\dfrac{64}{100}} = \dfrac{8}{10} = \dfrac{4}{5}$ because

$$\left(\frac{8}{10}\right)^2 = \frac{8}{10} \cdot \frac{8}{10} = \frac{64}{100}.$$

23. hypotenuse $= \sqrt{(\text{leg})^2 + (\text{other leg})^2}$

$$= \sqrt{4^2 + 4^2}$$

$$= \sqrt{16 + 16}$$

$$= \sqrt{32}$$

$$\approx 5.66$$

The hypotenuse is 5.66 centimeters.

24.
$$\frac{n}{12} = \frac{5}{8}$$

$$n \cdot 8 = 12 \cdot 5$$

$$8n = 60$$

$$\frac{8n}{8} = \frac{60}{8}$$

$$n = 7.5$$

25. Let x be the height of the tower.

$$\frac{x}{5\frac{3}{4}} = \frac{48}{4}$$

$$x \cdot 4 = 5\frac{3}{4} \cdot 48$$

$$4x = \frac{23}{4} \cdot 48$$

$$4x = 276$$

$$\frac{4x}{4} = \frac{276}{4}$$

$$x = 69$$

The tower is approximately 69 feet tall.

Cumulative Review Chapters 1–6

1. a. $12 - 9 = 3$ *Check*: $3 + 9 = 12$

   b. $22 - 7 = 15$ *Check*: $15 + 7 = 22$

   c. $35 - 35 = 0$ *Check*: $0 + 35 = 35$

   d. $70 - 0 = 70$ *Check*: $70 + 0 = 70$

2. a. $20 \cdot 0 = 0$

   b. $20 \cdot 1 = 20$

   c. $0 \cdot 20 = 0$

   d. $1 \cdot 20 = 20$

3. To round 248,982 to the nearest hundred, observe that the digit in the tens place is 8. Since this digit is at least 5, we add 1 to the digit in the hundreds place. The number 248,982 rounded to the nearest hundred is 249,000.

4. To round 248,982 to the nearest thousand, observe that the digit in the hundreds place is 9. Since this digit is at least 5, we add 1 to the digit in the thousands place. The number 248,982 rounded to the nearest thousand is 249,000.

5. a.
$$\begin{array}{r} \overset{4}{} \\ 25 \\ \times \quad 8 \\ \hline 200 \end{array}$$

   b.
$$\begin{array}{r} \overset{2\,3}{} \\ 246 \\ \times \quad 5 \\ \hline 1230 \end{array}$$

6.
$$\begin{array}{r} 373 \text{ R } 24 \\ 28\overline{)10,468} \\ \underline{-8\,4} \\ 2\,06 \\ \underline{-1\,96} \\ 108 \\ \underline{-84} \\ 24 \end{array}$$

7. $1 + (-10) + (-8) + 9 = -9 + (-8) + 9$
$$= -17 + 9$$
$$= -8$$

8. $-12(7) = -84$

9. $80 = 2 \cdot 40$
$$= 2 \cdot 2 \cdot 20$$
$$= 2 \cdot 2 \cdot 2 \cdot 10$$
$$= 2 \cdot 2 \cdot 2 \cdot 2 \cdot 5$$
$$= 2^4 \cdot 5$$

10. $3^2 - 1^2 = 9 - 1 = 8$

11. $\dfrac{12}{20} = \dfrac{4 \cdot 3}{4 \cdot 5} = \dfrac{3}{5}$

12. $9^2 \cdot 3 = 81 \cdot 3 = 243$

13. $-\dfrac{6}{13} \cdot -\dfrac{26}{30} = \dfrac{6 \cdot 26}{13 \cdot 30} = \dfrac{6 \cdot 13 \cdot 2}{13 \cdot 6 \cdot 5} = \dfrac{2}{5}$

14. $3\frac{3}{8} \cdot 4\frac{5}{9} = \frac{27}{8} \cdot \frac{41}{9} = \frac{9 \cdot 3 \cdot 41}{8 \cdot 9} = \frac{123}{8}$ or $15\frac{3}{8}$

15. $\frac{7}{8} + \frac{6}{8} + \frac{3}{8} = \frac{7+6+3}{8} = \frac{16}{8} = 2$

16. $\frac{7}{10} - \frac{3}{10} + \frac{4}{10} = \frac{7-3+4}{10} = \frac{8}{10} = \frac{4}{5}$

17. $7 = 7$
$14 = 2 \cdot 7$
$LCD = 2 \cdot 7 = 14$

18. $\begin{aligned}
\frac{17}{25} + \frac{3}{10} &= \frac{17}{25} \cdot \frac{2}{2} + \frac{3}{10} \cdot \frac{5}{5} \\
&= \frac{17 \cdot 2}{25 \cdot 2} + \frac{3 \cdot 5}{10 \cdot 5} \\
&= \frac{34}{50} + \frac{15}{50} \\
&= \frac{34+15}{50} \\
&= \frac{49}{50}
\end{aligned}$

19. $4 \cdot 5 = 20$, so
$\frac{3}{4} = \frac{3 \cdot 5}{4 \cdot 5} = \frac{15}{20}$

20. $\frac{10}{55} = \frac{5 \cdot 2}{5 \cdot 11} = \frac{2}{11}$
$\frac{6}{33} = \frac{3 \cdot 2}{3 \cdot 11} = \frac{2}{11}$
Yes, they are equivalent.

21. $\begin{aligned}
\frac{2}{3} - \frac{10}{11} &= \frac{2}{3} \cdot \frac{11}{11} - \frac{10}{11} \cdot \frac{3}{3} \\
&= \frac{22}{33} - \frac{30}{33} \\
&= \frac{22-30}{33} \\
&= -\frac{8}{33}
\end{aligned}$

22.
$$17\frac{5}{24} \qquad 17\frac{15}{72} \qquad 16\frac{87}{72}$$
$$-\ 9\frac{5}{9} \qquad -\ 9\frac{40}{72} \qquad -\ 9\frac{40}{72}$$
$$\qquad\qquad\qquad\qquad\qquad\qquad 7\frac{47}{72}$$

23. $\frac{5}{12} - \frac{1}{4} = \frac{5}{12} - \frac{3}{12} = \frac{5-3}{12} = \frac{2}{12} = \frac{1}{6}$
There is $\frac{1}{6}$ hour remaining.

24. $80 \div 8 \cdot 2 + 7 = 10 \cdot 2 + 7 = 20 + 7 = 27$

25.
$$2\frac{1}{3} \qquad\qquad 2\frac{8}{24}$$
$$+ 5\frac{3}{8} \qquad\qquad + 5\frac{9}{24}$$
$$\qquad\qquad\qquad\qquad 7\frac{17}{24}$$

26. $\begin{aligned}
\frac{\frac{3}{5} + \frac{4}{9} + \frac{11}{15}}{3} &= \left(\frac{27}{45} + \frac{20}{45} + \frac{33}{45} \right) \div 3 \\
&= \frac{80}{45} \cdot \frac{1}{3} \\
&= \frac{16}{9} \cdot \frac{1}{3} \\
&= \frac{16}{27}
\end{aligned}$

27. $\frac{3}{4} = \frac{3 \cdot 11}{4 \cdot 11} = \frac{33}{44}$
$\frac{9}{11} = \frac{9 \cdot 4}{11 \cdot 4} = \frac{36}{44}$
Since $33 < 36$, $\frac{33}{44} < \frac{36}{44}$ and $\frac{3}{4} < \frac{9}{11}$.

28. $\begin{aligned}
5y - 8y &= 24 \\
-3y &= 24 \\
\frac{-3y}{-3} &= \frac{24}{-3} \\
y &= -8
\end{aligned}$

29.
$$y - 5 = -2 - 6$$
$$y - 5 = -8$$
$$y - 5 + 5 = -8 + 5$$
$$y = -3$$

30.
$$3y - 6 = 7y - 6$$
$$3y - 6 + 6 = 7y - 6 + 6$$
$$3y = 7y$$
$$3y - 7y = 7y - 7y$$
$$-4y = 0$$
$$\frac{-4y}{-4} = \frac{0}{-4}$$
$$y = 0$$

31.
$$3a - 6 = a + 4$$
$$3a - 6 - a = a + 4 - a$$
$$2a - 6 = 4$$
$$2a - 6 + 6 = 4 + 6$$
$$2a = 10$$
$$\frac{2a}{2} = \frac{10}{2}$$
$$a = 5$$

32.
$$4(y + 1) - 3 = 21$$
$$4y + 4 - 3 = 21$$
$$4y + 1 = 21$$
$$4y + 1 - 1 = 21 - 1$$
$$4y = 20$$
$$\frac{4y}{4} = \frac{20}{4}$$
$$y = 5$$

33.
$$3(2x - 6) + 6 = 0$$
$$6x - 18 + 6 = 0$$
$$6x - 12 = 0$$
$$6x - 12 + 12 = 0 + 12$$
$$6x = 12$$
$$\frac{6x}{6} = \frac{12}{6}$$
$$x = 2$$

34. Seventy-five thousandths written in standard form is 0.075.

35. To round 736.2359 to the nearest tenth, observe that the digit in the hundredths place is 3. Since this digit is less than 5, we do not add 1 to the digit in the tenths place. The number 736.2359 rounded to the nearest tenth is 736.2.

36. To round 736.2359 to the nearest thousandth, observe that the digit in the ten-thousandths place is 9. Since this digit is at least 5, we add 1 to the digit in the thousandths place. The number 736.2359 rounded to the nearest thousandth is 736.236.

37.
$$\begin{array}{r} 23.850 \\ +\ 1.604 \\ \hline 25.454 \end{array}$$

38.
$$\begin{array}{r} 700.00 \\ -\ 18.76 \\ \hline 681.24 \end{array}$$

39.
$$\begin{array}{r} 0.0531 \\ \times\ \ \ \ 16 \\ \hline 3186 \\ 5310\ \ \\ \hline 0.8496 \end{array}$$

40.
$$\begin{array}{r} 0.375 \\ 8\overline{)\ 3.000} \\ -2\ 4\ \ \ \ \\ \hline 60\ \ \\ -56\ \ \\ \hline 40 \\ -40 \\ \hline 0 \end{array}$$

$$\frac{3}{8} = 0.375$$

41.

$$115\overline{)5.980} \quad \begin{array}{r} 0.052 \\ \end{array}$$

$$\begin{array}{r} 0.052 \\ 115\overline{)5.980} \\ \underline{-5\ 75} \\ 230 \\ \underline{-230} \\ 0 \end{array}$$

$-5.98 \div 115 = -0.052$

42. $7.9 = 7\dfrac{9}{10} = \dfrac{7 \cdot 10 + 9}{10} = \dfrac{79}{10}$

43. $-0.5(8.6 - 1.2) = -0.5(7.4) = -3.7$

44.

$$\dfrac{n}{4} = \dfrac{12}{16}$$
$$n \cdot 16 = 4 \cdot 12$$
$$16n = 48$$
$$\dfrac{16n}{16} = \dfrac{48}{16}$$
$$n = 3$$

45. $\dfrac{9}{20} = 0.450$

$\dfrac{4}{9} = 0.\overline{4} = 0.444...$

$0.456 = 0.456$

$\dfrac{4}{9}, \dfrac{9}{20}, 0.456$

46. $\dfrac{700 \text{ meters}}{5 \text{ seconds}} = \dfrac{5 \cdot 140 \text{ meters}}{5 \cdot 1 \text{ second}} = \dfrac{140 \text{ meters}}{1 \text{ second}}$

The unit rate is 140 meters/second.

47. The ratio of \$15 to \$10 is

$\dfrac{\$15}{\$10} = \dfrac{15}{10} = \dfrac{5 \cdot 3}{5 \cdot 2} = \dfrac{3}{2}.$

48. The ratio of 7 to 21 is $\dfrac{7}{21} = \dfrac{1 \cdot 7}{3 \cdot 7} = \dfrac{1}{3}.$

49. The ratio of 2.5 to 3.15 is

$\dfrac{2.5}{3.15} = \dfrac{2.5 \cdot 100}{3.15 \cdot 100} = \dfrac{250}{315} = \dfrac{50 \cdot 5}{63 \cdot 5} = \dfrac{50}{63}.$

50. The ratio of 900 to 9000 is

$\dfrac{900}{9000} = \dfrac{900 \cdot 1}{900 \cdot 10} = \dfrac{1}{10}.$

Chapter 7

Practice Problems

1. Since 27 students out of 100 students in a club are freshman, the fraction is $\dfrac{27}{100}$. Then
$$\dfrac{27}{100} = 27\%.$$

2. $\dfrac{31}{100} = 31\%$

3. $49\% = 49(0.01) = 0.49$

4. $3.1\% = 3.1(0.01) = 0.031$

5. $175\% \; 175(0.01) = 1.75$

6. $0.46\% = 0.46(0.01) = 0.0046$

7. $600\% = 600(0.01) = 6.00 \text{ or } 6$

8. $50\% = 50 \cdot \dfrac{1}{100} = \dfrac{50}{100} = \dfrac{1 \cdot 50}{2 \cdot 50} = \dfrac{1}{2}$

9. $2.3\% = 2.3 \cdot \dfrac{1}{100} = \dfrac{2.3}{100} = \dfrac{2.3 \cdot 10}{100 \cdot 10} = \dfrac{23}{1000}$

10. $150\% = 150 \cdot \dfrac{1}{100} = \dfrac{150}{100} = \dfrac{3 \cdot 50}{2 \cdot 50} = \dfrac{3}{2} \text{ or } 1\dfrac{1}{2}$

11. $66\dfrac{2}{3}\% = 66\dfrac{2}{3} \cdot \dfrac{1}{100}$
$$= \dfrac{200}{3} \cdot \dfrac{1}{100}$$
$$= \dfrac{2 \cdot 100}{3} \cdot \dfrac{1}{100}$$
$$= \dfrac{2}{3}$$

12. $12\% = 12 \cdot \dfrac{1}{100} = \dfrac{12}{100} = \dfrac{3 \cdot 4}{25 \cdot 4} = \dfrac{3}{25}$

13. $0.14 = 0.14(100\%) = 14.\% \text{ or } 14\%$

14. $1.75 = 1.75(100\%) = 175.\% \text{ or } 175\%$

15. $0.057 = 0.057(100\%) = 05.7\% \text{ or } 5.7\%$

16. $0.5 = 0.5(100\%) = 050.\% \text{ or } 50\%$

17. $\dfrac{1}{4} = \dfrac{1}{4} \cdot 100\% = \dfrac{1}{4} \cdot \dfrac{100}{1}\% = \dfrac{100}{4}\% = 25\%$

18. $\dfrac{9}{40} = \dfrac{9}{40} \cdot 100\%$
$$= \dfrac{9}{40} \cdot \dfrac{100}{1}\%$$
$$= \dfrac{900}{40}\%$$
$$= 22\dfrac{20}{40}\%$$
$$= 22\dfrac{1}{2}\%$$

19. $5\dfrac{1}{2} = \dfrac{11}{2}$
$$= \dfrac{11}{2} \cdot 100\%$$
$$= \dfrac{11}{2} \cdot \dfrac{100}{1}\%$$
$$= \dfrac{1100}{2}\%$$
$$= 550\%$$

20. $\dfrac{3}{17} = \dfrac{3}{17} \cdot 100\%$
$$= \dfrac{3}{17} \cdot \dfrac{100\%}{1}$$
$$= \dfrac{300}{17}\%$$
$$\approx 17.65\%$$

$$\begin{array}{r} 17.646 \approx 17.65 \\ 17{\overline{\smash{\big)}\,300.000}} \\ \underline{-17} \\ 130 \\ \underline{-119} \\ 11\,0 \\ \underline{-10\,2} \\ 80 \\ \underline{-68} \\ 120 \end{array}$$

Thus, $\dfrac{3}{17}$ is approximately 17.65%.

21. As a decimal $27.5\% = 27.5(0.01) = 0.275.$

As a fraction $27.5\% = 27.5 \cdot \dfrac{1}{100}$
$$= \dfrac{27.5}{100}$$
$$= \dfrac{27.5}{100} \cdot \dfrac{10}{10}$$
$$= \dfrac{275}{1000}$$
$$= \dfrac{11 \cdot 25}{40 \cdot 25}$$
$$= \dfrac{11}{40}.$$

Thus, 27.5% written as a decimal is 0.275, and written as a fraction is $\dfrac{11}{40}$.

22. $1\dfrac{3}{4} = \dfrac{7}{4}$
$$= \dfrac{7}{4} \cdot 100\%$$
$$= \dfrac{7}{4} \cdot \dfrac{100\%}{1}$$
$$= \dfrac{700}{4}\%$$
$$= 175\%$$

Thus, a "$1\dfrac{3}{4}$ times" increase is the same as a "175% increase."

Vocabulary and Readiness Check

1. Percent means "per hundred."

3. The % symbol is read as percent.

5. To write a percent as a *decimal*, drop the % symbol and multiply by 0.01.

Exercise Set 7.1

1. $\dfrac{96}{100} = 96\%$

3. 37 out of 100 adults preferred football.
$\dfrac{37}{100} = 37\%$

5. 37 of the adults preferred football, while 13 preferred soccer. Thus,
$37 + 13 = 50$ preferred football or soccer.
$\dfrac{50}{100} = 50\%$

7. $41\% = 41(0.01) = 0.41$

9. $6\% = 6(0.01) = 0.06$

11. $100\% = 100(0.01) = 1.00$ or 1

13. $73.6\% = 73.6(0.01) = 0.736$

15. $2.8\% = 2.8(0.01) = 0.028$

17. $0.6\% = 0.6(0.01) = 0.006$

19. $300\% = 300(0.01) = 3.00$ or 3

21. $32.58\% = 32.58(0.01) = 0.3258$

23. $8\% = 8 \cdot \dfrac{1}{100} = \dfrac{8}{100} = \dfrac{4 \cdot 2}{4 \cdot 25} = \dfrac{2}{25}$

25. $4\% = 4 \cdot \dfrac{1}{100} = \dfrac{4}{100} = \dfrac{1 \cdot 4}{4 \cdot 25} = \dfrac{1}{25}$

27. $4.5\% = 4.5 \cdot \dfrac{1}{100}$

$= \dfrac{4.5}{100}$

$= \dfrac{4.5 \cdot 10}{100 \cdot 10}$

$= \dfrac{45}{1000}$

$= \dfrac{5 \cdot 9}{5 \cdot 200}$

$= \dfrac{9}{200}$

29. $175\% = 175 \cdot \dfrac{1}{100} = \dfrac{175}{100} = \dfrac{7 \cdot 25}{4 \cdot 25} = \dfrac{7}{4}$ or $1\dfrac{3}{4}$

31. $6.25\% = 6.25 \cdot \dfrac{1}{100}$

$= \dfrac{6.25}{100}$

$= \dfrac{6.25 \cdot 100}{100 \cdot 100}$

$= \dfrac{625}{10,000}$

$= \dfrac{1 \cdot 625}{16 \cdot 625}$

$= \dfrac{1}{16}$

33. $10\dfrac{1}{3}\% = 10\dfrac{1}{3} \cdot \dfrac{1}{100} = \dfrac{31}{3} \cdot \dfrac{1}{100} = \dfrac{31}{300}$

35. $22\dfrac{3}{8}\% = 22\dfrac{3}{8} \cdot \dfrac{1}{100} = \dfrac{179}{8} \cdot \dfrac{1}{100} = \dfrac{179}{800}$

37. $0.006 = 0.006(100\%) = 0.6\%$

39. $0.22 = 0.22(100\%) = 22\%$

41. $5.3 = 5.3(100\%) = 530\%$

43. $0.056 = 0.056(100\%) = 5.6\%$

45. $0.2228 = 0.2228(100\%) = 22.28\%$

47. $3.00 = 3.00(100\%) = 300\%$

49. $0.7 = 0.7(100\%) = 70\%$

51. $\dfrac{7}{10} = \dfrac{7}{10} \cdot 100\% = \dfrac{700}{10}\% = 70\%$

53. $\dfrac{4}{5} = \dfrac{4}{5} \cdot 100\% = \dfrac{400}{5}\% = 80\%$

55. $\dfrac{34}{50} = \dfrac{34}{50} \cdot 100\% = \dfrac{3400}{50}\% = 68\%$

57. $\dfrac{3}{8} = \dfrac{3}{8} \cdot 100\% = \dfrac{300}{8}\% = \dfrac{75}{2}\% = 37\dfrac{1}{2}\%$

59. $\dfrac{1}{3} = \dfrac{1}{3} \cdot 100\% = \dfrac{100}{3}\% = 33\dfrac{1}{3}\%$

61. $4\dfrac{1}{2} = 4\dfrac{1}{2} \cdot 100\%$

$= \dfrac{9}{2} \cdot 100\%$

$= \dfrac{900}{2}\%$

$= 450\%$

63. $1\dfrac{9}{10} = 1\dfrac{9}{10} \cdot 100\%$

$= \dfrac{19}{10} \cdot 100\%$

$= \dfrac{1900}{10}\%$

$= 190\%$

65. $\dfrac{9}{11} = \dfrac{9}{11} \cdot 100\% = \dfrac{900}{11}\%$

$$
\begin{array}{r}
81.818 \approx 81.82 \\
11\overline{)\,900.000} \\
-88 \\
\hline
20 \\
-11 \\
\hline
9\,0 \\
-8\,8 \\
\hline
20 \\
-11 \\
\hline
90 \\
-88 \\
\hline
2
\end{array}
$$

$\dfrac{9}{11}$ is approximately 81.82%.

67. $\dfrac{4}{15} = \dfrac{4}{15} \cdot 100\% = \dfrac{400}{15}\%$

$$
\begin{array}{r}
26.666 \approx 26.67 \\
15\overline{)\,400.000} \\
-30 \\
\hline
100 \\
-90 \\
\hline
10\,0 \\
-9\,0 \\
\hline
1\,00 \\
-90 \\
\hline
10
\end{array}
$$

$\dfrac{4}{15}$ is approximately 26.67%.

69.

Percent	Decimal	Fraction
60%	0.6	$\dfrac{3}{5}$
$23\dfrac{1}{2}\%$	0.235	$\dfrac{47}{200}$
80%	0.8	$\dfrac{4}{5}$
$33\dfrac{1}{3}\%$	$0.333\overline{3}$	$\dfrac{1}{3}$
87.5%	0.875	$\dfrac{7}{8}$
7.5%	0.075	$\dfrac{3}{40}$

71.

Percent	Decimal	Fraction
200%	2	2
280%	2.8	$2\dfrac{4}{5}$
705%	7.05	$7\dfrac{1}{20}$
454%	4.54	$4\dfrac{27}{50}$

73. $38\% = 38(0.01) = 0.38$

$38\% = 38 \cdot \dfrac{1}{100} = \dfrac{38}{100} = \dfrac{19}{50}$

75. $25.2\% = 25.2(0.01) = 0.252$

$$
\begin{aligned}
25.2\% &= 25.2 \cdot \dfrac{1}{100} \\
&= \dfrac{25.2}{100} \\
&= \dfrac{25.2 \cdot 10}{100 \cdot 10} \\
&= \dfrac{252}{1000} \\
&= \dfrac{63}{250}
\end{aligned}
$$

77. $32.2\% = 32.2(0.01) = 0.322$

$$32.2\% = 32.2 \cdot \frac{1}{100}$$

$$= \frac{32.2}{100}$$

$$= \frac{32.2}{100} \cdot \frac{10}{10}$$

$$= \frac{322}{1000}$$

$$= \frac{161}{500}$$

79. $0.4\% = 0.4(0.01) = 0.004$

$$0.4\% = 0.4 \cdot \frac{1}{100}$$

$$= \frac{0.4}{100}$$

$$= \frac{0.4}{100} \cdot \frac{10}{10}$$

$$= \frac{4}{1000}$$

$$= \frac{1}{250}$$

81. $12\% = 12(0.01) = 0.12$

$$12\% = 12 \cdot \frac{1}{100} = \frac{12}{100} = \frac{3}{25}$$

83. $0.442 = 0.442(100\%) = 44.2\%$

85. $\dfrac{7}{1000} = \dfrac{7}{1000}(100\%)$

$$= \frac{700}{1000}\%$$

$$= \frac{7}{10}\%$$

$$= 0.7\%$$

87. $3.9\% = 3.9 \cdot \dfrac{1}{100} = \dfrac{3.9}{100} = \dfrac{3.9}{100} \cdot \dfrac{10}{10} = \dfrac{39}{1000}$

89. $\dfrac{3}{4} - \dfrac{1}{2} \cdot \dfrac{8}{9} = \dfrac{3}{4} - \dfrac{1 \cdot 8}{2 \cdot 9}$

$$= \frac{3}{4} - \frac{1 \cdot 2 \cdot 4}{2 \cdot 9}$$

$$= \frac{3}{4} - \frac{4}{9}$$

$$= \frac{3}{4} \cdot \frac{9}{9} - \frac{4}{9} \cdot \frac{4}{4}$$

$$= \frac{27}{36} - \frac{16}{36}$$

$$= \frac{27 - 16}{36}$$

$$= \frac{11}{36}$$

91. $6\dfrac{2}{3} - 4\dfrac{5}{6} = \dfrac{20}{3} - \dfrac{29}{6}$

$$= \frac{20}{3} \cdot \frac{2}{2} - \frac{29}{6}$$

$$= \frac{40}{6} - \frac{29}{6}$$

$$= \frac{11}{6}$$

$$= 1\frac{5}{6}$$

93. a. 52.8647% rounded to the nearest tenth of a percent is 52.9%.

b. 52.8647% rounded to the nearest hundredth of a percent is 52.86%.

95. a. $6.5\% = 6.5(0.01) = 0.065$
   INCORRECT

b. $7.8\% = 7.8(0.01) = 0.078$ CORRECT

c. $120\% = 120(0.01) = 1.20$ INCORRECT

d. $0.35\% = 0.35(0.01) = 0.0035$
   CORRECT

b and d are correct.

97. $45\% + 40\% + 11\% = 96\%$
$100\% - 96\% = 4\%$
4% of the U.S. population have AB blood type.

99. 3 of 4 equal parts are shaded, so
$$\frac{3}{4} = \frac{3}{4} \cdot 100\% = \frac{300}{4}\% = 75\%$$

101. A fraction written as a percent is greater than 100% when the numerator is <u>greater</u> than the denominator.

103. $\dfrac{21}{79} \approx 0.2658 \approx 0.266$

$0.2658(100\%) = 26.58\% \approx 26.6\%$

105. The longest bar corresponds to network systems and data communication analysts, so that is predicted to be the fastest growing occupation.

107. The percent change for physicians assistants is 49%, so $49\% = 49(0.01) = 0.49$.

109. answers may vary

Section 7.2

Practice Problems

1. 8 is <u>what percent</u> of 48?

$$8 = x \cdot 48$$

2. 2.6 is 40% of <u>what number</u>?

$$2.6 = 40\% \cdot x$$

3. <u>What number</u> is 90% of 0.045?

$$x = 90\% \cdot 0.045$$

4. 56% of 180 is <u>what number</u>?

$$56\% \cdot 180 = x$$

5. 12% of <u>what number</u> is 21?

$$12\% \cdot x = 21$$

6. <u>What percent</u> of 95 is 76?

$$x \cdot 95 = 76$$

7. <u>What number</u> is 25% of 90?

$$x = 25\% \cdot 90$$

$x = 25\% \cdot 90$
$x = 0.25 \cdot 90$
$x = 22.5$
Then 22.5 is 25% of 90.
Is this reasonable? To see, round 25% to 30%. Then 30% or 0.30(90) is 27. Our result is reasonable since 22.5 is close to 27.

8. 95% of 400 is <u>what number</u>?

$$95\% \cdot 400 = x$$

$0.95 \cdot 400 = x$
$380 = x$
Then 95% of 400 is 380. Is this result reasonable? To see, round 95% to 100%. Then 100% of 400 or 1.0(400) = 400, which is close to 380.

9. 15% of <u>what number</u> is 2.4?

$$15\% \cdot x = 2.4$$

$0.15 \cdot x = 2.4$
$\dfrac{0.15x}{0.15} = \dfrac{2.4}{0.15}$
$x = 16$

Then 15% of 16 is 2.4. Is this result reasonable? To see, round 15% to 20%. Then 20% of 16 or 0.20(16) = 3.2, which is close to 2.4.

10. 18 is $4\frac{1}{2}$% of <u>what number</u>?

$\downarrow\downarrow\ \downarrow\quad\downarrow\qquad\quad\downarrow$

$18 = 4\frac{1}{2}\% \cdot \qquad x$

$18 = 0.045x$

$\dfrac{18}{0.045} = \dfrac{0.045x}{0.045}$

$400 = x$

Then 18 is $4\frac{1}{2}$% of 400.

11. <u>What percent</u> of 90 is 27?

$\qquad\downarrow\qquad\downarrow\ \downarrow\ \downarrow$

$\qquad x\qquad\ \cdot\ 90\ = 27$

$90x = 27$

$\dfrac{90x}{90} = \dfrac{27}{90}$

$x = \dfrac{3}{10}$

or $x = 0.30$

Since we are looking for percent, we can write $\dfrac{3}{10}$ or 0.30 as a percent. $x = 30\%$

Then 30% of 90 is 27. To check, see that $30\% \cdot 90 = 27$.

12. 63 is <u>what percent</u> of 45?

$\downarrow\downarrow\qquad\downarrow\qquad\ \downarrow\ \downarrow$

$63 = \qquad x\qquad \cdot\ 45$

$63 = 45x$

$\dfrac{63}{45} = \dfrac{45x}{45}$

$\dfrac{7}{5} = x$

$1.4 = x$

$140\% = x$

Then 63 is 140% of 45.

Vocabulary and Readiness Check

1. The word <u>is</u> translates to "=."

3. In the statement "10% of 90 is 9," the number 9 is called the <u>amount</u>, 90 is called the <u>base</u>, and 10 is called the <u>percent</u>.

5. Any "percent greater than 100%" of "a number" = "a number <u>greater</u> than the original number."

Exercise Set 7.2

1. 18% of 81 is <u>what number</u>?

$\downarrow\quad\downarrow\ \downarrow\ \downarrow\qquad\quad\downarrow$

$18\%\ \cdot\ 81 = \qquad x$

3. 20% of <u>what number</u> is 105?

$\downarrow\quad\downarrow\qquad\quad\downarrow\qquad\ \downarrow\ \downarrow$

$20\%\ \cdot\qquad\quad x\qquad = 105$

5. 0.6 is 40% of <u>what number</u>?

$\downarrow\ \downarrow\ \downarrow\ \downarrow\qquad\quad\downarrow$

$0.6 =\ 40\% \cdot\qquad\quad x$

7. <u>What percent</u> of 80 is 3.8?

$\qquad\downarrow\qquad\quad\downarrow\ \downarrow\ \downarrow\ \downarrow$

$\qquad x\qquad\quad \cdot\ 80 =\ 3.8$

9. <u>What number</u> is 9% of 43?

$\qquad\downarrow\qquad\quad\downarrow\ \downarrow\ \downarrow\ \downarrow$

$\qquad x\qquad\quad = 9\%\ \cdot\ 43$

11. What percent of 250 is 150?

$$\downarrow \qquad \downarrow \downarrow \downarrow \downarrow$$
$$x \qquad \cdot 250 = 150$$

13. $10\% \cdot 35 = x$
$0.10 \cdot 35 = x$
$\qquad 3.5 = x$
10% of 35 is 3.5.

15. $x = 14\% \cdot 205$
$x = 0.14 \cdot 205$
$x = 28.7$
28.7 is 14% of 205.

17. $\quad 1.2 = 12\% \cdot x$
$\quad 1.2 = 0.12x$
$\dfrac{1.2}{0.12} = \dfrac{0.12x}{0.12}$
$\quad 10 = x$
1.2 is 12% of 10.

19. $8\dfrac{1}{2}\% \cdot x = 51$
$0.085x = 51$
$\dfrac{0.085x}{0.085} = \dfrac{51}{0.085}$
$\quad x = 600$

$8\dfrac{1}{2}\%$ of 600 is 51.

21. $x \cdot 80 = 88$
$\dfrac{x \cdot 80}{80} = \dfrac{88}{80}$
$\quad x = 1.1$
$\quad x = 110\%$
88 is 110% of 80.

23. $\quad 17 = x \cdot 50$
$\dfrac{17}{50} = \dfrac{x \cdot 50}{50}$
$0.34 = x$
$34\% = x$
17 is 34% of 50.

25. $\quad 0.1 = 10\% \cdot x$
$\quad 0.1 = 0.10x$
$\dfrac{0.1}{0.1} = \dfrac{0.1x}{0.1}$
$\quad 1 = x$
0.1 is 10% of 1.

27. $150\% \cdot 430 = x$
$1.5 \cdot 430 = x$
$\quad 645 = x$
150% of 430 is 645.

29. $\quad 82.5 = 16\dfrac{1}{2}\% \cdot x$
$\quad 82.5 = 0.165x$
$\dfrac{82.5}{0.165} = \dfrac{0.165x}{0.165}$
$\quad 500 = x$

82.5 is $16\dfrac{1}{2}\%$ of 500.

31. $\quad 2.58 = x \cdot 50$
$\dfrac{2.58}{50} = \dfrac{x \cdot 50}{50}$
$0.0516 = x$
$5.16\% = x$
2.58 is 5.16% of 50.

33. $x = 42\% \cdot 60$
$x = 0.42 \cdot 60$
$x = 25.2$
25.2 is 42% of 60.

35. $x \cdot 184 = 64.4$
$\dfrac{x \cdot 184}{184} = \dfrac{64.4}{184}$
$\quad x = 0.35$
$\quad x = 35\%$
35% of 184 is 64.4.

37. $120\% \cdot x = 42$
$1.20 \cdot x = 42$
$\dfrac{1.2x}{1.2} = \dfrac{42}{1.2}$
$\quad x = 35$
120% of 35 is 42.

39. $2.4\% \cdot 26 = x$
$0.024 \cdot 26 = x$
$0.624 = x$
2.4% of 26 is 0.624.

41. $x \cdot 600 = 3$
$\dfrac{x \cdot 600}{600} = \dfrac{3}{600}$
$x = 0.005$
$x = 0.5\%$
0.5% of 600 is 3.

43. $6.67 = 4.6\% \cdot x$
$6.67 = 0.046x$
$\dfrac{6.67}{0.046} = \dfrac{0.046x}{0.046}$
$145 = x$
6.67 is 4.6% of 145.

45. $1575 = x \cdot 2500$
$\dfrac{1575}{2500} = \dfrac{x \cdot 2500}{2500}$
$0.63 = x$
$63\% = x$
1575 is 63% of 2500.

47. $2 = x \cdot 50$
$\dfrac{2}{50} = \dfrac{x \cdot 50}{50}$
$0.04 = x$
$4\% = x$
2 is 4% of 50.

49. $\dfrac{27}{x} = \dfrac{9}{10}$
$27 \cdot 10 = x \cdot 9$
$270 = 9x$
$\dfrac{270}{9} = \dfrac{9x}{9}$
$30 = x$

51. $\dfrac{x}{5} = \dfrac{8}{11}$
$11 \cdot x = 8 \cdot 5$
$11x = 40$
$\dfrac{11x}{11} = \dfrac{40}{11}$
$x = 3\dfrac{7}{11}$

53. $\dfrac{17}{12} = \dfrac{x}{20}$

55. $\dfrac{8}{9} = \dfrac{14}{x}$

57. $5 \cdot n = 32$
$\dfrac{5 \cdot n}{5} = \dfrac{32}{5}$
$n = \dfrac{32}{5}$
Choice c is correct.

59. $0.06 = n \cdot 7$
$\dfrac{0.06}{7} = \dfrac{n \cdot 7}{7}$
$\dfrac{0.06}{7} = n$
Choice b is correct.

61. What number is thirty-three and one-third percent of twenty-four?

63. Since 100% of 98 is 98, the percent is 100%; a.

65. Since 55% is less than 100%, which is 1, 55% of 45 is less than 45; c.

67. Since 100% is 1, 100% of 45 is equal to 45; a.

69. Since 100% is 1, 100% of 45 is equal to 45; a.

71. answers may vary

73. $1.5\% \cdot 45,775 = x$
$0.015 \cdot 45,775 = x$
$686.625 = x$
1.5% of 45,775 is 686.625.

75. $22,113 = 180\% \cdot x$
$22,113 = 1.80 \cdot x$
$\dfrac{22,113}{1.8} = \dfrac{1.8 \cdot x}{1.8}$
$12,285 = x$
22,113 is 180% of 12,285.

The Bigger Picture

1. $\dfrac{2}{9} + \dfrac{1}{5} = \dfrac{2 \cdot 5}{9 \cdot 5} + \dfrac{1 \cdot 9}{5 \cdot 9} = \dfrac{10}{45} + \dfrac{9}{45} = \dfrac{19}{45}$

2. $42 \div 2 \cdot 3 = 21 \cdot 3 = 63$

3. $-0.03(0.7) = -0.021$

4. $\sqrt{49} + \sqrt{1} = 7 + 1 = 8$

5. $\dfrac{3}{8} = \dfrac{n}{128}$
$3 \cdot 128 = 8 \cdot n$
$384 = 8n$
$\dfrac{384}{8} = \dfrac{8n}{8}$
$48 = n$

6. $\dfrac{7.2}{n} = \dfrac{36}{8}$
$7.2 \cdot 8 = n \cdot 36$
$57.6 = 36n$
$\dfrac{57.6}{36} = \dfrac{36n}{36}$
$1.6 = n$

7. $215 = x \cdot 86$
$\dfrac{215}{86} = \dfrac{x \cdot 86}{86}$
$2.5 = x$
$250\% = x$
215 is 250% of 86.

8. $95\% \cdot 48 = x$
$0.95(48) = x$
$45.6 = x$
95% of 48 is 45.6 or $45\dfrac{3}{5}$.

9. $4.2 = x \cdot 15$
$\dfrac{4.2}{15} = \dfrac{x \cdot 15}{15}$
$0.28 = x$
$28\% = x$
4.2 is 28% of 15.

10. $93.6 = 52\% \cdot x$
$93.6 = 0.52 \cdot x$
$\dfrac{93.6}{0.52} = \dfrac{0.52x}{0.52}$
$180 = x$
93.6 is 52% of 180.

Section 7.3

Practice Problems

1.

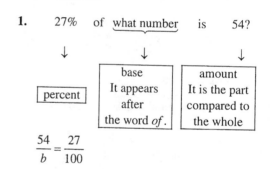

$\dfrac{54}{b} = \dfrac{27}{100}$

2. 30 is what percent of 90?

amount   percent   base

$\dfrac{30}{90} = \dfrac{p}{100}$

3. What number is 25% of 116?

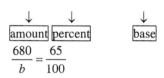

$$\frac{a}{116}=\frac{25}{100}$$

4. 680 is 65% of what number?

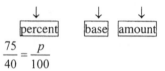

$$\frac{680}{b}=\frac{65}{100}$$

5. What percent of 40 is 75?

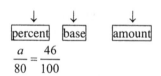

$$\frac{75}{40}=\frac{p}{100}$$

6. 46% of 80 is what number?

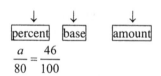

$$\frac{a}{80}=\frac{46}{100}$$

7. What number is 8% of 120?

$$\frac{a}{120}=\frac{8}{100}$$
$$a\cdot100=120\cdot8$$
$$100a=960$$
$$\frac{100a}{100}=\frac{960}{100}$$
$$a=9.6$$

9.6 is 8% of 120.

8. 65% of what number is 52?

9.

$$\frac{52}{b}=\frac{65}{100}$$
$$52\cdot100=b\cdot65$$
$$5200=65b$$
$$\frac{5200}{65}=\frac{65b}{65}$$
$$80=b$$

Thus, 65% of 80 is 52.

9. 15.4 is 5% of what number?

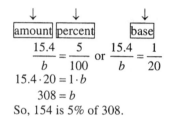

$$\frac{15.4}{b}=\frac{5}{100}\quad\text{or}\quad\frac{15.4}{b}=\frac{1}{20}$$
$$15.4\cdot20=1\cdot b$$
$$308=b$$

So, 154 is 5% of 308.

10. What percent of 40 is 8?

$$\frac{8}{40}=\frac{p}{100}\quad\text{or}\quad\frac{1}{5}=\frac{p}{100}$$
$$1\cdot100=5\cdot p$$
$$100=5p$$
$$\frac{100}{5}=\frac{5p}{5}$$
$$20=p$$

So, 20% of 40 is 8.

11. 414 is what percent of 180?

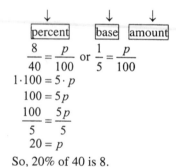

$$\frac{414}{180}=\frac{p}{100}$$
$$414\cdot100=180\cdot p$$
$$41,400=180p$$
$$\frac{41,400}{180}=\frac{180p}{180}$$
$$230=p$$

Then 414 is 230% of 180.

Vocabulary and Readiness Check

1. When translating the statement "20% of 15 is 3" to a proportion, the number 3 is called the <u>amount</u>, 15 is the <u>base</u>, and 20 is the <u>percent</u>.

3. In the question "What number is 25% of 200?", which part of the percent proportion is unknown? <u>amount</u>

Exercise Set 7.3

1. 98% of 45 is <u>what number</u>?

   percent base amount = a

   $$\frac{a}{45} = \frac{98}{100}$$

3. <u>What number</u> is 4% of 150?

   amount = a percent base

   $$\frac{a}{150} = \frac{4}{100}$$

5. 14.3 is 26% of <u>what number</u>?

   amount percent base = b

   $$\frac{14.3}{b} = \frac{26}{100}$$

7. 35% of <u>what number</u> is 84?

   percent base = b amount

   $$\frac{84}{b} = \frac{35}{100}$$

9. <u>What percent</u> of 400 is 70?

   percent = p base amount

   $$\frac{70}{400} = \frac{p}{100}$$

11. 8.2 is <u>what percent</u> of 82?

    amount percent = p base

    $$\frac{8.2}{82} = \frac{p}{100}$$

13. $$\frac{a}{65} = \frac{40}{100}$$
    $$a \cdot 100 = 65 \cdot 40$$
    $$100a = 2600$$
    $$\frac{100a}{100} = \frac{2600}{100}$$
    $$a = 26$$
    40% of 65 is 26.

15. $$\frac{a}{105} = \frac{18}{100}$$
    $$a \cdot 100 = 18 \cdot 105$$
    $$a \cdot 100 = 1890$$
    $$\frac{a \cdot 100}{100} = \frac{1890}{100}$$
    $$a = 18.9$$
    18.9 is 18% of 105.

17. $$\frac{90}{b} = \frac{15}{100}$$
    $$90 \cdot 100 = b \cdot 15$$
    $$9000 = 15b$$
    $$\frac{9000}{15} = \frac{15b}{15}$$
    $$600 = b$$
    15% of 600 is 90.

19.
$$\frac{7.8}{b} = \frac{78}{100}$$
$$7.8 \cdot 100 = 78 \cdot b$$
$$780 = 78 \cdot b$$
$$\frac{780}{78} = \frac{78 \cdot b}{78}$$
$$10 = b$$
7.8 is 78% of 10.

21.
$$\frac{42}{35} = \frac{p}{100}$$
$$42 \cdot 100 = 35 \cdot p$$
$$4200 = 35p$$
$$\frac{4200}{35} = \frac{35p}{35}$$
$$120 = p$$
42 is 120% of 35.

23.
$$\frac{14}{50} = \frac{p}{100}$$
$$14 \cdot 100 = 50 \cdot p$$
$$1400 = 50p$$
$$\frac{1400}{50} = \frac{50p}{50}$$
$$28 = p$$
14 is 28% of 50.

25.
$$\frac{3.7}{b} = \frac{10}{100}$$
$$3.7 \cdot 100 = b \cdot 10$$
$$370 = 10b$$
$$\frac{370}{10} = \frac{10b}{10}$$
$$37 = b$$
3.7 is 10% of 37.

27.
$$\frac{a}{70} = \frac{2.4}{100}$$
$$a \cdot 100 = 70 \cdot 2.4$$
$$100a = 168$$
$$\frac{100a}{100} = \frac{168}{100}$$
$$a = 1.68$$
1.68 is 2.4% of 70.

29.
$$\frac{160}{b} = \frac{16}{100}$$
$$160 \cdot 100 = b \cdot 16$$
$$16,000 = 16b$$
$$\frac{16,000}{16} = \frac{16b}{16}$$
$$1000 = b$$
160 is 16% of 1000.

31.
$$\frac{394.8}{188} = \frac{p}{100}$$
$$394.8 \cdot 100 = 188 \cdot p$$
$$39,480 = 188p$$
$$\frac{39,480}{188} = \frac{188p}{188}$$
$$210 = p$$
394.8 is 210% of 188.

33.
$$\frac{a}{62} = \frac{89}{100}$$
$$a \cdot 100 = 62 \cdot 89$$
$$100a = 5518$$
$$\frac{100a}{100} = \frac{5518}{100}$$
$$a = 55.18$$
55.18 is 89% of 62.

35.
$$\frac{2.7}{6} = \frac{p}{100}$$
$$2.7 \cdot 100 = 6 \cdot p$$
$$270 = 6p$$
$$\frac{270}{6} = \frac{6p}{6}$$
$$45 = p$$
45% of 6 is 2.7.

37.
$$\frac{105}{b} = \frac{140}{100}$$
$$105 \cdot 100 = b \cdot 140$$
$$10,500 = 140b$$
$$\frac{10,500}{140} = \frac{140b}{140}$$
$$75 = b$$
140% of 75 is 105.

39.
$$\frac{a}{48} = \frac{1.8}{100}$$
$$a \cdot 100 = 48 \cdot 1.8$$
$$100a = 86.4$$
$$\frac{100a}{100} = \frac{86.4}{100}$$
$$a = 0.864$$
1.8% of 48 is 0.864.

41.
$$\frac{4}{800} = \frac{p}{100}$$
$$4 \cdot 100 = 800 \cdot p$$
$$400 = 800p$$
$$\frac{400}{800} = \frac{800p}{800}$$
$$0.5 = p$$
0.5% of 800 is 4.

43.
$$\frac{3.5}{b} = \frac{2.5}{100}$$
$$3.5 \cdot 100 = b \cdot 2.5$$
$$350 = 2.5b$$
$$\frac{350}{2.5} = \frac{2.5b}{2.5}$$
$$140 = b$$
3.5 is 2.5% of 140.

45.
$$\frac{a}{48} = \frac{20}{100}$$
$$a \cdot 100 = 48 \cdot 20$$
$$100a = 960$$
$$\frac{100a}{100} = \frac{960}{100}$$
$$a = 9.6$$
20% of 48 is 9.6.

47.
$$\frac{2486}{2200} = \frac{p}{100}$$
$$2486 \cdot 100 = 2200 \cdot p$$
$$248,600 = 2200p$$
$$\frac{248,600}{2200} = \frac{2200p}{2200}$$
$$113 = p$$
2486 is 113% of 2200.

49.
$$-\frac{11}{16} + \left(-\frac{3}{16}\right) = \frac{-11-3}{16}$$
$$= \frac{-14}{16}$$
$$= -\frac{2 \cdot 7}{2 \cdot 8}$$
$$= -\frac{7}{8}$$

51.
$$3\frac{1}{2} - \frac{11}{30} = \frac{7}{2} - \frac{11}{30}$$
$$= \frac{7 \cdot 15}{2 \cdot 15} - \frac{11}{30}$$
$$= \frac{105}{30} - \frac{11}{30}$$
$$= \frac{105-11}{30}$$
$$= \frac{94}{30}$$
$$= 3\frac{4}{30}$$
$$= 3\frac{2}{15}$$

53.
$$\overset{1}{0.41}$$
$$\underline{0.29}$$
$$0.70$$

55.
$$2.38$$
$$\underline{-\,0.19}$$
$$2.19$$

57. answers may vary

59.
$$\frac{a}{64} = \frac{25}{100}$$
$$\frac{17}{64} \overset{?}{=} \frac{25}{100}$$
$$17 \cdot 100 \overset{?}{=} 25 \cdot 64$$
$$1700 = 1600 \quad \text{False}$$
The amount is not 17.

291

$$\frac{a}{64} = \frac{25}{100}$$
$$a \cdot 100 = 25 \cdot 64$$
$$100a = 1600$$
$$\frac{100a}{100} = \frac{1600}{100}$$
$$a = 16$$
25% of 64 is 16.

61. $$\frac{36}{12} = \frac{p}{100}$$
$$\frac{36}{12} \overset{?}{=} \frac{50}{100}$$
$$3 = \frac{1}{2} \quad \text{False}$$

The percent is not 50.

$$\frac{36}{12} = \frac{p}{100}$$
$$36 \cdot 100 = 12 \cdot p$$
$$3600 = 12p$$
$$\frac{3600}{12} = \frac{12p}{12}$$
$$300 = p$$

The percent is 300 (300%).

63. answers may vary

65. $$\frac{88,542}{110,736} = \frac{p}{100}$$
$$88,542 \cdot 100 = 110,736 \cdot p$$
$$8,854,200 = 110,736p$$
$$\frac{8,854,200}{110,736} = \frac{110,736p}{110,736}$$
$$79.958 \approx p$$

80.0% of 110,736 is 88,542.

The Bigger Picture

1. $\frac{2}{9} + \frac{1}{5} = \frac{2 \cdot 5}{9 \cdot 5} + \frac{1 \cdot 9}{5 \cdot 9} = \frac{10}{45} + \frac{9}{45} = \frac{10+9}{45} = \frac{19}{45}$

2. $42 \div 2 \cdot 3 = 21 \cdot 3 = 63$

3. $-0.03(0.7) = -0.021$

4. $7.9 + 0.1 = 8$

5. $$\frac{3}{8} = \frac{n}{128}$$
$$3 \cdot 128 = n \cdot 8$$
$$384 = 8n$$
$$\frac{384}{8} = \frac{8n}{8}$$
$$48 = n$$

6. $$\frac{7.2}{n} = \frac{36}{8}$$
$$7.2 \cdot 8 = 36 \cdot n$$
$$57.6 = 36n$$
$$\frac{57.6}{36} = \frac{36n}{36}$$
$$1.6 = n$$

7. $$\frac{215}{86} = \frac{p}{100}$$
$$215 \cdot 100 = p \cdot 86$$
$$21,500 = 86p$$
$$\frac{21,500}{86} = \frac{86p}{86}$$
$$250 = p$$

215 is 250% of 86.

8. $$\frac{a}{48} = \frac{95}{100}$$
$$a \cdot 100 = 95 \cdot 48$$
$$100a = 4560$$
$$\frac{100a}{100} = \frac{4560}{100}$$
$$a = 45.6 \text{ or } 45\frac{6}{10} = 45\frac{3}{5}$$

95% of 48 is 45.6 or $45\frac{3}{5}$.

9. $$\frac{4.2}{15} = \frac{p}{100}$$
$$4.2 \cdot 100 = 15 \cdot p$$
$$420 = 15 \cdot p$$
$$\frac{420}{15} = \frac{15}{15}p$$
$$28 = p$$

4.2 is 28% of 15.

10.
$$\frac{93.6}{b} = \frac{52}{100}$$
$$93.6 \cdot 100 = 52 \cdot b$$
$$9360 = 52b$$
$$\frac{9360}{52} = \frac{52b}{52}$$
$$180 = b$$
93.6 of 180 is 52%.

Integrated Review

1. $0.94 = 0.94(100\%) = 94\%$

2. $0.17 = 0.17(100\%) = 17\%$

3. $\frac{3}{8} = \frac{3}{8} \cdot 100\% = \frac{300}{8}\% = 37.5\%$

4. $\frac{7}{2} = \frac{7}{2} \cdot 100\% = \frac{700\%}{2} = 350\%$

5. $4.7 = 4.7(100\%) = 470\%$

6. $8 = 8(100\%) = 800\%$

7. $\frac{9}{20} = \frac{9}{20} \cdot 100\% = \frac{900}{20}\% = 45\%$

8. $\frac{53}{50} = \frac{53}{50} \cdot 100\% = \frac{5300}{50}\% = 106\%$

9. $6\frac{3}{4} = \frac{27}{4} = \frac{27}{4} \cdot 100\% = \frac{2700}{4}\% = 675\%$

10. $3\frac{1}{4} = \frac{13}{4} = \frac{13}{4} \cdot 100\% = \frac{1300}{4}\% = 325\%$

11. $0.02 = 0.02(100\%) = 2\%$

12. $0.06 = 0.06(100\%) = 6\%$

13. $71\% = 71(0.01) = 0.71$

14. $31\% = 31(0.01) = 0.31$

15. $3\% = 3(0.01) = 0.03$

16. $4\% = 4(0.01) = 0.04$

17. $224\% = 224(0.01) = 2.24$

18. $700\% = 700(0.01) = 7.0$

19. $2.9\% = 2.9(0.01) = 0.029$

20. $6.6\% = 6.6(0.01) = 0.066$

21. $7\% = 7(0.01) = 0.07$
$$7\% = 7 \cdot \frac{1}{100} = \frac{7}{100}$$

22. $5\% = 5(0.01) = 0.05$
$$5\% = 5 \cdot \frac{1}{100} = \frac{5}{100} = \frac{1}{20}$$

23. $6.8\% = 6.8(0.01) = 0.068$
$$6.8\% = 6.8 \cdot \frac{1}{100}$$
$$= \frac{6.8}{100}$$
$$= \frac{6.8 \cdot 10}{100 \cdot 10}$$
$$= \frac{68}{1000}$$
$$= \frac{17}{250}$$

24. $11.25\% = 11.25(0.01) = 0.1125$
$$11.25\% = 11.25 \cdot \frac{1}{100}$$
$$= \frac{11.25}{100}$$
$$= \frac{11.25 \cdot 100}{100 \cdot 100}$$
$$= \frac{1125}{10,000}$$
$$= \frac{9}{80}$$

25. $74\% = 74(0.01) = 0.74$
$$74\% = 74 \cdot \frac{1}{100} = \frac{74}{100} = \frac{37}{50}$$

26. $45\% = 45(0.01) = 0.45$

$45\% = 45 \cdot \dfrac{1}{100} = \dfrac{45}{100} = \dfrac{9}{20}$

27. $16\dfrac{1}{3}\% = 16.3\overline{3}(0.01) \approx 0.163$

$16\dfrac{1}{3}\% = \dfrac{49}{3}\% = \dfrac{49}{3} \cdot \dfrac{1}{100} = \dfrac{49}{300}$

28. $12\dfrac{2}{3}\% = 12.6\overline{6}(0.01) \approx 0.127$

$12\dfrac{2}{3}\% = \dfrac{38}{3}\% = \dfrac{38}{3} \cdot \dfrac{1}{100} = \dfrac{38}{300} = \dfrac{19}{150}$

29. $\dfrac{a}{90} = \dfrac{15}{100}$

$a \cdot 100 = 15 \cdot 90$

$100a = 1350$

$\dfrac{100a}{100} = \dfrac{1350}{100}$

$a = 13.5$

15% of 90 is 13.5.

30. $\dfrac{78}{b} = \dfrac{78}{100}$

$78 \cdot 100 = 78 \cdot b$

$7800 = 78b$

$\dfrac{7800}{78} = \dfrac{78b}{78}$

$100 = b$

78% of 100 is 78.

31. $\dfrac{297.5}{b} = \dfrac{85}{100}$

$297.5 \cdot 100 = 85 \cdot b$

$29,750 = 85b$

$\dfrac{29,750}{85} = \dfrac{85b}{85}$

$350 = b$

297.5 is 85% of 350.

32. $\dfrac{78}{65} = \dfrac{p}{100}$

$78 \cdot 100 = 65 \cdot p$

$7800 = 65p$

$\dfrac{7800}{65} = \dfrac{65p}{65}$

$120 = p$

78 is 120% of 65.

33. $\dfrac{23.8}{85} = \dfrac{p}{100}$

$23.8 \cdot 100 = 85 \cdot p$

$2380 = 85p$

$\dfrac{2380}{85} = \dfrac{85p}{85}$

$28 = p$

23.8 is 28% of 85.

34. $\dfrac{a}{200} = \dfrac{38}{100}$

$a \cdot 100 = 38 \cdot 200$

$100a = 7600$

$\dfrac{100a}{100} = \dfrac{7600}{100}$

$a = 76$

38% of 200 is 76.

35. $\dfrac{a}{85} = \dfrac{40}{100}$

$a \cdot 100 = 40 \cdot 85$

$100a = 3400$

$\dfrac{100a}{100} = \dfrac{3400}{100}$

$a = 34$

34 is 40% of 85.

36. $\dfrac{128.7}{99} = \dfrac{p}{100}$

$128.7 \cdot 100 = p \cdot 99$

$12,870 = 99p$

$\dfrac{12,870}{99} = \dfrac{99p}{99}$

$130 = p$

130% of 99 is 128.7.

37.
$$\frac{115}{250} = \frac{p}{100}$$
$$115 \cdot 100 = p \cdot 250$$
$$11,500 = 250p$$
$$\frac{11,500}{250} = \frac{250p}{250}$$
$$46 = p$$
46% of 250 is 115.

38.
$$\frac{a}{84} = \frac{45}{100}$$
$$a \cdot 100 = 45 \cdot 84$$
$$100a = 3780$$
$$\frac{100a}{100} = \frac{3780}{100}$$
$$a = 37.8$$
37.8 is 45% of 84.

39.
$$\frac{63}{b} = \frac{42}{100}$$
$$63 \cdot 100 = 42 \cdot b$$
$$6300 = 42b$$
$$\frac{6300}{42} = \frac{42b}{42}$$
$$150 = b$$
42% of 150 is 63.

40.
$$\frac{58.9}{b} = \frac{95}{100}$$
$$58.9 \cdot 100 = 95 \cdot b$$
$$5890 = 95b$$
$$\frac{5890}{95} = \frac{95b}{95}$$
$$62 = b$$
95% of 62 is 58.9.

Section 7.4

Practice Problems

1. *Method 1:*

What number is 4% of 837,000?

$$\begin{array}{ccccc} \downarrow & & \downarrow & \downarrow & \downarrow & \downarrow \\ x & & = & 4\% & \cdot & 837,000 \end{array}$$
$$x = 0.04 \cdot 837,000$$
$$x = 33,480$$
We predict 33,480 adults in Rhode Island to choose chocolate chip as their favorite ice

cream.
Method 2:

What number is 4% of 837,000?

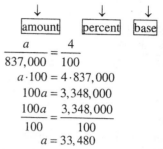

$$\frac{a}{837,000} = \frac{4}{100}$$
$$a \cdot 100 = 4 \cdot 837,000$$
$$100a = 3,348,000$$
$$\frac{100a}{100} = \frac{3,348,000}{100}$$
$$a = 33,480$$
We predict 33,480 adults in Rhode Island to choose Chocolate Chip as their favorite ice cream.

2. *Method 1:*

34,000 is what percent of 130,000?

$$\begin{array}{ccccc} \downarrow & \downarrow & & \downarrow & & \downarrow & \downarrow \\ 34,000 & = & & x & & \cdot & 130,000 \end{array}$$
$$34,000 = 130,000x$$
$$\frac{34,000}{130,000} = \frac{130,000x}{130,000}$$
$$0.26 \approx x$$
$$26\% = x$$
In Florida, 34,000 or 26% more new nurses were needed in 2006.
Method 2:

34,000 is what percent of 130,000?

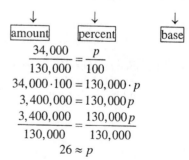

$$\frac{34,000}{130,000} = \frac{p}{100}$$
$$34,000 \cdot 100 = 130,000 \cdot p$$
$$3,400,000 = 130,000p$$
$$\frac{3,400,000}{130,000} = \frac{130,000p}{130,000}$$
$$26 \approx p$$
In Florida, 34,000 or 26% more new nurses were needed in 2006.

3. *Method 1*:
864 is 32% of <u>what number</u>?

$$\downarrow \downarrow \downarrow \quad \downarrow \qquad \downarrow$$
$$864 = 32\% \quad \cdot \qquad x$$
$$864 = 0.32 \cdot x$$
$$\frac{864}{0.32} = \frac{0.32x}{0.32}$$
$$2700 = x$$
There are 2700 students at Euclid
University.
Method 2:
864   is   32% of <u>what number</u>?

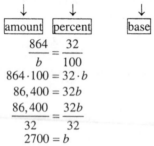
$$\frac{864}{b} = \frac{32}{100}$$
$$864 \cdot 100 = 32 \cdot b$$
$$86,400 = 32b$$
$$\frac{86,400}{32} = \frac{32b}{32}$$
$$2700 = b$$
There are 2700 students at Euclid
University.

4. <u>What number</u> is 20% of 202 million?

$$\downarrow \qquad \downarrow \downarrow \downarrow \qquad \downarrow$$
$$x \qquad = 20\% \cdot \qquad 202$$
$$x = 0.20 \cdot 202$$
$$x = 40.4 \text{ million}$$

a. The increase in the number of vehicles
on the road in 2004 is 40.4 million.

b. The total number of registered vehicles
on the road in 2004 was
202 million + 40.4 million = 242.4
million

Method 2:
<u>What number</u> is 20% of 202 million?

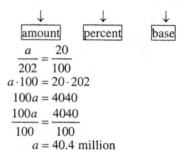
$$\frac{a}{202} = \frac{20}{100}$$
$$a \cdot 100 = 20 \cdot 202$$
$$100a = 4040$$
$$\frac{100a}{100} = \frac{4040}{100}$$
$$a = 40.4 \text{ million}$$

a. The increase in the number of vehicles
on the road in 2004 is 40.4 million.

b. The total number of registered vehicles
on the road in 2004 was
202 million + 40.4 million
= 242.4 million

5. Find the amount of increase by subtracting
the original number of attendants from the
new number of attendants.
amount of increase = 333 − 285 = 48
The amount of increase is 48 attendants.

$$\text{percent of increase} = \frac{\text{amount of increase}}{\text{original amount}}$$
$$= \frac{48}{285}$$
$$\approx 0.168$$
$$= 16.8\%$$
The number of attendants to the local play,
Peter Pan, increased by about 16.8%.

6. Find the amount of decrease by subtracting
18,483 from 20,200.
Amount of decrease = 20,200 − 18,483
$$= 1717$$
The amount of decrease is 1717.

$$\text{percent of decrease} = \frac{\text{amount of decrease}}{\text{original amount}}$$
$$= \frac{1717}{20,200}$$
$$= 0.085$$
$$= 8.5\%$$

The population decreased by 8.5%.

Exercise Set 7.4

1. 4% of 220 is what number?
Method 1:
$$4\% \cdot 220 = x$$
$$0.04 \cdot 220 = x$$
$$8.8 = x$$
The minimum weight resistance is 8.8 pounds.
Method 2:
$$\frac{a}{220} = \frac{4}{100}$$
$$a \cdot 100 = 220 \cdot 4$$
$$100a = 880$$
$$\frac{100a}{100} = \frac{880}{100}$$
$$a = 8.8$$
The minimum weight resistance is 8.8 pounds.

3. 24 is 1.5% of what number?
Method 1:
$$24 = 1.5\% \cdot x$$
$$24 = 0.015x$$
$$\frac{24}{0.015} = \frac{0.015x}{0.015}$$
$$1600 = x$$
1600 bolts were inspected.
Method 2:
$$\frac{24}{b} = \frac{1.5}{100}$$
$$24 \cdot 100 = 1.5 \cdot b$$
$$2400 = 1.5b$$
$$\frac{2400}{1.5} = \frac{1.5b}{1.5}$$
$$1600 = b$$
1600 bolts were inspected.

5. 378 is what percent of 2700?
Method 1:
$$378 = x \cdot 2700$$
$$\frac{378}{2700} = \frac{x \cdot 2700}{2700}$$
$$0.14 = x$$
$$14\% = x$$
The student spent 14% of last semester's college cost on books.
Method 2:
$$\frac{378}{2700} = \frac{p}{100}$$
$$378 \cdot 100 = 2700 \cdot p$$
$$37,800 = 2700p$$
$$\frac{37,800}{2700} = \frac{2700p}{2700}$$
$$14 = p$$

The student spent 14% of last semester's college cost on books.

7. 58% of 940 is what number?
Method 1:
$$58\% \cdot 940 = x$$
$$0.58 \cdot 940 = x$$
$$545.2 = x$$
545 films were rated R.
Method 2:
$$\frac{a}{940} = \frac{58}{100}$$
$$a \cdot 100 = 58 \cdot 940$$
$$100a = 54,520$$
$$\frac{100a}{100} = \frac{54,520}{100}$$
$$a = 545.2$$
545 films were rated R.

9. 35 is what percent of 535?
Method 1:
$$35 = x \cdot 535$$
$$\frac{35}{535} = \frac{x \cdot 535}{535}$$
$$0.065 \approx x$$
$$6.5\% \approx x$$
6.5% of the members of the 109th U.S. Congress attended a community college.

Method 2:

$$\frac{35}{535} = \frac{p}{100}$$
$$35 \cdot 100 = 535 \cdot p$$
$$3500 = 535p$$
$$\frac{3500}{535} = \frac{535p}{535}$$
$$6.5 \approx p$$

6.5% of the members of the 109th U.S. Congress attended a community college.

11. 8% of 6200 is what number?
Method 1:
$$8\% \cdot 6200 = x$$
$$0.08 \cdot 6200 = x$$
$$496 = x$$
The decrease in the number of chairs produced is 496. The new number of chairs produced each month is 6200 − 496 = 5704 chairs.
Method 2:
$$\frac{a}{6200} = \frac{8}{100}$$
$$100 \cdot a = 8 \cdot 6200$$
$$100a = 49,600$$
$$\frac{100a}{100} = \frac{49,600}{100}$$
$$a = 496$$
The decrease in the number of chairs produced is 496. The new number of chairs produced each month is 6200 − 496 = 5704 chairs.

13. What number is 49% of 63,000?
Method 1:
$$x = 49\% \cdot 63,000$$
$$x = 0.49 \cdot 63,000$$
$$x = 30,870$$
The number of people employed as physician assistants is expected to be 63,000 + 30,870 = 93,870.

Method 2:

$$\frac{a}{63,000} = \frac{49}{100}$$
$$a \cdot 100 = 63,000 \cdot 49$$
$$100a = 3,087,000$$
$$\frac{100a}{100} = \frac{3,087,000}{100}$$
$$a = 30,870$$

The number of people employed as physician assistants is expected to be 63,000 + 30,870 = 93,870.

15. 20 is what percent of 40?
Method 1:
$$20 = x \cdot 40$$
$$\frac{20}{40} = \frac{40x}{40}$$
$$0.5 = x$$
$$50\% = x$$
50% of the total calories come from fat.
Method 2:
$$\frac{20}{40} = \frac{p}{100}$$
$$20 \cdot 100 = p \cdot 40$$
$$2000 = 40p$$
$$\frac{2000}{40} = \frac{40p}{40}$$
$$50 = p$$

50% of the total calories come from fat.

17. 10 is what percent of 80?
Method 1:
$$10 = x \cdot 80$$
$$\frac{10}{80} = \frac{x \cdot 80}{80}$$
$$0.125 = x$$
$$12.5\% = x$$
12.5% of the total calories come from fat.

Method 2:

$$\frac{10}{80} = \frac{p}{100}$$

$$10 \cdot 100 = 80 \cdot p$$

$$1000 = 80p$$

$$\frac{1000}{80} = \frac{80p}{80}$$

$$12.5 = p$$

12.5% of the total calories come from fat.

19. 35 is what percent of 120?
Method 1:

$$35 = x \cdot 120$$

$$\frac{35}{120} = \frac{120x}{120}$$

$$0.292 \approx x$$

$$29.2\% \approx x$$

29.2% of the total calories come from fat.
Method 2:

$$\frac{35}{120} = \frac{p}{100}$$

$$35 \cdot 100 = p \cdot 120$$

$$3500 = 120p$$

$$\frac{3500}{120} = \frac{120p}{120}$$

$$29.2 \approx p$$

29.2% of the total calories come from fat.

21. 26,250 is 15% of what number?
Method 1:

$$26,250 = 15\% \cdot x$$

$$26,250 = 0.15 \cdot x$$

$$\frac{26,250}{0.15} = \frac{0.15 \cdot x}{0.15}$$

$$175,000 = x$$

The price of the home was $175,000.
Method 2:

$$\frac{26,250}{b} = \frac{15}{100}$$

$$26,250 \cdot 100 = b \cdot 15$$

$$2,625,000 = 15b$$

$$\frac{2,625,000}{15} = \frac{15b}{15}$$

$$175,000 = b$$

The price of the home was $175,000.

23. What number is 78% of 40?
Method 1:

$$x = 78\% \cdot 40$$

$$x = 0.78 \cdot 40$$

$$x = 31.2$$

The owner can bill 31.2 hours each week for a repairman.
Method 2:

$$\frac{a}{40} = \frac{78}{100}$$

$$a \cdot 100 = 78 \cdot 40$$

$$100a = 3120$$

$$\frac{100a}{100} = \frac{3120}{100}$$

$$a = 31.2$$

The owner can bill 31.2 hours each week for a repairman.

25. What number is 4.5% of 19,286?
Method 1:

$$x = 4.5\% \cdot 19,286$$

$$x = 0.045 \cdot 19,286$$

$$x = 867.87$$

The price of the car will increase by $867.87. The new price of that model will be $19,286 + $867.87 = $20,153.87.
Method 2:

$$\frac{a}{19,286} = \frac{4.5}{100}$$

$$a \cdot 100 = 4.5 \cdot 19,286$$

$$100a = 86,787$$

$$\frac{100a}{100} = \frac{86,787}{100}$$

$$a = 867.87$$

The price of the car will increase by $867.87. The new price of that model will be $19,286 + $867.87 = $20,153.87.

27. What number is 32% of 6950?
Method 1:

$$x = 32\% \cdot 6950$$

$$x = 0.32 \cdot 6950$$

$$x = 2224$$

The number of adults in Buckeye predicted to prefer vanilla ice cream is 2224.

Method 2:

$$\frac{a}{6950} = \frac{32}{100}$$

$$a \cdot 100 = 32 \cdot 6950$$

$$100a = 222,400$$

$$\frac{100a}{100} = \frac{222,400}{100}$$

$$a = 2224$$

The number of adults in Buckeye predicted to prefer vanilla ice cream is 2224.

29. What number is 80% of 35?

Method 1:

$x = 80\% \cdot 35$

$x = 0.80 \cdot 35$

$x = 28$

The increase is projected to be 28 million. The population is projected to be (35 + 28) million = 63 million.

Method 2:

$$\frac{a}{35} = \frac{80}{100}$$

$$a \cdot 100 = 80 \cdot 35$$

$$100a = 2800$$

$$\frac{100a}{100} = \frac{2800}{100}$$

$$a = 28$$

The increase is projected to be 28 million. The population is projected to be (35 + 28) million = 63 million.

	Original Amount	New Amount	Amount of Increase	Percent Increase
31.	50	80	80 − 50 = 30	$\frac{30}{50} = 0.6 = 60\%$
33.	65	117	117 − 65 = 52	$\frac{52}{65} = 0.8 = 80\%$

	Original Amount	New Amount	Amount of Decrease	Percent Decrease
35.	8	6	$8 - 6 = 2$	$\dfrac{2}{8} = 0.25 = 25\%$
37.	160	40	$160 - 40 = 120$	$\dfrac{120}{160} = 0.75 = 75\%$

39. $\text{percent decrease} = \dfrac{\text{amount of decrease}}{\text{original amount}}$

$$= \frac{150 - 84}{150}$$

$$= \frac{66}{150}$$

$$= 0.44$$

The decrease in calories is 44%.

41. $\text{percent decrease} = \dfrac{\text{amount of decrease}}{\text{original amount}}$

$$= \frac{10,845 - 10,700}{10,845}$$

$$= \frac{145}{10,845}$$

$$\approx 0.013$$

The decrease in cable TV systems was 1.3%.

43. $\text{percent increase} = \dfrac{\text{amount of increase}}{\text{original amount}}$

$$= \frac{444 - 174}{174}$$

$$= \frac{270}{174}$$

$$\approx 1.552$$

The increase in acres was 155.2%

45. $\text{percent increase} = \dfrac{\text{amount of increase}}{\text{original amount}}$

$$= \frac{464,000 - 455,000}{455,000}$$

$$= \frac{9000}{455,000}$$

$$\approx 0.0198$$

The increase in computer programmers is expected to be 2%.

47. percent increase $= \dfrac{\text{amount of increase}}{\text{original amount}}$

$= \dfrac{3769 - 3570}{3570}$

$= \dfrac{199}{3570}$

≈ 0.056

The increase in elementary and secondary teachers is expected to be 5.6%.

49. percent decrease $= \dfrac{\text{amount of decrease}}{\text{original amount}}$

$= \dfrac{7151 - 5713}{7151}$

$= \dfrac{1438}{7151}$

≈ 0.201

The decrease in cinema sites was 20.1%

51. percent increase $= \dfrac{\text{amount of increase}}{\text{original amount}}$

$= \dfrac{19.9 - 13.1}{13.1}$

$= \dfrac{6.8}{13.1}$

≈ 0.519

The increase in soft drink size was 51.9%.

53. percent increase $= \dfrac{\text{amount of increase}}{\text{original amount}}$

$= \dfrac{212,000 - 110,000}{110,000}$

$= \dfrac{102,000}{110,000}$

≈ 0.927

The increase in cellular subscribers was 92.7%.

55.
$$\begin{array}{r} 0.12 \\ \times\ \ 38 \\ \hline 96 \\ 360 \\ \hline 4.56 \end{array}$$

57.
$$\begin{array}{r} {\scriptstyle 1\,1} \\ 9.20 \\ +1.98 \\ \hline 11.18 \end{array}$$

59. $-\dfrac{3}{8} - \dfrac{5}{12} = -\dfrac{3}{8}\cdot\dfrac{3}{3} - \dfrac{5}{12}\cdot\dfrac{2}{2}$

$= -\dfrac{9}{24} - \dfrac{10}{24}$

$= -\dfrac{19}{24}$

61. $2\dfrac{4}{5} \div 3\dfrac{9}{10} = \dfrac{14}{5} \div \dfrac{39}{10}$

$= \dfrac{14}{5}\cdot\dfrac{10}{39}$

$= \dfrac{14\cdot 10}{5\cdot 39}$

$= \dfrac{14\cdot 2\cdot 5}{5\cdot 39}$

$= \dfrac{28}{39}$

63. The increased number is double the original number.

65. To find the percent increase, she should have divided by the original amount, which is 150.

percent increase $= \dfrac{30}{150} = 0.2 = 20\%$

67. False; the percents are different.
percent increase from 1980 to 1990 = 20%
percent decrease from 1990 to 2000
$= 16\dfrac{2}{3}\%$

Section 7.5

Practice Problems

1. sales tax = tax rate · purchase price
    $$= \quad 8.5\% \quad \cdot \quad \$59.90$$
    $$= 0.085 \cdot \$59.90$$
    $$\approx \$5.09$$
    The sales tax is $5.09.
    Total Price = purchase price + sales tax
    $$= \quad \$59.90 \quad + \quad \$5.09$$
    $$= \$64.99$$
    The sales tax on $59.90 is $5.09, and the total price is $64.99.

2. sales tax = tax rate · purchase price
    $$\$1665 = r \cdot \$18,500$$
    $$\frac{1665}{18,500} = \frac{r \cdot 18,500}{18,500}$$
    $$0.09 = r$$
    The sales tax rate is 9%.

3. commission = commission rate · sales
    $$= \quad 6.6\% \quad \cdot \quad \$47,632$$
    $$= 0.066 \cdot \$47,632$$
    $$\approx \$3143.712$$
    The sales representative's commission for the month is $3143.71.

4. commission = commission rate · sales
    $$\$645 = r \cdot \$4300$$
    $$\frac{645}{4300} = r$$
    $$0.15 = r$$
    $$15\% = r$$
    The commission rate is 15%.

5. amount of discount
    = discount rate · original price
    $$= \quad 35\% \quad \cdot \quad \$700$$
    $$= 0.35 \cdot \$700$$
    $$= \$245$$
    The discount is $245.
    sale price = original price − discount
    $$= \quad \$700 \quad - \quad \$245$$
    $$= \$455$$
    The sale price is $455.

Vocabulary and Readiness Check

1. <u>sales tax</u> = tax rate · purchase price.

3. <u>commission</u> = commission rate · sales.

5. <u>sale price</u> = original price − amount of discount.

Exercise Set 7.5

1. sales tax = 5% · $150 = 0.05 · $150 = $7.50
    The sales tax is $7.50.

3. sales tax = 7.5% · $799
    $$= 0.075 \cdot \$799$$
    $$\approx \$59.93$$
    total price = $799 + $59.93 = $858.93
    The total price of the camcorder is $858.93.

5. $335.30 = r · $4790
    $$\frac{\$335.30}{\$4790} = r$$
    $$0.07 = r$$
    The sales tax rate is 7%.

7. a. $10.20 = 8.5% · p
        $$\$10.2 = 0.085p$$
        $$\frac{\$10.2}{0.085} = p$$
        $$\$120 = p$$
        The purchase price of the table saw is $120.

    b.   total price = $120 + $10.20 = $130.20
           The total price of the table saw is $130.20.

9.   sales tax = $6.5\% \cdot \$1800$
              $= 0.065 \cdot \$1800$
              $= \$117$
   total price = $1800 + $117 = $1917
   The total price of the bracelet is $1917.

11. $\$24.25 = 5\% \cdot p$
      $\$24.25 = 0.05p$
      $\dfrac{\$24.25}{0.05} = p$
        $\$485 = p$
   The purchase price of the futon is $485.

13. $\$98.70 = r \cdot \$1645$
      $\dfrac{\$98.70}{\$1645} = r$
        $0.06 = r$
   The sales tax rate is 6%.

15. Purchase price = $210 + $15 + $5 = $230
     Sales tax = $7\% \cdot \$230$
             $= 0.07 \cdot \$230$
             $= \$16.10$
   Total price = $230 + $16.10 = $246.10
   The sales tax is $16.10 and the total price of the items is $246.10.

17. commission = $4\% \cdot \$1,329,401$
                $= 0.04 \cdot \$1,329,401$
                $= \$53,176.04$
   Her commission was $53,176.04.

19. $\$1380.40 = r \cdot \$9860$
      $\dfrac{\$1380.40}{\$9860} = r$
         $0.14 = r$
   The commission rate is 14%.

21. commission = $1.5\% \cdot \$325,900$
                $= 0.015 \cdot \$325,900$
                $= \$4888.50$
   His commission will be $4888.50.

23. $\$5565 = 3\% \cdot \text{sales}$

$\$5565 = 0.03 \cdot s$

$\dfrac{\$5565}{0.03} = s$

$\$185,500 = s$

The selling price of the house was $185,500.

	Original Price	Discount Rate	Amount of Discount	Sale Price
25.	$89	10%	$10\% \cdot \$89 = \$8.90$	$\$89 - \$8.90 = \$80.10$
27.	$196.50	50%	$50\% \cdot \$196.50 = \$98.25$	$\$196.50 - \$98.25 = \$98.25$
29.	$410	35%	$35\% \cdot \$410 = 143.50$	$\$410 - \$143.50 = \$266.50$
31.	$21,700	15%	$15\% \cdot \$21,700 = \$3255$	$\$21,700 - \$3255 = \$18,445$

33. discount = $15\% \cdot \$300 = 0.15 \cdot \$300 = \$45$

sale price = $\$300 - \$45 = \$255$

The discount is $45 and the sale price is $255.

	Purchase Price	Tax Rate	Sales Tax	Total Price
35.	$305	9%	$9\% \cdot \$305 = \$27.45$	$\$305 + \$27.45 = \$332.45$
37.	$56	5.5%	$5.5\% \cdot \$56 = \$3.08$	$\$56 + \$3.08 = \$59.08$

	Sale	Commission Rate	Commission
39.	$235,800	3%	$\$235,800 \cdot 3\% = \$7074$
41.	$17,900	$\dfrac{\$1432}{\$17,900} = 0.08 = 8\%$	$1432

43. $2000 \cdot \dfrac{3}{10} \cdot 2 = 600 \cdot 2 = 1200$

45. $400 \cdot \dfrac{3}{100} \cdot 11 = 12 \cdot 11 = 132$

47. $600 \cdot 0.04 \cdot \dfrac{2}{3} = 24 \cdot \dfrac{2}{3} = 16$

49. Round $68 to $70 and 9.5% to 10%.
$10\% \cdot \$70 = 0.10 \cdot \$70 = \$7$
$\$70 + \$7 = \$77$
The best estimate of the total price is $77; d.

	Bill Amount	10%	15%	20%
51.	$40.21 \approx \$40$	$4.00	$\$4 + \dfrac{1}{2}(\$4) = \$4 + \$2 = \$6.00$	$2(\$4) = \$8.00$
53.	$72.17 \approx \$72.00$	$7.20	$\$7.20 + \dfrac{1}{2}(\$7.20) = \$7.20 + \$3.60$ $= \$10.80$	$2(\$7.20) = \$14.40$

55. A discount of 60% is better; answers may vary.

57. $7.5\% \cdot \$24,966 = 0.075 \cdot \$24,966 = \$1872.45$
$\$24,966 + \$1872.45 = \$26,838.45$
The total price of the necklace is $26,838.45.

Section 7.6

Practice Problems

1. $I = P \cdot R \cdot T$
$I = \$875 \cdot 7\% \cdot 5$
$\quad = \$875 \cdot (0.07) \cdot 5$
$\quad = \$306.25$
The simple interest is $306.25.

2. $I = P \cdot R \cdot T$
$I = \$1500 \cdot 20\% \cdot \dfrac{9}{12}$
$\quad = \$1500 \cdot (0.20) \cdot \dfrac{9}{12}$
$\quad = \$225$
She paid $225 in interest.

3. $I = P \cdot R \cdot T$
$\quad = \$2100 \cdot 13\% \cdot \dfrac{6}{12}$
$\quad = \$2100 \cdot (0.13) \cdot \dfrac{6}{12}$
$\quad = \$136.50$
The interest is $136.50.

total amount = principal + interest
$$= \$2100 + \$136.50$$
$$= \$2236.50$$
After 6 months, the total amount paid will be $2236.50.

4. $P = \$3000$, $r = 4\% = 0.04$, $n = 1$, $t = 6$ years

$$A = P\left(1 + \frac{r}{n}\right)^{n \cdot t}$$

$$= \$3000\left(1 + \frac{0.04}{1}\right)^{1 \cdot 6}$$

$$= \$3000(1.04)^6$$

$$\approx \$3795.96$$

The total amount after 6 years is $3795.96.

5. $P = \$5500$, $r = 6\frac{1}{4}\% = 0.0625$, $n = 365$,

$t = 5$

$$A = P\left(1 + \frac{r}{n}\right)^{n \cdot t}$$

$$= \$5500\left(1 + \frac{0.0625}{365}\right)^{365 \cdot 5}$$

$$\approx \$7517.41$$

The total amount after 5 years is $7517.41.

Calculator Explorations

1. $A = \$600\left(1 + \frac{0.09}{4}\right)^{4 \cdot 5} \approx \$936.31$

3. $A = \$1200\left(1 + \frac{0.11}{1}\right)^{1 \cdot 20} \approx \$9674.77$

5. $A = \$500\left(1 + \frac{0.06}{4}\right)^{4 \cdot 4} \approx \$634.49$

Vocabulary and Readiness Check

1. To calculate <u>simple</u> interest, use $I = P \cdot R \cdot T$.

3. <u>Compound</u> interest is computed on not only the original principal, but on interest already earned in previous compounding periods.

5. <u>Total amount</u> (paid or received) = principal + interest.

Exercise Set 7.6

1. simple interest = principal · rate · time
$$= \$200 \cdot 8\% \cdot 2$$
$$= \$200 \cdot 0.08 \cdot 2$$
$$= \$32$$

3. simple interest = principal · rate · time
$$= \$160 \cdot 11.5\% \cdot 4$$
$$= \$160 \cdot 0.115 \cdot 4$$
$$= \$73.60$$

5. simple interest = principal · rate · time
$$= \$5000 \cdot 10\% \cdot 1\frac{1}{2}$$
$$= \$5000 \cdot 0.10 \cdot 1.5$$
$$= \$750$$

7. simple interest = principal · rate · time
$$= \$375 \cdot 18\% \cdot \frac{6}{12}$$
$$= \$375 \cdot 0.18 \cdot 0.5$$
$$= \$33.75$$

9. simple interest = principal · rate · time
$$= \$2500 \cdot 16\% \cdot \frac{21}{12}$$
$$= \$2500 \cdot 0.16 \cdot 1.75$$
$$= \$700$$

11. simple interest = principal · rate · time
$$= \$162,500 \cdot 12.5\% \cdot 5$$
$$= \$162,500 \cdot 0.125 \cdot 5$$
$$= \$101,562.50$$
The amount of interest is $101,562.50.

13. simple interest = principal · rate · time

$$= \$5000 \cdot 9\% \cdot \frac{15}{12}$$
$$= \$5000 \cdot 0.09 \cdot 1.25$$
$$= \$562.50$$
Total = \$5000 + \$562.50 = \$5562.50

15. Simple interest = principal · rate · time

$$= \$8500 \cdot 17\% \cdot 4$$
$$= \$8500 \cdot 0.17 \cdot 4$$
$$= \$5780$$
Total amount = \$8500 + \$5780 = \$14,280

17. $A = P\left(1 + \dfrac{r}{n}\right)^{n \cdot t}$

$$= 6150\left(1 + \frac{0.14}{2}\right)^{2 \cdot 15}$$
$$= 6150(1.07)^{30}$$
$$\approx 46,815.37$$
The total amount is \$46,815.37.

19. $A = P\left(1 + \dfrac{r}{n}\right)^{n \cdot t}$

$$= 1560\left(1 + \frac{0.08}{365}\right)^{365 \cdot 5}$$
$$= 1560\left(1 + \frac{0.08}{365}\right)^{1825}$$
$$\approx 2327.14$$
The total amount is \$2327.14.

21. $A = P\left(1 + \dfrac{r}{n}\right)^{n \cdot t}$

$$= 10,000\left(1 + \frac{0.09}{2}\right)^{2 \cdot 20}$$
$$= 10,000(1.045)^{40}$$
$$\approx 58,163.65$$
The total amount is \$58,163.65.

23. $A = P\left(1 + \dfrac{r}{n}\right)^{n \cdot t}$

$$= 2675\left(1 + \frac{0.09}{1}\right)^{1 \cdot 1}$$
$$= 2675(1.09)$$
$$= 2915.75$$
The total amount is \$2915.75.

25. $A = P\left(1 + \dfrac{r}{n}\right)^{n \cdot t}$

$$= 2000\left(1 + \frac{0.08}{1}\right)^{1 \cdot 5}$$
$$= 2000(1.08)^{5}$$
$$\approx 2938.66$$
The total amount is \$2938.66.

27. $A = P\left(1 + \dfrac{r}{n}\right)^{n \cdot t}$

$$= 2000\left(1 + \frac{0.08}{4}\right)^{4 \cdot 5}$$
$$= 2000(1.02)^{20}$$
$$\approx 2971.89$$
The total amount is \$2971.89.

29. perimeter = 10 + 6 + 10 + 6 = 32
The perimeter is 32 yards.

31. Perimeter = 7 + 7 + 7 + 7 + 7 = 35
The perimeter is 35 meters.

33. $\dfrac{x}{4} + \dfrac{x}{5} = \dfrac{x}{4} \cdot \dfrac{5}{5} + \dfrac{x}{5} \cdot \dfrac{4}{4} = \dfrac{5x}{20} + \dfrac{4x}{20} = \dfrac{9x}{20}$

35. $\left(\dfrac{2}{3}\right)\left(-\dfrac{1}{3}\right)-\left(\dfrac{9}{10}\right)\left(\dfrac{2}{5}\right)=-\dfrac{2}{9}-\dfrac{18}{50}$

$\qquad = -\dfrac{2}{9}\cdot\dfrac{50}{50}-\dfrac{18}{50}\cdot\dfrac{9}{9}$

$\qquad = -\dfrac{100}{450}-\dfrac{162}{450}$

$\qquad = -\dfrac{262}{450}$

$\qquad = -\dfrac{131}{225}$

37. answers may vary

39. answers may vary

Chapter 7 Vocabulary Check

1. In a mathematical statement, of usually means "multiplication."

2. In a mathematical statement, is means "equals."

3. Percent means "per hundred."

4. Compound interest is computed not only on the principal, but also on interest already earned in previous compounding periods.

5. In the percent proportion
$$\dfrac{\text{amount}}{\text{base}}=\dfrac{\text{percent}}{100}.$$

6. To write a decimal or fraction as a percent, multiply by 100%.

7. The decimal equivalent of the % symbol is 0.01.

8. The fraction equivalent of the % symbol is $\dfrac{1}{100}$.

9. The percent equation is
base · percent = amount.

10. Percent of decrease = $\dfrac{\text{amount of decrease}}{\text{original amount}}$.

11. Percent of increase = $\dfrac{\text{amount of increase}}{\text{original amount}}$.

12. Sales tax = tax rate · purchase price.

13. Total price = purchase price + sales tax.

14. Commission = commission rate · sales.

15. Amount of discount = discount rate · original price.

16. Sale price = original price − amount of discount.

Chapter 7 Review

1. $\dfrac{37}{100}=37\%$

37% of adults preferred pepperoni.

2. $\dfrac{77}{100}=77\%$

77% of free throws were made.

3. $26\%=26(0.01)=0.26$

4. $75\%=75(0.01)=0.75$

5. $3.5\%=3.5(0.01)=0.035$

6. $1.5\%=1.5(0.01)=0.015$

7. $275\%=275(0.01)=2.75$

8. $400\%=400(0.01)=4.00$ or 4

9. $47.85\%=47.85(0.01)=0.4785$

10. $85.34\%=85.34(0.01)=0.8534$

11. $1.6=1.6(100\%)=160\%$

12. $0.055=0.055(100\%)=5.5\%$

13. $0.076 = 0.076(100\%) = 7.6\%$

14. $0.085 = 0.085(100\%) = 8.5\%$

15. $0.71 = 0.71(100\%) = 71\%$

16. $0.65 = 0.65(100\%) = 65\%$

17. $6 = 6(100)\% = 600\%$

18. $9 = 9(100\%) = 900\%$

19. $7\% = 7\left(\dfrac{1}{100}\right) = \dfrac{7}{100}$

20. $15\% = 15\left(\dfrac{1}{100}\right) = \dfrac{15}{100} = \dfrac{3}{20}$

21. $25\% = 25\left(\dfrac{1}{100}\right) = \dfrac{25}{100} = \dfrac{1}{4}$

22. $8.5\% = 8.5\left(\dfrac{1}{100}\right)$
$= \dfrac{8.5}{100}$
$= \dfrac{8.5 \cdot 10}{100 \cdot 10}$
$= \dfrac{85}{1000}$
$= \dfrac{17}{200}$

23. $10.2\% = 10.2\left(\dfrac{1}{100}\right)$
$= \dfrac{10.2}{100}$
$= \dfrac{10.2 \cdot 10}{100 \cdot 10}$
$= \dfrac{102}{1000}$
$= \dfrac{51}{500}$

24. $16\dfrac{2}{3}\% = \dfrac{50}{3}\% = \dfrac{50}{3}\left(\dfrac{1}{100}\right) = \dfrac{50}{300} = \dfrac{1}{6}$

25. $33\dfrac{1}{3}\% = \dfrac{100}{3}\% = \dfrac{100}{3}\left(\dfrac{1}{100}\right) = \dfrac{100}{300} = \dfrac{1}{3}$

26. $110\% = 110\left(\dfrac{1}{100}\right) = \dfrac{110}{100} = 1\dfrac{10}{100} = 1\dfrac{1}{10}$

27. $\dfrac{2}{5} = \dfrac{2}{5} \cdot \dfrac{100}{1}\% = \dfrac{200}{5}\% = 40\%$

28. $\dfrac{7}{10} = \dfrac{7}{10} \cdot \dfrac{100}{1}\% = \dfrac{700}{10}\% = 70\%$

29. $\dfrac{7}{12} = \dfrac{7}{12} \cdot \dfrac{100}{1}\% = \dfrac{700}{12}\% = \dfrac{175}{3}\% = 58\dfrac{1}{3}\%$

30. $1\dfrac{2}{3} = \dfrac{5}{3} \cdot \dfrac{100}{1}\% = \dfrac{500}{3}\% = 166\dfrac{2}{3}\%$

31. $1\dfrac{1}{4} = \dfrac{5}{4} \cdot \dfrac{100}{1}\% = \dfrac{500}{4}\% = 125\%$

32. $\dfrac{3}{5} = \dfrac{3}{5} \cdot \dfrac{100}{1}\% = \dfrac{300}{5}\% = 60\%$

33. $\dfrac{1}{16} = \dfrac{1}{16} \cdot \dfrac{100}{1}\% = \dfrac{100}{16}\% = 6.25\%$

34. $\dfrac{5}{8} = \dfrac{5}{8} \cdot \dfrac{100}{1}\% = \dfrac{500}{8}\% = 62.5\%$

35. $1250 = 1.25\% \cdot x$
$1250 = 0.0125x$
$\dfrac{1250}{0.0125} = \dfrac{0.0125x}{0.0125}$
$100,000 = x$
1250 is 1.25% of 100,000.

36. $x = 33\frac{1}{3}\% \cdot 24,000$

$x = \frac{100}{3} \cdot \frac{1}{100} \cdot 24,000$

$x = \frac{1}{3} \cdot 24,000$

$x = 8000$

8000 is $33\frac{1}{3}\%$ of 24,000.

37. $124.2 = x \cdot 540$

$\frac{124.2}{540} = \frac{540x}{540}$

$0.23 = x$

$23\% = x$

124.2 is 23% of 540.

38. $22.9 = 20\% \cdot x$

$22.9 = 0.20 \cdot x$

$\frac{22.9}{0.20} = \frac{0.20x}{0.20}$

$114.50 = x$

22.9 is 20% of 114.50.

39. $x = 17\% \cdot 640$

$x = 0.17 \cdot 640$

$x = 108.8$

108.8 is 17% of 640.

40. $693 = x \cdot 462$

$\frac{693}{462} = \frac{462x}{462}$

$1.5 = x$

$150\% = x$

693 is 150% of 462.

41. $\frac{104.5}{b} = \frac{25}{100}$

$104.5 \cdot 100 = 25 \cdot b$

$10,450 = 25b$

$\frac{10,450}{25} = \frac{25b}{25}$

$418 = b$

104.5 is 25% of 418.

42. $\frac{16.5}{b} = \frac{5.5}{100}$

$16.5 \cdot 100 = 5.5 \cdot b$

$1650 = 5.5b$

$\frac{1650}{5.5} = \frac{5.5b}{5.5}$

$300 = b$

16.5 is 5.5% of 300.

43. $\frac{a}{532} = \frac{30}{100}$

$a \cdot 100 = 30 \cdot 532$

$100a = 15,960$

$\frac{100a}{100} = \frac{15,960}{100}$

$a = 159.6$

159.6 is 30% of 532.

44. $\frac{63}{35} = \frac{p}{100}$

$63 \cdot 100 = p \cdot 35$

$6300 = 35p$

$\frac{6300}{35} = \frac{35p}{35}$

$180 = p$

63 is 180% of 35.

45. $\frac{93.5}{85} = \frac{p}{100}$

$93.5 \cdot 100 = p \cdot 85$

$9350 = 85p$

$\frac{9350}{85} = \frac{85p}{85}$

$110 = p$

93.5 is 110% of 85.

46. $\frac{a}{500} = \frac{33}{100}$

$a \cdot 100 = 33 \cdot 500$

$100a = 16,500$

$\frac{100a}{100} = \frac{16,500}{100}$

$a = 165$

165 is 33% of 500.

47. 1320 is what percent of 2000?
Method 1:
$$1320 = x \cdot 2000$$
$$\frac{1320}{2000} = \frac{2000x}{2000}$$
$$0.66 = x$$
$$66\% = x$$
66% of people own microwaves.
Method 2:
$$\frac{1320}{2000} = \frac{p}{100}$$
$$1320 \cdot 100 = p \cdot 2000$$
$$132,000 = 2000p$$
$$\frac{132,000}{2000} = \frac{2000p}{2000}$$
$$66 = p$$
66% of people own microwaves.

48. 2000 is what percent of 12,360?
Method 1:
$$2000 = x \cdot 12,360$$
$$\frac{2000}{12,360} = \frac{12,360x}{12,360}$$
$$0.16 \approx x$$
$$16\% \approx x$$
16% of freshmen are enrolled in prealgebra.
Method 2:
$$\frac{200}{12,360} = \frac{p}{100}$$
$$2000 \cdot 100 = 12,360 \cdot p$$
$$200,000 = 12,360p$$
$$\frac{200,000}{12,360} = \frac{12,360p}{12,360}$$
$$16 \approx p$$
16% of freshman are enrolled in prealgebra.

49. percent decrease $= \dfrac{\text{amount of decrease}}{\text{original amount}}$
$$= \frac{675 - 534}{675}$$
$$= \frac{141}{675}$$
$$\approx 0.209$$
$$\approx 20.9\%$$
Violent crime decreased 20.9%.

50. percent increase $= \dfrac{\text{amount of increase}}{\text{original amount}}$
$$= \frac{33 - 16}{16}$$
$$= \frac{17}{16}$$
$$= 1.0625$$
$$= 106.25\%$$
The charge will increase 106.25%.

51. Amount of decrease
= percent decrease · original amount
$$= 4\% \cdot \$215,000$$
$$= 0.04 \cdot \$215,000$$
$$= \$8600$$
Total amount $= \$215.00 - \$8600$
$$= \$206.400$$
$206,400 is expected to be collected next year.

52. Amount of increase
= percent increase · original amount
$$= 15\% \cdot \$11.50$$
$$= 0.15 \cdot \$11.50$$
$$\approx \$1.73$$
Total amount $= \$11.50 + \$1.73 = \$13.23$
The new hourly rate is $13.23.

53. sales tax = tax rate · purchase price
$$= 5.5\% \cdot \$250$$
$$= 0.055 \cdot \$250$$
$$= \$13.75$$
Total amount $= \$250 + \$13.75 = \$263.75$
The total price for the coat is $263.75.

54. sales tax = tax rate · purchase price
$$= 4.5\% \cdot \$25.50$$
$$= 0.045 \cdot \$25.50$$
$$\approx \$1.15$$
The sales tax is $1.15.

55. commission = commission rate · sales
$$= 5\% \cdot \$100,000$$
$$= 0.05 \cdot \$100,000$$
$$= \$5000$$
His commission is $5000.

56. commission = commission rate · sales

$$= 7.5\% \cdot \$4005$$
$$= 0.075 \cdot \$4005$$
$$\approx \$300.38$$

Her commission is $300.38.

57. Amount of discount

$$= \text{discount} \cdot \text{original price}$$
$$= 30\% \cdot \$3000$$
$$= 0.30 \cdot \$3000$$
$$= \$900$$

Sale price = $3000 − $900 = $2100
The amount of discount is $900; the sale price is $2100.

58. Amount of discount

$$= \text{discount} \cdot \text{original price}$$
$$= 10\% \cdot \$90$$
$$= 0.10 \cdot \$90$$
$$= \$9.00$$

Sale price = $90 − $9 = $81
The amount of discount is $9; the sale price is $81.

59. $I = P \cdot R \cdot T$

$$= \$4000 \cdot 12\% \cdot \frac{4}{12}$$
$$= \$4000 \cdot 0.12 \cdot \frac{1}{3}$$
$$= \$160$$

The simple interest is $160.

60. $I = P \cdot R \cdot T$

$$= \$650 \cdot 20\% \cdot \frac{3}{12}$$
$$= \$6500 \cdot 0.20 \cdot 0.25$$
$$= \$325$$

The simple interest is $325.

61. $A = P\left(1 + \dfrac{r}{n}\right)^{n \cdot t}$

$$= 5500\left(1 + \frac{0.12}{1}\right)^{1 \cdot 15}$$
$$= 5500(1.12)^{15}$$
$$\approx 30,104.61$$

The total amount is $30,104.61.

62. $A = P\left(1 + \dfrac{r}{n}\right)^{n \cdot t}$

$$= 6000\left(1 + \frac{0.11}{2}\right)^{2 \cdot 10}$$
$$= 6000(1.055)^{20}$$
$$\approx 17,506.54$$

The total amount is $17,506.54.

63. $A = P\left(1 + \dfrac{r}{n}\right)^{n \cdot t}$

$$= 100\left(1 + \frac{0.12}{4}\right)^{4 \cdot 5}$$
$$= 100(1.03)^{20}$$
$$\approx 180.61$$

The total amount is $180.61.

64. $A = P\left(1 + \dfrac{r}{n}\right)^{n \cdot t}$

$$= 1000\left(1 + \frac{0.18}{4}\right)^{4 \cdot 20}$$
$$= 1000(1.045)^{80}$$
$$\approx 33,830.10$$

The total amount is $33,830.10.

65. $3.8\% = 3.8(0.01) = 0.038$

66. $124.5\% = 124.5(0.01) = 1.245$

67. $0.54 = 0.54(100\%) = 54\%$

68. $95.2 = 95.2(100\%) = 9520\%$

69. $47\% = 47\left(\dfrac{1}{100}\right) = \dfrac{47}{100}$

70. $5.6\% = 5.6\left(\dfrac{1}{100}\right)$

$\qquad = \dfrac{5.6}{100}$

$\qquad = \dfrac{5.6 \cdot 10}{100 \cdot 10}$

$\qquad = \dfrac{56}{1000}$

$\qquad = \dfrac{7}{125}$

71. $\dfrac{3}{8} = \dfrac{3}{8} \cdot \dfrac{100}{1}\% = \dfrac{300}{8}\% = 37\dfrac{1}{2}\%$

72. $\dfrac{6}{5} = \dfrac{6}{5} \cdot \dfrac{100}{1}\% = \dfrac{600}{5}\% = 120\%$

73. $\quad 43 = 16\% \cdot x$

$\qquad 43 = 0.16x$

$\qquad \dfrac{43}{0.16} = \dfrac{0.16x}{0.16}$

$\qquad 268.75 = x$

43 is 16% of 268.75.

74. $\quad 27.5 = x \cdot 25$

$\qquad \dfrac{27.5}{25} = \dfrac{25x}{25}$

$\qquad 1.1 = x$

$\qquad 110\% = x$

27.5 is 110% of 25.

75. $x = 36\% \cdot 1968$

$\quad x = 0.36 \cdot 1968$

$\quad x = 708.48$

708.48 is 36% of 1968.

76. $\quad 67 = x \cdot 50$

$\qquad \dfrac{67}{50} = \dfrac{50x}{50}$

$\qquad 1.34 = x$

$\qquad 134\% = x$

67 is 134% of 50.

77. $\qquad \dfrac{75}{25} = \dfrac{p}{100}$

$\qquad 75 \cdot 100 = p \cdot 25$

$\qquad 7500 = 25p$

$\qquad \dfrac{7500}{25} = \dfrac{25p}{25}$

$\qquad 300 = p$

75 is 300% of 25.

78. $\qquad \dfrac{a}{240} = \dfrac{16}{100}$

$\qquad a \cdot 100 = 16 \cdot 240$

$\qquad 100a = 3840$

$\qquad \dfrac{100a}{100} = \dfrac{3840}{100}$

$\qquad a = 38.4$

38.4 is 16% of 240.

79. $\qquad \dfrac{28}{b} = \dfrac{5}{100}$

$\qquad 28 \cdot 100 = 5 \cdot b$

$\qquad 2800 = 5b$

$\qquad \dfrac{2800}{5} = \dfrac{5b}{5}$

$\qquad 560 = b$

28 is 5% of 560.

80. $\qquad \dfrac{52}{16} = \dfrac{p}{100}$

$\qquad 52 \cdot 100 = p \cdot 16$

$\qquad 5200 = 16p$

$\qquad \dfrac{5200}{16} = \dfrac{16p}{16}$

$\qquad 325 = p$

52 is 325% of 16.

81. $\dfrac{78}{300} = 0.26 = 26\%$

26% of the soft drinks have been sold.

82. $\$96{,}950 \cdot 7\% = \$96{,}950 \cdot 0.07 = \$6786.50$

The house has lost $6786.50 in value.

83. Sales tax $= 8.75\% \cdot \$568$
$$= 0.0875 \cdot \$568$$
$$= \$49.70$$
Total price $= \$568 + \$49.70 = \$617.70$
The total price is $617.70.

84. Amount of discount $= 15\% \cdot \$23.00$
$$= 0.15 \cdot \$23.00$$
$$= \$3.45$$

85. commission = commission rate $\cdot$ sales
$$\$1.60 = r \cdot \$12.80$$
$$\frac{\$1.60}{\$12.80} = r$$
$$0.125 = r$$
$$12.5\% = r$$
His rate of commission is 12.5%.

86. Simple interest = principal $\cdot$ rate $\cdot$ time
$$= \$1400 \cdot 13\% \cdot \frac{6}{12}$$
$$= \$1400 \cdot 0.13 \cdot 0.5$$
$$= \$91$$
Total amount $= \$1400 + \$91 = \$1491$
The total amount is $1491.

87. Simple interest = principal $\cdot$ rate $\cdot$ time
$$= \$5500 \cdot 12.5\% \cdot 9$$
$$= \$5500 \cdot 0.125 \cdot 9$$
$$= \$6187.50$$
Total amount $= \$5500 + \$6187.50 = \$11,687.50$
The total amount is $11,687.50.

Chapter 7 Test

1. $85\% = 85(0.01) = 0.85$

2. $500\% = 500(0.01) = 5$

3. $0.6\% = 0.6(0.01) = 0.006$

4. $0.056 = 0.056(100\%) = 5.6\%$

5. $6.1 = 6.1(100\%) = 610\%$

6. $0.35 = 0.35(100\%) = 35\%$

7. $120\% = 120\left(\dfrac{1}{100}\right) = \dfrac{120}{100} = 1\dfrac{1}{5}$

8. $38.5\% = 38.5\left(\dfrac{1}{100}\right) = \dfrac{38.5}{100} = \dfrac{385}{1000} = \dfrac{77}{200}$

9. $0.2\% = 0.2\left(\dfrac{1}{100}\right) = \dfrac{0.2}{100} = \dfrac{2}{1000} = \dfrac{1}{500}$

10. $\dfrac{11}{20} = \dfrac{11}{20} \cdot \dfrac{100}{1}\% = \dfrac{1100}{20}\% = 55\%$

11. $\dfrac{3}{8} = \dfrac{3}{8} \cdot \dfrac{100}{1}\% = \dfrac{300}{8}\% = 37.5\%$

12. $1\dfrac{3}{4} = \dfrac{7}{4} \cdot \dfrac{100}{1}\% = \dfrac{700}{4}\% = 175\%$

13. $\dfrac{1}{5} = \dfrac{1}{5} \cdot \dfrac{100}{1}\% = \dfrac{100}{5}\% = 20\%$

14. $64\% = 64\left(\dfrac{1}{100}\right) = \dfrac{64}{100} = \dfrac{16}{25}$

15. *Method 1*:
$x = 42\% \cdot 80$
$x = 0.42 \cdot 80$
$x = 33.6$
33.6 is 42% of 80.
Method 2:
$$\frac{a}{80} = \frac{42}{100}$$
$$a \cdot 100 = 42 \cdot 80$$
$$100a = 3360$$
$$\frac{100a}{100} = \frac{3360}{100}$$
$$a = 33.6$$
33.6 is 42% of 80.

16. *Method 1:*
$$0.6\% \cdot x = 7.5$$
$$0.006x = 7.5$$
$$\frac{0.006x}{0.006} = \frac{7.5}{0.006}$$
$$x = 1250$$
0.6% of 1250 is 7.5.
Method 2:
$$\frac{7.5}{b} = \frac{0.6}{100}$$
$$7.5 \cdot 100 = b \cdot 0.6$$
$$750 = 0.6b$$
$$\frac{750}{0.6} = \frac{0.6b}{0.6}$$
$$1250 = b$$
0.6% of 1250 is 7.5.

17. *Method 1:*
$$567 = x \cdot 756$$
$$\frac{567}{756} = \frac{x \cdot 756}{756}$$
$$0.75 = x$$
$$75\% = x$$
567 is 75% of 756.
Method 2:
$$\frac{567}{756} = \frac{p}{100}$$
$$567 \cdot 100 = 756 \cdot p$$
$$56,700 = 756p$$
$$\frac{56,700}{756} = \frac{756p}{756}$$
$$75 = p$$
567 is 75% of 756.

18. 12% of 320 is what number?
$$12\% \cdot 320 = x$$
$$0.12 \cdot 320 = x$$
$$38.4 = x$$
There are 38.4 pounds of copper.

19. 20% of what number is $11,350?
$$20\% \cdot x = \$11,350$$
$$0.20x = \$11,350$$
$$\frac{0.2x}{0.2} = \frac{\$11,350}{0.2}$$
$$x = \$56,750$$
The value of the potential crop is $56,750.

20. tax = 1.25% · $354 = 0.0125 · $354 ≈ $4.43
total amount = $354 + $4.43 = $358.43
The total amount of the stereo is $358.43.

21. percent increase $= \dfrac{\text{amount of increase}}{\text{original amount}}$
$$= \frac{26,460 - 25,200}{25,200}$$
$$= \frac{1260}{25,200}$$
$$= 0.05$$
The increase in population was 5%.

22. Amount of discount $= 15\% \cdot \$120$
$$= 0.15 \cdot \$120$$
$$= \$18$$
Sale price = $120 − $18 = $102
The amount of the discount is $18; the sale price is $102.

23. commission $= 4\% \cdot \$9875$
$$= 0.04 \cdot \$9875$$
$$= \$395$$
His commission was $395.

24.
$$\$1.53 = \text{rate} \cdot \$152.99$$
$$\frac{\$1.53}{\$152.99} = r$$
$$0.01 \approx r$$
$$1\% \approx r$$
The sales tax rate is 1%.

25. simple interest = principal · rate · time
$$= \$2000 \cdot 9.25\% \cdot 3\frac{1}{2}$$
$$= \$2000 \cdot 0.0925 \cdot 3.5$$
$$= \$647.5$$

26.
$$A = P\left(1 + \frac{r}{n}\right)^{n \cdot t}$$
$$= 1365\left(1 + \frac{0.08}{1}\right)^{1 \cdot 5}$$
$$= 1365(1.08)^5$$
$$\approx 2005.63$$

The total amount of \$2005.63.

27. Simple interest = principal · rate · time
$$= \$400 \cdot 13.5\% \cdot \frac{6}{12}$$
$$= \$400 \cdot 0.135 \cdot 0.5$$
$$= \$27$$

Total amount = \$400 + \$27 = \$427

28. percent decrease $= \dfrac{\text{amount of decrease}}{\text{original amount}}$
$$= \frac{125,587 - 118,346}{125,587}$$
$$= \frac{7241}{125,587}$$
$$\approx 0.0576$$
$$\approx 5.8\%$$

The number of crimes has decreased 5.8%.

Cumulative Review Chapters 1–7

1.
$$\begin{array}{r} 236 \\ \times\;\;\;86 \\ \hline 1\;416 \\ 18\;880 \\ \hline 20,296 \end{array}$$

2.
$$\begin{array}{r} 409 \\ \times\;\;\;76 \\ \hline 2\;454 \\ 28\;630 \\ \hline 31,084 \end{array}$$

3. $-3 - 7 = -3 + (-7) = -10$

4. $8 - (-2) = 8 + 2 = 10$

5.
$$x - 2 = -1$$
$$x - 2 + 2 = -1 + 2$$
$$x = 1$$

6.
$$x + 4 = 3$$
$$x + 4 - 4 = 3 - 4$$
$$x = -1$$

7.
$$3(2x - 6) + 6 = 0$$
$$3 \cdot 2x - 3 \cdot 6 + 6 = 0$$
$$6x - 18 + 6 = 0$$
$$6x - 12 = 0$$
$$6x - 12 + 12 = 0 + 12$$
$$6x = 12$$
$$\frac{6x}{6} = \frac{12}{6}$$
$$x = 2$$

8.
$$5(x - 2) = 3x$$
$$5 \cdot x - 5 \cdot 2 = 3x$$
$$5x - 10 = 3x$$
$$5x - 5x - 10 = 3x - 5x$$
$$-10 = -2x$$
$$\frac{-10}{-2} = \frac{-2x}{-2}$$
$$5 = x$$

9. $3 = \dfrac{3}{1} \cdot \dfrac{7}{7} = \dfrac{3 \cdot 7}{1 \cdot 7} = \dfrac{21}{7}$

10. $8 = \dfrac{8}{1} \cdot \dfrac{5}{5} = \dfrac{8 \cdot 5}{1 \cdot 5} = \dfrac{40}{5}$

11. $-\dfrac{10}{27} = -\dfrac{2 \cdot 5}{3 \cdot 3 \cdot 3}$

Since 10 and 27 have no common factors, $-\dfrac{10}{27}$ is already in simplest form.

12. $\dfrac{10y}{32} = \dfrac{5 \cdot 2 \cdot y}{16 \cdot 2} = \dfrac{5y}{16}$

13. $-\dfrac{5}{16} \div -\dfrac{3}{4} = -\dfrac{5}{16} \cdot -\dfrac{4}{3} = \dfrac{5 \cdot 4}{4 \cdot 4 \cdot 3} = \dfrac{5}{12}$

14. $\dfrac{-2}{5} \div \dfrac{7}{10} = -\dfrac{2}{5} \cdot \dfrac{10}{7} = -\dfrac{2 \cdot 2 \cdot 5}{5 \cdot 7} = -\dfrac{4}{7}$

15. $y - x = -\dfrac{8}{10} - \left(-\dfrac{3}{10}\right)$

$= -\dfrac{8}{10} + \dfrac{3}{10}$

$= -\dfrac{5}{10}$

$= -\dfrac{1}{2}$

16. $2x + 3y = 2\left(\dfrac{2}{5}\right) + 3\left(\dfrac{-1}{5}\right) = \dfrac{4}{5} + \dfrac{-3}{5} = \dfrac{1}{5}$

17. $-\dfrac{3}{4} - \dfrac{1}{14} + \dfrac{6}{7} = -\dfrac{3}{4} \cdot \dfrac{7}{7} - \dfrac{1}{14} \cdot \dfrac{2}{2} + \dfrac{6}{7} \cdot \dfrac{4}{4}$

$= -\dfrac{21}{28} - \dfrac{2}{28} + \dfrac{24}{28}$

$= \dfrac{1}{28}$

18. $\dfrac{2}{9} + \dfrac{7}{15} - \dfrac{1}{3} = \dfrac{2}{9} \cdot \dfrac{5}{5} + \dfrac{7}{15} \cdot \dfrac{3}{3} - \dfrac{1}{3} \cdot \dfrac{15}{15}$

$= \dfrac{10}{45} + \dfrac{21}{45} - \dfrac{15}{45}$

$= \dfrac{16}{45}$

19. $\dfrac{\frac{1}{2} + \frac{3}{8}}{\frac{3}{4} - \frac{1}{6}} = \dfrac{\frac{1}{2} \cdot \frac{4}{4} + \frac{3}{8}}{\frac{3}{4} \cdot \frac{3}{3} - \frac{1}{6} \cdot \frac{2}{2}}$

$= \dfrac{\frac{4}{8} + \frac{3}{8}}{\frac{9}{12} - \frac{2}{12}}$

$= \dfrac{\frac{7}{8}}{\frac{7}{12}}$

$= \dfrac{7}{8} \div \dfrac{7}{12}$

$= \dfrac{7}{8} \cdot \dfrac{12}{7}$

$= \dfrac{7 \cdot 4 \cdot 3}{4 \cdot 2 \cdot 7}$

$= \dfrac{3}{2}$

20. $\dfrac{\frac{2}{3} + \frac{1}{6}}{\frac{3}{4} - \frac{3}{5}} = \dfrac{\frac{2}{3} \cdot \frac{2}{2} + \frac{1}{6}}{\frac{3}{4} \cdot \frac{5}{5} - \frac{3}{5} \cdot \frac{4}{4}}$

$= \dfrac{\frac{4}{6} + \frac{1}{6}}{\frac{15}{20} - \frac{12}{20}}$

$= \dfrac{\frac{5}{6}}{\frac{3}{20}}$

$= \dfrac{5}{6} \div \dfrac{3}{20}$

$= \dfrac{5}{6} \cdot \dfrac{20}{3}$

$= \dfrac{5 \cdot 2 \cdot 10}{2 \cdot 3 \cdot 3}$

$= \dfrac{50}{9}$

21. $\dfrac{x}{2} = \dfrac{x}{3} + \dfrac{1}{2}$

$6\left(\dfrac{x}{2}\right) = 6\left(\dfrac{x}{3} + \dfrac{1}{2}\right)$

$3x = 6 \cdot \dfrac{x}{3} + 6 \cdot \dfrac{1}{2}$

$3x = 2x + 3$

$3x - 2x = 2x + 3 - 2x$

$x = 3$

22.
$$\frac{x}{2}+\frac{1}{5}=3-\frac{x}{5}$$
$$10\left(\frac{x}{2}+\frac{1}{5}\right)=10\left(3-\frac{x}{5}\right)$$
$$10\cdot\frac{x}{2}+10\cdot\frac{1}{5}=10\cdot3-10\cdot\frac{x}{5}$$
$$5x+2=30-2x$$
$$5x+2+2x=30-2x+2x$$
$$7x+2=30$$
$$7x+2-2=30-2$$
$$7x=28$$
$$\frac{7x}{7}=\frac{28}{7}$$
$$x=4$$

23. a. $\quad 4\frac{2}{9}=\frac{9\cdot4+2}{9}=\frac{36+2}{9}=\frac{38}{9}$

b. $\quad 1\frac{8}{11}=\frac{11\cdot1+8}{11}=\frac{11+8}{11}=\frac{19}{11}$

24. a. $\quad 3\frac{2}{5}=\frac{5\cdot3+2}{5}=\frac{15+2}{5}=\frac{17}{5}$

b. $\quad 6\frac{2}{7}=\frac{7\cdot6+2}{7}=\frac{42+2}{7}=\frac{44}{7}$

25. $\quad 0.125=\frac{125}{1000}=\frac{1}{8}$

26. $\quad 0.85=\frac{85}{100}=\frac{17}{20}$

27. $\quad -105.083=-105\frac{83}{1000}$

28. $\quad 17.015=17\frac{15}{1000}=17\frac{3}{200}$

29.
$$\begin{array}{r} 85.00 \\ -\ 17.31 \\ \hline 67.69 \end{array}$$

30.
$$\begin{array}{r} 38.00 \\ -\ 10.06 \\ \hline 27.94 \end{array}$$

31. $7.68\times10=76.8$

32. $12.483\times100=1248.3$

33. $(-76.3)(1000)=-76,300$

34. $-853.75\times10=-8537.5$

35. $x\div y=2.5\div0.05$

$$0.05\overline{)2.5}\quad\text{becomes}\quad 5\overline{)\begin{array}{r}50\\250\\-25\\\hline00\end{array}}$$

36. $\quad \dfrac{x}{100}=4.75$

$$\frac{470}{100}\overset{?}{=}4.75$$
$$4.7=4.75\quad\text{False}$$
No, 470 is not a solution.

37. 55, 67, 75, 86, 91, 91
$$\text{median}=\frac{75+86}{2}=\frac{161}{2}=80.5$$

38. $\quad \text{mean}=\dfrac{36+40+86+30}{4}=\dfrac{192}{4}=48$

39. $\quad \dfrac{2.5}{3.15}=\dfrac{2500}{3150}=\dfrac{50}{63}$

40. $\quad \dfrac{5.8}{7.6}=\dfrac{58}{76}=\dfrac{29}{38}$

41.
$$16\overline{)\begin{array}{r}0.21\\3.36\\-32\\\hline16\\-16\\\hline0\end{array}}$$

The unit price is $0.21/ounce.

42.
$$\begin{array}{r} 2.25 \\ 40\overline{)90.00} \\ \underline{-80} \\ 10\ 0 \\ \underline{-8\ 0} \\ 2\ 00 \\ \underline{-2\ 00} \\ 0 \end{array}$$

The price is \$2.25 per tile, or \$2.25 per square foot.

43.
$$\frac{4.1}{7} \stackrel{?}{=} \frac{2.9}{5}$$
$$4.1(5) \stackrel{?}{=} 2.9(7)$$
$$20.5 \neq 20.3$$
No, it is not a true proportion.

44.
$$\frac{6.3}{9} \stackrel{?}{=} \frac{3.5}{5}$$
$$6.3(5) \stackrel{?}{=} 3.5(9)$$
$$31.5 = 31.5$$
Yes, it is a true proportion.

45.
$$\frac{5 \text{ miles}}{2 \text{ inches}} = \frac{x \text{ miles}}{7 \text{ inches}}$$
$$5 \cdot 7 = 2 \cdot x$$
$$35 = 2x$$
$$\frac{35}{2} = \frac{2x}{2}$$
$$17.5 = x$$
17.5 miles corresponds to 7 inches.

46.
$$\frac{7 \text{ problems}}{6 \text{ minutes}} = \frac{x \text{ problems}}{30 \text{ minutes}}$$
$$7 \cdot 30 = 6 \cdot x$$
$$210 = 6x$$
$$\frac{210}{6} = \frac{6x}{6}$$
$$35 = x$$
The student can complete 35 problems in 30 minutes.

47. $1.9\% = 1.9\left(\dfrac{1}{100}\right) = \dfrac{1.9}{100} = \dfrac{19}{1000}$

48. $2.3\% = 2.3\left(\dfrac{1}{100}\right) = \dfrac{2.3}{100} = \dfrac{23}{1000}$

49. $33\dfrac{1}{3}\% = \dfrac{100}{3}\left(\dfrac{1}{100}\right) = \dfrac{100}{300} = \dfrac{1}{3}$

50. $108\% = 108\left(\dfrac{1}{100}\right) = \dfrac{108}{100} = 1\dfrac{2}{25}$

Chapter 8

Section 8.1

Practice Problems

1. **a.** Sweden has $1\frac{1}{2}$ symbols and each symbol represents 50 billion kilowatt-hours, so Sweden generated approximately $1\frac{1}{2}(50) = 75$ billion kilowatt-hours of nuclear energy.

   b. Sweden and Germany, together, have $7\frac{1}{2}$ symbols, so they generated approximately $7\frac{1}{2}(50) = 375$ billion kilowatt-hours of nuclear energy.

2. **a.** The height of the bar for insects is 45, so approximately 45 endangered species are insects.

   b. The shortest bar corresponds to arachnids, so arachnids have the fewest endangered species.

3.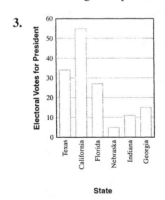

4. The height of the bar for 80–89 is 12, so 12 students scored 80–89 on the test.

5. The height of the bar for 40–49 is 1, for 50–59 is 3, for 60–69 is 2, for 70–79 is 10. So, $1 + 3 + 2 + 10 = 16$ students scored less than 80 on the test.

6.

Class Interval (Credit Card Balances)	Tally	Class Frequency (Number of Months)
$0–$49	|||	3
$50–$99	||||	4
$100–$149	||	2
$150–$199			1
$200–$249			1
$250–$299			1

7.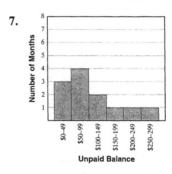

8. **a.** The lowest point on the graph corresponds to January, so the average daily temperature is the lowest during January.

   b. The point on the graph that corresponds to 25 is December, so the average daily temperature is 25°F in December.

c. The points on the graph that are greater than 70 are June, July, and August. So, the average daily temperature is greater than 70°F in June, July, and August.

Vocabulary and Readiness Check

1. A <u>bar</u> graph presents data using vertical or horizontal bars.

3. A <u>line</u> graph displays information with a line that connects data points.

Exercise Set 8.1

1. Washington has the greatest number of apples, so the greatest quantity of apples was produced in the state of Washington.

3. New York is represented by 2.5 apples, and each apple represents 10 million bushels, so there were approximately 2.5(10) = 25 million bushels grown.

5. New York is represented by 2.5 apples. From the pictograph, California, with 1.5 apples, and Pennsylvania, with 1 apple, together produced as many apples as New York.

7. New York produced approximately 25 million bushels in 2005. Therefore, the percent increase in apple production from 2001 to 2005 is $\frac{25-23.8}{23.8} \approx 0.05$ or about 5%.

9. The year 2002 has 7.5 flames and each flame represents 12,000 wildfires, so there were approximately 7.5(12,000) = 90,000 wildfires in 2002.

11. 2000 has 10 flames and each flame represents 12,000 wildfires, so there were approximately 10(12,000) = 120,000 wildfires in 2000.

13. 2004 has 6.5 flames and 2005 has 5.5 flames, which is one fewer. Thus, the decrease in the number of wildfires from 2004 to 2005 was 12,000.

15. 2003 has 7 flames, 2004 has 6.5 flames, and 2005 has 5.5 flames. The average is $\frac{7+6.5+5.5}{3} = 6\frac{1}{3}$. Each flame represents 12,000 wildfires, so the average annual number of wildfires from 2003 to 2005 is $6\frac{1}{3}(12,000) = 76,000.$

17. The longest bar corresponds to September, so the month in which most hurricanes made landfall is September.

19. The length of the bar for August is 27, so approximately 27 hurricanes made landfall in August.

21. Two of the six hurricanes that made landfall in July did so in 2005. The fraction is $\frac{2}{6} = \frac{1}{3}.$

23. The longest bar corresponds to Tokyo, Japan, and the length of the bar is 35.2 million. So, the city with the largest population is Tokyo, Japan and its population is about 35.2 million or 35,200,000.

25. The longest bar corresponding to a city in the United States is the bar for New York. The population is approximately 21.9 million or 21,900,000.

27. The bar corresponding to Seoul, South Korea has length 22.1 and the bar corresponding to Sao Paolo, Brazil has length 20.1. Thus, Seoul, South Korea is about 22.1 − 20.1 = 2 million larger than Sao Paolo, Brazil.

29.

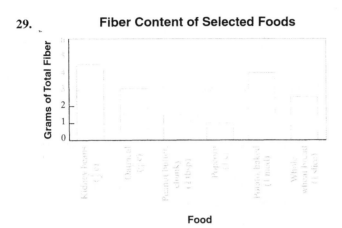

Fiber Content of Selected Foods

31.

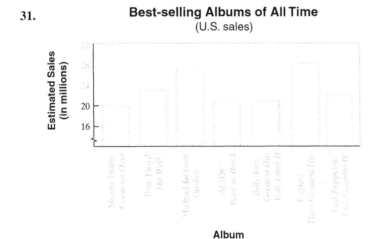

Best-selling Albums of All Time
(U.S. sales)

33. The height of the bar for 100–149 miles per week is 15, so 15 of the adults drive 100–149 miles per week.

35. 29 of the adults drive 0–49 miles per week, 17 of the adults drive 50–99 miles per week, and 15 of the adults drive 100–149 miles per week, so 29 + 17 + 15 = 61 of the adults drive fewer than 150 miles per week.

37. 15 of the adults drive 100–149 miles per week and 9 of the adults drive 150–199 miles per week, so 15 + 9 = 24 of the adults drive 100–199 miles per week.

39. 21 of the adults drive 250–299 miles per week and 9 of the adults drive 200–249 miles per week, so 21 − 9 = 12 more adults drive 250–299 miles per week than 200–249 miles per week.

41. 9 of the 100 adults surveyed drive

150–199 miles per week, so the ratio is $\dfrac{9}{100}$.

43. The tallest bar corresponds to 45–54, so the most householders are in the 45–54 age range.

45. According to the bar graph, approximately 21 million householders will be 55–64 years old.

47. The sum of the first three bars is $21 + 17 + 6 = 44$, so approximately 44 million householders will be 44 years old or younger.

49. The height of the bar for 45–54 is 25 and the height of the bar for 55–64 is 21. So approximately $25 - 21 = 4$ million more householders will be 45–54 years old than 55–64 years old.

	Class Interval (Scores)	Tally	Class Frequency (Number of Games)
51.	70–79	|	1
53.	90–99	ⅢⅢ Ⅲ	8

	Class Interval (Account Balances)	Tally	Class Frequency (Number of People)
55.	$0–$99	ⅢⅢ |	6
57.	$200–$299	ⅢⅢ |	6
59.	$400–$499	||	2

61.

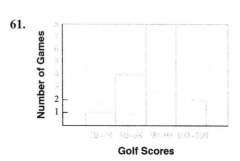

Golf Scores

63. The point on the graph corresponding to 1994 is 2.7, so the average number of goals per game in 1994 was 2.7.

65. The highest point on the graph corresponds to 1982, so the average number of goals per game was the greatest in 1982.

67. The graph decreases between 2002 and 2006, so the average number of goals per game decreased from 2002 to 2006.

69. The dots for 1990 and 2006 are below the 2.5-level, so the average number of goals per game was less than 2.5 in 1990 and 2006.

71. 30% of 12 is 0.30 · 12 = 3.6.

73. 10% of 62 is 0.10 · 62 = 6.2

75. $\dfrac{1}{4} = \dfrac{1 \cdot 25}{4 \cdot 25} = \dfrac{25}{100} = 25\%$

77. $\dfrac{17}{50} = \dfrac{17 \cdot 2}{50 \cdot 2} = \dfrac{34}{100} = 34\%$

79. The point on the high temperature graph corresponding to Thursday is 83, so the high temperature reading on Thursday was 83°F.

81. The lowest point on the graph of low temperatures corresponds to Sunday. The low temperature on Sunday was 68°F.

83. The difference between the graphs is the greatest for Tuesday. The high temperature was 86°F and the low temperature was 73°F, so the difference is 86 − 73 = 13°F.

85. answers may vary

Section 8.2

Practice Problems

1. Eight of the 100 adults prefer golf. The ratio is

$$\dfrac{\text{adults preferring golf}}{\text{total adults}} = \dfrac{8}{100} = \dfrac{2}{25}$$

2. Add the percents corresponding to three computers and four or more computers. 6% + 3% = 9%

3. amount = percent · base
= 0.03 · 299,800,000
= 8,994,000

Thus, 8,994,000 Americans have four or more working computers at home.

4.

Year	Percent	Degrees in Sector
Freshmen	30%	30% of 360° = 0.30(360°) = 108°
Sophomores	27%	27% of 360° = 0.27(360°) = 97.2°
Juniors	25%	25% of 360° = 0.25(360°) = 90°
Seniors	18%	18% of 360° = 0.18(360°) = 64.8°

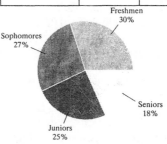

Vocabulary and Readiness Check

1. In a <u>circle</u> graph, each section (shaped like a piece of pie) shows a category and the relative size of the category.

3. The number of degrees in a whole circle is <u>360</u>.

Exercise Set 8.2

1. The largest sector corresponds to the category "parent or guardian's home," thus most of the students live in a parent or guardian's home.

3. 180 of the 700 total students live in campus housing.
$$\frac{180}{700} = \frac{9}{35}$$
The ratio is $\frac{9}{35}$.

5. 180 of the students live in campus housing while 320 live in a parent or guardian's home.
$$\frac{180}{320} = \frac{9}{16}$$
The ratio is $\frac{9}{16}$.

7. The largest sector corresponds to Asia. Thus, the largest continent is Asia.

9. $30\% + 7\% = 37\%$
37% of the land on Earth is accounted for by Europe and Asia.

11. Asia accounts for 30% of the land on Earth.
30% of $57,000,000 = 0.30 \cdot 57,000,000$
$= 17,100,000$
Asia is 17,100,000 square miles.

13. Australia accounts for 5% of the land on Earth.
5% of $57,000,000 = 0.05 \cdot 57,000,000$
$= 2,850,000$
Australia is 2,850,000 square miles.

15. Add the percent for adult's fiction (33%) to the percent for children's fiction (22%).
$33\% + 22\% = 55\%$
Thus, 55% of books are classified as some type of fiction.

17. The second-largest sector corresponds to nonfiction, so the second-largest category of books is nonfiction.

19. Nonfiction accounts for 25% of the books.
25% of $25,600 = 0.25 \cdot 125,600 = 31,400$
The library has 31,400 nonfiction books.

21. Children's fiction accounts for 22% of the books.
22% of $125,600 = 0.22 \cdot 125,600 = 27,632$
The library has 27,632 children's fiction books.

23. Reference or other accounts for 17% + 3% = 20% of the books.
20% of 125,600 = 0.20 · 125,600 = 25,120
The library has 25,120 reference or other books.

25.

Energy Source	Percent	Degrees in Sector
Coal	50.6%	50.6% of 360° = 0.506(360°) ≈ 182°
Natural Gas	15.8%	15.8% of 360° = 0.158(360°) ≈ 57°
Hydroelectric	8.2%	8.2% of 360° = 0.082(360°) ≈ 30°
Nuclear	20.6%	20.6% of 360° = 0.206(360°) ≈ 74°
Petroleum	1.6%	1.6% of 360° = 0.016(360°) ≈ 6°
Other	3.2%	3.2% of 360° = 0.032(360°) ≈ 12°

27.

2006 Hybrid Sales by Make of Car		
Company	**Percent**	**Degrees in Sector**
Toyota	63%	63% of 360° = 0.63(360°) ≈ 227°
Honda	17%	17% of 360° = 0.17(360°) ≈ 61°
Lexus	10%	10% of 360° = 0.10(360°) = 36°
Ford/Mercury	10%	10% of 360° = 0.10(360°) = 36°

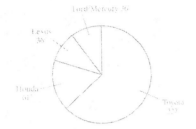

29. $20 = 2 \cdot 10 = 2 \cdot 2 \cdot 5 = 2^2 \cdot 5$

31. $40 = 2 \cdot 20 = 2 \cdot 2 \cdot 10 = 2 \cdot 2 \cdot 2 \cdot 5 = 2^3 \cdot 5$

33. $85 = 5 \cdot 17$

35. answers may vary

37. Pacific Ocean:
$49\% \cdot 264{,}489{,}800 = 129{,}600{,}002$ square kilometers

39. Indian Ocean:
$21\% \cdot 264{,}489{,}800 = 55{,}542{,}858$ square kilometers

41. $21.5\% \cdot 2800 = 602$ respondents

43. $59.8\% + 18.7\% = 78.5\%$
$0.785 \cdot 2800 = 2198$ respondents

45. $\dfrac{\text{number of respondents who spend } \$0-\$15}{\text{number of respondents who spend } \$15-\$175} = \dfrac{602}{1674}$

$$= \frac{2 \cdot 301}{2 \cdot 837}$$
$$= \frac{301}{837}$$

47. no; answers may vary

Section 8.3

Practice Problems

1.

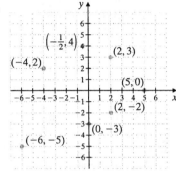

2. Point *A* has coordinates (5, 0).
   Point *B* has coordinates (−5, 4)
   Point *C* has coordinates (−3, −4).
   Point *D* has coordinates (0, −1).
   Point *E* has coordinates (2, 1).

3. $x + 3y = -12$
   $0 + 3(-4) \stackrel{?}{=} -12$
   $-12 = -12$ True
   Yes, (0, −4) is a solution of $x + 3y = -12$.

4.

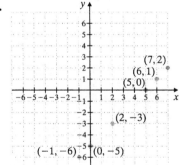

5. **a.** $y = -5x$
   $y = -5(5)$
   $y = -25$
   The ordered-pair solution is (5, −25).

b. $y = -5x$
   $0 = -5x$
   $\dfrac{0}{-5} = \dfrac{-5x}{-5}$
   $0 = x$
   The ordered-pair solution is (0, 0).

c. $y = -5x$
   $y = -5(-3)$
   $y = 15$
   The ordered-pair solution is (−3, 15).

6. **a.** $y = 5x + 2$
   $y = 5(0) + 2$
   $y = 0 + 2$
   $y = 2$
   The ordered-pair solution is (0, 2).

b. $y = 5x + 2$
   $-3 = 5x + 2$
   $-3 - 2 = 5x + 2 - 2$
   $-5 = 5x$
   $\dfrac{-5}{5} = \dfrac{5x}{5}$
   $-1 = x$
   The ordered-pair solution is (−1, −3).

Vocabulary and Readiness Check

1. In the ordered pair (−1, 2), the <u>x</u>-value is −1 and the <u>y</u>-value is 2.

3. The axes divide the plane into <u>four</u> regions, called quadrants.

5. The process of locating a point on the rectangular coordinate system is called <u>plotting</u> the point.

7. A <u>plane</u> is a flat surface that extends indefinitely in all directions.

Exercise Set 8.3

1.

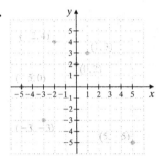

3.

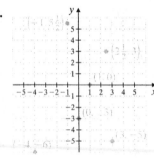

5. Point *A* has coordinates (0, 0).

Point *B* has coordinates $\left(3\frac{1}{2}, 0\right)$.

Point *C* has coordinates (3, 2).
Point *D* has coordinates (−1, 3).
Point *E* has coordinates (−2, −2).
Point *F* has coordinates (0, −1).
Point *G* has coordinates (2, −1).

7. $y = -20x$
$0 \stackrel{?}{=} -20(0)$
$0 = 0$ True
Yes, (0, 0) is a solution of $y = -20x$.

9. $x - y = 3$
$1 - 2 \stackrel{?}{=} 3$
$-1 = 3$ False
No, (1, 2) is not a solution of $x - y = 3$.

11. $y = 2x + 1$
$-3 \stackrel{?}{=} 2(-2) + 1$
$-3 \stackrel{?}{=} -4 + 1$
$-3 = -3$ True
Yes, (−2, −3) is a solution of $y = 2x + 1$.

13. $x = -3y$
$6 \stackrel{?}{=} -3(-2)$
$6 = 6$ True
Yes, (6, −2) is a solution of $x = -3y$.

15. $3y + 2x = 10$
$3 \cdot 0 + 2 \cdot 5 \stackrel{?}{=} 10$
$0 + 10 \stackrel{?}{=} 10$
$10 = 10$ True
Yes, (5, 0) is a solution of $3y + 2x = 10$.

17. $x - 5y = -1$
$3 - 5(1) \stackrel{?}{=} -1$
$3 - 5 \stackrel{?}{=} -1$
$-2 = -1$ False
No, (3, 1) is not a solution of $x - 5y = -1$.

19.

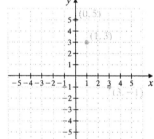

21.

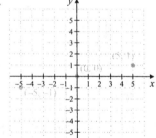

23.

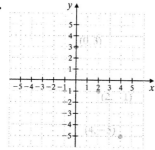

25. $y = -9x$
$y = -9(1)$
$y = -9$
The solution is $(1, -9)$.
$y = -9x$
$y = -9(0)$
$y = 0$
The solution is $(0, 0)$.
$y = -9x$
$-18 = -9x$
$\dfrac{-18}{-9} = \dfrac{-9x}{-9}$
$2 = x$
The solution is $(2, -18)$.

27. $x - y = 14$
$2 - y = 14$
$-2 + 2 - y = -2 + 14$
$-y = 12$
$y = -12$
The solution is $(2, -12)$.
$x - y = 14$
$x - (-8) = 14$
$x + 8 = 14$
$x + 8 - 8 = 14 - 8$
$x = 6$
The solution is $(6, -8)$.
$x - y = 14$
$0 - y = 14$
$-y = 14$
$y = -14$
The solution is $(0, -14)$.

29. $x + y = -2$
$-2 + y = -2$
$-2 + 2 + y = -2 + 2$
$y = 0$
The solution is $(-2, 0)$.
$x + y = -2$
$1 + y = -2$
$1 - 1 + y = -2 - 1$
$y = -3$
The solution is $(1, -3)$.
$x + y = -2$
$x + 5 = -2$
$x + 5 - 5 = -2 - 5$
$x = -7$
The solution is $(-7, 5)$.

31. $y = x + 5$
$y = 1 + 5$
$y = 6$
The solution is $(1, 6)$.
$y = x + 5$
$7 = x + 5$
$7 - 5 = x + 5 - 5$
$2 = x$
The solution is $(2, 7)$.
$y = x + 5$
$y = 3 + 5$
$y = 8$
The solution is $(3, 8)$.

33. $y = 3x - 5$
$y = 3 \cdot 1 - 5$
$y = 3 - 5$
$y = -2$
The solution is $(1, -2)$.
$y = 3x - 5$
$y = 3 \cdot 2 - 5$
$y = 6 - 5$
$y = 1$
The solution is $(2, 1)$.

$$y = 3x - 5$$
$$4 = 3x - 5$$
$$4 + 5 = 3x - 5 + 5$$
$$9 = 3x$$
$$\frac{9}{3} = \frac{3x}{3}$$
$$3 = x$$

The solution is (3, 4).

35. $y = -x$
$$0 = -x$$
$$0 = x$$

The solution is (0, 0).
$$y = -x$$
$$y = -2$$

The solution is (2, −2).
$$y = -x$$
$$2 = -x$$
$$-2 = x$$

The solution is (−2, 2).

37. $x + 2y = -8$
$$4 + 2y = -8$$
$$4 - 4 + 2y = -8 - 4$$
$$2y = -12$$
$$\frac{2y}{2} = \frac{-12}{2}$$
$$y = -6$$

The solution is (4, −6).
$$x + 2y = -8$$
$$x + 2(-3) = -8$$
$$x - 6 = -8$$
$$x - 6 + 6 = -8 + 6$$
$$x = -2$$

The solution is (−2, −3).
$$x + 2y = -8$$
$$0 + 2y = -8$$
$$2y = -8$$
$$\frac{2y}{2} = \frac{-8}{2}$$
$$y = -4$$

The solution is (0, −4).

39.
$$\begin{array}{r} 5.6 \\ -\ 3.9 \\ \hline 1.7 \end{array}$$

41.
$$\begin{array}{r} 5.6 \\ \times\ 3.9 \\ \hline 5\ 04 \\ 16\ 80 \\ \hline 21.80 \end{array}$$

43. $(0.236)(-100) = -23.6$

45. To plot (a, b), start at the origin and move a units to the right and b units up. The point will be in quadrant I. Thus, (a, b) is in quadrant I is a true statement.

47. To plot $(0, b)$, start at the origin and move 0 units to the right and b units up. The point will lie on the y-axis. Thus, $(0, b)$ lies on the y-axis is a true statement.

49. To plot $(0, -b)$, start at the origin and move 0 units to the right and b units down. The point will be on the y-axis. Thus, $(0, -b)$ lies on the x-axis is a false statement.

51. To plot $(-a, b)$, start at the origin and move a units to the left and b units up. The point will lie in quadrant II. Thus, $(-a, b)$ lies in quadrant III is a false statement.

53. To plot $(4, -3)$, start at the origin and move 4 units to the right, so the point $(4, -3)$ is plotted to the right of the y-axis.

55.

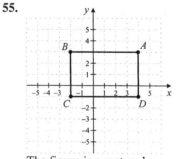

The figure is a rectangle.

332

57. $P = 2l + 2w$
Let $l = 6$ and $w = 4$.
$P = 2 \cdot 6 + 2 \cdot 4 = 12 + 8 = 20$
The perimeter is 20 units.

Integrated Review

1. Customer service representatives has 10 figures and each figure represents 50,000 workers. Thus, the increase in the number of customer service representatives is approximately $10 \cdot 50,000 = 500,000$.

2. Post-secondary teachers has 11 figures and each figure represents 50,000 workers. Thus, the increase in the number of post-secondary teachers is approximately $11 \cdot 50,000 = 550,000$.

3. Retail salespeople has the greatest number of figures, so the greatest increase is expected for retail salespeople.

4. Waitstaff has the least number of figures, so the least increase is expected for waitstaff.

5. The tallest bar corresponds to Oroville, CA. Thus, the U.S. dam with the greatest height is the Oroville Dam, which is approximately 755 feet.

6. The bar whose height is between 625 and 650 feet corresponds to New Bullards Bar, CA. Thus, the U.S. dam with a height between 625 and 650 feet is the New Bullards Bar Dam, which is approximately 635 feet.

7. From the graph, the Hoover Dam is approximately 725 feet and the Glen Canyon Dam is approximately 710 feet. The Hoover Dam is $725 - 710 = 15$ feet higher than the Glen Canyon Dam.

8. There are 4 bars that are taller than the 700-feet level, so there are 4 dams with heights over 700 feet.

9. The highest points on the graph correspond to Thursday and Saturday. Thus, the highest temperature occurs on Thursday and Saturday and is 100°F.

10. The lowest point on the graph corresponds to Monday. Thus, the lowest temperature occurs on Monday, and is 82°F.

11. The points on the graph that are lower than the 90 level correspond to Sunday, Monday, and Tuesday. Thus, the temperature was less than 90°F on Sunday, Monday, and Tuesday.

12. The points on the graph that are higher than the 90 level correspond to Wednesday, Thursday, Friday, and Saturday. Thus, the temperature was greater than 90°F on Wednesday, Thursday, Friday, and Saturday.

13. The sector corresponding to whole milk is 35%.
35% of $200 = 0.35 \cdot 200 = 70$
Thus, 70 quart containers of whole milk are sold.

14. The sector corresponding to skim milk is 26%.
26% of $200 = 0.26 \cdot 200 = 52$
Thus, 52 quart containers of skim milk are sold.

15. The sector corresponding to buttermilk is 1%.
1% of $200 = 0.01 \cdot 200 = 2$
Thus, 2 quart containers of buttermilk are sold.

16. The sector corresponding to Flavored reduced fat and skim milk is 3%.
3% of $200 = 0.03 \cdot 200 = 6$
Thus, 6 quart containers of flavored reduced fat and skim milk are sold.

	Class Intervals (Scores	Tally	Class Frequency (Number of Quizzes)
17.	50–59	||	2
18.	60–69	|	1
19.	70–79	|||	3
20.	80–89	ⵜﾞ |	6
21.	90–99	ⵜﾞ	5

22.

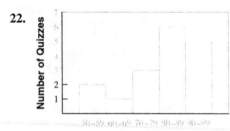

Quiz Scores

23.

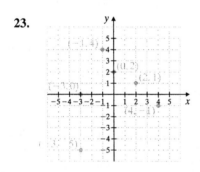

24. $x = 3y$
$1 \overset{?}{=} 3 \cdot 3$
$1 = 9$ False
No, $(1, 3)$ is not a solution of $x = 3y$.

25. $x + y = -6$
$-2 + (-4) \overset{?}{=} -6$
$-6 = -6$ True
Yes, $(-2, -4)$ is a solution of $x + y = -6$.

26. $x - y = 6$
$0 - y = 6$
$-y = 6$
$y = -6$
$(0, -6)$ is a solution.
$x - y = 6$
$x - 0 = 6$
$x = 6$
$(6, 0)$ is a solution.
$x - y = 6$
$2 - y = 6$
$2 - 2 - y = 6 - 2$
$-y = 4$
$y = -4$
$(2, -4)$ is a solution.

Section 8.4

Practice Problems

1.

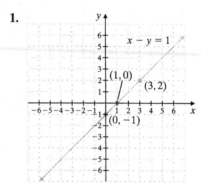

2. $y - x = 4$
Find any 3 ordered-pair solutions.
Let $x = 0$.
$y - x = 4$
$y - 0 = 4$
$y = 4$
$(0, 4)$
Let $x = 2$.
$y - x = 4$
$y - 2 = 4$
$y - 2 + 2 = 4 + 2$
$y = 6$
$(2, 6)$

Let $y = 0$.

$y - x = 4$

$0 - x = 4$

$x = -4$

$(-4, 0)$

Plot $(0, 4)$, $(2, 6)$, and $(-4, 0)$. Then draw the line through them.

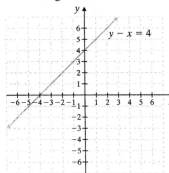

3. $y = -3x + 2$

Find any 3 ordered-pair solutions.

Let $x = 0$.

$y = -3x + 2$

$y = -3 \cdot 0 + 2$

$y = 2$

$(0, 2)$

Let $x = 1$.

$y = -3x + 2$

$y = -3(1) + 2$

$y = -3 + 2$

$y = -1$

$(1, -1)$

Let $x = -1$.

$y = -3x + 2$

$y = -3(-1) + 2$

$y = 3 + 2$

$y = 5$

$(-1, 5)$

Plot $(0, 2)$, $(1, -1)$, and $(-1, 5)$. Then draw the line through them.

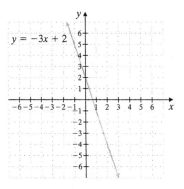

4. $y = -4$

No matter what x-value we choose, y is always -4.

x	y
-2	-4
0	-4
2	-4

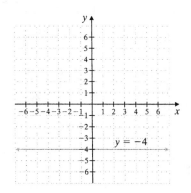

5. $x = 5$

No matter what y-value we choose, x is always 5.

x	y
5	-2
5	0
5	2

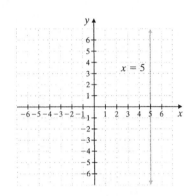

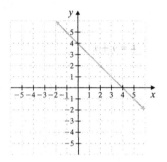

$$x = 5$$

Vocabulary and Readiness Check

1. A <u>linear</u> equation in two variables can be written in the form $ax + by = c$.

3. The graph of the equation $y = -2$ is a <u>horizontal</u> line.

Exercise Set 8.4

1. $x + y = 4$
Find any 3 ordered-pair solutions.
Let $x = 0$.
$x + y = 4$
$0 + y = 4$
$y = 4$
$(0, 4)$
Let $x = 2$.
$x + y = 4$
$2 + y = 4$
$2 - 2 + y = 4 - 2$
$y = 2$
$(2, 2)$
Let $x = 4$.
$x + y = 4$
$4 + y = 4$
$4 - 4 + y = 4 - 4$
$y = 0$
$(4, 0)$
Plot $(0, 4)$, $(2, 2)$, and $(4, 0)$. Then draw the line through them.

3. $x - y = -6$
Find any 3 ordered-pair solutions.
Let $x = 0$.
$x - y = -6$
$0 - y = -6$
$y = 6$
$(0, 6)$
Let $x = -2$.
$x - y = -6$
$-2 - y = -6$
$2 - 2 - y = 2 - 6$
$-y = -4$
$y = 4$
$(-2, 4)$
Let $y = 0$.
$x - y = -6$
$x - 0 = -6$
$x = -6$
$(-6, 0)$
Plot $(0, 6)$, $(-2, 4)$, and $(-6, 0)$. Then draw the line through them.

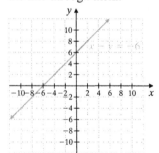

5. $y = 4x$

Find any 3 ordered-pair solutions.
Let $x = 0$.
$y = 4x$
$y = 4(0)$
$y = 0$
$(0, 0)$
Let $x = 1$.
$y = 4x$
$y = 4(1)$
$y = 4$
$(1, 4)$
Let $x = -1$.
$y = 4x$
$y = 4(-1)$
$y = -4$
$(-1, -4)$
Plot $(0, 0)$, $(1, 4)$ and $(-1, -4)$. Then draw the line through them.

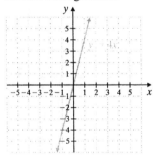

7. $y = 2x - 1$

Find any 3 ordered-pair solutions.
Let $x = 0$.
$y = 2x - 1$
$y = 2 \cdot 0 - 1$
$y = -1$
$(0, -1)$
Let $x = 1$.
$y = 2x - 1$
$y = 2 \cdot 1 - 1$
$y = 2 - 1$
$y = 1$
$(1, 1)$

Let $x = 2$.
$y = 2x - 1$
$y = 2 \cdot 2 - 1$
$y = 4 - 1$
$y = 3$
$(2, 3)$
Plot $(0, -1)$, $(1, 1)$, and $(2, 3)$. Then draw the line through them.

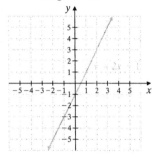

9. $x = -3$

No matter what y-value we choose, x is always -3.

x	y
-3	-4
-3	0
-3	4

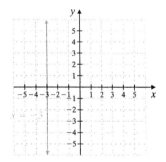

11. $y = -3$

No matter what x-value we choose, y is always -3.

x	y
-2	-3
0	-3
2	-3

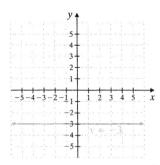

13. $x = 0$

No matter what y-value we choose, x is always 0.

x	y
0	-3
0	0
0	3

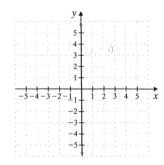

15. $y = -2x$

Find any 3 ordered-pair solutions.
Let $x = 0$.
$y = -2 \cdot 0$
$y = 0$
$(0, 0)$
Let $x = 2$.
$y = -2x$
$y = -2 \cdot 2$
$y = -4$
$(2, -4)$
Let $x = -2$.
$y = -2x$
$y = -2(-2)$
$y = 4$
$(-2, 4)$
Plot $(0, 0)$, $(2, -4)$, and $(-2, 4)$. Then draw the line through them.

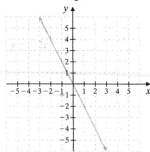

17. $y = -2$

No matter what x-value we choose, y is always -2.

x	y
-4	-2
0	-2
4	-2

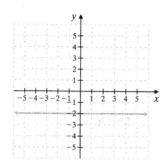

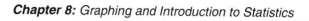
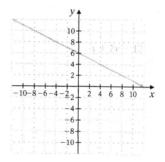

19. $x + 2y = 12$

Find any 3 ordered-pair solutions.

Let $x = 0$.

$x + 2y = 12$

$0 + 2y = 12$

$\dfrac{2y}{2} = \dfrac{12}{2}$

$y = 6$

$(0, 6)$

Let $y = 0$.

$x + 2y = 12$

$x + 2 \cdot 0 = 12$

$x + 0 = 12$

$x = 12$

$(12, 0)$

Let $x = 6$.

$x + 2y = 12$

$6 + 2y = 12$

$6 - 6 + 2y = 12 - 6$

$2y = 6$

$\dfrac{2y}{2} = \dfrac{6}{2}$

$y = 3$

$(6, 3)$

Plot $(0, 6)$, $(12, 0)$, and $(6, 3)$. Then draw the line through them.

21. $x = 6$

No matter what y-value we choose, x is always 6.

x	y
6	-5
6	0
6	5

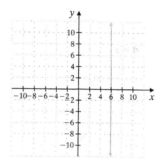

23. $y = x - 3$

Find any 3 ordered-pair solutions.

Let $x = 0$.

$y = x - 3$

$y = 0 - 3$

$y = -3$

$(0, -3)$

Let $x = -3$.

$y = x - 3$

$y = -3 - 3$

$y = -6$

$(-3, -6)$

Let $y = 0$.
$$y = x - 3$$
$$0 = x - 3$$
$$0 + 3 = x - 3 + 3$$
$$3 = x$$
$(3, 0)$

Plot $(0, -3)$, $(-3, -6)$, and $(3, 0)$. Then draw the line through them.

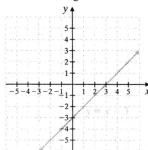

25. $x = y - 4$

Find any 3 ordered-pair solutions.
Let $y = 0$.
$$x = y - 4$$
$$x = 0 - 4$$
$$x = -4$$
$(-4, 0)$
Let $y = 4$.
$$x = y - 4$$
$$x = 4 - 4$$
$$x = 0$$
$(0, 4)$
Let $y = 6$.
$$x = y - 4$$
$$x = 6 - 4$$
$$x = 2$$
$(2, 6)$

Plot $(-4, 0)$, $(0, 4)$, and $(2, 6)$. Then draw the line through them.

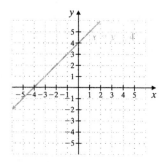

27. $x + 3 = 0$ or $x = -3$

No matter what y-value we choose, x is always -3.

x	y
-3	-2
-3	0
-3	2

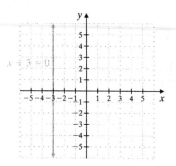

29. $y = -\dfrac{1}{4}x$

Find any 3 ordered-pair solutions.
Let $x = -4$.
$$y = -\frac{1}{4}x$$
$$y = -\frac{1}{4}(-4)$$
$$y = 1$$
$(-4, 1)$

Let $x = 0$.

$y = -\dfrac{1}{4}x$

$y = -\dfrac{1}{4} \cdot 0$

$y = 0$

$(0, 0)$

Let $x = 4$.

$y = -\dfrac{1}{4}x$

$y = -\dfrac{1}{4} \cdot 4$

$y = -1$

$(4, -1)$

Plot $(-4, 1)$, $(0, 0)$, and $(4, -1)$. Then draw the line through them.

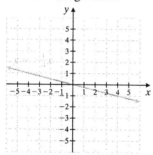

31. $y = \dfrac{1}{3}x$

Find any 3 ordered-pair solutions.

Let $x = -3$.

$y = \dfrac{1}{3}x$

$y = \dfrac{1}{3}(-3)$

$y = -1$

$(-3, -1)$

Let $x = 0$.

$y = \dfrac{1}{3}x$

$y = \dfrac{1}{3} \cdot 0$

$y = 0$

Let $x = 3$.

$y = \dfrac{1}{3}x$

$y = \dfrac{1}{3} \cdot 3$

$y = 1$

$(3, 1)$

Plot $(-3, -1)$, $(0, 0)$, and $(3, 1)$. Then draw the line through them.

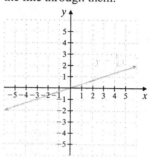

33. $y = 4x + 2$

Find any 3 ordered-pair solutions.

Let $x = -1$.

$y = 4x + 2$

$y = 4(-1) + 2$

$y = -4 + 2$

$y = -2$

$(-1, -2)$

Let $x = 0$.

$y = 4x + 2$

$y = 4(0) + 2$

$y = 0 + 2$

$y = 2$

$(0, 2)$

Let $x = 1$.

$y = 4x + 2$

$y = 4(1) + 2$

$y = 4 + 2$

$y = 6$

$(1, 6)$

Plot $(-1, -2)$, $(0, 2)$ and $(1, 6)$. Then draw the line through them.

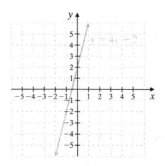

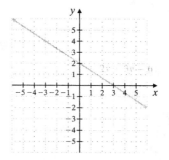

35. $2x + 3y = 6$
Find any 3 ordered-pair solutions.
Let $x = 0$.
$$2x + 3y = 6$$
$$2 \cdot 0 + 3y = 6$$
$$3y = 6$$
$$\frac{3y}{3} = \frac{6}{3}$$
$$y = 2$$
$(0, 2)$
Let $y = 0$.
$$2x + 3y = 6$$
$$2x + 3 \cdot 0 = 6$$
$$2x = 6$$
$$\frac{2x}{2} = \frac{6}{2}$$
$$x = 3$$
$(3, 0)$
Let $x = 6$.
$$2x + 3y = 6$$
$$2 \cdot 6 + 3y = 6$$
$$12 + 3y = 6$$
$$12 - 12 + 3y = 6 - 12$$
$$3y = -6$$
$$\frac{3y}{3} = \frac{-6}{3}$$
$$y = -2$$
$(6, -2)$
Plot $(0, 2)$, $(3, 0)$, and $(6, -2)$. Then draw the line through them.

37. $x = -3.5$
No matter what y-value we choose, x is always -3.5.

x	y
-3.5	-3
-3.5	0
-3.5	3

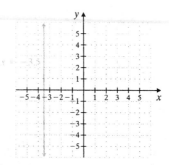

39. $3x - 4y = 24$
Find any 3 ordered-pair solutions.
Let $x = 0$.
$$3x - 4y = 24$$
$$3 \cdot 0 - 4y = 24$$
$$-4y = 24$$
$$\frac{-4y}{-4} = \frac{24}{-4}$$
$$y = -6$$
$(0, -6)$

Let $y = 0$.

$$3x - 4y = 24$$
$$3x - 4 \cdot 0 = 24$$
$$3x = 24$$
$$\frac{3x}{3} = \frac{24}{3}$$
$$x = 8$$

$(8, 0)$

Let $x = 4$.

$$3x - 4y = 24$$
$$3 \cdot 4 - 4y = 24$$
$$12 - 4y = 24$$
$$12 - 12 - 4y = 24 - 12$$
$$-4y = 12$$
$$\frac{-4y}{-4} = \frac{12}{-4}$$
$$y = -3$$

$(4, -3)$

Plot $(0, -6)$, $(8, 0)$, and $(4, -3)$. Then draw the line through them.

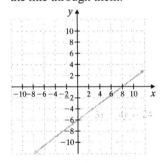

41. $(0.5)\left(-\dfrac{1}{8}\right) = \left(\dfrac{1}{2}\right)\left(-\dfrac{1}{8}\right) = -\dfrac{1 \cdot 1}{2 \cdot 8} = -\dfrac{1}{16}$

43. $\dfrac{3}{4} \div \left(-\dfrac{19}{20}\right) = \dfrac{3}{4} \cdot \left(-\dfrac{20}{19}\right)$

$$= -\frac{3 \cdot 20}{4 \cdot 19}$$
$$= -\frac{3 \cdot 4 \cdot 5}{4 \cdot 19}$$
$$= -\frac{15}{19}$$

45. $\dfrac{2}{11} - \dfrac{x}{11} = \dfrac{2 - x}{11}$

47.

x	$y = |x|$
-3	$|-3| = 3$
-2	$|-2| = 2$
-1	$|-1| = 1$
0	$|0| = 0$
1	$|1| = 1$
2	$|2| = 2$
3	$|3| = 3$

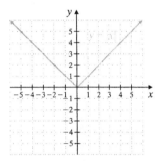

49. answers may vary

51. Since the line corresponding to DVD is increasing from left to right, U.S. spending on DVD home entertainment is increasing.

53. The DVD line is between 20 and 25 above 2005. Approximately $23.3 billion was spent on DVD home entertainment in 2005.

55. The lines meet between years 2001 and 2002, so the amount of spending on each were equal between 2001 and 2002.

57. The drawn line crosses the $40 billion line between 2009 and 2010, so the amount spent on DVD home entertainment is expected to be $40 billion between 2009 and 2010.

Section 8.5

Practice Problems

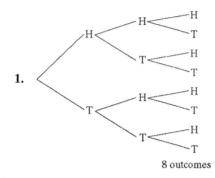

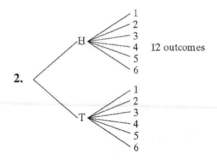

8 outcomes

12 outcomes

3. The probability of an event is a measure of the likelihood of it occurring.

5. A probability of 0 means that an event won't occur.

Exercise Set 8.5

1. 12 outcomes

3. Red, Blue, Yellow — 3 outcomes

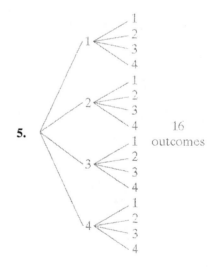

5. 16 outcomes

3. The possibilities are:
H, H, H,   H, H, T,   H, T, H,   H, T, T,
T, H, H,   T, H, T,   T, T, H,   T, T, T
T, H, T is one of the 8 possible outcomes, so
the probability is $\dfrac{1}{8}$.

4. A 2 or a 5 are two of the six possible outcomes.
The probability is $\dfrac{2}{6} = \dfrac{1}{3}$.

5. A blue is 2 out of the 4 possible marbles.
The probability is $\dfrac{2}{4} = \dfrac{1}{2}$.

Vocabulary and Readiness Check

1. A possible result of an experiment is called an outcome.

7.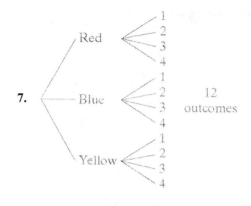

9.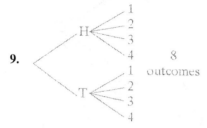

11. A 5 is one of the six possible outcomes. The probability is $\frac{1}{6}$.

13. A 1 or a 6 are two of the six possible outcomes. The probability is $\frac{2}{6} = \frac{1}{3}$.

15. Three of the six possible outcomes are even. The probability is $\frac{3}{6} = \frac{1}{2}$.

17. Four of the six possible outcomes are numbers greater than 2. The probability is $\frac{4}{6} = \frac{2}{3}$.

19. A 2 is one of three possible outcomes. The probability is $\frac{1}{3}$.

21. A 1, a 2, or a 3 are three of three possible outcomes. The probability is $\frac{3}{3} = 1$.

23. An odd number is a 1 or a 3, which are two of three possible outcomes. The probability is $\frac{2}{3}$.

25. One of the seven marbles is red. The probability is $\frac{1}{7}$.

27. Two of the seven marbles are yellow. The probability is $\frac{2}{7}$.

29. Four of the seven marbles are either green or red. The probability is $\frac{4}{7}$.

31. The blood pressure was higher for 38 of the 200 people. The probability is $\frac{38}{200} = \frac{19}{100}$.

33. The blood pressure did not change for 10 of the 200 people. The probability is $\frac{10}{200} = \frac{1}{20}$.

35. $\frac{1}{2} + \frac{1}{3} = \frac{1}{2} \cdot \frac{3}{3} + \frac{1}{3} \cdot \frac{2}{2} = \frac{3}{6} + \frac{2}{6} = \frac{3+2}{6} = \frac{5}{6}$

37. $\frac{1}{2} \cdot \frac{1}{3} = \frac{1 \cdot 1}{2 \cdot 3} = \frac{1}{6}$

39. $5 \div \frac{3}{4} = \frac{5}{1} \div \frac{3}{4} = \frac{5}{1} \cdot \frac{4}{3} = \frac{5 \cdot 4}{1 \cdot 3} = \frac{20}{3}$ or $6\frac{2}{3}$

41. One of the 52 cards is the king of hearts. The probability is $\frac{1}{52}$.

43. Four of the 52 cards are kings. The probability is $\frac{4}{52} = \frac{1}{13}$.

45. Thirteen of the 52 cards are hearts. The probability is $\frac{13}{52} = \frac{1}{4}$.

47. Twenty six of the cards are in black ink. The probability is $\frac{26}{52} = \frac{1}{2}$.

Tree diagram for 49.–51.

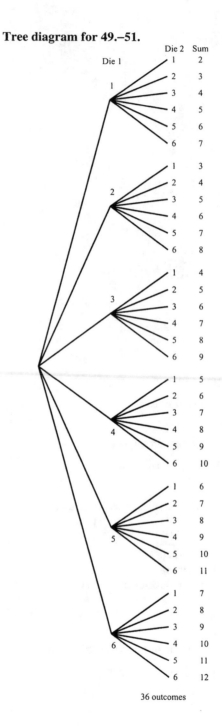

36 outcomes

49. Five of the 36 sums are 6. The probability is $\frac{5}{36}$.

51. None of the 36 sums are 13. The probability is $\frac{0}{36} = 0$.

53. answers may vary

Chapter 8 Vocabulary Check

1. A bar graph presents data using vertical or horizontal bars.

2. The possible results of an experiment are the outcomes.

3. A pictograph is a graph in which pictures or symbols are used to visually present data.

4. A line graph displays information with a line that connects data points.

5. In the ordered pair (a, b), the x-value is a and the y-value is b.

6. A tree diagram is one way to picture and count outcomes.

7. An experiment is an activity being considered, such as tossing a coin or rolling a die.

8. In a circle graph, each section (shaped like a piece of pie) shows a category and the relative size of the category.

9. The probability of an event is $\frac{\text{number of ways that event can occur}}{\text{number of possible outcomes}}$.

10. A histogram is a special bar graph in which the width of each bar represents a class interval and the height of each bar represents the class frequency.

11. The point of intersection of the x-axis and the y-axis in the rectangular coordinate system is called the origin.

12. The axis divide the plane into 4 regions, called quadrants.

13. The process of locating a point on the rectangular coordinate system is called plotting the point.

14. A linear equation in two variables can be written in the form $ax + by = c$.

Chapter 8 Review

1. Midwest has 5 houses, and each house represents 800,000 homes, so there were $5(800,000) = 4,000,000$ housing starts in the Midwest.

2. Northeast has 3 houses, and each house represents 800,000 homes, so there were $3(800,000) = 2,400,000$ housing starts in the Northeast.

3. South has the greatest number of houses, so the most housing starts were in the South.

4. Northeast has the least number of houses, so the fewest housing starts were in the Northeast.

5. Each house represents 800,000 homes, so look for the regions with $\frac{6,400,000}{800,000} = 8$ or more houses. The South and West had 6,400,000 or more housing starts.

6. Each house represents 800,000 homes, so look for the regions with fewer than $\frac{6,400,000}{800,000} = 8$ houses. The Northeast and Midwest had fewer than 6,400,000 housing starts.

7. The height of the bar representing 1960 is 7.5. Thus, approximately 7.5% of persons completed four or more years of college in 1960.

8. The tallest bar corresponds to 2003. Thus, the greatest percent of persons completing four or more years of college was in 2003.

9. The bars whose height is at a level of 15 or more are 1980, 1990, 2000, and 2003. Thus, 15% or more persons completed four or more years of college in 1980, 1990, 2000, and 2003.

10. answers may vary

11. The point on the graph corresponding to 1998 is 68. Thus, there were approximately 68 medal events at the Winter Olympics of 1998.

12. The point on the graph corresponding to 2002 is 78. Thus, there were approximately 78 medal events at the Winter Olympics of 2002.

13. The point on the graph corresponding to 2006 is 84. Thus, there were approximately 84 medal events at the Winter Olympics of 2006.

14. The points on the graph corresponding to 2006 and 2002 are 84 and 78, respectively. Thus, there were $84 - 78 = 6$ more medal events at the Winter Olympics of 2006 than the Winter Olympics of 2002.

15. The points on the graph corresponding to 2006 and 1992 are 84 and 57, respectively. Thus, there were $84 - 57 = 27$ more medal events at the Winter Olympics of 2006 than the Winter Olympics of 1992.

16. The height of the bar corresponding to 21–25 is 4. Thus, 4 employees work 21–25 hours per week.

17. The height of the bar corresponding to 41–45 is 1. Thus, 1 employee works 41–45 hours per week.

18. Add the heights of the bars corresponding to 36–40 and 41–45. Thus, $8 + 1 = 9$ employees work 36 or more hours per week.

19. Add the heights of the bars corresponding to 16–20, 21–25, and 26–30. Thus, $6 + 4 + 8 = 18$ employees work 30 hours or less per week.

	Class Interval (Temperatures)	Tally	Class Frequency (Number of Months)
20.	80°–89°	JHÍ	5
21.	90°–99°	|||	3
22.	100°–109°	||||	4

23.

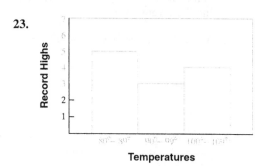

Temperatures

24. The largest sector corresponds to the category "Mortgage payment," thus the largest budget item is mortgage payment.

25. The smallest sector corresponds to the category "Utilities," thus the smallest budget item is utilities.

26. Add the amounts for mortgage payment and utilities. Thus, $975 + $250 = $1225 is budgeted for the mortgage payment and utilities.

27. Add the amounts for savings and contributions. Thus, $400 + $300 = $700 is budgeted for savings and contributions.

28. $\dfrac{\text{mortgage payment}}{\text{total}} = \dfrac{\$975}{\$4000}$

$= \dfrac{39 \cdot 25}{160 \cdot 25}$

$= \dfrac{39}{160}$

The ratio is $\dfrac{39}{160}$.

29. $\dfrac{\text{food}}{\text{total}} = \dfrac{\$700}{\$4000} = \dfrac{7 \cdot 100}{40 \cdot 100} = \dfrac{7}{40}$

The ratio is $\dfrac{7}{40}$.

30. $\dfrac{\text{car expenses}}{\text{food}} = \dfrac{\$500}{\$700} = \dfrac{5 \cdot 100}{7 \cdot 100} = \dfrac{5}{7}$

The ratio is $\dfrac{5}{7}$.

31. The sector corresponding to 70 mph is 40%.
40% of 50 = 0.40 · 50 = 20
Thus, 20 states have a rural interstate highway speed limit of 70 mph.

32. The sector corresponding to 75 mph is 22%.
22% of 50 = 0.22 · 50 = 11
Thus, 11 states have a rural interstate highway speed limit of 75 mph.

33. The sector corresponding to 65 mph is 34%.
34% of 50 = 0.34 · 50 = 17
Thus, 17 states have a rural interstate highway speed limit of 65 mph.

34. The sectors corresponding to 60 mph and 80 mph are 2% and 2%, respectively.
4% of 50 = 0.04 · 50 = 2
Thus, 2 states have a rural interstate highway speed limit of 60 mph or 80 mph.

35. $x = -6y$
$0 = -6y$
$\dfrac{0}{-6} = \dfrac{-6y}{-6}$
$0 = y$
The solution is (0, 0).
$x = -6y$
$x = -6(-1)$
$x = 6$
The solution is (6, –1).
$x = -6y$
$-6 = -6y$
$\dfrac{-6}{-6} = \dfrac{-6y}{-6}$
$1 = y$
The solution is (–6, 1).

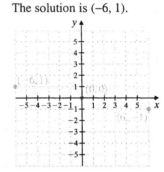

36. $y = 3x - 2$
$y = 3 \cdot 0 - 2$
$y = 0 - 2$
$y = -2$
The solution is (0, –2).
$y = 3x - 2$
$y = 3(-1) - 2$
$y = -3 - 2$
$y = -5$
The solution is (–1, –5).
$y = 3x - 2$
$y = 3 \cdot 2 - 2$
$y = 6 - 2$
$y = 4$
The solution is (2, 4).

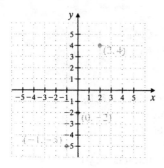

37. $x + y = -4$
$-1 + y = -4$
$1 - 1 + y = 1 - 4$
$y = -3$
The solution is $(-1, -3)$.
$x + y = -4$
$x + 0 = -4$
$x = -4$
The solution is $(-4, 0)$.
$x + y = -4$
$-5 + y = -4$
$5 - 5 + y = 5 - 4$
$y = 1$
The solution is $(-5, 1)$.

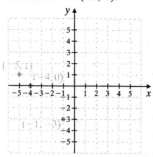

38. $x - y = 3$
$4 - y = 3$
$-4 + 4 - y = -4 + 3$
$-y = -1$
$y = 1$
The solution is $(4, 1)$.

$x - y = 3$
$0 - y = 3$
$-y = 3$
$y = -3$
The solution is $(0, -3)$.
$x - y = 3$
$x - 3 = 3$
$x - 3 + 3 = 3 + 3$
$x = 6$
The solution is $(6, 3)$.

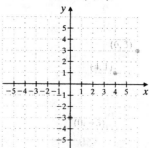

39. $y = 3x$
$y = 3 \cdot 1$
$y = 3$
The solution is $(1, 3)$.
$y = 3x$
$y = 3(-2)$
$y = -6$
The solution is $(-2, -6)$.
$y = 3x$
$0 = 3x$
$\dfrac{0}{3} = \dfrac{3x}{3}$
$0 = x$
The solution is $(0, 0)$.

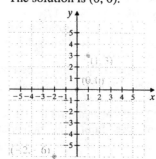

40.　$x = y + 6$
$1 = y + 6$
$1 - 6 = y + 6 - 6$
$-5 = y$
The solution is $(1, -5)$.
$x = y + 6$
$6 = y + 6$
$6 - 6 = y + 6 - 6$
$0 = y$
The solution is $(6, 0)$.
$x = y + 6$
$x = -4 + 6$
$x = 2$
The solution is $(2, -4)$.

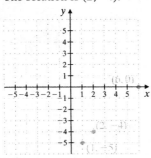

41.　$x = -5$
No matter what y-value we choose, x is always -5.

x	y
-5	-2
-5	0
-5	2

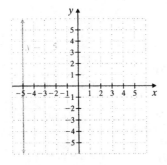

42.　$y = 0$
No matter what x-value we choose, y is always 0.

x	y
-2	0
0	0
2	0

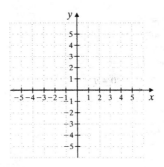

43.　$x + y = 11$
Find any 3 ordered-pair solutions.
Let $x = 0$.
$x + y = 11$
$0 + y = 11$
$y = 11$
$(0, 11)$
Let $y = 0$.
$x + y = 11$
$x + 0 = 11$
$x = 11$
$(11, 0)$
Let $x = 5$.
$x + y = 11$
$5 + y = 11$
$5 + y = 11$
$5 - 5 + y = 11 - 5$
$y = 6$
$(5, 6)$
Plot $(0, 11)$, $(11, 0)$, and $(5, 6)$. Then draw the line through them.

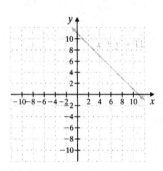

44. $x - y = 11$

Find any 3 ordered-pair solutions.

Let $x = 0$.

$x - y = 11$

$0 - y = 11$

$-y = 11$

$y = -11$

$(0, -11)$

Let $y = 0$.

$x - y = 11$

$x - 0 = 11$

$x = 11$

$(11, 0)$

Let $x = 5$.

$x - y = 11$

$5 - y = 11$

$5 - 5 - y = 11 - 5$

$-y = 6$

$y = -6$

$(5, -6)$

Plot $(0, -11)$, $(11, 0)$, and $(5, -6)$. Then draw the line through them.

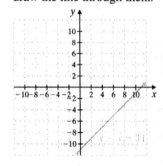

45. $y = 4x - 2$

Find any 3 ordered-pair solutions.

Let $x = 0$.

$y = 4x - 2$

$y = 4 \cdot 0 - 2$

$y = 0 - 2$

$y = -2$

$(0, -2)$

Let $y = 0$.

$y = 4x - 2$

$0 = 4x - 2$

$0 + 2 = 4x - 2 + 2$

$2 = 4x$

$\dfrac{2}{4} = \dfrac{4x}{4}$

$\dfrac{1}{2} = x$

$\left(\dfrac{1}{2}, 0\right)$

Let $x = 1$.

$y = 4x - 2$

$y = 4 \cdot 1 - 2$

$y = 4 - 2$

$y = 2$

$(1, 2)$

Plot $(0, -2)$, $\left(\dfrac{1}{2}, 0\right)$, and $(1, 2)$. Then draw the line through them.

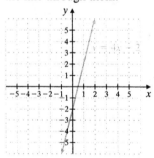

46. $y = 5x$

Find any 3 ordered-pair solutions.

Let $x = 0$.

$y = 5x$

$y = 5 \cdot 0$

$y = 0$

$(0, 0)$

Let $x = 1$.
$y = 5x$
$y = 5(1)$
$y = 5$
$(1, 5)$
Let $x = -1$.
$y = 5x$
$y = 5(-1)$
$y = -5$
$(-1, -5)$
Plot $(0, 0)$, $(1, 5)$, and $(-1, -5)$. Then draw the line through them.

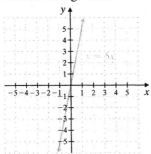

47. $x = -2y$
Find any 3 ordered-pair solutions.
Let $y = 0$.
$x = -2y$
$x = -2 \cdot 0$
$x = 0$
$(0, 0)$
Let $y = -1$.
$x = -2y$
$x = -2(-1)$
$x = 2$
$(2, -1)$
Let $y = 1$.
$x = -2y$
$x = -2(1)$
$x = -2$
$(-2, 1)$
Plot $(0, 0)$, $(2, -1)$, and $(-2, 1)$. Then draw the line through them.

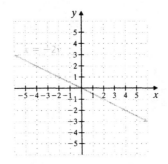

48. $x + y = -1$
Find any 3 ordered-pair solutions.
Let $x = 0$.
$x + y = -1$
$0 + y = -1$
$y = -1$
$(0, -1)$
Let $y = 0$.
$x + y = -1$
$x + 0 = -1$
$x = -1$
$(-1, 0)$
Let $x = -2$.
$x + y = -1$
$-2 + y = -1$
$-2 + 2 + y = -1 + 2$
$y = 1$
$(-2, 1)$
Plot $(0, -1)$, $(-1, 0)$, and $(-2, 1)$. Then draw the line through them.

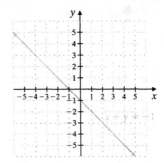

49. $2x - 3y = 12$
Find any 3 ordered-pair solutions.
Let $x = 0$.
$$2x - 3y = 12$$
$$2 \cdot 0 - 3y = 12$$
$$0 - 3y = 12$$
$$-3y = 12$$
$$\frac{-3y}{-3} = \frac{12}{-3}$$
$$y = -4$$
$(0, -4)$
Let $y = 0$.
$$2x - 3y = 12$$
$$2x - 3 \cdot 0 = 12$$
$$2x - 0 = 12$$
$$2x = 12$$
$$\frac{2x}{2} = \frac{12}{2}$$
$$x = 6$$
$(6, 0)$
Let $x = 3$.
$$2x - 3y = 12$$
$$2 \cdot 3 - 3y = 12$$
$$6 - 3y = 12$$
$$6 - 6 - 3y = 12 - 6$$
$$-3y = 6$$
$$\frac{-3y}{-3} = \frac{6}{-3}$$
$$y = -2$$
$(3, -2)$
Plot $(0, -4)$, $(6, 0)$, and $(3, -2)$. Then draw the line through them.

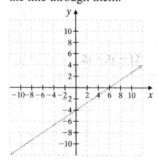

50. $x = \frac{1}{2}y$

Find any 3 ordered-pair solutions.
Let $y = 0$.
$$x = \frac{1}{2}y$$
$$x = \frac{1}{2} \cdot 0$$
$$x = 0$$
$(0, 0)$
Let $y = 2$.
$$x = \frac{1}{2}y$$
$$x = \frac{1}{2} \cdot 2$$
$$x = 1$$
$(1, 2)$
Let $y = 4$.
$$x = \frac{1}{2}y$$
$$x = \frac{1}{2} \cdot 4$$
$$x = 2$$
$(2, 4)$
Plot $(0, 0)$, $(1, 2)$, and $(2, 4)$. Then draw the line through them.

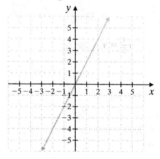

51.

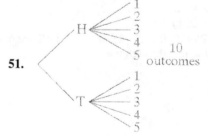

52.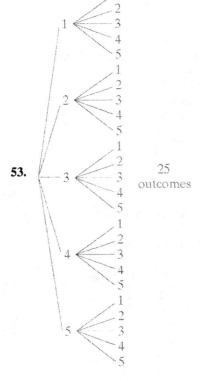

53.

54.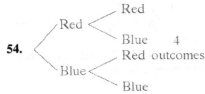

55.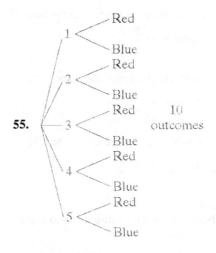

56. One of the six possible outcomes is 4. The probability is $\frac{1}{6}$.

57. One of the six possible outcomes is 3. The probability is $\frac{1}{6}$.

58. One of the five possible outcomes is 4. The probability is $\frac{1}{5}$.

59. One of the five possible outcomes is 3. The probability is $\frac{1}{5}$.

60. Three of the five possible outcomes are a 1, 3, or 5. The probability is $\frac{3}{5}$.

61. Two of the five possible outcomes are a 2 or a 4. The probability is $\frac{2}{5}$.

62. Two of the eight marbles are blue. The probability is $\frac{2}{8} = \frac{1}{4}$.

63. Three of the eight marbles are yellow. The probability is $\dfrac{3}{8}$.

64. Two of the eight marbles are red. The probability is $\dfrac{2}{8} = \dfrac{1}{4}$.

65. One of the eight marbles is green. The probability is $\dfrac{1}{8}$.

66. $x = -4$

No matter what y-value we choose, x is always -4.

x	y
-4	-2
-4	0
-4	2

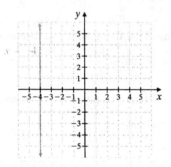

67. $y = 3$

No matter what x-value we choose, y is always 3.

x	y
-2	3
0	3
2	3

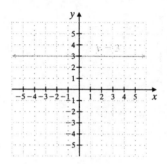

68. $x - 2y = 8$

Find any 3 ordered-pair solutions.

Let $x = 0$.

$x - 2y = 8$

$0 - 2y = 8$

$-2y = 8$

$\dfrac{-2y}{-2} = \dfrac{8}{-2}$

$y = -4$

$(0, -4)$

Let $x = 4$.

$x - 2y = 8$

$4 - 2y = 8$

$-4 + 4 - 2y = -4 + 8$

$-2y = 4$

$\dfrac{-2y}{-2} = \dfrac{4}{-2}$

$y = -2$

$(4, -2)$

Let $y = 0$.

$x - 2y = 8$

$x - 2 \cdot 0 = 8$

$x - 0 = 8$

$x = 8$

$(8, 0)$

Plot $(0, -4)$, $(4, -2)$, and $(8, 0)$. Then draw a line through them.

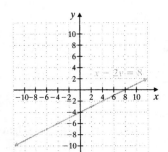

69. $2x + y = 6$
Find any 3 ordered-pair solutions.
Let $x = 0$.
$$2x + y = 6$$
$$2 \cdot 0 + y = 6$$
$$0 + y = 6$$
$$y = 6$$
$(0, 6)$
Let $y = 0$.
$$2x + y = 6$$
$$2x + 0 = 6$$
$$2x = 6$$
$$\frac{2x}{2} = \frac{6}{2}$$
$$x = 3$$
$(3, 0)$
Let $x = -1$.
$$2x + y = 6$$
$$2(-1) + y = 6$$
$$-2 + y = 6$$
$$2 - 2 + y = 2 + 6$$
$$y = 8$$
$(-1, 8)$
Plot $(0, 6)$, $(3, 0)$, and $(-1, 8)$. Then draw the line through them.

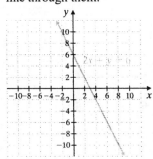

70. $x = y + 3$
Find any 3 ordered-pair solutions.
Let $x = 0$.
$$x = y + 3$$
$$0 = y + 3$$
$$0 - 3 = y + 3 - 3$$
$$-3 = y$$
$(0, -3)$
Let $y = 0$.
$$x = y + 3$$
$$x = 0 + 3$$
$$x = 3$$
$(3, 0)$
Let $y = -1$.
$$x = y + 3$$
$$x = -1 + 3$$
$$x = 2$$
$(2, -1)$
Plot $(0, -3)$, $(3, 0)$, and $(2, -1)$. Then draw the line through them.

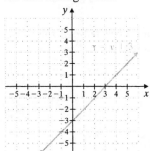

71. $x + y = -4$
Find any 3 ordered-pair solutions.
Let $x = 0$.
$$x + y = -4$$
$$0 + y = -4$$
$$y = -4$$
$(0, -4)$
Let $y = 0$.
$$x + y = -4$$
$$x + 0 = -4$$
$$x = -4$$
$(-4, 0)$

Let $x = -2$.
$$x + y = -4$$
$$-2 + y = -4$$
$$2 - 2 + y = 2 - 4$$
$$y = -2$$
$(-2, -2)$
Plot $(0, -4)$, $(-4, 0)$, and $(-2, -2)$. Then draw the line through them.

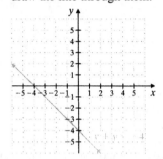

72. $y = \dfrac{3}{4}x$

Find any 3 ordered-pair solutions.
Let $x = 0$.
$$y = \dfrac{3}{4}x$$
$$y = \dfrac{3}{4} \cdot 0$$
$$y = 0$$
$(0, 0)$
Let $x = 4$.
$$y = \dfrac{3}{4}x$$
$$y = \dfrac{3}{4} \cdot 4$$
$$y = 3$$
$(4, 3)$

Let $x = -4$.
$$y = \dfrac{3}{4}x$$
$$y = \dfrac{3}{4}(-4)$$
$$y = -3$$
$(-4, -3)$
Plot $(0, 0)$, $(4, 3)$, and $(-4, -3)$. Then draw the line through them.

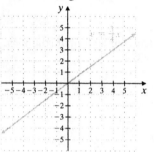

73. $y = -\dfrac{3}{4}x$

Find any 3 ordered-pair solutions.
Let $x = 0$.
$$y = -\dfrac{3}{4}x$$
$$y = -\dfrac{3}{4} \cdot 0$$
$$y = 0$$
$(0, 0)$
Let $x = 4$.
$$y = -\dfrac{3}{4}x$$
$$y = -\dfrac{3}{4} \cdot 4$$
$$y = -3$$
$(4, -3)$
Let $x = -4$.
$$y = -\dfrac{3}{4}x$$
$$y = -\dfrac{3}{4} \cdot (-4)$$
$$y = 3$$
$(-4, 3)$

Plot (0, 0), (4, −3), and (−4, 3). Then draw the line through them.

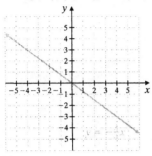

Chapter 8 Test

1. There are $4\frac{1}{2}$ dollar symbols for the second week. Each dollar symbol corresponds to $50.

   $$4\frac{1}{2} \cdot \$50 = \frac{9}{2} \cdot \$50 = \frac{\$450}{2} = \$225$$

   $225 was collected during the second week.

2. Week 3 has the greatest number of dollar symbols. So the most money was collected during the 3rd week. The 3rd week has 7 dollar symbols and each dollar symbol corresponds to $50, so 4 · $50 = $350 was collected during week 3.

3. There are a total of 22 dollar symbols and each dollar symbol corresponds to $50, so a total of 22 · $50 = $1100 was collected.

4. Look for the bars whose height is greater than 9. June, August, and September normally have more than 9 centimeters.

5. The shortest bar corresponds to February. The normal monthly rainfall in February in Chicago is 3 centimeters.

6. The bars corresponding to March and November have a height of 7. Thus, during March and November, 7 centimeters of precipitation normally occurs.

7.

Countries with the Highest Newspaper Circulations

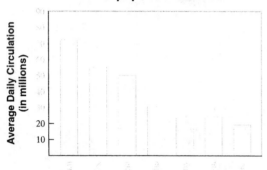

8. The point on the graph corresponding to 2002 is at about 1.5. Thus, the annual inflation rate in 2002 was about 1.5%.

9. The line graph is above the 3 level for 1990, 1991, and 2000. Thus the inflation rate was greater than 3% in 1990, 1991, and 2000.

10. Look for the years where the line graph is increasing. During 1994–1995, 1995–1996, 1998–1999, 1999–2000, and 2002–2003, the inflation rate was increasing.

11. $\dfrac{\text{number who prefer rock music}}{\text{total number}} = \dfrac{85}{200} = \dfrac{17}{40}$

The ratio is $\dfrac{17}{40}$.

12. $\dfrac{\text{number who prefer country}}{\text{number who prefer jazz}} = \dfrac{62}{44} = \dfrac{31}{22}$

The ratio is $\dfrac{31}{22}$.

13. Services employed 31% of the labor force.
$$31\% \text{ of } 132,000,000 = 0.31 \times 132,000,000$$
$$= 40,920,000$$
40,920,000 people were employed in the service industry.

14. Government employed 16% of the labor force.
$$16\% \text{ of } 132,000,000 = 0.16 \times 132,000,000$$
$$= 21,120,000$$
21,120,000 people were employed by government.

15. The height of the bar for 5'8"– 5'11" is 9. Thus, there are 9 students who are 5'8"–5'11" tall.

16. Add the heights of the bars for 5'0"–5'3"
and 5'4"–5'7". There are $5 + 6 = 11$
students who are 5'7" tall or shorter.

17.

Class Interval (Scores)	Tally	Class Frequency (Number of Students)
40–49	|	1
50–59	|||	3
60–69	||||	4
70–79	ⅢⅠ	5
80–89	ⅢⅠ|||	8
90–99	||||	4

18.

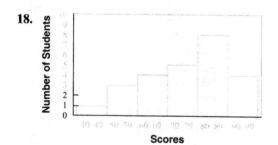

19. Point *A* has coordinates (4, 0).

20. Point *B* has coordinates (0, –3).

21. Point *C* has coordinates (–3, 4).

22. Point *D* has coordinates (–2, –1).

23. $x = -6y$
$0 = -6y$
$\dfrac{0}{-6} = \dfrac{-6y}{-6}$
$0 = y$
$(0, 0)$

$x = -6y$
$x = -6(1)$
$x = -6$
$(-6, 1)$
$x = -6y$
$12 = -6y$
$\dfrac{12}{-6} = \dfrac{-6y}{-6}$
$-2 = y$
$(12, -2)$

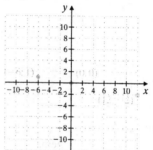

24. $y = 7x - 4$
$y = 7 \cdot 2 - 4$
$y = 14 - 4$
$y = 10$
$(2, 10)$
$y = 7x - 4$
$y = 7(-1) - 4$
$y = -7 - 4$
$y = -11$
$(-1, -11)$
$y = 7x - 4$
$y = 7 \cdot 0 - 4$
$y = 0 - 4$
$y = -4$
$(0, -4)$

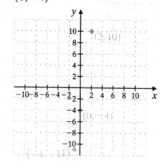

25. $y + x = -4$
Find any 3 ordered-pair solutions.
Let $x = 0$.
$y + x = -4$
$y + 0 = -4$
$y = -4$
$(0, -4)$
Let $y = 0$.
$y + x = -4$
$0 + x = -4$
$x = -4$
$(-4, 0)$
Let $x = -2$.
$y + x = -4$
$y + (-2) = -4$
$y + (-2) + 2 = -4 + 2$
$y = -2$
$(-2, -2)$
Plot $(0, -4)$, $(-4, 0)$, and $(-2, -2)$. Then draw the line through them.

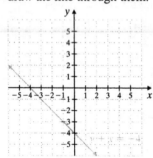

26. $y = -4$
No matter what x-value we choose, y is always -4.

x	y
-2	-4
0	-4
2	-4

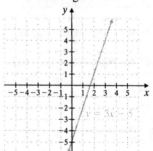

27. $y = 3x - 5$
Find any 3 ordered-pair solutions.
Let $x = 0$.
$y = 3x - 5$
$y = 3 \cdot 0 - 5$
$y = 0 - 5$
$y = -5$
$(0, -5)$
Let $x = 1$.
$y = 3x - 5$
$y = 3 \cdot 1 - 5$
$y = 3 - 5$
$y = -2$
$(-1, -2)$
Let $x = 2$.
$y = 3x - 5$
$y = 3 \cdot 2 - 5$
$y = 6 - 5$
$y = 1$
$(2, 1)$
Plot $(0, -5)$, $(1, -2)$, and $(2, 1)$. Then draw the line through them.

28. $x = 5$

No matter what y-value we choose, x is always 5.

x	y
5	-2
5	0
5	2

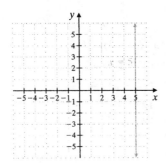

29. $y = -\dfrac{1}{2}x$

Find any 3 ordered-pair solutions.
Let $x = 0$.

$$y = -\frac{1}{2}x$$

$$y = -\frac{1}{2} \cdot 0$$

$$y = 0$$

$(0, 0)$
Let $x = -2$.

$$y = -\frac{1}{2}x$$

$$y = -\frac{1}{2} \cdot (-2)$$

$$y = 1$$

$(-2, 1)$
Let $x = 2$.

$$y = -\frac{1}{2}x$$

$$y = -\frac{1}{2} \cdot 2$$

$$y = -1$$

$(2, -1)$

Plot $(0, 0)$, $(-2, 1)$, and $(2, -1)$. Then draw the line through them.

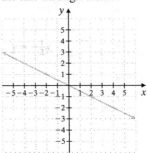

30. $3x - 2y = 12$

Find any 3 ordered-pair solutions.
Let $x = 0$.

$$3x - 2y = 12$$

$$3 \cdot 0 - 2y = 12$$

$$-2y = 12$$

$$\frac{-2y}{-2} = \frac{12}{-2}$$

$$y = -6$$

$(0, -6)$
Let $y = 0$.

$$3x - 2y = 12$$

$$3x - 2 \cdot 0 = 12$$

$$3x = 12$$

$$\frac{3x}{3} = \frac{12}{3}$$

$$x = 4$$

$(4, 0)$
Let $x = 2$.

$$3x - 2y = 12$$

$$3 \cdot 2 - 2y = 12$$

$$6 - 2y = 12$$

$$6 - 6 - 2y = 12 - 6$$

$$-2y = 6$$

$$\frac{-2y}{-2} = \frac{6}{-2}$$

$$y = -3$$

$(2, -3)$
Plot $(0, -6)$, $(4, 0)$, and $(2, -3)$. Then draw the line through them.

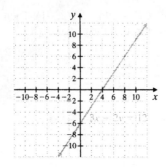

31.

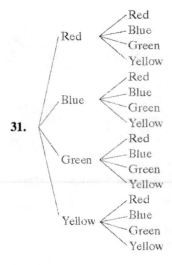

32.

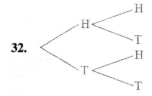

33. One of the ten possible outcomes is a 6. The probability is $\dfrac{1}{10}$.

34. Two of the ten possible outcomes are a 3 or a 4. The probability is $\dfrac{2}{10} = \dfrac{1}{5}$.

Cumulative Review Chapters 1–8

1. $\quad 4^3 + [3^2 - (10 \div 2)] - 7 \cdot 3$

$\quad = 4^3 + [3^2 - 5] - 7 \cdot 3$

$\quad = 4^3 + (9 - 5) - 7 \cdot 3$

$\quad = 4^3 + 4 - 7 \cdot 3$

$\quad = 64 + 4 - 7 \cdot 3$

$\quad = 64 + 4 - 21$

$\quad = 68 - 21$

$\quad = 47$

2. $\quad 7^2 - [5^3 + (6 \div 3)] + 4 \cdot 2$

$\quad = 7^2 - [5^3 + 2] + 4 \cdot 2$

$\quad = 7^2 - (125 + 2) + 4 \cdot 2$

$\quad = 7^2 - 127 + 4 \cdot 2$

$\quad = 49 - 127 + 4 \cdot 2$

$\quad = 49 - 127 + 8$

$\quad = -78 + 8$

$\quad = -70$

3. $x - y = -3 - 9 = -3 + (-9) = -12$

4. $x - y = 7 - (-2) = 7 + 2 = 9$

5. $\quad 3y - 7y = 12$

$\quad (3 - 7)y = 12$

$\quad -4y = 12$

$\quad \dfrac{-4y}{-4} = \dfrac{12}{-4}$

$\quad y = -3$

6. $\quad 2x - 6x = 24$

$\quad (2 - 6)x = 24$

$\quad -4x = 24$

$\quad \dfrac{-4x}{-4} = \dfrac{24}{-4}$

$\quad x = -6$

7. $\dfrac{x}{6} + 1 = \dfrac{4}{3}$

$$6\left(\dfrac{x}{6} + 1\right) = 6\left(\dfrac{4}{3}\right)$$

$$6 \cdot \dfrac{x}{6} + 6 \cdot 1 = 6 \cdot \dfrac{4}{3}$$

$$x + 6 = 8$$

$$x + 6 - 6 = 8 - 6$$

$$x = 2$$

8. $\dfrac{7}{2} + \dfrac{a}{4} = 1$

$$4\left(\dfrac{7}{2} + \dfrac{a}{4}\right) = 4(1)$$

$$4 \cdot \dfrac{7}{2} + 4 \cdot \dfrac{a}{4} = 4 \cdot 1$$

$$14 + a = 4$$

$$14 - 14 + a = 4 - 14$$

$$a = -10$$

9.
$$\begin{array}{r} 2\frac{1}{3} \\ + 5\frac{3}{8} \\ \hline \end{array} \qquad \begin{array}{r} 2\frac{8}{24} \\ + 5\frac{9}{24} \\ \hline 7\frac{17}{24} \end{array}$$

$$2\frac{1}{3} + 5\frac{3}{8} = 7\frac{17}{24}$$

10.
$$\begin{array}{r} 3\frac{2}{5} \\ + 4\frac{3}{4} \\ \hline \end{array} \qquad \begin{array}{r} 3\frac{8}{20} \\ + 4\frac{15}{20} \\ \hline 7\frac{23}{20} = 7 + 1\frac{3}{20} = 8\frac{3}{20} \end{array}$$

$$3\frac{2}{5} + 4\frac{3}{4} = 8\frac{3}{20}$$

11. $5.9 = 5\dfrac{9}{10}$

12. $2.8 = 2\dfrac{8}{10} = 2\dfrac{4}{5}$

13.
$$\begin{array}{r} 3.500 \\ - 0.068 \\ \hline 3.432 \end{array}$$

14.
$$\begin{array}{r} 7.400 \\ - 0.073 \\ \hline 7.327 \end{array}$$

15.
0.0531	4 decimal places
16	

$$\begin{array}{r} 3186 \\ + 5310 \\ \hline 0.8496 \end{array} \quad \text{4 decimal places}$$

16.
0.147	3 decimal places
0.2	1 decimal place
0.0294	$3 + 1 = 4$ decimal places

17.
$$\begin{array}{r} 0.052 \\ 115\overline{)\,5.980} \\ -5\ 75 \\ \hline 230 \\ -230 \\ \hline 0 \end{array}$$

$$-5.98 \div 115 = -0.052$$

18.
$$\begin{array}{r} 0.136 \\ 205\overline{)\,27.880} \\ -20\ 5 \\ \hline 7\ 38 \\ -6\ 15 \\ \hline 1\ 230 \\ -1\ 230 \\ \hline 0 \end{array}$$

$$27.88 \div 205 = 0.136$$

19. $(-1.3)^2 + 2.4 = 1.69 + 2.4 = 4.09$

20. $(-2.7)^2 = (-2.7)(-2.7) = 7.29$

21. $\dfrac{1}{4} = \dfrac{1 \cdot 25}{4 \cdot 25} = \dfrac{25}{100} = 0.25$

22. $\dfrac{3}{8} = \dfrac{3 \cdot 125}{8 \cdot 125} = \dfrac{375}{1000} = 0.375$

23.
$$5(x - 0.36) = -x + 2.4$$
$$5 \cdot x - 5 \cdot 0.36 = -x + 2.4$$
$$5x - 1.8 = -x + 2.4$$
$$5x + x - 1.8 = -x + x + 2.4$$
$$6x - 1.8 = 2.4$$
$$6x - 1.8 + 1.8 = 2.4 + 1.8$$
$$6x = 4.2$$
$$\dfrac{6x}{6} = \dfrac{4.2}{6}$$
$$x = 0.7$$

24.
$$4(0.35 - x) = x - 7$$
$$4 \cdot 0.35 - 4 \cdot x = x - 7$$
$$1.4 - 4x = x - 7$$
$$1.4 - 4x - x = x - x - 7$$
$$1.4 - 5x = -7$$
$$1.4 - 1.4 - 5x = -7 - 1.4$$
$$-5x = -8.4$$
$$\dfrac{-5x}{-5} = \dfrac{-8.4}{-5}$$
$$x = 1.68$$

25. $\sqrt{80} \approx 8.944$

26. $\sqrt{60} \approx 7.746$

27. The ratio of 21 to 29 is $\dfrac{21}{29}$.

28. The ratio of 7 to 15 is $\dfrac{7}{15}$.

29.
$$\begin{array}{r} 22.5 \\ 15\overline{)\,337.5} \\ \underline{-30} \\ 37 \\ \underline{-30} \\ 7\,5 \\ \underline{-7\,5} \\ 0 \end{array}$$

$$\dfrac{337.5 \text{ miles}}{15 \text{ gallons}} = \dfrac{22.5 \text{ miles}}{1 \text{ gallon}}$$

The unit rate is 22.5 miles/gallon.

30.
$$\begin{array}{r} 0.53 \\ 3\overline{)\,1.59} \\ \underline{-1\,5} \\ 09 \\ \underline{-9} \\ 0 \end{array}$$

$$\dfrac{\$1.59}{3 \text{ ounces}} = \dfrac{\$0.53}{1 \text{ ounce}}$$

The unit rate is \$0.53 per ounce.

31.
$$\dfrac{51}{34} = \dfrac{-3}{x}$$
$$51 \cdot x = 34 \cdot (-3)$$
$$51x = -102$$
$$\dfrac{51x}{51} = \dfrac{-102}{51}$$
$$x = -2$$

Check:
$$\dfrac{51}{34} = \dfrac{-3}{x}$$
$$\dfrac{51}{34} \overset{?}{=} \dfrac{-3}{-2}$$
$$51(-2) \overset{?}{=} 34(-3)$$
$$-102 = -102 \quad \text{True}$$

The solution is -2.

32.
$$\frac{8}{5} = \frac{x}{10}$$
$$8 \cdot 10 = 5 \cdot x$$
$$80 = 5x$$
$$\frac{80}{5} = \frac{5x}{5}$$
$$16 = x$$

Check:
$$\frac{8}{5} = \frac{x}{10}$$
$$\frac{8}{5} \stackrel{?}{=} \frac{16}{10}$$
$$8 \cdot 10 \stackrel{?}{=} 5 \cdot 16$$
$$80 = 80 \quad \text{True}$$

The solution is 16.

33. $\dfrac{12 \text{ feet}}{19 \text{ feet}} = \dfrac{12}{19}$

34. $\dfrac{4}{9}$

35. $4.6\% = 4.6(0.01) = 0.046$

36. $32\% = 32(0.01) = 0.32$

37. $0.74\% = 0.74(0.01) = 0.0074$

38. $2.7\% = 2.7(0.01) = 0.027$

39. 35% of $60 = 0.35 \cdot 60 = 21$
21 is 35% of 60.

40. 40% of $36 = 0.40 \cdot 36 = 14.4$
14.4 is 40% of 36.

41.
$$20.8 = 40\% \cdot x$$
$$20.8 = 0.4x$$
$$\frac{20.8}{0.4} = \frac{0.4x}{0.4}$$
$$52 = x$$
20.8 is 40% of 52.

42.
$$9.5 = 25\% \cdot x$$
$$9.5 = 0.25x$$
$$\frac{9.5}{0.25} = \frac{0.25x}{0.25}$$
$$38 = x$$
9.5 is 25% of 38.

43.
$$\$406 \cdot r = \$34.51$$
$$406r = 34.51$$
$$\frac{406r}{406} = \frac{34.51}{406}$$
$$r = 0.085 \text{ or } 8.5\%$$
The sales tax rate is 8.5%.

44.
$$\$2 \cdot r = \$0.13$$
$$2r = 0.13$$
$$\frac{2r}{2} = \frac{0.13}{2}$$
$$r = 0.065 \text{ or } 6.5\%$$
The sales tax rate is 6.5%.

45.
$$A = P\left(1 + \frac{r}{n}\right)^{n \cdot t}$$
$$= \$4000 \cdot \left(1 + \frac{0.053}{4}\right)^{4 \cdot 10}$$
$$= \$4000(1.01325)^{40}$$
$$\approx \$6772.12$$

46.
$$\frac{\text{total}}{12 \text{ months}} = \frac{\$1600 + \$128.60}{12}$$
$$= \frac{\$1728.60}{12}$$
$$= \$144.05$$
The monthly payment is \$144.05.

47. List the numbers in order from least to greatest.
55, 67, 75, 86, 91, 91
Since there are an even number of values,
the median is $\dfrac{75 + 86}{2} = \dfrac{161}{2} = 80.5$.

48. The numbers are listed in order. Since there are an even number of values, the median is
$$\frac{47+50}{2} = \frac{97}{2} = 48.5.$$

49. Two of the six possible outcomes are a 3 or a 4. The probability is $\frac{2}{6} = \frac{1}{3}$.

50. Three of the six possible outcomes are even. The probability is $\frac{3}{6} = \frac{1}{2}$.

Chapter 9

Practice Problems

1. Figure (a) is part of a line with one endpoint. It is ray AB or $\overrightarrow{AB}$.
   Figure (b) has two endpoints. It is line segment RS or $\overline{RS}$.
   Figure (c) extends indefinitely in two directions. It is line EF or $\overleftrightarrow{EF}$.
   Figure (d) has two rays with a common endpoint. It is $\angle HVT$ or $\angle TVH$ or $\angle V$.

2. Two other ways to name $\angle z$ are $\angle RTS$ and $\angle STR$.

3. a. $\angle R$ is an obtuse angle. It measures between 90° and 180°.

   b. $\angle N$ is a straight angle.

   c. $\angle M$ is an acute angle. It measures between 0° and 90°.

   d. $\angle Q$ is a right angle.

4. The complement of a 29° angle is an angle that measures $90° - 29° = 61°$.

5. The supplement of a 67° angle is an angle that measures $180° - 67° = 113°$.

6. a. $m\angle y = m\angle ADC - m\angle BDC$
   $= 141° - 97°$
   $= 44°$

   b. $m\angle x = 79° - 51° = 28°$

   c. Since the measures of both $\angle x$ and $\angle y$ are between 0° and 90°, they are acute angles.

7. Since $\angle a$ and the angle marked 109° are vertical angles, they have the same measure; so $m\angle a = 109°$.
   Since $\angle a$ and $\angle b$ are adjacent angles, their measures have a sum of 180°. So $m\angle b = 180° - 109° = 71°$.
   Since $\angle b$ and $\angle c$ are vertical angles, they have the same measure; so $m\angle c = 71°$.

8. $\angle w$ and $\angle x$ are vertical angles. $\angle w$ and $\angle y$ are corresponding angles, as are $\angle x$ and $\angle d$. So all of these angles have the same measure:
   $m\angle x = m\angle y = m\angle d = m\angle w = 45°$.
   $\angle w$ and $\angle a$ are adjacent angles, as are $\angle w$ and $\angle b$, so
   $m\angle a = m\angle b = 180° - 45° = 135°$.
   $\angle a$ and $\angle c$ are corresponding angles so $m\angle c = m\angle a = 135°$. $\angle c$ and $\angle z$ are vertical angles, so $m\angle z = m\angle c = 135°$.

Vocabulary and Readiness Check

1. A <u>plane</u> is a flat surface that extends indefinitely.

3. <u>Space</u> extends in all directions indefinitely.

5. A <u>ray</u> is part of a line with one endpoint.

7. A <u>straight</u> angle measures 180°.

9. An <u>acute</u> angle measures between 0° and 90°.

11. <u>Parallel</u> lines never meet and <u>intersecting</u> lines meet at a point.

13. An angle can be measured in <u>degrees</u>.

15. When two lines intersect, four angles are formed, called <u>vertical</u> angles.

Exercise Set 9.1

1. The figure extends indefinitely in two directions. It is line CD, line l, or $\overleftrightarrow{CD}$

3. The figure has two endpoints. It is line segment MN or $\overline{MN}$.

5. The figure has two rays with a common endpoint. It is an angle, which can be named $\angle GHI$, $\angle IHG$, or $\angle H$.

7. The figure has one endpoint and extends indefinitely in one direction. It is ray UW or $\overrightarrow{UW}$.

9. Two other ways to name $\angle x$ are $\angle CPR$ and $\angle RPC$.

11. Two other ways to name $\angle z$ are $\angle TPM$ and $\angle MPT$.

13. $\angle S$ is a straight angle.

15. $\angle R$ is a right angle.

17. $\angle Q$ is an obtuse angle.

19. $\angle P$ is an acute angle.

21. The complement of an angle that measures $23°$ is an angle that measures $90° - 23° = 67°$.

23. The supplement of an angle that measures $17°$ is an angle that measures $180° - 17° = 163°$.

25. The complement of an angle that measures $58°$ is an angle that measures $90° - 58° = 32°$.

27. The supplement of an angle that measures $150°$ is an angle that measures $180° - 150° = 30°$.

29. $52° + 38° = 90°$, so $\angle PNQ$ and $\angle QNR$ are complementary. $60° + 30° = 90°$, so $\angle MNP$ and $\angle RNO$ are complementary.

31. $45° + 135° = 180°$, so there are 4 pairs of supplementary angles: $\angle SPT$ and $\angle RPS$, $\angle SPT$ and $\angle QPT$, $\angle QPR$ and $\angle RPS$, $\angle QPR$ and $\angle QPT$.

33. $m\angle x = 74° - 47° = 27°$

35. $m\angle x = 42° + 90° = 132°$

37. $\angle x$ and the angle marked $150°$ are supplementary, so $m\angle x = 180° - 150° = 30°$. $\angle y$ and the angle marked $150°$ are vertical angles, so $m\angle y = 150°$. $\angle z$ and $\angle x$ are vertical angles so $m\angle z = m\angle x = 30°$.

39. $\angle x$ and the angle marked $103°$ are supplementary, so $m\angle x = 180° - 103° = 77°$. $\angle y$ and the angle marked $103°$ are vertical angles, so $m\angle y = 103°$. $\angle x$ and $\angle z$ are vertical angles, so $m\angle z = m\angle x = 77°$.

41. $\angle x$ and the angle marked $80°$ are supplementary, so $m\angle x = 180° - 80° = 100°$. $\angle y$ and the angle marked $80°$ are alternate interior angles, so $m\angle y = 80°$. $\angle x$ and $\angle z$ are corresponding angles, so $m\angle z = m\angle x = 100°$.

43. $\angle x$ and the angle marked $46°$ are supplementary, so $m\angle x = 180° - 46° = 134°$. $\angle y$ and the angle marked $46°$ are corresponding angles, so $m\angle y = 46°$. $\angle x$ and $\angle z$ are corresponding angles, so $m\angle z = m\angle x = 134°$.

45. $\angle x$ can also be named $\angle ABC$ or $\angle CBA$.

47. $\angle z$ can also be named $\angle DBE$ or $\angle EBD$.

49. $m\angle ABC = 15°$

51. $m\angle CBD = 50°$

53. $m\angle DBA = m\angle DBC + m\angle CBA$
$$= 50° + 15°$$
$$= 65°$$

55. $m\angle CBE = m\angle CBD + m\angle DBE$
$$= 50° + 45°$$
$$= 95°$$

57. $\dfrac{7}{8} + \dfrac{1}{4} = \dfrac{7}{8} + \dfrac{2}{8} = \dfrac{9}{8}$ or $1\dfrac{1}{8}$

59. $\dfrac{7}{8} \cdot \dfrac{1}{4} = \dfrac{7}{32}$

61. $3\dfrac{1}{3} - 2\dfrac{1}{2} = \dfrac{10}{3} - \dfrac{5}{2}$
$$= \dfrac{20}{6} - \dfrac{15}{6}$$
$$= \dfrac{5}{6}$$

63. $3\dfrac{1}{3} \div 2\dfrac{1}{2} = \dfrac{10}{3} \div \dfrac{5}{2} = \dfrac{10}{3} \cdot \dfrac{2}{5} = \dfrac{20}{15} = \dfrac{4}{3}$ or $1\dfrac{1}{3}$

65. The supplement of an angle that measures 125.2° is an angle with measure $180° - 125.2° = 54.8°$.

67. False, since $100° + 80° = 180° \neq 90°$.

69. True; since 120° is less than 180°, it is possible to find the supplement of a 120° angle.

71. $\angle a$ and the angle marked 60° are alternate interior angles, so $m\angle a = 60°$. The sum of $\angle a$, $\angle b$, and the angle marked 70° is a straight angle, so $m\angle b = 180° - 60° - 70° = 50°$. $\angle d$ and the angle marked 70° are alternate interior angles, so $m\angle d = 70°$. $\angle c$ and $\angle d$ are supplementary, so $m\angle c = 180° - 70° = 110°$. $\angle e$ and the angle marked 60° are supplementary, so $m\angle e = 180° - 60° = 120°$.

73. no; answers may vary

75. no; answers may vary

Section 9.2

Practice Problems

1. a. Perimeter $= 10$ m $+ 10$ m $+ 18$ m $+ 18$ m
$$= 56 \text{ meters}$$

   b. Perimeter
$$= 125 \text{ ft} + 125 \text{ ft} + 50 \text{ ft} + 50 \text{ ft}$$
$$= 350 \text{ ft}$$

2. $P = 2 \cdot l + 2 \cdot w$
$$= 2 \cdot 32 \text{ cm} + 2 \cdot 15 \text{ cm}$$
$$= 64 \text{ cm} + 30 \text{ cm}$$
$$= 94 \text{ centimeters}$$

3. $P = 4 \cdot s = 4 \cdot 4 \text{ ft} = 16 \text{ feet}$

4. $P = a + b + c$
$$= 6 \text{ cm} + 10 \text{ cm} + 8 \text{ cm}$$
$$= 24 \text{ centimeters}$$

5. Perimeter $= 6$ km $+ 4$ km $+ 9$ km $+ 4$ km
$$= 23 \text{ kilometers}$$

6. The unmarked horizontal side has length 20 m $- 15$ m $= 5$ m. The unmarked vertical side has length 31 m $- 6$ m $= 25$ m.
$$P = 15 \text{ m} + 31 \text{ m} + 20 \text{ m} + 6 \text{ m} + 5 \text{ m} + 25 \text{ m}$$
$$= 102 \text{ meters}$$

7. $P = 2 \cdot l + 2 \cdot w$
   $= 2 \cdot 120 \text{ feet} + 2 \cdot 60 \text{ feet}$
   $= 240 \text{ feet} + 120 \text{ feet}$
   $= 360 \text{ feet}$
   cost $= \$1.90$ per foot $\cdot 360$ feet $= \$684$
   The cost of the fencing is $84.

8. $C = \pi \cdot d = \pi \cdot 20 \text{ yd} = 20\pi \text{ yd} \approx 62.8 \text{ yd}$
   The exact circumference of the watered
   region is 20π yards, which is approximately
   62.8 yards.

Vocabulary and Readiness Check

1. The <u>perimeter</u> of a polygon is the sum of the
   lengths of its sides.

3. The exact ratio of circumference to diameter
   is <u>π</u>.

5. Both $\dfrac{22}{7}$ (or 3.14) and $3.14 \left(\text{or } \dfrac{22}{7} \right)$ are

   approximations for π.

Exercise Set 9.2

1. $P = 2 \cdot l + 2 \cdot w$
   $= 2 \cdot 17 \text{ ft} + 2 \cdot 15 \text{ ft}$
   $= 34 \text{ ft} + 30 \text{ ft}$
   $= 64 \text{ ft}$
   The perimeter is 64 feet.

3. $P = 35 \text{ cm} + 25 \text{ cm} + 35 \text{ cm} + 25 \text{ cm}$
   $= 120 \text{ cm}$
   The perimeter is 120 centimeters.

5. $P = a + b + c$
   $= 5 \text{ in.} + 7 \text{ in.} + 9 \text{ in.}$
   $= 21 \text{ in.}$
   The perimeter is 21 inches.

7. Sum the lengths of the sides.
   $P = 10 \text{ ft} + 8 \text{ ft} + 8 \text{ ft} + 15 \text{ ft} + 7 \text{ ft}$
   $= 48 \text{ ft}$
   The perimeter is 48 feet.

9. All sides of a regular polygon have the same
   length, so the perimeter is the number of
   sides multiplied by the length of a side.
   $P = 3 \cdot 14 \text{ in.} = 42 \text{ in.}$
   The perimeter is 42 inches.

11. All sides of a regular polygon have the same
   length, so the perimeter is the number of
   sides multiplied by the length of a side.
   $P = 5 \cdot 31 \text{ cm} = 155 \text{ cm}$
   The perimeter is 155 centimeters.

13. Sum the lengths of the sides.
   $P = 5 \text{ ft} + 3 \text{ ft} + 2 \text{ ft} + 7 \text{ ft} + 4 \text{ ft}$
   $= 21 \text{ ft}$
   the perimeter is 21 feet.

15. $P = 4 \cdot s = 4 \cdot 90 \text{ ft} = 360 \text{ ft}$
   360 feet of lime powder will be deposited.

17. $P = 2 \cdot l + 2 \cdot w$
   $= 2 \cdot 120 \text{ yd} + 2 \cdot 53 \text{ yd}$
   $= 240 \text{ yd} + 106 \text{ yd}$
   $= 346 \text{ yd}$
   The perimeter of the football field is 346
   yards.

19. $P = 2 \cdot l + 2 \cdot w$
   $= 2 \cdot 8 \text{ ft} + 2 \cdot 3 \text{ ft}$
   $= 16 \text{ ft} + 6 \text{ ft}$
   $= 22 \text{ ft}$
   22 feet of stripping is needed.

21. The amount of stripping needed is 22 feet.
   22 feet $\cdot \$2.50$ per foot $= \$55$
   The total cost of the stripping is $55.

23. All sides of a regular polygon have the same
   length, so the perimeter is the number of
   sides multiplied by the length of a side.
   $P = 8 \cdot 9 \text{ in.} = 72 \text{ in.}$
   The perimeter is 72 inches.

25. $P = 4 \cdot s = 4 \cdot 7 \text{ in.} = 28 \text{ in.}$
   The perimeter is 28 inches.

27. $P = 2 \cdot l + 2 \cdot w$
$\quad = 2 \cdot 11 \text{ ft} + 2 \cdot 10 \text{ ft}$
$\quad = 22 \text{ ft} + 20 \text{ ft}$
$\quad = 42 \text{ ft}$
42 ft · $0.86 per foot = $36.12
The cost is $36.12.

29. The unmarked vertical side has length
28 m − 20 m = 8 m.
The unmarked horizontal side has length
20 m − 17 m = 3m.
$P = 17 \text{ m} + 8 \text{ m} + 3 \text{ m} + 20 \text{ m} + 20 \text{ m} + 28 \text{ m}$
$\quad = 96 \text{ m}$
The perimeter is 96 meters.

31. The unmarked horizontal side has length
(3 + 6 + 4) ft = 13 ft.
$P = (3 + 5 + 6 + 5 + 4 + 15 + 13 + 15) \text{ ft} = 66 \text{ ft}$
The perimeter is 66 feet.

33. The unmarked vertical side has length
5 cm + 14 cm = 19 cm.
The unmarked horizontal side has length
18 cm − 9 cm = 9 cm.
$P = 18 \text{ cm} + 19 \text{ cm} + 9 \text{ cm} + 14 \text{ cm} + 9 \text{ cm} + 5 \text{ cm}$
$\quad = 74 \text{ cm}$
The perimeter is 74 centimeters.

35. $C = \pi \cdot d = \pi \cdot 17 \text{ cm} = 17\pi \text{ cm} \approx 53.38 \text{ cm}$

37. $C = 2 \cdot \pi \cdot r$
$\quad = 2 \cdot \pi \cdot 8 \text{ mi}$
$\quad = 16\pi \text{ mi}$
$\quad \approx 50.24 \text{ mi}$

39. $C = \pi \cdot d = \pi \cdot 26 \text{ m} = 26\pi \text{ m} \approx 81.64 \text{ m}$

41. $\pi \cdot d = \pi \cdot 15 \text{ ft} = 15\pi \text{ ft} \approx 47.1$
He needs 15π feet of netting or 47.1 feet.

43. $C = \pi \cdot d = \pi \cdot 4000 \text{ ft} = 4000\pi \text{ ft} \approx 12,560 \text{ ft}$
The distance around is about 12,560 feet.

45. Sum the lengths of the sides.
$P = 9 \text{ mi} + 6 \text{ mi} + 11 \text{ mi} + 4.7 \text{ mi}$
$\quad = 30.7 \text{ mi}$
The perimeter is 30.7 miles.

47. $C = \pi \cdot d = \pi \cdot 14 \text{ cm} = 14\pi \text{ cm} \approx 43.96 \text{ cm}$

49. $P = 5 \cdot 8 \text{ mm} = 40 \text{ mm}$
The perimeter is 40 millimeters.

51. The unmarked vertical side has length
(22 − 8) ft = 14 ft.
The unmarked horizontal side has length
(20 − 7) ft = 13 ft.
$P = (7 + 8 + 13 + 14 + 20 + 22) \text{ ft} = 84 \text{ ft}$
The perimeter is 84 feet.

53. $5 + 6 \cdot 3 = 5 + 18 = 23$

55. $(20 - 16) \div 4 = 4 \div 4 = 1$

57. $(18 + 8) - (12 + 4) = 26 - 16 = 10$

59. $(72 \div 2) \cdot 6 = 36 \cdot 6 = 216$

61. a. The first age category that 8-year-old children fit into is "Under 9," thus the minimum width is 30 yards and the minimum length is 40 yards.

   b. $P = 2 \cdot l + 2 \cdot w$
$\quad = 2 \cdot 40 \text{ yd} + 2 \cdot 30 \text{ yd}$
$\quad = 80 \text{ yd} + 60 \text{ yd}$
$\quad = 140 \text{ yd}$
The perimeter of the field is 140 yards.

63. The square's perimeter is 4 · 3 in. = 12 in.
The circle's circumference is
$\pi \cdot 4$ in. ≈ 12.56 in.
So the circle has the greater distance around;
b.

65. a. Smaller circle:
$C = 2 \cdot \pi \cdot r$
$\quad = 2 \cdot \pi \cdot 10 \text{ m}$
$\quad = 20\pi \text{ m}$
$\quad \approx 62.8 \text{ m}$
Larger circle:
$C = 2 \cdot \pi \cdot r$
$\quad = 2 \cdot \pi \cdot 20 \text{ m}$
$\quad = 40\pi \text{ m}$
$\quad \approx 125.6 \text{ m}$

b. Yes, when the radius of a circle is doubled, the circumference is also doubled.

67. The total length of the two straight sections is $2 \cdot 22$ m $= 44$ m. The total length of the two curved sections is the circumference of a circle of radius 5 m.
$$C = 2 \cdot \pi \cdot r = 2 \cdot \pi \cdot 5 \text{ m} = 10\pi \text{ m}$$
The perimeter of the track is the sum of these.
$$P = 44 \text{ m} + 10\pi \text{ m} = (44 + 10\pi) \text{ m} \approx 75.4 \text{ m}$$

69. The three linear sides have lengths of 7 feet, 5 feet, and 7 feet. The length of the curved side is half of the circumference of a circle with diameter 5 feet, or
$$\frac{1}{2}\pi \cdot d = \frac{1}{2}\pi \cdot 5 = 2.5\pi \approx 7.9 \text{ feet}$$
$$7 \text{ ft} + 5 \text{ ft} + 7 \text{ ft} + 7.9 \text{ ft} = 26.9 \text{ ft}$$
The perimeter of the window is 26.9 feet.

Section 9.3

Practice Problems

1. $A = \dfrac{1}{2}bh$
$$= \frac{1}{2} \cdot 12 \text{ in.} \cdot 8\frac{1}{4} \text{ in.}$$
$$= \frac{1}{2} \cdot 12 \text{ in.} \cdot \frac{33}{4} \text{ in.}$$
$$= 49\frac{1}{2} \text{ sq in.}$$

The area is $49\dfrac{1}{2}$ square inches.

2. $A = \dfrac{1}{2}(b + B)h$
$$= \frac{1}{2}(5 \text{ yd} + 11 \text{ yd})(6.1 \text{ yd})$$
$$= 48.8 \text{ sq yd}$$
The area is 48.8 square yards.

3. Split the rectangle into two pieces, a top rectangle with dimensions 12 m by 24 m, and a bottom rectangle with dimensions 6 m by 18 m. The area of the figure is the sum of the areas of these.
$$A = 12 \text{ m} \cdot 24 \text{ m} + 6 \text{ m} \cdot 18 \text{ m}$$
$$= 288 \text{ sq m} + 108 \text{ sq m}$$
$$= 396 \text{ sq m}$$
The area is 396 square meters.

4. $A = \pi r^2$
$$= \pi \cdot (7 \text{ cm})^2$$
$$= 49\pi \text{ sq cm}$$
$$\approx 153.86 \text{ sq cm}$$

The area is 49π square centimeters, which is approximately 153.86 square centimeters.

5. $V = lwh = 7 \text{ ft} \cdot 3 \text{ ft} \cdot 4 \text{ ft} = 84 \text{ cu ft}$
The volume of the box is 84 cubic feet.
SA
$$= 2lh + 2wh + 2lw$$
$$= 2(7 \text{ ft})(4 \text{ ft}) + 2(3 \text{ ft})(4 \text{ ft}) + 2(7 \text{ ft})(3 \text{ ft})$$
$$= 56 \text{ sq ft} + 24 \text{ sq ft} + 42 \text{ sq ft}$$
$$= 122 \text{ sq ft}$$

The surface area of the box is 122 square feet.

6. $V = \dfrac{4}{3}\pi r^3 = \dfrac{4}{3}\pi \cdot \left(\dfrac{1}{2} \text{ cm}\right)^3$
$$= \frac{4}{3}\pi \cdot \frac{1}{8} \text{ cu cm}$$
$$= \frac{1}{6}\pi \text{ cu cm}$$
$$\approx \frac{1}{6} \cdot \frac{22}{7} \text{ cu cm}$$
$$= \frac{11}{21} \text{ cu cm}$$

The volume is $\dfrac{1}{6}\pi$ cubic centimeter, which is approximately $\dfrac{11}{21}$ cubic centimeter.

$$SA = 4\pi r^2 = 4\pi \cdot \left(\frac{1}{2}\ \text{cm}\right)^2$$
$$= 4\pi \cdot \frac{1}{4}\ \text{sq cm}$$
$$= \pi\ \text{sq cm}$$
$$\approx \frac{22}{7}\ \text{sq cm or } 3\frac{1}{7}\ \text{sq cm}$$

The surface area is π square centimeters, which is approximately $3\frac{1}{7}$ square centimeters.

7. $V = \pi r^2 h = \pi \cdot (5\ \text{in.})^2 \cdot 9\ \text{in.}$
$$= \pi \cdot 25\ \text{sq in.} \cdot 9\ \text{in.}$$
$$= 225\pi\ \text{cu in.}$$
$$\approx 706.5\ \text{cu in.}$$

The volume is 225π cubic inches, which is approximately 706.5 cubic inches.

8. $V = \frac{1}{3}s^2 h = \frac{1}{3} \cdot (3\ \text{m})^2 \cdot 5.1\ \text{m}$
$$= \frac{1}{3} \cdot 9\ \text{sq m} \cdot 5.1\ \text{m}$$
$$= 15.3\ \text{cu m}$$

The volume is 15.3 cubic meters.

Vocabulary and Readiness Check

1. The <u>surface area</u> of a polyhedron is the sum of the areas of its faces.

3. <u>Area</u> measures the amount of surface enclosed by a region.

5. Area is measured in <u>square</u> units.

Exercise Set 9.3

1. $A = l \cdot w = 3.5\ \text{m} \cdot 2\ \text{m} = 7\ \text{sq m}$

3. $A = \frac{1}{2} \cdot b \cdot h$
$$= \frac{1}{2} \cdot 6\frac{1}{2}\ \text{yd} \cdot 3\ \text{yd}$$
$$= \frac{1}{2} \cdot \frac{13}{2}\ \text{yd} \cdot 3\ \text{yd}$$
$$= \frac{39}{4}\ \text{sq yd}$$
$$= 9\frac{3}{4}\ \text{sq yd}$$

5. $A = \frac{1}{2} \cdot b \cdot h = \frac{1}{2} \cdot 6\ \text{yd} \cdot 5\ \text{yd} = 15\ \text{sq yd}$

7. $r = d \div 2 = (3\ \text{in.}) \div 2 = 1.5\ \text{in.}$
$$A = \pi r^2$$
$$= \pi(1.5\ \text{in.})^2$$
$$= 2.25\pi\ \text{sq in.}$$
$$\approx 7.065\ \text{sq in.}$$

9. $A = b \cdot h = 7\ \text{ft} \cdot 5.25\ \text{ft} = 36.75\ \text{sq ft}$

11. $A = \frac{1}{2}(b + B) \cdot h$
$$= \frac{1}{2}(5\ \text{m} + 9\ \text{m}) \cdot 4\ \text{m}$$
$$= \frac{1}{2} \cdot 14\ \text{m} \cdot 4\ \text{m}$$
$$= 28\ \text{sq m}$$

13. $A = \frac{1}{2}(b + B) \cdot h$
$$= \frac{1}{2}(7\ \text{yd} + 4\ \text{yd}) \cdot 4\ \text{yd}$$
$$= \frac{1}{2}(11\ \text{yd}) \cdot 4\ \text{yd}$$
$$= 22\ \text{sq yd}$$

15. $A = b \cdot h$

$\quad = 7 \text{ ft} \cdot 5\frac{1}{4} \text{ ft}$

$\quad = 7 \text{ ft} \cdot \frac{21}{4} \text{ ft}$

$\quad = \frac{147}{4} \text{ sq ft}$

$\quad = 36\frac{3}{4} \text{ sq ft}$

17. $A = b \cdot h$

$\quad = 5 \text{ in.} \cdot 4\frac{1}{2} \text{ in.}$

$\quad = 5 \text{ in.} \cdot \frac{9}{2} \text{ in.}$

$\quad = \frac{45}{2} \text{ sq in.}$

$\quad = 22\frac{1}{2} \text{ sq in.}$

19. The base of the triangle is

$7 \text{ cm} - 1\frac{1}{2} \text{ cm} - 1\frac{1}{2} \text{ cm} = 4 \text{ cm}$, so its area

is $\frac{1}{2} \cdot 4 \text{ cm} \cdot 2 \text{ cm} = 4 \text{ sq cm}$.

The area of the rectangle is
7 cm · 3 cm = 21 sq cm.
The area of the figure is the sum of these.
$A = 4 \text{ sq cm} + 21 \text{ sq cm} = 25 \text{ sq cm}$

21. The figure can be divided into two
rectangles, one measuring 10 mi by 5 mi
and one measuring 12 mi by 3 mi. The area
of the figure is the sum of the areas of the
two rectangles.
$A = 10 \text{ mi} \cdot 5 \text{ mi} + 12 \text{ mi} \cdot 3 \text{ mi}$

$\quad = 50 \text{ sq mi} + 36 \text{ sq mi}$

$\quad = 86 \text{ sq mi}$

23. The top of the figure is a square with sides
of length 3 cm, so its area is

$s^2 = (3 \text{ cm})^2 = 9 \text{ sq cm}$.

The bottom of the figure is a parallelogram
with area $b \cdot h = 3 \text{ cm} \cdot 5 \text{ cm} = 15 \text{ sq cm}$.
The area of the figure is the sum of these.
$A = 9 \text{ sq cm} + 15 \text{ sq cm} = 24 \text{ sq cm}$

25. $A = \pi r^2$

$\quad = \pi (6 \text{ in.})^2$

$\quad = 36\pi \text{ sq in.}$

$\quad \approx 113\frac{1}{7} \text{ sq in.}$

27. $V = l \cdot w \cdot h = 6 \text{ in.} \cdot 4 \text{ in.} \cdot 3 \text{ in.} = 72 \text{ cu in.}$
SA
$\quad = 2lh + 2wh + 2lw$
$\quad = 2 \cdot 6 \text{ in.} \cdot 3 \text{ in.} + 2 \cdot 4 \text{ in.} \cdot 3 \text{ in.} + 2 \cdot 6 \text{ in.} \cdot 4 \text{ in.}$
$\quad = 36 \text{ sq in.} + 24 \text{ sq in.} + 48 \text{ sq in.}$
$\quad = 108 \text{ sq in.}$

29. $V = s^3 = (8 \text{ cm})^3 = 512 \text{ cu cm}$

$SA = 6s^2$

$\quad = 6(8 \text{ cm})^2$

$\quad = 6 \cdot 64 \text{ sq cm}$

$\quad = 384 \text{ sq cm}$

31. $V = \frac{1}{3} \cdot \pi \cdot r^2 \cdot h$

$\quad = \frac{1}{3} \cdot \pi (2 \text{ yd})^2 (3 \text{ yd})$

$\quad = 4\pi \text{ cu yd}$

$\quad \approx 4 \cdot \frac{22}{7} \text{ cu yd}$

$\quad = \frac{88}{7} \text{ cu yd}$

$\quad = 12\frac{4}{7} \text{ cu yd}$

$SA = \pi \cdot r\sqrt{r^2 + h^2} + \pi r^2$

$\quad = \pi \cdot 2 \text{ yd}\sqrt{(2 \text{ yd})^2 + (3 \text{ yd})^2} + \pi (2 \text{ yd})^2$

$\quad = 2\pi\sqrt{13} \text{ sq yd} + 4\pi \text{ sq yd}$

$\quad \approx 35.23 \text{ sq yd}$

33. $r = \dfrac{1}{2} \cdot d = \dfrac{1}{2} \cdot 10 \text{ in.} = 5 \text{ in.}$

$$V = \dfrac{4}{3}\pi r^3$$
$$= \dfrac{4}{3}\pi(5 \text{ in.})^3$$
$$= \dfrac{4}{3}\pi \cdot 125 \text{ cu in.}$$
$$= \dfrac{500}{3}\pi \text{ cu in.}$$
$$\approx \dfrac{500}{3} \cdot \dfrac{22}{7} \text{ cu in.} = 523\dfrac{17}{21} \text{ cu in.}$$

$$SA = 4\pi r^2$$
$$= 4\pi(5 \text{ in.})^2$$
$$= 4\pi \cdot 25 \text{ sq in.}$$
$$= 100\pi \text{ sq in.}$$
$$\approx 100 \cdot \dfrac{22}{7} \text{ sq in.} = 314\dfrac{2}{7} \text{ sq in.}$$

35. $r = \dfrac{1}{2}d = \dfrac{1}{2} \cdot 2 \text{ in.} = 1 \text{ in.}$

$$V = \pi \cdot r^2 \cdot h$$
$$= \pi(1 \text{ in.})^2 \cdot 9 \text{ in}$$
$$= 9\pi \text{ cu in.}$$
$$\approx 9 \cdot \dfrac{22}{7} \text{ cu in.} = 28\dfrac{2}{7} \text{ cu in.}$$

37. $V = \dfrac{1}{3}s^2 h$

$$= \dfrac{1}{3}(5 \text{ cm})^2 (9 \text{ cm})$$
$$= \dfrac{1}{3} \cdot 25 \text{ sq cm} \cdot 9 \text{ cm}$$
$$= 75 \text{ cu cm}$$

39. $V = s^3 = \left(1\dfrac{1}{3} \text{ in.}\right)^3$

$$= \left(\dfrac{4}{3} \text{ in.}\right)^3$$
$$= \dfrac{64}{27} \text{ cu in.}$$
$$= 2\dfrac{10}{27} \text{ cu in.}$$

41. $V = lwh = (2 \text{ ft})(1.4 \text{ ft})(3 \text{ ft}) = 8.4 \text{ cu ft}$

SA
$$= 2lh + 2wh + 2lw$$
$$= 2(2 \text{ ft})(3 \text{ ft}) + 2(1.4 \text{ ft})(3 \text{ ft}) + 2(2 \text{ ft})(1.4 \text{ ft})$$
$$= 12 \text{ sq ft} + 8.4 \text{ sq ft} + 5.6 \text{ sq ft}$$
$$= 26 \text{ sq ft}$$

43. $A = l \cdot w = 505 \text{ ft} \cdot 225 \text{ ft} = 113{,}625 \text{ sq ft}$
The area of the flag is 113,625 square feet.

45. $A = l \cdot w = 7 \text{ ft} \cdot 6 \text{ ft} = 42 \text{ sq ft}$
$4 \cdot 42 \text{ sq ft} = 168 \text{ sq ft}$
Four panels require 168 square feet of material.

47. $V = \dfrac{1}{3} \cdot s^2 \cdot h$

$$= \dfrac{1}{3} \cdot (12 \text{ cm})^2 \cdot 20 \text{ cm}$$
$$= 960 \text{ cu cm}$$

49. The land is in the shape of a trapezoid.

$$A = \dfrac{1}{2}(b + B) \cdot h$$
$$= \dfrac{1}{2}(90 \text{ ft} + 140 \text{ ft}) \cdot 80 \text{ ft}$$
$$= \dfrac{1}{2} \cdot 230 \text{ ft} \cdot 80 \text{ ft}$$
$$= 9200 \text{ sq ft}$$

There are 9200 square feet of land in the plot.

51. $V = \dfrac{4}{3} \cdot \pi \cdot r^3$

$= \dfrac{4}{3}\pi(7 \text{ in.})^3$

$= \dfrac{1372}{3}\pi \text{ cu in. or } 457\dfrac{1}{3}\pi \text{ cu in.}$

$SA = 4 \cdot \pi \cdot r^2 = 4\pi(7 \text{ in.})^2 = 196\pi \text{ sq in.}$

53. $A = \dfrac{1}{2}(b + B)h$

$= \dfrac{1}{2}(25 \text{ ft} + 36 \text{ ft}) \cdot 12\dfrac{1}{2} \text{ ft}$

$= \dfrac{1}{2} \cdot 61 \text{ ft} \cdot 12\dfrac{1}{2} \text{ ft}$

$= 381\dfrac{1}{4} \text{ sq ft}$

To the nearest square foot, the area is 381 square feet.

55. $r = \dfrac{1}{2} \cdot d = \dfrac{1}{2} \cdot 6 \text{ in.} = 3 \text{ in.}$

$V = \dfrac{4}{3} \cdot \pi \cdot r^3$

$= \dfrac{4}{3}\pi \cdot (3 \text{ in.})^3$

$= 36\pi \text{ cu in.}$

$\approx 113.04 \text{ cu in.}$

57. $r = \dfrac{1}{2} \cdot d = \dfrac{1}{2} \cdot 4 \text{ ft} = 2 \text{ ft}$

$A = \pi r^2 = \pi(2 \text{ ft})^2 = \pi \cdot 4 \text{ sq ft}$

The area of the pizza is 4π square feet, or approximately $4 \cdot 3.14 = 12.56$ square feet.

59. $r = \dfrac{1}{2} \cdot d = \dfrac{1}{2} \cdot 3 \text{ m} = 1.5 \text{ m}$

$V = \dfrac{4}{3} \cdot \pi \cdot r^3$

$= \dfrac{4}{3}\pi(1.5 \text{ m})^3$

$= 4.5\pi \text{ cu m}$

$\approx 4.5 \cdot 3.14 \text{ cu m} = 14.13 \text{ cu m}$

61. $A = l \cdot w = 16 \text{ ft} \cdot 10\dfrac{1}{2} \text{ ft}$

$= 16 \text{ ft} \cdot \dfrac{21}{2} \text{ ft}$

$= 168 \text{ sq ft}$

The area of the wall is 168 square feet.

63. $V = \dfrac{1}{3} \cdot s^2 \cdot h$

$= \dfrac{1}{3}(5 \text{ in.})^2 \cdot \dfrac{13}{10} \text{ in.}$

$= \dfrac{1}{3} \cdot 25 \cdot \dfrac{13}{10} \text{ cu in.}$

$= 10\dfrac{5}{6} \text{ cu in.}$

65. $5^2 = 5 \cdot 5 = 25$

67. $3^2 = 3 \cdot 3 = 9$

69. $1^2 + 2^2 = 1 \cdot 1 + 2 \cdot 2 = 1 + 4 = 5$

71. $4^2 + 2^2 = 4 \cdot 4 + 2 \cdot 2 = 16 + 4 = 20$

73. A fence goes around the edge of a yard, thus the situation involves perimeter.

75. Carpet covers the entire floor of a room, so the situation involves area.

77. Paint covers the surface of the wall, thus the situation involves area.

79. A wallpaper border goes around the edge of a room, so the situation involves perimeter.

81. Note that the dimensions given are the diameters of the pizzas.
12-inch pizza:

$r = \dfrac{1}{2} \cdot d = \dfrac{1}{2} \cdot 12 \text{ in.} = 6 \text{ in.}$

$A = \pi \cdot r^2 = \pi(6 \text{ in.})^2 = 36\pi \text{ sq in.}$

Price per square inch $= \dfrac{\$10}{36\pi \text{ sq in.}}$

$\approx \$0.0884$

8-inch pizzas:

$r = \frac{1}{2} \cdot d = \frac{1}{2} \cdot 8$ in. $= 4$ in.

$A = \pi \cdot r^2 = \pi(4 \text{ in.})^2 = 16\pi$ sq in.

$2 \cdot A = 2 \cdot 16\pi$ sq in. $= 32\pi$ sq in.

Price per square inch: $\dfrac{\$9}{32\pi \text{ sq in.}} \approx \$0.0895$

Since the price per square inch for the 12-inch pizza is less, the 12-inch pizza is the better deal.

83. $V = \pi \cdot r^2 \cdot h$

$\quad = \pi \cdot (16.3 \text{ ft})^2 \cdot 32 \text{ ft}$

$\quad = 8502.08\pi$ cu ft

$\quad \approx 26{,}696.5$ cu ft

The volume of the tank is about 26,696.5 cubic feet.

85. no; answers may vary

87. The area of the shaded region is the area of the square minus the area of the circle.
Square:

$A = s^2 = (6 \text{ in.})^2 = 36$ sq in.

Circle:

$r = \frac{1}{2} \cdot d = \frac{1}{2}(6 \text{ in.}) = 3$ in.

$A = \pi \cdot r^2$

$\quad = \pi(3 \text{ in.})^2$

$\quad = 9\pi$ sq in.

$\quad \approx 28.26$ sq in.

36 sq in. − 28.26 sq in. = 7.74 sq in.
The shaded region has area of approximately 7.74 square inches.

89. The skating area is a rectangle with a half circle on each end.
Rectangle:
$A = l \cdot w = 22$ m $\cdot$ 10 m $= 220$ sq m
Half circles:

$A = 2 \cdot \frac{1}{2} \cdot \pi \cdot r^2$

$\quad = \pi(5 \text{ m})^2$

$\quad = 25\pi$ sq m ≈ 78.5 sq m

220 sq m + 78.5 sq m = 298.5 sq m
The skating surface has area of 298.5 square meters.

91. no; answers may vary

Integrated Review

1. The supplement of a 27° angle measures
$180° − 27° = 153°$.
The complement of a 27° angle measures
$90° − 27° = 63°$.

2. $\angle x$ and the angle marked 105° are supplementary angles, so
$m\angle x = 180° − 105° = 75°$.
$\angle y$ and the angle marked 105° are vertical angles, so $m\angle y = 105°$.
$\angle z$ and the angle marked 105° are supplementary angles, so
$m\angle z = 180° − 105° = 75°$.

3. $\angle x$ and the angle marked 52° are supplementary angles, so
$m\angle x = 180° − 52° = 128°$.
$\angle y$ and the angle marked 52° are corresponding angles, so $m\angle y = 52°$.
$\angle z$ and $\angle y$ are supplementary angles, so
$m\angle z = 180° − m\angle y = 180° − 52° = 128°$.

4. The sum of the measures of the angles of a triangle is 180°.
$m\angle x = 180° − 90° − 38° = 52°$

5. $d = 2 \cdot r = 2 \cdot 2.3$ in. $= 4.6$ in.

6. $r = \frac{1}{2} \cdot d$

$= \frac{1}{2} \cdot 8\frac{1}{2}$ in.

$= \frac{1}{2} \cdot \frac{17}{2}$ in.

$= \frac{17}{4}$ in.

$= 4\frac{1}{4}$ in.

7. $P = 4 \cdot s = 4 \cdot 5 \text{ m} = 20 \text{ m}$

$A = s^2 = (5 \text{ m})^2 = 25 \text{ sq m}$

8. $P = a + b + c = 4 \text{ ft} + 5 \text{ ft} + 3 \text{ ft} = 12 \text{ ft}$

$A = \frac{1}{2} \cdot b \cdot h = \frac{1}{2} \cdot 3 \text{ ft} \cdot 4 \text{ ft} = 6 \text{ sq ft}$

9. $C = 2 \cdot \pi \cdot r = 2 \cdot \pi \cdot 5 \text{ cm} = 10\pi \text{ cm} \approx 31.4 \text{ cm}$

$A = \pi \cdot r^2$

$= \pi(5 \text{ cm})^2$

$= 25\pi \text{ sq cm} \approx 78.5 \text{ sq cm}$

10. $P = 11 \text{ mi} + 5 \text{ mi} + 11 \text{ mi} + 5 \text{ mi} = 32 \text{ mi}$

$A = l \cdot w = 11 \text{ mi} \cdot 4 \text{ mi} = 44 \text{ sq mi}$

11. The unmarked horizontal side has length
$17 \text{ cm} - 8 \text{ cm} = 9 \text{ cm}$.
The unmarked vertical side has length
$7 \text{ cm} + 3 \text{ cm} = 10 \text{ cm}$.
$P = (8 + 3 + 9 + 7 + 17 + 10) \text{ cm} = 54 \text{ cm}$
The figure is made up of two rectangles, one
with dimensions 10 cm by 8 cm, the other
with dimensions 7 cm by 9 cm.
$A = 10 \text{ cm} \cdot 8 \text{ cm} + 7 \text{ cm} \cdot 9 \text{ cm} = 143 \text{ cm}$

12. $P = 2 \cdot l + 2 \cdot w$

$= 2(17 \text{ ft}) + 2(14 \text{ ft})$

$= 34 \text{ ft} + 28 \text{ ft}$

$= 62 \text{ ft}$

$A = l \cdot w = 17 \text{ ft} \cdot 14 \text{ ft} = 238 \text{ sq ft}$

13. $V = s^3 = (4 \text{ in.})^3 = 64 \text{ cu in.}$

$SA = 6 \cdot s^2$

$= 6(4 \text{ in.})^2$

$= 6 \cdot 16 \text{ sq in.}$

$= 96 \text{ sq in.}$

14. $V = l \cdot w \cdot h = 3 \text{ ft} \cdot 2 \text{ ft} \cdot 5.1 \text{ ft} = 30.6 \text{ cu ft}$

SA

$= 2lh + 2wh + 2lw$

$= 2(3 \text{ ft})(5.1 \text{ ft}) + 2(2 \text{ ft})(5.1 \text{ ft}) + 2(3 \text{ ft})(2 \text{ ft})$

$= 30.6 \text{ sq ft} + 20.4 \text{ sq ft} + 12 \text{ sq ft}$

$= 63 \text{ sq ft}$

15. $V = \frac{1}{3}s^2 h = \frac{1}{3}(10 \text{ cm})^2 (12 \text{ cm}) = 400 \text{ cu cm}$

16. $r = \frac{1}{2} \cdot d = \frac{1}{2} \cdot 3 \text{ mi} = \frac{3}{2} \text{ mi}$

$V = \frac{4}{3}\pi r^3$

$= \frac{4}{3}\pi \left(\frac{3}{2} \text{ mi}\right)^3$

$= \frac{9}{2}\pi \text{ cu mi}$

$= 4\frac{1}{2}\pi \text{ cu mi}$

$\approx \frac{9}{2} \cdot \frac{22}{7} \text{ cu mi} = 14\frac{1}{7} \text{ cu mi}$

Section 9.4

Practice Problems

1. $6 \text{ ft} = \frac{6 \text{ ft}}{1} \cdot 1$

$= \frac{6 \text{ ft}}{1} \cdot \frac{12 \text{ in.}}{1 \text{ ft}}$

$= 6 \cdot 12 \text{ in.}$

$= 72 \text{ in.}$

2. $8 \text{ yd} = \frac{8 \text{ yd}}{1} \cdot 1 = \frac{8 \text{ yd}}{1} \cdot \frac{3 \text{ ft}}{1 \text{ yd}} = 8 \cdot 3 \text{ ft} = 24 \text{ ft}$

3. $18 \text{ in.} = \dfrac{18 \text{ in.}}{1} \cdot \dfrac{1 \text{ ft}}{12 \text{ in.}} = \dfrac{18}{12} \text{ ft} = 1.5 \text{ ft}$

4. $68 \text{ in.} = \dfrac{68 \text{ in.}}{1} \cdot \dfrac{1 \text{ ft}}{12 \text{ in.}} = \dfrac{68}{12} \text{ ft}$

$$
\begin{array}{r}
5 \\
12\overline{)\ 68} \\
\underline{-60} \\
8
\end{array}
$$

Thus, 68 in. = 5 ft 8 in.

5. $5 \text{ yd} = \dfrac{5 \text{ yd}}{1} \cdot \dfrac{3 \text{ ft}}{1 \text{ yd}} = 15 \text{ ft}$

5 yd 2 ft = 15 ft + 2 ft = 17 ft

6.
$$
\begin{array}{r}
4 \text{ ft } \ 8 \text{ in.} \\
+ \ 8 \text{ ft } 11 \text{ in.} \\
\hline
12 \text{ ft } 19 \text{ in.}
\end{array}
$$
Since 19 inches is the same as 1 ft 7 in., we have
12 ft 19 in. = 12 ft + 1 ft 7 in. = 13 ft 7 in.

7.
$$
\begin{array}{r}
4 \text{ ft } \ 7 \text{ in.} \\
\times \qquad\qquad 4 \\
\hline
16 \text{ ft } 28 \text{ in.}
\end{array}
$$
Since 28 in. is the same as 2 ft 4 in., we simplify as
16 ft 28 in. = 16 ft + 2 ft 4 in. = 18 ft 4 in.

8.
$$
\begin{array}{rcr}
5 \text{ ft } 8 \text{ in.} & \rightarrow & 4 \text{ ft } 20 \text{ in.} \\
- 1 \text{ ft } 9 \text{ in.} & & - 1 \text{ ft } \ 9 \text{ in.} \\
\hline
& & 3 \text{ ft } 11 \text{ in.}
\end{array}
$$
The remaining board length is 3 ft 11 in.

9. $2.5 \text{ m} = \dfrac{2.5 \text{ m}}{1} \cdot \dfrac{1000 \text{ mm}}{1 \text{ m}} = 2500 \text{ mm}$

10. 3500 m = 3.500 km or 3.5 km

11. 640 m = 0.64 km
$$
\begin{array}{r}
2.10 \text{ km} \\
- 0.64 \text{ km} \\
\hline
1.46 \text{ km}
\end{array}
$$

2.1 km = 2100 m
$$
\begin{array}{r}
2100 \text{ m} \\
- \ 640 \text{ m} \\
\hline
1460 \text{ m}
\end{array}
$$

12.
$$
\begin{array}{r}
18.3 \text{ hm} \\
\times \ \ 5 \\
\hline
91.5 \text{ hm}
\end{array}
$$

13. 0.8 m = 80 cm
80 cm + 45 cm = 125 cm
The scarf will be 125 cm or 1.25 m.

Vocabulary and Readiness Check

1. The basic unit of length in the metric system is the <u>meter</u>.

3. A meter is slightly longer than a <u>yard</u>.

5. One yard equals 3 <u>feet</u>.

7. One mile equals 5280 <u>feet</u>.

Exercise Set 9.4

1. $60 \text{ in.} = \dfrac{60 \text{ in.}}{1} \cdot \dfrac{1 \text{ ft}}{12 \text{ in.}} = \dfrac{60}{12} \text{ ft} = 5 \text{ ft}$

3. $12 \text{ yd} = \dfrac{12 \text{ yd}}{1} \cdot \dfrac{3 \text{ ft}}{1 \text{ yd}} = 12 \cdot 3 \text{ ft} = 36 \text{ ft}$

5. $42{,}240 \text{ ft} = \dfrac{42{,}240 \text{ ft}}{1} \cdot \dfrac{1 \text{ mi}}{5280 \text{ ft}}$
$\qquad\qquad = \dfrac{42{,}240}{5280} \text{ mi}$
$\qquad\qquad = 8 \text{ mi}$

7. $8\dfrac{1}{2} \text{ ft} = \dfrac{8.5 \text{ ft}}{1} \cdot \dfrac{12 \text{ in.}}{1 \text{ ft}} = 8.5 \cdot 12 \text{ in.} = 102 \text{ in.}$

9. $10 \text{ ft} = \dfrac{10 \text{ ft}}{1} \cdot \dfrac{1 \text{ yd}}{3 \text{ ft}} = \dfrac{10}{3} \text{ yd} = 3\dfrac{1}{3} \text{ yd}$

11. $6.4 \text{ mi} = \dfrac{6.4 \text{ mi}}{1} \cdot \dfrac{5280 \text{ ft}}{1 \text{ mi}}$
$= 6.4 \cdot 5280 \text{ ft}$
$= 33,792 \text{ ft}$

13. $162 \text{ in.} = \dfrac{162 \text{ in.}}{1} \cdot \dfrac{1 \text{ ft}}{12 \text{ in.}} \cdot \dfrac{1 \text{ yd}}{3 \text{ ft}}$
$= \dfrac{162}{36} \text{ yd}$
$= 4.5 \text{ yd}$

15. $3 \text{ in.} = \dfrac{3 \text{ in.}}{1} \cdot \dfrac{1 \text{ ft}}{12 \text{ in.}} = \dfrac{3}{12} \text{ ft} = 0.25 \text{ ft}$

17. $40 \text{ ft} = \dfrac{40 \text{ ft}}{1} \cdot \dfrac{1 \text{ yd}}{3 \text{ ft}} = \dfrac{40}{3} \text{ yd}$

$$
\begin{array}{r}
13 \text{ yd } 1 \text{ ft} \\
3\overline{)\,40} \\
\underline{-3} \\
10 \\
\underline{-9} \\
1
\end{array}
$$

19. $85 \text{ in.} = \dfrac{85 \text{ in.}}{1} \cdot \dfrac{1 \text{ ft}}{12 \text{ in.}} = \dfrac{85}{12} \text{ ft}$

$$
\begin{array}{r}
7 \text{ ft } 1 \text{ in.} \\
12\overline{)\,85} \\
\underline{-84} \\
1
\end{array}
$$

21. $10,000 \text{ ft} = \dfrac{10,000 \text{ ft}}{1} \cdot \dfrac{1 \text{ mi}}{5280 \text{ ft}} = \dfrac{10,000}{5280} \text{ mi}$

$$
\begin{array}{r}
1 \text{ mi } 4720 \text{ ft} \\
5280\overline{)\,10,000} \\
\underline{-5280} \\
4720
\end{array}
$$

23. $5 \text{ ft } 2 \text{ in.} = \dfrac{5 \text{ ft}}{1} \cdot \dfrac{12 \text{ in.}}{1 \text{ ft}} + 2 \text{ in.}$
$= 60 \text{ in.} + 2 \text{ in.}$
$= 62 \text{ in.}$

25. $8 \text{ yd } 2 \text{ ft} = \dfrac{8 \text{ yd}}{1} \cdot \dfrac{3 \text{ ft}}{1 \text{ yd}} + 2 \text{ ft}$
$= 24 \text{ ft} + 2 \text{ ft}$
$= 26 \text{ ft}$

27. $2 \text{ yd } 1 \text{ ft} = \dfrac{2 \text{ yd}}{1} \cdot \dfrac{3 \text{ ft}}{1 \text{ yd}} + 1 \text{ ft}$
$= 6 \text{ ft} + 1 \text{ ft}$
$= 7 \text{ ft}$
$7 \text{ ft} = \dfrac{7 \text{ ft}}{1} \cdot \dfrac{12 \text{ in.}}{1 \text{ ft}} = 7 \cdot 12 \text{ in.} = 84 \text{ in.}$

29. $3 \text{ ft } 10 \text{ in.} + 7 \text{ ft } 4 \text{ in.} = 10 \text{ ft } 14 \text{ in.}$
$= 10 \text{ ft} + 1 \text{ ft } 2 \text{ in.}$
$= 11 \text{ ft } 2 \text{ in.}$

31. $12 \text{ yd } 2 \text{ ft} + 9 \text{ yd } 2 \text{ ft} = 21 \text{ yd } 4 \text{ ft}$
$= 21 \text{ yd} + 1 \text{ yd } 1 \text{ ft}$
$= 22 \text{ yd } 1 \text{ ft}$

33.
$$
\begin{array}{r}
22 \text{ ft } 8 \text{ in.} \\
-\ 16 \text{ ft } 3 \text{ in.} \\
\hline
6 \text{ ft } 5 \text{ in.}
\end{array}
$$

35.
$$
\begin{array}{rcr}
18 \text{ ft } 3 \text{ in.} & \rightarrow & 17 \text{ ft } 15 \text{ in.} \\
-\ 10 \text{ ft } 9 \text{ in.} & & -\ 10 \text{ ft }\ \ 9 \text{ in.} \\
\hline
 & & 7 \text{ ft }\ \ 6 \text{ in.}
\end{array}
$$

37. $28 \text{ ft } 8 \text{ in.} \div 2 = 14 \text{ ft } 4 \text{ in.}$

39.
$$
\begin{array}{r}
16 \text{ yd }\ 2 \text{ ft} \\
\times 5 \\
\hline
80 \text{ yd } 10 \text{ ft} = 80 \text{ yd} + 3 \text{ yd } 1 \text{ ft} = 83 \text{ yd } 1 \text{ ft}
\end{array}
$$

41. $60 \text{ m} = \dfrac{60 \text{ m}}{1} \cdot \dfrac{100 \text{ cm}}{1 \text{ m}} = 6000 \text{ cm}$

43. $40 \text{ mm} = \dfrac{40 \text{ mm}}{1} \cdot \dfrac{1 \text{ cm}}{10 \text{ mm}} = 4.0 \text{ cm}$

45. $500 \text{ m} = \dfrac{500 \text{ m}}{1} \cdot \dfrac{1 \text{ km}}{1000 \text{ m}}$
$= \dfrac{500}{1000} \text{ km}$
$= 0.5 \text{ km}$

47. $1700 \text{ mm} = \dfrac{1700 \text{ mm}}{1} \cdot \dfrac{1 \text{ m}}{1000 \text{ mm}} = 1.7 \text{ m}$

49. $1500 \text{ cm} = \dfrac{1500 \text{ cm}}{1} \cdot \dfrac{1 \text{ m}}{100 \text{ cm}} = \dfrac{1500}{100} \text{ m} = 15 \text{ m}$

51. $0.42 \text{ km} = \dfrac{0.42 \text{ km}}{1} \cdot \dfrac{100,000 \text{ cm}}{1 \text{ km}}$
$= 42,000 \text{ cm}$

53. $7 \text{ km} = \dfrac{7 \text{ km}}{1} \cdot \dfrac{1000 \text{ m}}{1 \text{ km}} = 7000 \text{ m}$

55. $8.3 \text{ cm} = \dfrac{8.3 \text{ cm}}{1} \cdot \dfrac{10 \text{ mm}}{1 \text{ cm}} = 83 \text{ mm}$

57. $20.1 \text{ mm} = \dfrac{20.1 \text{ mm}}{1} \cdot \dfrac{1 \text{ dm}}{100 \text{ mm}} = \dfrac{20.1}{100} \text{ dm} = 0.201 \text{ dm}$

59. $0.04 \text{ m} = \dfrac{0.04 \text{ m}}{1} \cdot \dfrac{1000 \text{ mm}}{1 \text{ m}} = 40 \text{ mm}$

61. $\begin{array}{r} 8.60 \text{ m} \\ + \ 0.34 \text{ m} \\ \hline 8.94 \text{ m} \end{array}$

63. $\begin{array}{r} 2.9 \text{ m} \\ + \ 40.0 \text{ mm} \\ \hline \end{array} \quad \begin{array}{r} 2.90 \text{ m} \\ + \ 0.04 \text{ m} \\ \hline 2.94 \text{ m} \end{array} \text{ or } \begin{array}{r} 2900 \text{ mm} \\ + \ 40 \text{ mm} \\ \hline 2940 \text{ mm} \end{array}$

65. $\begin{array}{r} 24.8 \text{ mm} \\ - \ 1.19 \text{ cm} \\ \hline \end{array} \quad \begin{array}{r} 24.8 \text{ mm} \\ - \ 11.9 \text{ mm} \\ \hline 12.9 \text{ mm} \end{array} \text{ or } \begin{array}{r} 2.48 \text{ cm} \\ - \ 1.19 \text{ cm} \\ \hline 1.29 \text{ cm} \end{array}$

67. $\begin{array}{r} 15 \text{ km} \\ - \ 2360 \text{ m} \\ \hline \end{array} \quad \begin{array}{r} 15.00 \text{ km} \\ - \ 2.36 \text{ km} \\ \hline 12.64 \text{ km} \end{array} \text{ or } \begin{array}{r} 15,000 \text{ m} \\ - \ 2360 \text{ m} \\ \hline 12,640 \text{ m} \end{array}$

69. $18.3 \text{ m} \times 3 = 54.9 \text{ m}$

71. $6.2 \text{ km} \div 4 = 1.55 \text{ km}$

		Yards	Feet	Inches
73.	Chrysler Building in New York City	$348\frac{2}{3}$	1046	12,552
75.	Python length	$11\frac{2}{3}$	35	420

		meters	millimeters	kilometers	centimeters
77.	Length of elephant	5	5000	0.005	500
79.	Tennis ball diameter	0.065	65	0.000065	6.5
81.	Distance from London to Paris	342,000	342,000,000	342	34,200,000

83.   6 ft 10 in.
    + 3 ft 8 in.
    9 ft 18 in. = 9 ft + 1 ft 6 in. = 10 ft 6 in.
    The bamboo is 10 ft 6 in. now.

85.   6000 ft
    − 900 ft
    5100 ft
    The Grand Canyon of the Colorado River is 5100 feet deeper than the Grand Canyon of the Yellowstone River.

87. 8 ft 11 in. ÷ 22.5 in. = 107 in. ÷ 22.5 in. ≈ 4.8
    Robert is 4.8 times as tall as Gul.

89.   80 mm   80.0 mm
    − 5.33 cm   − 53.3 mm
                26.7 mm
    The ice must be 26.7 mm thicker before skating is allowed.

91. 1 ft 9 in. × 9 = 9 ft 81 in.
                   = 9 ft + 6 ft 9 in.
                   = 15 ft 9 in.
    The stacks extend 15 ft 9 in. from the wall.

93.

$$\begin{array}{r} 3.35 \\ 20\overline{)\,67.00} \\ \underline{-60} \\ 7\,0 \\ \underline{-6\,0} \\ 1\,00 \\ \underline{-1\,00} \\ 0 \end{array}$$

Each piece will be 3.35 meters long.

95. $79 \text{ ft} \times 4 = 316 \text{ ft}$
4 trucks are 316 feet long.

$$\begin{array}{r} 105 \\ 3\overline{)\,316} \\ \underline{-3} \\ 01 \\ \underline{-0} \\ 16 \\ \underline{-15} \\ 1 \end{array}$$

The four trucks are $105\frac{1}{3}$ yards long.

97. $0.21 = \dfrac{21}{100}$

99. $\dfrac{13}{100} = 0.13$

101. $\dfrac{1}{4} = \dfrac{1}{4} \cdot \dfrac{25}{25} = \dfrac{25}{100} = 0.25$

103. No, the width of a twin-size bed being 20 meters is not reasonable.

105. Yes, glass for a drinking glass being 2 millimeters thick is reasonable.

107. No, the distance across the Colorado River being 50 kilometers is not reasonable.

109. 5 yd 2 in. is close to 5 yd. 7 yd 30 in. is close to 7 yd 36 in. = 8 yd.
Estimate: 5 yd + 8 yd = 13 yd.

111. answers may vary; for example,

$$4 \text{ ft} = \frac{4 \text{ ft}}{1} \cdot \frac{1 \text{ yd}}{3 \text{ ft}} = \frac{4}{3} \text{ yd} = 1\frac{1}{3} \text{ yd}$$

$$4 \text{ ft} = \frac{4 \text{ ft}}{1} \cdot \frac{12 \text{ in.}}{1 \text{ ft}} = 48 \text{ in.}$$

113. answers may vary

115. $18.3 \text{ m} \times 18.3 \text{ m} = 334.89 \text{ sq m}$
The area of the sign is 334.89 square meters.

Section 9.5

Practice Problems

1.
$$\begin{aligned} 6500 \text{ lb} &= \frac{6500 \text{ lb}}{1} \cdot \frac{1 \text{ ton}}{2000 \text{ lb}} \\ &= \frac{6500}{2000} \text{ tons} \\ &= \frac{13}{4} \text{ tons or } 3\frac{1}{4} \text{ tons} \end{aligned}$$

2.
$$\begin{aligned} 72 \text{ oz} &= \frac{72 \text{ oz}}{1} \cdot \frac{1 \text{ lb}}{16 \text{ oz}} \\ &= \frac{72}{16} \text{ lb} \cdot \frac{9}{2} \text{ lb or } 4\frac{1}{2} \text{ lb} \end{aligned}$$

3. $47 \text{ oz} = 47 \text{ oz} \cdot \dfrac{1 \text{ lb}}{16 \text{ oz}} = \dfrac{47}{16} \text{ lb}$

$$\begin{array}{r} 2 \text{ lb } 15 \text{ oz} \\ 16\overline{)\,47} \\ \underline{-32} \\ 15 \end{array}$$

Thus, $47 \text{ oz} = 2 \text{ lb } 15 \text{ oz}$

4.
$$\begin{array}{r} 8 \text{ tons } 100 \text{ lb} \\ -\ 5 \text{ tons } 1200 \text{ lb} \end{array} \rightarrow \begin{array}{r} 7 \text{ tons } 2100 \text{ lb} \\ -\ 5 \text{ tons } 1200 \text{ lb} \\ \hline 2 \text{ tons } 900 \text{ lb} \end{array}$$

5.
$$\begin{array}{r} 1 \text{ lb } \quad 6 \text{ oz} \\ 4\overline{)\,5 \text{ lb } \quad 8 \text{ oz}} \\ \underline{-\ 4 \text{ lb}} \\ 1 \text{ lb} = \underline{16 \text{ oz}} \\ 24 \text{ oz} \end{array}$$

6.

batch weight	5 lb 14 oz
+ container weight → +	6 oz
total weight	5 lb 20 oz

5 lb 20 oz = 5 lb + 1 lb 4 oz = 6 lb 4 oz
The total weight is 6 lb 4 oz.

7. $3.41 \text{ g} = \dfrac{3.41 \text{ g}}{1} \cdot \dfrac{1000 \text{ mg}}{1 \text{ g}} = 3410 \text{ mg}$

8. $56.2 \text{ cg} = 5\underset{\smile}{6}.2 \text{ cg} = 0.562 \text{ g}$

9. 3.1 dg = 0.31 g   or   2.5 g = 25 dg

$$\begin{array}{r} 2.50 \text{ g} \\ -\ 0.31 \text{ g} \\ \hline 2.19 \text{ g} \end{array} \qquad \begin{array}{r} 25.0 \text{ dg} \\ -\ 3.1 \text{ dg} \\ \hline 21.9 \text{ dg} \end{array}$$

10.
$$\begin{array}{r} 22.9 \quad \approx 23 \\ 24\overline{)550.0 \text{ kg}} \\ \underline{-48} \\ 70 \\ \underline{-48} \\ 22\ 0 \\ \underline{-21\ 6} \\ 4 \end{array}$$

Each bag weighs about 23 kg.

Vocabulary and Readiness Check

1. <u>Mass</u> is a measure of the amount of substance in an object. This measure does not change.

3. The basic unit of mass in the metric system is the <u>gram</u>.

5. One ton equals <u>2000</u> pounds.

Exercise Set 9.5

1. $2 \text{ lb} = \dfrac{2 \text{ lb}}{1} \cdot \dfrac{16 \text{ oz}}{1 \text{ lb}} = 2 \cdot 16 \text{ oz} = 32 \text{ oz}$

3. $5 \text{ tons} = \dfrac{5 \text{ tons}}{1} \cdot \dfrac{2000 \text{ lb}}{1 \text{ ton}}$
$= 5 \cdot 2000 \text{ lb}$
$= 10,000 \text{ lb}$

5. $18,000 \text{ lb} = \dfrac{18,000 \text{ lb}}{1} \cdot \dfrac{1 \text{ ton}}{2000 \text{ lb}}$
$= \dfrac{18,000}{2000} \text{ tons}$
$= 9 \text{ tons}$

7. $60 \text{ oz} = \dfrac{60 \text{ oz}}{1} \cdot \dfrac{1 \text{ lb}}{16 \text{ oz}}$
$= \dfrac{60}{16} \text{ lb}$
$= \dfrac{15}{4} \text{ lb}$
$= 3\dfrac{3}{4} \text{ lb}$

9. $3500 \text{ lb} = \dfrac{3500 \text{ lb}}{1} \cdot \dfrac{1 \text{ ton}}{2000 \text{ lb}}$
$= \dfrac{3500}{2000} \text{ tons}$
$= \dfrac{7}{4} \text{ tons}$
$= 1\dfrac{3}{4} \text{ tons}$

11. $12.75 \text{ lb} = \dfrac{12.75 \text{ lb}}{1} \cdot \dfrac{16 \text{ oz}}{1 \text{ lb}}$
$= 12.75 \cdot 16 \text{ oz}$
$= 204 \text{ oz}$

13. $4.9 \text{ tons} = \dfrac{4.9 \text{ tons}}{1} \cdot \dfrac{2000 \text{ lb}}{1 \text{ ton}}$
$= 4.9 \cdot 2000 \text{ lb}$
$= 9800 \text{ lb}$

15. $4\dfrac{3}{4} \text{ lb} = \dfrac{19}{4} \text{ lb}$
$= \dfrac{\frac{19}{4} \text{ lb}}{1} \cdot \dfrac{16 \text{ oz}}{1 \text{ lb}}$
$= \dfrac{19}{4} \cdot 16 \text{ oz}$
$= 76 \text{ oz}$

17. $2950 \text{ lb} = \dfrac{2950 \text{ lb}}{1} \cdot \dfrac{1 \text{ ton}}{2000 \text{ lb}}$

$= \dfrac{2950}{2000} \text{ tons}$

$= \dfrac{59}{40} \text{ tons}$

$= 1.475 \text{ tons}$

$\approx 1.5 \text{ tons}$

19. $\dfrac{4}{5} \text{ oz} = \dfrac{\frac{4}{5}}{1} \cdot \dfrac{1 \text{ lb}}{16 \text{ oz}} = \dfrac{4}{5} \cdot \dfrac{1}{16} \text{ lb} = \dfrac{1}{20} \text{ lb}$

21. $5\dfrac{3}{4} \text{ lb} = \dfrac{23}{4} \text{ lb}$

$= \dfrac{\frac{23}{4} \text{ lb}}{1} \cdot \dfrac{16 \text{ oz}}{1 \text{ lb}}$

$= \dfrac{23}{4} \cdot 16 \text{ oz}$

$= 92 \text{ oz}$

23. $10 \text{ lb } 1 \text{ oz} = 10 \cdot 16 \text{ oz} + 1 \text{ oz}$

$= 160 \text{ oz} + 1 \text{ oz}$

$= 161 \text{ oz}$

25. $89 \text{ oz} = \dfrac{89 \text{ oz}}{1} \cdot \dfrac{1 \text{ lb}}{16 \text{ oz}} = \dfrac{89}{16} \text{ lb}$

$$\begin{array}{r} 5 \text{ lb } 9 \text{ oz} \\ 16\overline{\smash{)}89} \\ \underline{-80} \\ 9 \end{array}$$

$89 \text{ oz} = 5 \text{ lb } 9 \text{ oz}$

27. $34 \text{ lb } 12 \text{ oz} + 18 \text{ lb } 14 \text{ oz} = 52 \text{ lb } 26 \text{ oz}$

$= 52 \text{ lb} + 1 \text{ lb } 10 \text{ oz}$

$= 53 \text{ lb } 10 \text{ oz}$

29. $3 \text{ tons } 1820 \text{ lb} + 4 \text{ tons } 930 \text{ lb}$

$= 7 \text{ tons } 2750 \text{ lb}$

$= 7 \text{ tons} + 1 \text{ ton } 750 \text{ lb}$

$= 8 \text{ tons } 750 \text{ lb}$

31.
$$\begin{array}{r} 5 \text{ tons } 1050 \text{ lb} \\ - 2 \text{ tons } 875 \text{ lb} \\ \hline 3 \text{ tons } 175 \text{ lb} \end{array}$$

33.
$$\begin{array}{r} 12 \text{ lb } 4 \text{ oz} \\ - 3 \text{ lb } 9 \text{ oz} \end{array} \qquad \begin{array}{r} 11 \text{ lb } 20 \text{ oz} \\ - 3 \text{ lb } 9 \text{ oz} \\ \hline 8 \text{ lb } 11 \text{ oz} \end{array}$$

35. $5 \text{ lb } 3 \text{ oz} \times 6 = 30 \text{ lb } 18 \text{ oz}$

$= 30 \text{ lb} + 1 \text{ lb } 2 \text{ oz}$

$= 31 \text{ lb } 2 \text{ oz}$

37. $6 \text{ tons } 1500 \text{ lb} \div 5$

$= \dfrac{6}{5} \text{ tons } 300 \text{ lb}$

$= 1\dfrac{1}{5} \text{ tons } 300 \text{ lb}$

$= 1 \text{ ton} + \dfrac{2000 \text{ lb}}{5} + 300 \text{ lb}$

$= 1 \text{ ton} + 400 \text{ lb} + 300 \text{ lb}$

$= 1 \text{ ton } 700 \text{ lb}$

39. $500 \text{ g} = \dfrac{500 \text{ g}}{1} \cdot \dfrac{1 \text{ kg}}{1000 \text{ g}} = \dfrac{500}{1000} \text{ kg} = 0.5 \text{ kg}$

41. $4 \text{ g} = \dfrac{4 \text{ g}}{1} \cdot \dfrac{1000 \text{ mg}}{1 \text{ g}}$

$= 4 \cdot 1000 \text{ mg}$

$= 4000 \text{ mg}$

43. $25 \text{ kg} = \dfrac{25 \text{ kg}}{1} \cdot \dfrac{1000 \text{ g}}{1 \text{ kg}}$

$= 25 \cdot 1000 \text{ g}$

$= 25,000 \text{ g}$

45. $48 \text{ mg} = \dfrac{48 \text{ mg}}{1} \cdot \dfrac{1 \text{ g}}{1000 \text{ mg}}$

$= \dfrac{48}{1000} \text{ g}$

$= 0.048 \text{ g}$

47. $6.3 \text{ g} = \dfrac{6.3 \text{ g}}{1} \cdot \dfrac{1 \text{ kg}}{1000 \text{ g}}$

$= \dfrac{6.3}{1000} \text{ kg}$

$= 0.0063 \text{ kg}$

49. $15.14 \text{ g} = \dfrac{15.14 \text{ g}}{1} \cdot \dfrac{1000 \text{ mg}}{1 \text{ g}}$

$\qquad = 15.14 \cdot 1000 \text{ mg}$

$\qquad = 15,140 \text{ mg}$

51. $6.25 \text{ kg} = \dfrac{6.25 \text{ kg}}{1} \cdot \dfrac{1000 \text{ g}}{1 \text{ kg}}$

$\qquad = 6.25 \cdot 1000 \text{ g}$

$\qquad = 6250 \text{ g}$

53. $35 \text{ hg} = \dfrac{35 \text{ hg}}{1} \cdot \dfrac{10,000 \text{ cg}}{1 \text{ hg}}$

$\qquad = 35 \cdot 10,000 \text{ cg}$

$\qquad = 350,000 \text{ cg}$

55.
$$\begin{array}{r} 3.8 \text{ mg} \\ + 9.7 \text{ mg} \\ \hline 13.5 \text{ mg} \end{array}$$

57. $205 \text{ mg} + 5.61 \text{ g} = 0.205 \text{ g} + 5.61 \text{ g} = 5.815 \text{ g}$
or
$\quad 205 \text{ mg} + 5.61 \text{ g} = 205 \text{ mg} + 5610 \text{ mg}$
$\qquad\qquad\qquad\qquad\quad = 5815 \text{ mg}$

59. $9 \text{ g} - 7150 \text{ mg}$

$$\begin{array}{r} 9000 \text{ mg} \\ - 7150 \text{ mg} \\ \hline 1850 \text{ mg} \end{array} \quad \text{or} \quad \begin{array}{r} 9.000 \text{ g} \\ - 7.150 \text{ g} \\ \hline 1.850 \text{ g} \end{array}$$

61. $1.61 \text{ kg} - 250 \text{ g} = 1.61 \text{ kg} - 0.250 \text{ kg} = 1.36 \text{ kg}$
or
$\quad 1.61 \text{ kg} - 250 \text{ g} = 1610 \text{ g} - 250 \text{ g} = 1360 \text{ g}$

63.
$$\begin{array}{r} 5.2 \text{ kg} \\ \times 2.6 \\ \hline 13.52 \text{ kg} \end{array}$$

65. $17 \text{ kg} \div 8 = \dfrac{17}{8} \text{ kg}$

$$
\begin{array}{r}
2.125 \\
8\overline{)\ 17.000} \\
\underline{-16} \\
1\ 0 \\
\underline{-8} \\
20 \\
\underline{-16} \\
40 \\
\underline{-40} \\
0
\end{array}
$$

$17 \text{ kg} \div 8 = 2.125 \text{ kg}$

	Object	Tons	Pounds	Ounces
67.	Statue of Liberty—weight of copper sheeting	100	200,000	3,200,000
69.	A 12-inch cube of osmium (heaviest metal)	$\frac{269}{400}$ or 0.6725	1345	21,520

	Object	Grams	Kilograms	Milligrams	Centigrams
71.	Capsule of Amoxicillin (Antibiotic)	0.5	0.0005	500	50
73.	A six-year-old boy	21,000	21	21,000,000	2,100,000

75.

$$
\begin{array}{r}
336 \\
\times\ 24 \\
\hline
1344 \\
6720 \\
\hline
8064
\end{array}
$$

$8064 \text{ g} = \dfrac{8064 \text{ g}}{1} \cdot \dfrac{1 \text{ kg}}{1000 \text{ g}} = \dfrac{8064}{1000} \text{ kg} = 8.064 \text{ kg}$

24 cans weigh 8.064 kg.

77. $0.09 \text{ g} = \dfrac{0.09 \text{ g}}{1} \cdot \dfrac{1000 \text{ mg}}{1 \text{ g}}$

$\phantom{0.09 \text{ g}} = 0.09 \cdot 1000 \text{ mg}$

$\phantom{0.09 \text{ g}} = 90 \text{ mg}$

$90 \text{ mg} - 60 \text{ mg} = 30 \text{ mg}$

The extra-strength tablet contains 30 mg more medication.

79.
$$\begin{array}{r} 1 \text{ lb } 10 \text{ oz} \\ + 3 \text{ lb } 14 \text{ oz} \\ \hline \end{array}$$
4 lb 24 oz = 4 lb + 1 lb 8 oz = 5 lb 8 oz
The total amount of rice is 5 lb 8 oz.

81.
$$\begin{array}{r} 64 \text{ lb } 8 \text{ oz} \\ - 28 \text{ lb } 10 \text{ oz} \\ \hline \end{array} \qquad \begin{array}{r} 63 \text{ lb } 24 \text{ oz} \\ - 28 \text{ lb } 10 \text{ oz} \\ \hline 35 \text{ lb } 14 \text{ oz} \end{array}$$

Carla's zucchini was 35 lb 14 oz lighter than the record weight.

83.
$$\begin{array}{r} 7 \text{ lb } 8 \text{ oz} \\ - \qquad 8.6 \text{ oz} \\ \hline \end{array} \qquad \begin{array}{r} 6 \text{ lb } 24 \text{ oz} \\ - \qquad 8.6 \text{ oz} \\ \hline 6 \text{ lb } 15.4 \text{ oz} \end{array}$$

This is 6 lb 15.4 oz lighter.

85. $3 \times 16 = 48$
3 cartons contain 48 boxes of fruit.
$3 \text{ mg} \times 48 = 144 \text{ mg}$
3 cartons contain 144 mg of preservatives.

87. $26 \text{ g} \times 12 = 312 \text{ g}$
$$\frac{312 \text{ g}}{1} \cdot \frac{1 \text{ kg}}{1000 \text{ g}} = 0.312 \text{ kg of packaging}$$
in a carton
$6.432 \text{ kg} - 0.312 \text{ kg} = 6.12 \text{ kg}$
The actual weight of the oatmeal is 6.12 kg.

89. $3 \text{ lb } 4 \text{ oz} \times 10 = 30 \text{ lb } 40 \text{ oz}$
$$\begin{aligned} &= 30 \text{ lb} + 2 \text{ lb } 8 \text{ oz} \\ &= 32 \text{ lb } 8 \text{ oz} \end{aligned}$$
Each box weighs 32 lb 8 oz.
$32 \text{ lb } 8 \text{ oz} \times 4 = 128 \text{ lb } 32 \text{ oz}$
$$\begin{aligned} &= 128 \text{ lb} + 2 \text{ lb} \\ &= 130 \text{ lb} \end{aligned}$$
4 boxes of meat weigh 130 lb.

91.
$$\begin{array}{r} 55 \text{ lb } 4 \text{ oz} \\ - 2 \text{ lb } 8 \text{ oz} \\ \hline \end{array} \qquad \begin{array}{r} 54 \text{ lb } 20 \text{ oz} \\ - 2 \text{ lb } \ 8 \text{ oz} \\ \hline 52 \text{ lb } 12 \text{ oz} \end{array}$$

$$\begin{array}{r} 52 \text{ lb } 12 \text{ oz} \\ \times \qquad 4 \\ \hline 208 \text{ lb } 48 \text{ oz} \end{array} = 208 \text{ lb} + 3 \text{ lb} = 211 \text{ lb}$$
4 cartons contain 211 lb of pineapple.

93. $\dfrac{4}{25} = \dfrac{4}{25} \cdot \dfrac{4}{4} = \dfrac{16}{100} = 0.16$

95. $\dfrac{7}{8} = \dfrac{7}{8} \cdot \dfrac{125}{125} = \dfrac{875}{1000} = 0.875$

97. No, a pill containing 2 kg of medication is not reasonable.

99. Yes, a bag of flour weighing 4.5 kg is reasonable.

101. No, a professor weighing less than 150 g is not reasonable.

103. answers may vary

105. True, a kilogram is 1000 grams.

107. answers may vary

Section 9.6

Practice Problems

1. $43 \text{ pt} = \dfrac{43 \text{ pt}}{1} \cdot \dfrac{1 \text{ qt}}{2 \text{ pt}} = \dfrac{43}{2} \text{ qt} = 21\dfrac{1}{2} \text{ qt}$

2. $26 \text{ qt} = \dfrac{26 \text{ qt}}{1} \cdot \dfrac{4 \text{ c}}{1 \text{ qt}} = 26 \cdot 4 \text{ c} = 104 \text{ c}$

3.
$$\begin{array}{r} 1 \text{ gal } 1 \text{ qt} \\ - \qquad 2 \text{ qt} \\ \hline \end{array} \rightarrow \begin{array}{r} 5 \text{ qt} \\ - 2 \text{ qt} \\ \hline 3 \text{ qt} \end{array}$$

4.
$$\begin{array}{r} 15 \text{ gal } 3 \text{ qt} \\ + 4 \text{ gal } 3 \text{ qt} \\ \hline 19 \text{ gal } 6 \text{ qt} \end{array} = 19 \text{ gal} + 1 \text{ gal } 2 \text{ qt} = 20 \text{ gal } 2$$

qt
The total amount of oil will be 20 gal 2 qt.

5. $2100 \text{ ml} = \dfrac{2100 \text{ ml}}{1} \cdot \dfrac{1 \text{ L}}{1000 \text{ ml}}$

$= \dfrac{2100}{1000} \text{ L}$

$= 2.1 \text{ L}$

6. $2.13 \text{ dal} = \dfrac{2.13 \text{ dal}}{1} \cdot \dfrac{10 \text{ L}}{1 \text{ dal}}$

$= 2.13 \cdot 10 \text{ L}$

$= 21.3 \text{ L}$

7. $1250 \text{ ml} = 1.250 \text{ L}$ $\qquad 2.9 \text{ L} = 2900 \text{ ml}$

$\qquad\quad 1.25 \text{ L} \qquad\qquad\qquad 1250 \text{ ml}$

$\qquad + \; 2.9 \;\; \text{L} \qquad\qquad\quad + \; 2900 \text{ ml}$

$\qquad\quad \overline{4.15 \text{ L}} \qquad\qquad\qquad \overline{4150 \text{ ml}}$

The total is 4.15 L or 4150 ml.

8. 28.6 L

$\underline{\times \quad 85}$

$\quad 143\,0$

$\quad 2288\,0$

$\overline{\;2431.0 \text{ L}\;}$

Thus, 2431 L can be pumped in 85 minutes.

Vocabulary and Readiness Check

1. Units of <u>capacity</u> are generally used to measure liquids.

3. One cup equals 8 <u>fluid ounces</u>.

5. One pint equals 2 <u>cups</u>.

7. One gallon equals 4 <u>quarts</u>.

Exercise Set 9.6

1. $32 \text{ fl oz} = \dfrac{32 \text{ fl oz}}{1} \cdot \dfrac{1 \text{ c}}{8 \text{ fl oz}} = \dfrac{32}{8} \text{ c} = 4 \text{ c}$

3. $8 \text{ qt} = \dfrac{8 \text{ qt}}{1} \cdot \dfrac{2 \text{ pt}}{1 \text{ qt}} = 8 \cdot 2 \text{ pt} = 16 \text{ pt}$

5. $14 \text{ qt} = \dfrac{14 \text{ qt}}{1} \cdot \dfrac{1 \text{ gal}}{4 \text{ qt}} = \dfrac{14}{4} \text{ gal} = 3\dfrac{1}{2} \text{ gal}$

7. $80 \text{ fl oz} = \dfrac{80 \text{ fl oz}}{1} \cdot \dfrac{1 \text{ c}}{8 \text{ fl oz}} = \dfrac{80}{8} \text{ c} = 10 \text{ c}$

$10 \text{ c} = \dfrac{10 \text{ c}}{1} \cdot \dfrac{1 \text{ pt}}{2 \text{ c}} = \dfrac{10}{2} \text{ pt} = 5 \text{ pt}$

9. $2 \text{ qt} = \dfrac{2 \text{ qt}}{1} \cdot \dfrac{2 \text{ pt}}{1 \text{ qt}} \cdot \dfrac{2 \text{ c}}{1 \text{ pt}} = 2 \cdot 2 \cdot 2 \text{ c} = 8 \text{ c}$

11. $120 \text{ fl oz} = \dfrac{120 \text{ fl oz}}{1} \cdot \dfrac{1 \text{ c}}{8 \text{ fl oz}} \cdot \dfrac{1 \text{ pt}}{2 \text{ c}} \cdot \dfrac{1 \text{ qt}}{2 \text{ pt}}$

$= \dfrac{120}{8 \cdot 2 \cdot 2} \text{ qt}$

$= \dfrac{15}{4} \text{ qt}$

$= 3\dfrac{3}{4} \text{ qt}$

13. $42 \text{ c} = \dfrac{42 \text{ c}}{1} \cdot \dfrac{1 \text{ qt}}{4 \text{ c}} = \dfrac{42}{4} \text{ qt} = 10\dfrac{1}{2} \text{ qt}$

15. $4\dfrac{1}{2} \text{ pt} = \dfrac{9}{2} \text{ pt} = \dfrac{\frac{9}{2} \text{ pt}}{1} \cdot \dfrac{2 \text{ c}}{1 \text{ pt}} = \dfrac{9}{2} \cdot 2 \text{ c} = 9 \text{ c}$

17. $5 \text{ gal } 3 \text{ qt} = \dfrac{5 \text{ gal}}{1} \cdot \dfrac{4 \text{ qt}}{1 \text{ gal}} + 3 \text{ qt}$

$= 5 \cdot 4 \text{ qt} + 3 \text{ qt}$

$= 20 \text{ qt} + 3 \text{ qt}$

$= 23 \text{ qt}$

19. $\dfrac{1}{2} \text{ c} = \dfrac{\frac{1}{2} \text{ c}}{1} \cdot \dfrac{1 \text{ pt}}{2 \text{ c}} = \dfrac{1}{2} \cdot \dfrac{1}{2} \text{ pt} = \dfrac{1}{4} \text{ pt}$

21. $58 \text{ qt} = 56 \text{ qt} + 2 \text{ qt}$

$= \dfrac{56 \text{ qt}}{1} \cdot \dfrac{1 \text{ gal}}{4 \text{ qt}} + 2 \text{ qt}$

$= \dfrac{56}{4} \text{ gal} + 2 \text{ qt}$

$= 14 \text{ gal } 2 \text{ qt}$

23. $39 \text{ pt} = 38 \text{ pt} + 1 \text{ pt}$

$= \dfrac{38 \text{ pt}}{1} \cdot \dfrac{1 \text{ qt}}{2 \text{ pt}} + 1 \text{ pt}$

$= 19 \text{ qt} + 1 \text{ pt}$

$= 16 \text{ qt} + 3 \text{ qt} + 1 \text{ pt}$

$= \dfrac{16 \text{ qt}}{1} \cdot \dfrac{1 \text{ gal}}{4 \text{ qt}} + 3 \text{ qt} + 1 \text{ pt}$

$= 4 \text{ gal} + 3 \text{ qt} + 1 \text{ pt}$

$= 4 \text{ gal } 3 \text{ qt } 1 \text{ pt}$

25. $2\dfrac{3}{4} \text{ gal} = \dfrac{11}{4} \text{ gal}$

$= \dfrac{\frac{11}{4} \text{ gal}}{1} \cdot \dfrac{4 \text{ qt}}{1 \text{ gal}} \cdot \dfrac{2 \text{ pt}}{1 \text{ qt}}$

$= \dfrac{11}{4} \cdot 4 \cdot 2 \text{ pt}$

$= 22 \text{ pt}$

27. $5 \text{ gal } 3 \text{ qt}$
    $\underline{+\ 7 \text{ gal } 3 \text{ qt}}$
    $12 \text{ gal } 6 \text{ qt} = 12 \text{ gal} + 1 \text{ gal } 2 \text{ qt}$
    $\phantom{12 \text{ gal } 6 \text{ qt}} = 13 \text{ gal } 2 \text{ qt}$

29. $1 \text{ c } 5 \text{ fl oz} + 2 \text{ c } 7 \text{ fl oz} = 3 \text{ c } 12 \text{ fl oz}$
$\phantom{1 \text{ c } 5 \text{ fl oz} + 2 \text{ c } 7 \text{ fl oz}} = 3 \text{ c} + 1 \text{ c } 4 \text{ fl oz}$
$\phantom{1 \text{ c } 5 \text{ fl oz} + 2 \text{ c } 7 \text{ fl oz}} = 4 \text{ c } 4 \text{ fl oz}$

31. 3 gal $2 \text{ gal } 4 \text{ qt}$
    $\underline{-\ 1 \text{ gal } 3 \text{ qt}}$ $\underline{-\ 1 \text{ gal } 3 \text{ qt}}$
    $\phantom{-\ 1 \text{ gal } 3 \text{ qt}}$ $1 \text{ gal } 1 \text{ qt}$

33. $3 \text{ gal } 1 \text{ qt}$ $2 \text{ gal } 5 \text{ qt}$ $2 \text{ gal } 4 \text{ qt } 2 \text{ pt}$
   $\underline{-\ 1 \text{ qt } 1 \text{ pt}}$ $\underline{-\ 1 \text{ qt } 1 \text{ pt}}$ $\underline{-\ 1 \text{ qt } 1 \text{ pt}}$
   $\phantom{-\ 1 \text{ qt } 1 \text{ pt}}$ $\phantom{-\ 1 \text{ qt } 1 \text{ pt}}$ $2 \text{ gal } 3 \text{ qt } 1 \text{ pt}$

35. $8 \text{ gal } 2 \text{ qt} \times 2 = 16 \text{ gal } 4 \text{ qt}$
$\phantom{8 \text{ gal } 2 \text{ qt} \times 2} = 16 \text{ gal} + 1 \text{ gal}$
$\phantom{8 \text{ gal } 2 \text{ qt} \times 2} = 17 \text{ gal}$

37. $9 \text{ gal } 2 \text{ qt} \div 2 = (8 \text{ gal } 4 \text{ qt} + 2 \text{ qt}) \div 2$
$\phantom{9 \text{ gal } 2 \text{ qt} \div 2} = 8 \text{ gal } 6 \text{ qt} \div 2$
$\phantom{9 \text{ gal } 2 \text{ qt} \div 2} = 4 \text{ gal } 3 \text{ qt}$

39. $5 \text{L} = \dfrac{5 \text{L}}{1} \cdot \dfrac{1000 \text{ ml}}{1 \text{ L}} = 5000 \text{ ml}$

41. $0.16 \text{ L} = \dfrac{0.16 \text{ L}}{1} \cdot \dfrac{1 \text{ kl}}{1000 \text{ L}}$

$= \dfrac{0.16}{1000} \text{ kl}$

$= 0.00016 \text{ kl}$

43. $5600 \text{ ml} = \dfrac{5600 \text{ ml}}{1} \cdot \dfrac{1 \text{ L}}{1000 \text{ ml}}$

$= \dfrac{5600}{1000} \text{ L}$

$= 5.6 \text{ L}$

45. $3.2 \text{ L} = \dfrac{3.2 \text{ L}}{1} \cdot \dfrac{100 \text{ cl}}{1 \text{ L}} = 3.2 \cdot 1000 \text{ cl} = 320 \text{ cl}$

47. $410 \text{ L} = \dfrac{410 \text{ L}}{1} \cdot \dfrac{1 \text{ kl}}{1000 \text{ L}} = \dfrac{410}{1000} \text{ kl} = 0.41 \text{ kl}$

49. $64 \text{ ml} = \dfrac{64 \text{ ml}}{1} \cdot \dfrac{1 \text{ L}}{1000 \text{ ml}}$

$= \dfrac{64}{1000} \text{ L}$

$= 0.064 \text{ L}$

51. $0.16 \text{ kl} = \dfrac{0.16 \text{ kl}}{1} \cdot \dfrac{1000 \text{ L}}{1 \text{ kl}}$

$= 0.16 \cdot 1000 \text{ L}$

$= 160 \text{ L}$

53. $3.6 \text{ L} = \dfrac{3.6 \text{ L}}{1} \cdot \dfrac{1000 \text{ ml}}{1 \text{ L}}$

$= 3.6 \cdot 1000 \text{ ml}$

$= 3600 \text{ ml}$

55. $3.4 \text{ L} + 15.9 \text{ L} = 19.3 \text{ L}$

57. $2700 \text{ ml} + 1.8 \text{ L} = 2.7 \text{ L} + 1.8 \text{ L} = 4.5 \text{ L}$
or
$ 2700 \text{ ml} + 1.8 \text{ L} = 2700 \text{ ml} + 1800 \text{ ml}$
$\phantom{57. 2700 \text{ ml} + 1.8 \text{ L}} = 4500 \text{ ml}$

59.

$$\begin{array}{ll} 8.6\text{ L} & 8600\text{ ml} \\ -\,190\text{ ml} & -\,190\text{ ml} \end{array} \text{ or } \begin{array}{l} 8.60\ \ \text{L} \\ -\,0.190\text{ L} \end{array}$$

$$\begin{array}{ll} & 8410\text{ ml} & 8.41\ \ \text{L} \end{array}$$

61. $17,500\text{ ml} - 0.9\text{ L} = 17,500\text{ ml} - 900\text{ ml}$
$$\phantom{17,500\text{ ml} - 0.9\text{ L} } = 16,600\text{ ml}$$
or
$$17,500\text{ ml} - 0.9\text{ L} = 17.5\text{ L} - 0.9\text{ L} = 16.6\text{ L}$$

63. $480\text{ ml} \times 8 = 3840\text{ ml}$

65. $81.2\text{ L} \div 0.5 = 81.2\text{ L} \div \dfrac{1}{2}$
$$\phantom{81.2\text{ L} \div 0.5 } = 81.2\text{ L} \cdot 2$$
$$\phantom{81.2\text{ L} \div 0.5 } = 162.4\text{ L}$$

	Capacity	Cups	Gallons	Quarts	Pints
67.	An average-size bath of water	336	21	84	168
69.	Your kidneys filter about this amount of blood every minute	4	$\frac{1}{4}$	1	2

71.

$$\begin{array}{ll} 2\text{ L} & 2.000\text{ L} \\ -\,410\text{ ml} & -\,0.410\text{ L} \end{array}$$
$$\phantom{2\text{ L} \quad } \overline{1.590\text{ L}}$$

There was 1.59 L left in the bottle.

73. $354\text{ ml} + 18.6\text{ L} = 0.354\text{ L} + 18.6\text{ L} = 18.954\text{ L}$
There were 18.954 liters of gasoline in her tank.

75. $\dfrac{1}{30}\text{ gal} = \dfrac{\frac{1}{30}\text{ gal}}{1} \cdot \dfrac{4\text{ qt}}{1\text{ gal}} \cdot \dfrac{2\text{ pt}}{1\text{ qt}} \cdot \dfrac{2\text{ c}}{1\text{ pt}} \cdot \dfrac{8\text{ fl oz}}{1\text{ c}}$

$$\phantom{\dfrac{1}{30}\text{ gal}} = \dfrac{1}{30} \cdot 128\text{ fl oz}$$
$$\phantom{\dfrac{1}{30}\text{ gal}} \approx 4.3\text{ fl oz}$$

$\dfrac{1}{30}$ gal is about 4.3 fluid ounces.

77. 5 pt 1 c + 2 pt 1 c = 7 pt 2 c
$$= 7 \text{ pt} + 1 \text{ pt}$$
$$= 8 \text{ pt}$$
$$= \frac{8 \text{ pt}}{1} \cdot \frac{1 \text{ qt}}{2 \text{ pt}}$$
$$= \frac{8}{2} \text{ qt}$$
$$= \frac{4 \text{ qt}}{1} \cdot \frac{1 \text{ gal}}{4 \text{ qt}}$$
$$= \frac{4}{4} \text{ gal}$$
$$= 1 \text{ gal}$$

Yes, the liquid can be poured into the container without causing it to overflow.

79. $44.3 \overline{)14.0}$ becomes

$$
\begin{array}{r}
0.3160 \approx 0.6 \\
443\overline{)140.0000} \\
-132\;9 \\
\hline
7\;10 \\
-4\;43 \\
\hline
2\;670 \\
-2\;658 \\
\hline
120 \\
-0 \\
\hline
120
\end{array}
$$

$$\frac{\$14}{44.3 \text{ L}} \approx \frac{\$0.316}{1 \text{ L}}$$

The price was $0.316 per liter.

81. $\dfrac{20}{25} = \dfrac{4 \cdot 5}{5 \cdot 5} = \dfrac{4}{5}$

83. $\dfrac{27}{45} = \dfrac{3 \cdot 9}{5 \cdot 9} = \dfrac{3}{5}$

85. $\dfrac{72}{80} = \dfrac{8 \cdot 9}{8 \cdot 10} = \dfrac{9}{10}$

87. No, a 2 L dose of cough medicine is not reasonable.

89. No, a tub filled with 3000 ml of hot water is not reasonable.

91. answers may vary

93. answers may vary

95. answers may vary

97. B indicates 1.5 cc.

99. D indicates 2.7 cc.

101. B indicates 54 u or 0.54 cc.

103. D indicates 86 u or 0.86 cc.

Section 9.7

Practice Problems

1. $1.5 \text{ cm} = \dfrac{1.5 \text{ cm}}{1} \cdot \dfrac{1 \text{ in.}}{2.54 \text{ cm}}$
$$= \dfrac{1.5}{2.54} \text{ in.} \approx 0.59 \text{ in.}$$

2. $8 \text{ oz} \approx \dfrac{8 \text{ oz}}{1} \cdot \dfrac{28.35 \text{ g}}{1 \text{ oz}} = 8 \cdot 28.35 \text{ g} = 226.8 \text{ g}$

3. $237 \text{ ml} \approx \dfrac{237 \text{ ml}}{1} \cdot \dfrac{1 \text{ fl oz}}{29.57 \text{ ml}}$
$$= \dfrac{237}{29.57} \text{ fl oz}$$
$$\approx 8 \text{ fl oz}$$

4. $F = \dfrac{9}{5} \cdot C + 32 = \dfrac{9}{5} \cdot 60 + 32 = 108 + 32 = 140$

Thus, 60°C is equivalent to 140°F.

5. $F = 1.8 \cdot C + 32$
$$= 1.8 \cdot 32 + 32$$
$$= 57.6 + 32$$
$$= 89.6$$

Therefore, 32°C is the same as 89.6°F.

6. $C = \dfrac{5}{9} \cdot (F - 32) = \dfrac{5}{9} \cdot (68 - 32) = \dfrac{5}{9} \cdot (36) = 20$

Therefore, 68°F is the same temperature as 20°C.

7. $C = \dfrac{5}{9} \cdot (F - 32) = \dfrac{5}{9} \cdot (113 - 32) = \dfrac{5}{9} \cdot (81) = 45$

    Therefore, 113°F is 45°C.

8. $C = \dfrac{5}{9} \cdot (F - 32) = \dfrac{5}{9} \cdot (102.8 - 32) = \dfrac{5}{9} \cdot (70.8) = 39.3$

    Albert's temperature is 39.3°C.

Exercise Set 9.7

1. $756 \text{ ml} \approx \dfrac{756 \text{ ml}}{1} \cdot \dfrac{1 \text{ fl oz}}{29.57 \text{ ml}}$
    $= \dfrac{756}{29.57} \text{ fl oz}$
    $\approx 25.57 \text{ fl oz}$

3. $86 \text{ in.} = \dfrac{86 \text{ in.}}{1} \cdot \dfrac{2.54 \text{ cm}}{1 \text{ in.}}$
    $= 86 \cdot 2.54 \text{ cm}$
    $= 218.44 \text{ cm}$

5. $1000 \text{ g} \approx \dfrac{1000 \text{ g}}{1} \cdot \dfrac{0.04 \text{ oz}}{1 \text{ g}}$
    $= 1000 \cdot 0.04 \text{ oz}$
    $= 40 \text{ oz}$

7. $93 \text{ km} \approx \dfrac{93 \text{ km}}{1} \cdot \dfrac{0.62 \text{ mi}}{1 \text{ km}}$
    $= 93 \cdot 0.62 \text{ mi}$
    $= 57.66 \text{ mi}$

9. $14.5 \text{ L} \approx \dfrac{14.5 \text{ L}}{1} \cdot \dfrac{0.26 \text{ gal}}{1 \text{ L}} \approx 3.77 \text{ gal}$

11. $30 \text{ lb} \approx \dfrac{30 \text{ lb}}{1} \cdot \dfrac{0.45 \text{ kg}}{1 \text{ lb}} = 30 \cdot 0.45 \text{ kg} = 13.5 \text{ kg}$

		Meters	Yards	Centimeters	Feet	Inches
13.	The Height of a Woman	1.5	$1\frac{2}{3}$	150	5	60
15.	Leaning Tower of Pisa	54.864	60	5486.4	180	2160

17. $10 \text{ cm} = \dfrac{10 \text{ cm}}{1} \cdot \dfrac{1 \text{ in.}}{2.54 \text{ cm}} \approx 3.94 \text{ in.}$
The balance beam is approximately 3.94 inches wide.

19. $50 \text{ mph} \approx \dfrac{50 \text{ mph}}{1} \cdot \dfrac{1.61 \text{ km}}{1 \text{ mi}} = 80.5 \text{ kph}$
The speed limit is approximately 80.5 kilometers per hour.

21. $200 \text{ mg} = 0.2 \text{ g} \approx \dfrac{0.2 \text{ g}}{1} \cdot \dfrac{0.04 \text{ oz}}{1 \text{ g}} = 0.008 \text{ oz}$

23. $100 \text{ kg} \approx \dfrac{100 \text{ kg}}{1} \cdot \dfrac{2.2 \text{ lb}}{1 \text{ kg}}$
$= 100 \cdot 2.2 \text{ lb}$
$= 220 \text{ lb}$
$15 \text{ stone } 10 \text{ lb}$
$= \dfrac{15 \text{ stone}}{1} \cdot \dfrac{14 \text{ lb}}{1 \text{ stone}} + 10 \text{ lb}$
$= 15 \cdot 14 \text{ lb} + 10 \text{ lb}$
$= 210 \text{ lb} + 10 \text{ lb}$
$= 220 \text{ lb}$
Yes; the stamp is approximately correct.

25. $4500 \text{ km} = \dfrac{4500 \text{ km}}{1} \cdot \dfrac{0.62 \text{ mi}}{1 \text{ km}} = 2790 \text{ mi}$
The trip is about 2790 miles.

27. $3\dfrac{1}{2} \text{ in.} = \dfrac{3\frac{1}{2} \text{ in.}}{1} \cdot \dfrac{2.54 \text{ cm}}{1 \text{ in.}} = 8.89 \text{ cm}$
$8.89 \text{ cm} = \dfrac{8.89 \text{ cm}}{1} \cdot \dfrac{10 \text{ mm}}{1 \text{ cm}}$
$= 88.9 \text{ mm} \approx 90 \text{ mm}$
The width is approximately 90 mm.

29. $1.5 \text{ lb} - 1.25 \text{ l} = 0.25 \text{ lb}$
$0.25 \text{ lb} \approx \dfrac{0.25 \text{ lb}}{1} \cdot \dfrac{0.45 \text{ kg}}{1 \text{ lb}} \cdot \dfrac{1000 \text{ g}}{1 \text{ kg}}$
$\approx 112.5 \text{ g}$
The difference is approximately 112.5 g.

31. $167 \text{ kmh} \approx \dfrac{167 \text{ kmh}}{1} \cdot \dfrac{0.62 \text{ mi}}{1 \text{ km}} \approx 104 \text{ mph}$
The sneeze is approximately 104 miles per hour.

33. $8 \text{ m} \approx \dfrac{8 \text{ m}}{1} \cdot \dfrac{3.28 \text{ ft}}{1 \text{ m}} \approx 26.24 \text{ ft}$
The base diameter is approximately 26.24 ft.

35. $4.5 \text{ km} \approx \dfrac{4.5 \text{ km}}{1} \cdot \dfrac{0.62 \text{ mi}}{1 \text{ km}} \approx 3 \text{ mi}$
The track is approximately 3 mi.

37. One dose every 4 hours results in
$\dfrac{24}{4} = 6$ doses per day and $6 \times 7 = 42$ doses per week.
$5 \text{ ml} \times 42 = 210 \text{ ml}$
$210 \text{ ml} \approx \dfrac{210 \text{ ml}}{1} \cdot \dfrac{1 \text{ fl oz}}{29.57 \text{ ml}} \approx 7.1 \text{ fl oz}$
8 fluid ounces of medicine should be purchased.

39. This math book has a height of about 28 cm; b.

41. A liter has greater capacity than a quart; b.

43. A kilogram weighs greater than a pound; c

45. An $8\dfrac{1}{2}$ ounce glass of water has a capacity of about 250 ml $\left(\dfrac{1}{4} \text{ L}\right)$; d.

47. The weight of an average man is about 70 kg
$(70 \text{ kg} \approx 2.2 \cdot 70 \text{ lb} = 154 \text{ lb})$; d

49. $C = \dfrac{5}{9}(F - 32) = \dfrac{5}{9}(77 - 32) = \dfrac{5}{9}(45) = 25$
$77°F$ is $25°C$.

51. $C = \dfrac{5}{9}(F - 32) = \dfrac{5}{9}(104 - 32) = \dfrac{5}{9}(72) = 40$
$104°F$ is $40°C$.

53. $F = \dfrac{9}{5}C + 32 = \dfrac{9}{5}(50) + 32 = 90 + 32 = 122$
50°C is 122°F.

55. $F = \dfrac{9}{5}C + 32 = \dfrac{9}{5}(115) + 32 = 207 + 32 = 239$
115°C is 239°F.

57. $C = \dfrac{5}{9}(F - 32)$
$= \dfrac{5}{9}(20 - 32)$
$= \dfrac{5}{9}(-12) \approx -6.7$
20°F is −6.7°C.

59. $C = \dfrac{5}{9}(F - 32)$
$= \dfrac{5}{9}(142.1 - 32)$
$= \dfrac{5}{9}(110.1)$
≈ 61.2
142.1°F is 61.2°C.

61. $F = 1.8C + 32$
$= 1.8(92) + 32$
$= 165.6 + 32$
$= 197.6$
92°C is 197.6°F.

63. $F = 1.8C + 32$
$= 1.8(12.4) + 32$
$= 22.32 + 32$
≈ 54.3
12.4°C is 54.3°F.

65. $C = \dfrac{5}{9}(F - 32)$
$= \dfrac{5}{9}(134 - 32)$
$= \dfrac{5}{9}(102)$
≈ 56.7
134°F is 56.7°C.

67. $F = 1.8C + 32$
$= 1.8(27) + 32$
$= 48.6 + 32$
$= 80.6$
27°C is 80.6°F.

69. $C = \dfrac{5}{9}(F - 32) = \dfrac{5}{9}(70 - 32) = \dfrac{5}{9}(38) \approx 21.1$
70°F is 21.1°C.

71. $F = 1.8C + 32$
$= 1.8(118) + 32$
$= 212.4 + 32$
$= 244.4$
118°C is 244.4°F.

73. $F = 1.8C + 32$
$= 1.8(4000) + 32$
$= 7200 + 32$
$= 7232$
4000°C is 7232°F.

75. $6 \cdot 4 + 5 \div 1 = 24 + 5 \div 1 = 24 + 5 = 29$

77. $3[(1 + 5) \cdot (8 - 6)] = 3(6 \cdot 2) = 3(12) = 36$

79. Yes, a 72°F room feels comfortable.

81. No, a fever of 40°F is not reasonable.

83. No, an overcoat is not needed when the temperature is 30°C.

85. Yes, a fever of 40°C is reasonable.

87. $\text{BSA} = \sqrt{\dfrac{90 \times 182}{3600}} \approx 2.13$
The BSA is approximately 2.13 sq m.

89. $40 \text{ in.} = \dfrac{40 \text{ in.}}{1} \cdot \dfrac{2.54 \text{ cm}}{1 \text{ in.}} = 101.6 \text{ cm}$
$\text{BSA} = \sqrt{\dfrac{50 \times 101.6}{3600}} \approx 1.19$
The BSA is approximately 1.19 sq m.

91. $60 \text{ in.} = \dfrac{60 \text{ in.}}{1} \cdot \dfrac{2.54 \text{ cm}}{1 \text{ in.}} = 152.4 \text{ cm}$

$150 \text{ lb} \approx \dfrac{150 \text{ lb}}{1} \cdot \dfrac{0.45 \text{ kg}}{1 \text{ lb}} = 67.5 \text{ kg}$

$\text{BSA} \approx \sqrt{\dfrac{67.5 \times 152.4}{3600}} \approx 1.69$

The BSA is approximately 1.69 sq m.

93. $C = \dfrac{5}{9}(F - 32)$

$\quad = \dfrac{5}{9}(918,000,000 - 32)$

$\quad = \dfrac{5}{9}(917,999,968)$

$\quad \approx 510,000,000$

918,000,000°F is approximately 510,000,000°C.

95. answers may vary

Chapter 9 Vocabulary Check

1. Weight is a measure of the pull of gravity.

2. Mass is a measure of the amount of substance in an object. This measure does not change.

3. The basic unit of length in the metric system is the meter.

4. To convert from one unit of length to another, unit fractions may be used.

5. A gram is the basic unit of mass in the metric system.

6. The liter is the basic unit of capacity in the metric system.

7. A line segment is a piece of a line with two endpoints.

8. Two angles that have a sum of 90° are called complementary angles.

9. A line is a set of points extending indefinitely in two directions.

10. The perimeter of a polygon is the distance around the polygon.

11. An angle is made up of two rays that share the same endpoint. The common endpoint is called the vertex.

12. Area measures the amount of surface of a region.

13. A ray is a part of a line with one endpoint. A ray extends indefinitely in one direction.

14. A line that intersects two or more lines at different points is called a transversal.

15. An angle that measures 180° is called a straight angle.

16. The measure of the space of a solid is called its volume.

17. When two lines intersect, four angles are formed. Two of these angles that are opposite each other are called vertical angles.

18. Two of the angles from #17 that share a common side are called adjacent angles.

19. An angle whose measure is between 90° and 180° is called an obtuse angle.

20. An angle that measures 90° is called a right angle.

21. An angle whose measure is between 0° and 90° is called an acute angle.

22. Two angles that have a sum of 180° are called supplementary angles.

23. The surface area of a polyhedron is the sum of the areas of the faces of the polyhedron.

Chapter 9 Review

1. $\angle A$ is a right angle.

2. $\angle B$ is a straight angle.

3. $\angle C$ is an acute angle.

4. $\angle D$ is an obtuse angle.

5. The complement of a 25° angle has measure $90° - 25° = 65°$.

6. The supplement of a 105° angle has measure $180° - 105° = 75°$.

7. $m\angle x = 90° - 32° = 58°$

8. $m\angle x = 180° - 82° = 98°$

9. $m\angle x = 105° - 15° = 90°$

10. $m\angle x = 45° - 20° = 25°$

11. $47° + 133° = 180°$, so $\angle a$ and $\angle b$ are supplementary. So are $\angle b$ and $\angle c$, $\angle c$ and $\angle d$, and $\angle d$ and $\angle a$.

12. $47° + 43° = 90°$, so $\angle x$ and $\angle w$ are complementary. Also, $58° + 32° = 90°$, so $\angle y$ and $\angle z$ are complementary.

13. $\angle x$ and the angle marked 100° are vertical angles, so $m\angle x = 100°$.
    $\angle x$ and $\angle y$ are adjacent angles, so $m\angle y = 180° - 100° = 80°$.
    $\angle y$ and $\angle z$ are vertical angles, so $m\angle z = m\angle y = 80°$.

14. $\angle x$ and the angle marked 25° are adjacent angles, so $m\angle x = 180° - 25° = 155°$.
    $\angle x$ and $\angle y$ are vertical angles, so $m\angle y = m\angle x = 155°$.
    $\angle z$ and the angle marked 25° are vertical angles, so $m\angle z = 25°$.

15. $\angle x$ and the angle marked 53° are vertical angles, so $m\angle x = 53°$.
    $\angle x$ and $\angle y$ are alternate interior angles, so $m\angle y = m\angle x = 53°$.
    $\angle y$ and $\angle z$ are adjacent angles, so $m\angle z = 180° - m\angle y = 180° - 53° = 127°$.

16. $\angle x$ and the angle marked 42° are vertical angles, so $m\angle x = 42°$.
    $\angle x$ and $\angle y$ are alternate interior angles, so $m\angle y = m\angle x = 42°$.
    $\angle y$ and $\angle z$ are adjacent angles, so $m\angle z = 180° - m\angle y = 180° - 42° = 138°$.

17. $P = 23 \text{ m} + 11\frac{1}{2} \text{ m} + 23 \text{ m} + 11\frac{1}{2} \text{ m} = 69 \text{ m}$

18. $P = 11 \text{ cm} + 7.6 \text{ cm} + 12 \text{ cm} = 30.6 \text{ cm}$

19. The unmarked vertical side has length $8 \text{ m} - 5 \text{ m} = 3 \text{ m}$. The unmarked horizontal side has length $10 \text{ m} - 7 \text{ m} = 3 \text{ m}$.
    $P = (7 + 3 + 3 + 5 + 10 + 8) \text{ m} = 36 \text{ m}$

20. The unmarked vertical side has length $5 \text{ ft} + 4 \text{ ft} + 11 \text{ ft} = 20 \text{ ft}$.
    $P = (22 + 20 + 22 + 11 + 3 + 4 + 3 + 5) \text{ ft}$
    $\quad = 90 \text{ ft}$

21. $P = 2 \cdot l + 2 \cdot w = 2 \cdot 10 \text{ ft} + 2 \cdot 6 \text{ ft} = 32 \text{ ft}$

22. $P = 4 \cdot s = 4 \cdot 110 \text{ ft} = 440 \text{ ft}$

23. $C = \pi \cdot d$
    $\quad = \pi \cdot 1.7 \text{ in.}$
    $\quad \approx 3.14 \cdot 1.7 \text{ in.}$
    $\quad = 5.338 \text{ in.}$

24. $C = 2 \cdot \pi \cdot r$
    $\quad = 2 \cdot \pi \cdot 5 \text{ yd}$
    $\quad = \pi \cdot 10 \text{ yd}$
    $\quad \approx 3.14 \cdot 10 \text{ yd}$
    $\quad = 31.4 \text{ yd}$

25. $A = \frac{1}{2} \cdot (b + B) \cdot h$

$\quad = \frac{1}{2} \cdot (12 \text{ ft} + 36 \text{ ft}) \cdot 10 \text{ ft}$

$\quad = \frac{1}{2} \cdot 48 \text{ ft} \cdot 10 \text{ ft}$

$\quad = 240 \text{ sq ft}$

26. $A = \frac{1}{2} \cdot b \cdot h = \frac{1}{2} \cdot 20 \text{ m} \cdot 14 \text{ m} = 140 \text{ sq m}$

27. $A = l \cdot w = 40 \text{ cm} \cdot 15 \text{ cm} = 600 \text{ sq cm}$

28. $A = b \cdot h = 21 \text{ yd} \cdot 9 \text{ yd} = 189 \text{ sq yd}$

29. $A = \pi \cdot r^2$

$\quad = \pi \cdot (7 \text{ ft})^2$

$\quad = 49\pi \text{ sq ft} \approx 153.86 \text{ sq ft}$

30. $A = s^2 = (9.1 \text{ m})^2 = 82.81 \text{ sq m}$

31. $A = \frac{1}{2} \cdot b \cdot h = \frac{1}{2} \cdot 34 \text{ in.} \cdot 7 \text{ in.} = 119 \text{ sq in.}$

32. $A = \frac{1}{2} \cdot (b + B) \cdot h$

$\quad = \frac{1}{2} \cdot (64 \text{ cm} + 32 \text{ cm}) \cdot 26 \text{ cm}$

$\quad = \frac{1}{2} \cdot 96 \text{ cm} \cdot 26 \text{ cm}$

$\quad = 1248 \text{ sq cm}$

33. The unmarked horizontal side has length
$13 \text{ m} - 3 \text{ m} = 10 \text{ m}$. The unmarked vertical
side has length $12 \text{ m} - 4 \text{ m} = 8 \text{ m}$. The area
is the sum of the areas of the two rectangles.
$A = 12 \text{ m} \cdot 10 \text{ m} + 8 \text{ m} \cdot 3 \text{ m}$

$\quad = 120 \text{ sq m} + 24 \text{ sq m}$

$\quad = 144 \text{ sq m}$

34. $A = l \cdot w = 36 \text{ ft} \cdot 12 \text{ ft} = 432 \text{ sq ft}$
The area of the driveway is 432 square feet.

35. $A = 10 \text{ ft} \cdot 13 \text{ ft} = 130 \text{ sq ft}$
130 square feet of carpet are needed.

36. $V = s^3$

$\quad = \left(2\frac{1}{2} \text{ in.}\right)^3$

$\quad = \left(\frac{5}{2} \text{ in.}\right)^3$

$\quad = \frac{125}{8} \text{ cu in.}$

$\quad = 15\frac{5}{8} \text{ cu in.}$

$SA = 6s^2$

$\quad = 6\left(2\frac{1}{2} \text{ in.}\right)^2$

$\quad = 6\left(\frac{5}{2} \text{ in.}\right)^2$

$\quad = 6\left(\frac{25}{4}\right) \text{ sq in.}$

$\quad = \frac{75}{2} \text{ sq in.}$

$\quad = 37\frac{1}{2} \text{ sq in.}$

37. $V = l \cdot w \cdot h = 2 \text{ ft} \cdot 7 \text{ ft} \cdot 6 \text{ ft} = 84 \text{ cu ft}$
$SA = 2lh + 2wh + 2lw$

$\quad = 2 \cdot 7 \text{ ft} \cdot 6 \text{ ft} + 2 \cdot 2 \cdot 6 \text{ ft} + 2 \cdot 7 \text{ ft} \cdot 2 \text{ ft}$

$\quad = 84 \text{ sq ft} + 24 \text{ sq ft} + 28 \text{ sq ft}$

$\quad = 136 \text{ sq ft}$

38. $V = \pi \cdot r^2 \cdot h$

$\quad = \pi \cdot (20 \text{ cm})^2 \cdot 50 \text{ cm}$

$\quad = 20,000\pi \text{ cu cm}$

$\quad \approx 62,800 \text{ cu cm}$

39. $V = \frac{4}{3} \cdot \pi \cdot r^3$

$\quad = \frac{4}{3} \cdot \pi \cdot \left(\frac{1}{2} \text{ km}\right)^3$

$\quad = \frac{1}{6}\pi \text{ cu km}$

$\quad \approx \frac{11}{21} \text{ cu km}$

40.
$$V = \frac{1}{3} \cdot s^2 \cdot h$$
$$= \frac{1}{3} \cdot (2 \text{ ft})^2 \cdot 2 \text{ ft}$$
$$= \frac{8}{3} \text{ cu ft}$$
$$= 2\frac{2}{3} \text{ cu ft}$$

The volume of the pyramid is $2\frac{2}{3}$ cubic feet.

41.
$$V = \pi \cdot r^2 \cdot h$$
$$= \pi \cdot (3.5 \text{ in.})^2 \cdot 8 \text{ in.}$$
$$= 98\pi \text{ cu in.}$$
$$\approx 307.72 \text{ cu in.}$$
The volume of the can is about 307.72 cubic inches.

42. Find the volume of each drawer.
$$V = l \cdot w \cdot h$$
$$= \left(2\frac{1}{2} \text{ ft}\right) \cdot \left(1\frac{1}{2} \text{ ft}\right) \cdot \left(\frac{2}{3} \text{ ft}\right)$$
$$= \frac{5}{2} \cdot \frac{3}{2} \cdot \frac{2}{3} \text{ cu ft}$$
$$= \frac{5}{2} \text{ cu ft}$$

The three drawers have volume
$3 \cdot \frac{5}{2} = \frac{15}{2} = 7\frac{1}{2}$ cubic feet.

43. $r = d \div 2 = 1 \text{ ft} \div 2 = 0.5 \text{ ft}$
$V = \pi \cdot r^2 \cdot h = \pi \cdot (0.5 \text{ ft})^2 \cdot 2 \text{ ft} = 0.5\pi \text{ cu ft}$

44. $108 \text{ in.} = \frac{108 \text{ in.}}{1} \cdot \frac{1 \text{ ft}}{12 \text{ in.}} = \frac{108}{12} \text{ ft} = 9 \text{ ft}$

45. $72 \text{ ft} = \frac{72 \text{ ft}}{1} \cdot \frac{1 \text{ yd}}{3 \text{ ft}} = \frac{72}{3} \text{ yd} = 24 \text{ yd}$

46. $1.5 \text{ mi} = \frac{1.5 \text{ mi}}{1} \cdot \frac{5280 \text{ ft}}{1 \text{ mi}}$
$$= 1.5 \cdot 5280 \text{ ft}$$
$$= 7920 \text{ ft}$$

47. $\frac{1}{2} \text{ yd} = \frac{\frac{1}{2} \text{ yd}}{1} \cdot \frac{3 \text{ ft}}{1 \text{ yd}} \cdot \frac{12 \text{ in.}}{1 \text{ ft}}$
$$= \frac{1}{2} \cdot 3 \cdot 12 \text{ in.}$$
$$= 18 \text{ in.}$$

48. $52 \text{ ft} = 51 \text{ ft} + 1 \text{ ft}$
$$= \frac{51 \text{ ft}}{1} \cdot \frac{1 \text{ yd}}{3 \text{ ft}} + 1 \text{ ft}$$
$$= \frac{51}{3} \text{ yd} + 1 \text{ ft}$$
$$= 17 \text{ yd } 1 \text{ ft}$$

49. $46 \text{ in.} = 36 \text{ in.} + 10 \text{ in.}$
$$= \frac{36 \text{ in.}}{1} \cdot \frac{1 \text{ ft}}{12 \text{ in.}} + 10 \text{ in.}$$
$$= \frac{36}{12} \text{ ft} + 10 \text{ in.}$$
$$= 3 \text{ ft } 10 \text{ in.}$$

50. $42 \text{ m} = \frac{42 \text{ m}}{1} \cdot \frac{100 \text{ cm}}{1 \text{ m}}$
$$= 42 \cdot 100 \text{ cm}$$
$$= 4200 \text{ cm}$$

51. $82 \text{ cm} = \frac{82 \text{ cm}}{1} \cdot \frac{10 \text{ mm}}{1 \text{ cm}}$
$$= 82 \cdot 10 \text{ mm}$$
$$= 820 \text{ mm}$$

52. $12.18 \text{ mm} = \frac{12.18 \text{ mm}}{1} \cdot \frac{1 \text{ m}}{1000 \text{ mm}}$
$$= \frac{12.18}{1000} \text{ m}$$
$$= 0.01218 \text{ m}$$

53. $2.31 \text{ m} = \frac{2.31 \text{ m}}{1} \cdot \frac{1 \text{ km}}{1000 \text{ m}}$
$$= \frac{2.31}{1000} \text{ km}$$
$$= 0.00231 \text{ km}$$

54.
$$\begin{array}{r} 4 \text{ yd } 2 \text{ ft} \\ + 16 \text{ yd } 2 \text{ ft} \\ \hline 20 \text{ yd } 4 \text{ ft} \end{array} = 20 \text{ yd} + 1 \text{ yd } 1 \text{ ft} = 21 \text{ yd } 1 \text{ ft}$$

55.
$$7 \text{ ft } 4 \text{ in.} \div 2 = (6 \text{ ft} + 1 \text{ ft } 4 \text{ in.}) \div 2$$
$$= (6 \text{ ft} + 16 \text{ in.}) \div 2$$
$$= \frac{6}{2} \text{ ft} + \frac{16}{2} \text{ in.}$$
$$= 3 \text{ ft } 8 \text{ in.}$$

56. 8 cm = 80 mm   15 mm = 1.5 cm

	80 mm			8.0 cm
+	15 mm	or	+	1.5 cm
	95 mm			9.5 cm

57. 4 m = 400 cm   126 cm = 1.26 m

	400 cm			4.00 m
−	126 cm	or	−	1.26 m
	274 cm			2.74 m

58.

	333 yd 1 ft		332 yd 4 ft
−	163 yd 2 ft	−	163 yd 2 ft
			169 yd 2 ft

The amount of material that remains is 169 yd 2 ft.

59.

	5 ft	2 in.
×		50
250 ft	100 in.	= 250 ft + 96 in. + 4 in.

$$= 250 \text{ ft} + 8 \text{ ft } 4 \text{ in.}$$
$$= 258 \text{ ft } 4 \text{ in.}$$
The sashes require 258 ft 4 in. of material.

60.

	217 km
×	2
	434 km

$$434 \text{ km} \div 4 = \frac{434}{4} \text{ km} = 108.5 \text{ km}$$
Each must drive 108.5 km.

61.

	0.8 m			0.8 m
×	30 cm		×	0.3 m
				0.24 sq m

The area is 0.24 sq m.

62. $66 \text{ oz} = \dfrac{66 \text{ oz}}{1} \cdot \dfrac{1 \text{ lb}}{16 \text{ oz}} = \dfrac{66}{16} \text{ lb} = 4.125 \text{ lb}$

63.
$$2.3 \text{ tons} = \frac{2.3 \text{ tons}}{1} \cdot \frac{2000 \text{ lb}}{1 \text{ ton}}$$
$$= 2.3 \cdot 2000 \text{ lb}$$
$$= 4600 \text{ lb}$$

64.
$$52 \text{ oz} = 48 \text{ oz} + 4 \text{ oz}$$
$$= \frac{48 \text{ oz}}{1} \cdot \frac{1 \text{ lb}}{16 \text{ oz}} + 4 \text{ oz}$$
$$= \frac{48}{16} \text{ lb} + 4 \text{ oz}$$
$$= 3 \text{ lb } 4 \text{ oz}$$

65.
$$10,300 \text{ lb} = 10,000 \text{ lb} + 300 \text{ lb}$$
$$= \frac{10,000 \text{ lb}}{1} \cdot \frac{1 \text{ ton}}{2000 \text{ lb}} + 300 \text{ lb}$$
$$= \frac{10,000}{200} \text{ tons} + 300 \text{ lb}$$
$$= 5 \text{ tons } 300 \text{ lb}$$

66.
$$27 \text{ mg} = \frac{27 \text{ mg}}{1} \cdot \frac{1 \text{ g}}{1000 \text{ mg}}$$
$$= \frac{27}{1000} \text{ g}$$
$$= 0.027 \text{ g}$$

67.
$$40 \text{ kg} = \frac{40 \text{ kg}}{1} \cdot \frac{1000 \text{ g}}{1 \text{ kg}}$$
$$= 40 \cdot 1000 \text{ g}$$
$$= 40,000 \text{ g}$$

68.
$$2.1 \text{ hg} = \frac{2.1 \text{ hg}}{1} \cdot \frac{10 \text{ dag}}{1 \text{ hg}}$$
$$= 2.1 \cdot 10 \text{ dag}$$
$$= 21 \text{ dag}$$

69.
$$0.03 \text{ mg} = \frac{0.03 \text{ mg}}{1} \cdot \frac{1 \text{ dg}}{100 \text{ mg}}$$
$$= \frac{0.03}{100} \text{ dg}$$
$$= 0.0003 \text{ dg}$$

70.

	6 lb	5 oz		5 lb	21 oz
−	2 lb	12 oz	−	2 lb	12 oz
				3 lb	9 oz

71.
$$\begin{array}{r} 8 \text{ lb } 6 \text{ oz} \\ \times \quad\quad 4 \\ \hline 32 \text{ lb } 24 \text{ oz} \end{array} = 32 \text{ lb} + 1 \text{ lb } 8 \text{ oz} = 33 \text{ lb } 8 \text{ oz}$$

72.
$$\begin{array}{r} 4.3 \text{ mg} \\ \times \quad 5 \\ \hline 21.5 \text{ mg} \end{array}$$

73.
$$\begin{array}{cc} 4.8 \text{ kg} = 4800 \text{ g} & 4200 \text{ g} = 4.2 \text{ kg} \\ \begin{array}{r} 4800 \text{ g} \\ - \ 4200 \text{ g} \\ \hline 600 \text{ g} \end{array} \quad \text{or} & \begin{array}{r} 4.8 \text{ kg} \\ - \ 4.2 \text{ kg} \\ \hline 0.6 \text{ kg} \end{array} \end{array}$$

74.
$$\begin{array}{r} 1 \text{ lb } 12 \text{ oz} \\ + \ 2 \text{ lb } \ 8 \text{ oz} \\ \hline 3 \text{ lb } 20 \text{ oz} \end{array} = 3 \text{ lb} + 1 \text{ lb } 4 \text{ oz} = 4 \text{ lb } 4 \text{ oz}$$
The total weight was 4 lb 4 oz.

75.
$$\begin{aligned} 38 \text{ tons } 300 \text{ lb} \div 4 &= \frac{38}{4} \text{ tons } \frac{300}{4} \text{ lb} \\ &= 9\frac{1}{2} \text{ tons } 75 \text{ lb} \\ &= 9 \text{ tons} + \frac{1}{2} \text{ ton} + 75 \text{ lb} \\ &= 9 \text{ tons} + 1000 \text{ lb} + 75 \text{ lb} \\ &= 9 \text{ tons } 1075 \text{ lb} \end{aligned}$$
They each receive 9 tons 1075 lb.

76. $20 \text{ pt} = \dfrac{20 \text{ pt}}{1} \cdot \dfrac{1 \text{ qt}}{2 \text{ pt}} = \dfrac{20}{2} \text{ qt} = 10 \text{ qt}$

77. $40 \text{ fl oz} = \dfrac{40 \text{ fl oz}}{1} \cdot \dfrac{1 \text{ c}}{8 \text{ fl oz}} = \dfrac{40}{8} \text{ c} = 5 \text{ c}$

78.
$$\begin{aligned} 3 \text{ qt } 1 \text{ pt} &= \frac{3 \text{ qt}}{1} \cdot \frac{2 \text{ pt}}{1 \text{ qt}} + 1 \text{ pt} \\ &= 3 \cdot 2 \text{ pt} + 1 \text{ pt} \\ &= 6 \text{ pt} + 1 \text{ pt} \\ &= 7 \text{ pt} \end{aligned}$$

79. $18 \text{ qt} = \dfrac{18 \text{ qt}}{1} \cdot \dfrac{2 \text{ pt}}{1 \text{ qt}} \cdot \dfrac{2 \text{ c}}{1 \text{ pt}} = 18 \cdot 2 \cdot 2 \text{ c} = 72 \text{ c}$

80.
$$\begin{aligned} 9 \text{ pt} &= 8 \text{ pt} + 1 \text{ pt} \\ &= \frac{8 \text{ pt}}{1} \cdot \frac{1 \text{ qt}}{2 \text{ pt}} + 1 \text{ pt} \\ &= \frac{8}{2} \text{ qt} + 1 \text{ pt} \\ &= 4 \text{ qt } 1 \text{ pt} \end{aligned}$$

81.
$$\begin{aligned} 15 \text{ qt} &= 12 \text{ qt} + 3 \text{ qt} \\ &= \frac{12 \text{ qt}}{1} \cdot \frac{1 \text{ gal}}{4 \text{ qt}} + 3 \text{ qt} \\ &= \frac{12}{4} \text{ gal} + 3 \text{ qt} \\ &= 3 \text{ gal } 3 \text{ qt} \end{aligned}$$

82.
$$\begin{aligned} 3.8 \text{ L} &= \frac{3.8 \text{ L}}{1} \cdot \frac{1000 \text{ ml}}{1 \text{ L}} \\ &= 3.8 \cdot 1000 \text{ ml} \\ &= 3800 \text{ ml} \end{aligned}$$

83. $14 \text{ hl} = \dfrac{14 \text{ hl}}{1} \cdot \dfrac{1 \text{ kl}}{10 \text{ hl}} = \dfrac{14}{10} \text{ kl} = 1.4 \text{ kl}$

84.
$$\begin{aligned} 30.6 \text{ L} &= \frac{30.6 \text{ L}}{1} \cdot \frac{100 \text{ cl}}{1 \text{ L}} \\ &= 30.6 \cdot 100 \text{ cl} \\ &= 3060 \text{ cl} \end{aligned}$$

85.
$$\begin{array}{r} 1 \text{ qt } 1 \text{ pt} \\ + \ 3 \text{ qt } 1 \text{ pt} \\ \hline 4 \text{ qt } 2 \text{ pt} \end{array} = 4 \text{ qt} + 1 \text{ qt} = 1 \text{ gal } 1 \text{ qt}$$

86.
$$\begin{array}{r} 3 \text{ gal } 2 \text{ qt} \\ \times \quad\quad 2 \\ \hline 6 \text{ gal } 4 \text{ qt} \end{array} = 6 \text{ gal} + 1 \text{ gal} = 7 \text{ gal}$$

87.
$$\begin{array}{cc} 0.946 \text{ L} = 946 \text{ ml} & 210 \text{ ml} = 0.21 \text{ L} \\ \begin{array}{r} 946 \text{ ml} \\ - \ 210 \text{ ml} \\ \hline 736 \text{ ml} \end{array} \quad \text{or} & \begin{array}{r} 0.946 \text{ L} \\ - \ 0.210 \text{ L} \\ \hline 0.736 \text{ L} \end{array} \end{array}$$

88.
$$\begin{array}{cc} 6.1 \text{ L} = 6100 \text{ ml} & 9400 \text{ ml} = 9.4 \text{ L} \\ \begin{array}{r} 6100 \text{ ml} \\ + \ 9400 \text{ ml} \\ \hline 15,500 \text{ ml} \end{array} \quad \text{or} & \begin{array}{r} 6.1 \text{ L} \\ + \ 9.4 \text{ L} \\ \hline 15.5 \text{ L} \end{array} \end{array}$$

89.

$$\begin{array}{r} 4 \text{ gal } 2 \text{ qt} \\ -1 \text{ gal } 3 \text{ qt} \end{array} \qquad \begin{array}{r} 3 \text{ gal } 6 \text{ qt} \\ -1 \text{ gal } 3 \text{ qt} \\ \hline 2 \text{ gal } 3 \text{ qt} \end{array}$$

There are 2 gal 3 qt of tea remaining.

90. 1 c 4 fl oz $\div 2 = (8$ fl oz $+ 4$ fl oz$) \div 2$
$$= 12 \text{ fl oz} \div 2$$
$$= 6 \text{ fl oz}$$
Use 6 fl oz of stock for half of a recipe.

91. 85 ml $\times 8 \times 16 = 10,880$ ml
$$\frac{10,880 \text{ ml}}{1} \cdot \frac{1 \text{ L}}{1000 \text{ ml}} = \frac{10,880}{1000} \text{ L} = 10.88 \text{ L}$$
There are 10.88 L of polish in 8 boxes.

92. 6 L $+ 1300$ ml $+ 2.6$ L $= 6$ L $+ 1.3$ L $+ 2.6$ L
$$= 9.9 \text{ L}$$
Since 9.9 L is less than 10 L, yes it will fit.

93. 7 m $\approx \dfrac{7 \text{ m}}{1} \cdot \dfrac{3.28 \text{ ft}}{1 \text{ m}} = 22.96$ ft

94. 11.5 yd $\approx \dfrac{11.5 \text{ yd}}{1} \cdot \dfrac{1 \text{ m}}{1.09 \text{ yd}} \approx 10.55$ m

95. 17.5 L $\approx \dfrac{17.5 \text{ L}}{1} \cdot \dfrac{0.26 \text{ gal}}{1 \text{ L}} = 4.55$ gal

96. 7.8 L $\approx \dfrac{7.8 \text{ L}}{1} \cdot \dfrac{1.06 \text{ qt}}{1 \text{ L}} = 8.268$ qt

97. 15 oz $\approx \dfrac{15 \text{ oz}}{1} \cdot \dfrac{28.35 \text{ g}}{1 \text{ oz}} = 425.25$ g

98. 23 lb $\approx \dfrac{23 \text{ lb}}{1} \cdot \dfrac{0.45 \text{ kg}}{1 \text{ lb}} = 10.35$ kg

99. 1.2 mm $\times 50 = 60$ mm
$$60 \text{ mm} = \frac{60 \text{ mm}}{1} \cdot \frac{1 \text{ cm}}{10 \text{ mm}} = 6 \text{ cm}$$
$$6 \text{ cm} = \frac{6 \text{ cm}}{1} \cdot \frac{1 \text{ in.}}{2.54 \text{ cm}} \approx 2.36 \text{ in.}$$
The height of the stack is approximately 2.36 in.

100. 82 kg $\approx \dfrac{82 \text{ kg}}{1} \cdot \dfrac{2.20 \text{ lb}}{1 \text{ kg}} = 180.4$ lb
The person weighs approximately 180.4 lb.

101. $F = 1.8C + 32$
$$= 1.8(42) + 32$$
$$= 75.6 + 32$$
$$= 107.6$$
42°C is 107.6°F.

102. $F = 1.8C + 32$
$$= 1.8(160) + 32$$
$$= 288 + 32$$
$$= 320$$
160°C is 320°F.

103. $C = \dfrac{5}{9}(F - 32) = \dfrac{5}{9}(41.3 - 32) = \dfrac{5}{9}(9.3) \approx 5.2$
41.3°F is 5.2°C.

104. $C = \dfrac{5}{9}(F - 32) = \dfrac{5}{9}(80 - 32) = \dfrac{5}{9}(48) \approx 26.7$
80°F is 26.7°C.

105. $C = \dfrac{5}{9}(F - 32) = \dfrac{5}{9}(35 - 32) = \dfrac{5}{9}(3) \approx 1.7$
35°F is 1.7°C.

106. $F = 1.8C + 32$
$$= 1.8(165) + 32$$
$$= 297 + 32$$
$$= 329$$
165°C is 329°F.

107. The supplement of a 72° angle is an angle that measures 180° − 72° = 108°.

108. The complement of a 1° angle is an angle that measures 90° − 1° = 89°.

109. $\angle x$ and the angle marked 85° are adjacent angles, so $m\angle x = 180° - 85° = 95°$.

110. Let $\angle y$ be the angle corresponding to $\angle x$ at the bottom intersection. Then $\angle y$ and the angle marked 123° are adjacent angles, so $m\angle x = m\angle y = 180° - 123° = 57°$.

111. $P = 7 \text{ in.} + 11.2 \text{ in.} + 9.1 \text{ in.} = 27.3 \text{ in.}$

112. The unmarked horizontal side has length 40 ft − 22 ft − 11 ft = 7 ft.
$P = (22 + 15 + 7 + 15 + 11 + 42 + 40 + 42) \text{ ft}$
$\quad = 194 \text{ ft}$

113. The unmarked horizontal side has length 43 m − 13 m = 30 m. The unmarked vertical side has length 42 m − 14 m = 28 m. The area is the sum of the areas of the two rectangles.
$A = 28 \text{ m} \cdot 13 \text{ m} + 42 \text{ m} \cdot 30 \text{ m}$
$\quad = 364 \text{ sq m} + 1260 \text{ sq m}$
$\quad = 1624 \text{ sq m}$

114. $A = \pi \cdot r^2$
$\quad = \pi \cdot (3 \text{ m})^2$
$\quad = 9\pi \text{ sq m} \approx 28.26 \text{ sq m}$

115. $V = \dfrac{1}{3} \cdot \pi \cdot r^2 \cdot h$
$\quad = \dfrac{1}{3} \cdot \pi \cdot \left(5\dfrac{1}{4} \text{ in.}\right)^2 \cdot 12 \text{ in.}$
$\quad = \dfrac{1}{3} \cdot \pi \cdot \left(\dfrac{21}{4} \text{ in.}\right)^2 \cdot 12 \text{ cu in.}$
$\quad = \dfrac{441}{4}\pi \text{ cu in.}$
$\quad \approx \dfrac{441}{4} \cdot \dfrac{22}{7} \text{ cu in.} = 346\dfrac{1}{2} \text{ cu in.}$

116. $V = l \cdot w \cdot h = 5 \text{ in.} \cdot 4 \text{ in.} \cdot 7 \text{ in.} = 140 \text{ cu in.}$
SA
$= 2lh + 2wh + 2lw$
$= 2 \cdot 7 \text{ in.} \cdot 5 \text{ in.} + 2 \cdot 4 \text{ in.} \cdot 5 \text{ in.} + 2 \cdot 7 \text{ in.} \cdot 4 \text{ in.}$
$= 70 \text{ sq in.} + 40 \text{ sq in.} + 56 \text{ sq in.}$
$= 166 \text{ sq in.}$

117. $6.25 \text{ ft} = \dfrac{6.25 \text{ ft}}{1} \cdot \dfrac{12 \text{ in.}}{1 \text{ ft}} = 75 \text{ in.}$

118. $8200 \text{ lb} = 8000 \text{ lb} + 200 \text{ lb}$
$\quad = \dfrac{8000 \text{ lb}}{1} \cdot \dfrac{1 \text{ ton}}{2000 \text{ lb}} + 200 \text{ lb}$
$\quad = 4 \text{ tons } 200 \text{ lb}$

119. $5 \text{ m} = \dfrac{5 \text{ m}}{1} \cdot \dfrac{100 \text{ cm}}{1 \text{ m}} = 500 \text{ cm}$

120. $286 \text{ mm} = \dfrac{286 \text{ mm}}{1} \cdot \dfrac{1 \text{ km}}{1,000,000 \text{ mm}}$
$\quad = 0.000286 \text{ km}$

121. $1400 \text{ mg} = \dfrac{1400 \text{ mg}}{1} \cdot \dfrac{1 \text{ g}}{1000 \text{ mg}} = 1.4 \text{ g}$

122. $6.75 \text{ gal} = \dfrac{6.75 \text{ gal}}{1} \cdot \dfrac{4 \text{ qt}}{1 \text{ gal}} = 27 \text{ qt}$

123. $F = 1.8C + 32$
$\quad = 1.8(86) + 32$
$\quad = 154.8 + 32$
$\quad = 186.8$
86°C is 186.8°F.

124. $C = \dfrac{5}{9}(F - 32) = \dfrac{5}{9}(51.8 - 32) = \dfrac{5}{9}(19.8) = 11$
51.8°F is 11°C.

125. 9.3 km = 9300 m   183 m = 0.183 km

$\quad\begin{array}{r} 9300 \text{ m} \\ -\ 183 \text{ m} \\ \hline 9117 \text{ m} \end{array}$ or $\begin{array}{r} 9.300 \text{ km} \\ -\ 0.183 \text{ km} \\ \hline 9.117 \text{ km} \end{array}$

126. 35 L = 35,000 ml   700 ml = 0.7 L

$\quad\begin{array}{r} 35,000 \text{ ml} \\ +\ \ 700 \text{ ml} \\ \hline 35,700 \text{ ml} \end{array}$ $\quad\begin{array}{r} 35.0 \text{ L} \\ +\ 0.7 \text{ L} \\ \hline 35.7 \text{ L} \end{array}$

127. $\begin{array}{r} 3 \text{ gal } 3 \text{ qt} \\ +\ 4 \text{ gal } 2 \text{ qt} \\ \hline 7 \text{ gal } 5 \text{ qt} \end{array} = 7 \text{ gal} + 1 \text{ gal } 1 \text{ qt} = 8 \text{ gal } 1 \text{ qt}$

128.

$$
\begin{array}{r}
3.2 \text{ kg} \\
\times \quad 4 \\
\hline
12.8 \text{ kg}
\end{array}
$$

Chapter 9 Test

1. The complement of an angle that measures
   78° is an angle that measures
   $90° - 78° = 12°$.

2. The supplement of a 124° angle is an angle
   that measures $180° - 124° = 56°$.

3. $m\angle x = 90° - 40° = 50°$

4. $\angle x$ and the angle marked 62° are adjacent
   angles, so $m\angle x = 180° - 62° = 118°$.
   $\angle y$ and the angle marked 62° are vertical
   angles, so $m\angle y = 62°$.
   $\angle x$ and $\angle z$ are vertical angles, so
   $m\angle z = m\angle x = 118°$.

5. $\angle x$ and the angle marked 73° are vertical
   angles, so $m\angle x = 73°$.
   $\angle x$ and $\angle y$ are alternate interior angles, so
   $m\angle y = m\angle x = 73°$.
   $\angle x$ and $\angle z$ are corresponding angles, so
   $m\angle z = m\angle x = 73°$.

6. $d = 2 \cdot r = 2 \cdot 3.1 \text{ m} = 6.2 \text{ m}$

7. $r = d \div 2 = 20 \text{ in.} \div 2 = 10 \text{ in.}$

8. Circumference:
   $C = 2 \cdot \pi \cdot r$
   $\quad = 2 \cdot \pi \cdot 9 \text{ in.}$
   $\quad = 18\pi \text{ in.}$
   $\quad \approx 56.52 \text{ in.}$
   Area:
   $A = \pi r^2$
   $\quad = \pi (9 \text{ in.})^2$
   $\quad = 81\pi \text{ sq in.}$
   $\quad \approx 254.34 \text{ sq in.}$

9. $P = 2 \cdot l + 2 \cdot w$
   $\quad = 2(7 \text{ yd}) + 2(5.3 \text{ yd})$
   $\quad = 14 \text{ yd} + 10.6 \text{ yd}$
   $\quad = 24.6 \text{ yd}$
   $A = l \cdot w = 7 \text{ yd} \cdot 5.3 \text{ yd} = 37.1 \text{ sq yd}$

10. The unmarked vertical side has length
    11 in. − 7 in. = 4 in. The unmarked
    horizontal side has length
    23 in. − 6 in. = 17 in.
    $P = (6 + 4 + 17 + 7 + 23 + 11) \text{ in.} = 68 \text{ in.}$
    Extending the unmarked vertical side
    downward divides the region into two
    rectangles. The region's area is the sum of
    the areas of these:
    $A = 11 \text{ in.} \cdot 6 \text{ in.} + 7 \text{ in.} \cdot 17 \text{ in.}$
    $\quad = 66 \text{ sq in.} + 119 \text{ sq in.}$
    $\quad = 185 \text{ sq in.}$

11. $V = \pi \cdot r^2 \cdot h$
    $\quad = \pi \cdot (2 \text{ in.})^2 \cdot 5 \text{ in.}$
    $\quad = 20\pi \text{ cu in.}$
    $\quad \approx 20 \cdot \dfrac{22}{7} \text{ cu in.} = 62\dfrac{6}{7} \text{ cu in.}$

12. $V = l \cdot w \cdot h = 5 \text{ ft} \cdot 3 \text{ ft} \cdot 2 \text{ ft} = 30 \text{ cu ft}$

13. $P = 4 \cdot s = 4 \cdot 4 \text{ in.} = 16 \text{ in.}$
    The perimeter of the frame is 16 inches.

14. $V = l \cdot w \cdot h = 3 \text{ ft} \cdot 3 \text{ ft} \cdot 2 \text{ ft} = 18 \text{ cu ft}$
    18 cubic feet of soil are needed.

15. $P = 2 \cdot l + 2 \cdot w$
    $\quad = 2 \cdot 18 \text{ ft} + 2 \cdot 13 \text{ ft}$
    $\quad = 36 \text{ ft} + 26 \text{ ft}$
    $\quad = 62 \text{ ft}$
    $\text{cost} = P \cdot \$1.87 \text{ per ft}$
    $\quad\quad\; = 62 \text{ ft} \cdot \$1.87 \text{ per ft}$
    $\quad\quad\; = \$115.94$
    62 feet of baseboard are needed, at a total
    cost of $115.94.

16.

$$12\overline{)280}$$

$$\begin{array}{r} 23 \\ 12\overline{)280} \\ \underline{-24} \\ 40 \\ \underline{-36} \\ 4 \end{array}$$

280 inches = 23 ft 4 in.

17. $2\frac{1}{2}$ gal $= \dfrac{2\frac{1}{2} \text{ gal}}{1} \cdot \dfrac{4 \text{ qt}}{1 \text{ gal}} = 2\frac{1}{2} \cdot 4 \text{ qt} = 10 \text{ qt}$

18. $30 \text{ oz} = \dfrac{30 \text{ oz}}{1} \cdot \dfrac{1 \text{ lb}}{16 \text{ oz}} = \dfrac{30}{16} \text{ lb} = 1.875 \text{ lb}$

19. $2.8 \text{ tons} = \dfrac{2.8 \text{ tons}}{1} \cdot \dfrac{2000 \text{ lb}}{1 \text{ ton}}$
$= 2.8 \cdot 2000 \text{ lb}$
$= 5600 \text{ lb}$

20. $38 \text{ pt} = \dfrac{38 \text{ pt}}{1} \cdot \dfrac{1 \text{ qt}}{2 \text{ pt}} \cdot \dfrac{1 \text{ gal}}{4 \text{ qt}}$
$= \dfrac{38}{8} \text{ gal}$
$= \dfrac{19}{4} \text{ gal}$
$= 4\dfrac{3}{4} \text{ gal}$

21. $40 \text{ mg} = \dfrac{40 \text{ mg}}{1} \cdot \dfrac{1 \text{ g}}{1000 \text{ mg}}$
$= \dfrac{40}{1000} \text{ g}$
$= 0.04 \text{ g}$

22. $2.4 \text{ kg} = \dfrac{2.4 \text{ kg}}{1} \cdot \dfrac{1000 \text{ g}}{1 \text{ kg}}$
$= 2.4 \cdot 1000 \text{ g}$
$= 2400 \text{ g}$

23. $3.6 \text{ cm} = \dfrac{3.6 \text{ cm}}{1} \cdot \dfrac{10 \text{ mm}}{1 \text{ cm}}$
$= 3.6 \cdot 10 \text{ mm}$
$= 36 \text{ mm}$

24. $4.3 \text{ dg} = \dfrac{4.3 \text{ dg}}{1} \cdot \dfrac{1 \text{ g}}{10 \text{ dg}} = \dfrac{4.3}{10} \text{ g} = 0.43 \text{ g}$

25. $0.83 \text{ L} = \dfrac{0.83 \text{ L}}{1} \cdot \dfrac{1000 \text{ ml}}{1 \text{ L}}$
$= 0.83 \cdot 1000$
$= 830 \text{ ml}$

26.

$$\begin{array}{r} 3 \text{ qt } 1 \text{ pt} \\ + 2 \text{ qt } 1 \text{ pt} \\ \hline 5 \text{ qt } 2 \text{ pt} \end{array} = 4 \text{ qt} + 1 \text{ qt} + 2 \text{ pt}$$
$$= 1 \text{ gal} + 1 \text{ qt} + 1 \text{ qt}$$
$$= 1 \text{ gal } 2 \text{ qt}$$

27.

$$\begin{array}{r} 8 \text{ lb } 6 \text{ oz} \\ - 4 \text{ lb } 9 \text{ oz} \end{array} \rightarrow \begin{array}{r} 7 \text{ lb } 22 \text{ oz} \\ - 4 \text{ lb } 9 \text{ oz} \\ \hline 3 \text{ lb } 13 \text{ oz} \end{array}$$

28. $2 \text{ ft } 9 \text{ in.} \times 3 = 6 \text{ ft } 27 \text{ in.}$
$= 6 \text{ ft} + 2 \text{ ft } 3 \text{ in.}$
$= 8 \text{ ft } 3 \text{ in.}$

29. $5 \text{ gal } 2 \text{ qt} \div 2 = 4 \text{ gal } 6 \text{ qt} \div 2$
$= \dfrac{4}{2} \text{ gal } \dfrac{6}{2} \text{ qt}$
$= 2 \text{ gal } 3 \text{ qt}$

30. $8 \text{ cm} = 80 \text{ mm}$ $14 \text{ mm} = 1.4 \text{ cm}$

$$\begin{array}{r} 80 \text{ mm} \\ - 14 \text{ mm} \\ \hline 66 \text{ mm} \end{array} \quad \text{or} \quad \begin{array}{r} 8.0 \text{ cm} \\ - 1.4 \text{ cm} \\ \hline 6.6 \text{ cm} \end{array}$$

31. $1.8 \text{ km} = 1800 \text{ m}$ $456 \text{ m} = 0.456 \text{ km}$

$$\begin{array}{r} 1800 \text{ m} \\ + 456 \text{ m} \\ \hline 2256 \text{ m} \end{array} \quad \text{or} \quad \begin{array}{r} 1.800 \text{ km} \\ + 0.456 \text{ km} \\ \hline 2.256 \text{ km} \end{array}$$

32. $C = \dfrac{5}{9}(F - 32)$
$= \dfrac{5}{9}(84 - 32)$
$= \dfrac{5}{9}(52)$
≈ 28.9
84°F is 28.9°C.

33. $F = 1.8C + 32$
$\quad\quad = 1.8(12.6) + 32$
$\quad\quad = 22.68 + 32$
$\quad\quad = 54.68$
12.6°C is 54.68°F

34. $8.4 \text{ m} \cdot \dfrac{2}{3} = \dfrac{8.4}{1} \cdot \dfrac{2}{3} \text{ m} = 5.6 \text{ m}$
The trees will be 5.6 m tall.

35.
$\begin{array}{r} 20 \text{ gal} \\ -15 \text{ gal } 1 \text{ qt} \\ \hline \end{array}$
$\begin{array}{r} 19 \text{ gal } 4 \text{ qt} \\ -15 \text{ gal } 1 \text{ qt} \\ \hline 4 \text{ gal } 3 \text{ qt} \end{array}$

Thus, 4 gal 3 qt remains in the container.

36. 88 m + 340 cm = 88 m + 3.40 m = 91.4 m
The span is 91.4 meters

37.
$\begin{array}{r} 2 \text{ ft } 9 \text{ in.} \\ \times \quad\quad 6 \\ \hline \end{array}$
12 ft 54 in. = 12 ft + 4 ft 6 in. = 16 ft 6 in.
Thus, 16 ft 6 in. of material is needed.

Cumulative Review Chapters 1–9

1. $\quad 3a - 6 = a + 4$
$\quad 3a - a - 6 = a - a + 4$
$\quad\quad 2a - 6 = 4$
$\quad\quad 2a - 6 + 6 = 4 + 6$
$\quad\quad\quad 2a = 10$
$\quad\quad\quad \dfrac{2a}{2} = \dfrac{10}{2}$
$\quad\quad\quad\quad a = 5$

2. $\quad 2x + 1 = 3x - 5$
$\quad 2x - 2x + 1 = 3x - 2x - 5$
$\quad\quad 1 = x - 5$
$\quad\quad 1 + 5 = x - 5 + 5$
$\quad\quad 6 = x$

3. a. $\left(\dfrac{2}{5}\right)^4 = \dfrac{2}{5} \cdot \dfrac{2}{5} \cdot \dfrac{2}{5} \cdot \dfrac{2}{5} = \dfrac{2^4}{5^4} = \dfrac{16}{625}$

b. $\left(-\dfrac{1}{4}\right)^2 = \left(-\dfrac{1}{4}\right)\left(-\dfrac{1}{4}\right) = \dfrac{1}{16}$

4. a. $\left(-\dfrac{1}{3}\right)^3 = \left(-\dfrac{1}{3}\right)\left(-\dfrac{1}{3}\right)\left(-\dfrac{1}{3}\right) = -\dfrac{1}{27}$

b. $\left(\dfrac{3}{7}\right)^2 = \dfrac{3}{7} \cdot \dfrac{3}{7} = \dfrac{9}{49}$

5.
$\begin{array}{r} 2\dfrac{4}{5} \\ 5 \\ +1\dfrac{1}{2} \\ \hline \end{array}$
$\begin{array}{r} 2\dfrac{8}{10} \\ 5 \\ +1\dfrac{5}{10} \\ \hline 8\dfrac{13}{10} = 8 + 1\dfrac{3}{10} = 9\dfrac{3}{10} \end{array}$

6.
$\begin{array}{r} 2\dfrac{1}{3} \\ 4\dfrac{2}{5} \\ +3 \\ \hline \end{array}$
$\begin{array}{r} 2\dfrac{5}{15} \\ 4\dfrac{6}{15} \\ +3 \\ \hline 9\dfrac{11}{15} \end{array}$

7. $11.1x - 6.3 + 8.9x - 4.6$
$= 11.1x + 8.9x - 6.3 - 4.6$
$= 20x - 10.9$

8. $2.5y + 3.7 - 1.3y - 1.9$
$= 2.5y - 1.3y + 3.7 - 1.9$
$= 1.2y + 1.8$

9. $\dfrac{5.68 + (0.9)^2 \div 100}{0.2} = \dfrac{5.69 + 0.81 \div 100}{0.2}$
$\quad\quad = \dfrac{5.69 + 0.0081}{0.2}$
$\quad\quad = \dfrac{5.6981}{0.2}$
$\quad\quad = 28.4905$

10. $\dfrac{0.12 + 0.96}{0.5} = \dfrac{1.08}{0.5} = 2.16$

11.
$$9\overline{)\,7.00}\;\;\substack{0.77...}$$
$$\underline{-63}$$
$$70$$
$$\underline{-63}$$
$$7$$

Thus $0.\overline{7} = \dfrac{7}{9}$.

12.
$$5\overline{)\,2.0}\;\;\substack{0.4}$$
$$\underline{-2\;0}$$
$$0$$

Thus $0.43 > \dfrac{2}{5}$.

13.
$$0.5y + 2.3 = 1.65$$
$$0.5y + 2.3 - 2.3 = 1.65 - 2.3$$
$$0.5y = -0.65$$
$$\frac{0.5y}{0.5} = \frac{-0.65}{0.5}$$
$$y = -1.3$$

14.
$$0.4x - 9.3 = 2.7$$
$$0.4x - 9.3 + 9.3 = 2.7 + 9.3$$
$$0.4x = 12$$
$$\frac{0.4x}{0.4} = \frac{12}{0.4}$$
$$x = 30$$

15. Use $a^2 + b^2 = c^2$ where $a = b = 300$.
$$300^2 + 300^2 = c^2$$
$$90,000 + 90,000 = c^2$$
$$180,000 = c^2$$
$$\sqrt{180,000} = c$$
$$424 \approx c$$
The length of the diagonal is approximately 424 feet.

16. Use $a^2 + b^2 = c^2$ where $a = 200$ and $b = 125$.
$$200^2 + 125^2 = c^2$$
$$40,000 + 15,625 = c^2$$
$$55,625 = c^2$$
$$\sqrt{55,625} = c$$
$$236 \approx c$$
The length of the diagonal is approximately 236 feet.

17. a. $\dfrac{\text{width}}{\text{length}} = \dfrac{5 \text{ feet}}{7 \text{ feet}} = \dfrac{5}{7}$

    b. $P = 2 \cdot l + 2 \cdot w$
$$= 2(7 \text{ feet}) + 2(5 \text{ feet})$$
$$= 14 \text{ feet} + 10 \text{ feet}$$
$$= 24 \text{ feet}$$
$$\dfrac{\text{length}}{\text{perimeter}} = \dfrac{7 \text{ feet}}{24 \text{ feet}} = \dfrac{7}{24}$$

18. a. $P = 4s = 4(9 \text{ in.}) = 36 \text{ in.}$
$$\dfrac{\text{side}}{\text{perimeter}} = \dfrac{9 \text{ inches}}{36 \text{ inches}} = \dfrac{9}{36} = \dfrac{1}{4}$$

    b. $A = s^2 = (9 \text{ in.})^2 = 81 \text{ sq in.}$
$$\dfrac{\text{perimeter}}{\text{area}} = \dfrac{36 \text{ inches}}{81 \text{ sq inches}} = \dfrac{36}{81} = \dfrac{4}{9}$$

19. $\dfrac{2160 \text{ dollars}}{12 \text{ weeks}} = \dfrac{180 \text{ dollars}}{1 \text{ week}}$

20. $\dfrac{8 \text{ chaperones}}{40 \text{ students}} = \dfrac{1 \text{ chaperone}}{5 \text{ students}}$

21.
$$\frac{1.6}{1.1} = \frac{x}{0.3}$$
$$1.6 \cdot 0.3 = 1.1 \cdot x$$
$$0.48 = 1.1x$$
$$\frac{0.48}{1.1} = \frac{1.1x}{1.1}$$
$$0.44 \approx x$$

22.
$$\frac{2.4}{3.5} = \frac{0.7}{x}$$
$$2.4 \cdot x = 3.5 \cdot 0.7$$
$$2.4x = 2.45$$
$$\frac{2.4x}{2.4} = \frac{2.45}{2.4}$$
$$x \approx 1.02$$

23. Let x be the dose for a 140-lb woman.
$$\frac{4 \text{ cc}}{25 \text{ lb}} = \frac{x \text{ cc}}{140 \text{ lb}}$$
$$\frac{4}{25} = \frac{x}{140}$$
$$4 \cdot 140 = 25 \cdot x$$
$$560 = 25x$$
$$\frac{560}{25} = \frac{25x}{25}$$
$$22.4 = x$$
The dose is 22.4 cc.

24. Let x be the amount for 5 pie crusts.
$$\frac{3 \text{ c}}{2 \text{ crusts}} = \frac{x \text{ c}}{5 \text{ crusts}}$$
$$\frac{3}{2} = \frac{x}{5}$$
$$3 \cdot 5 = 2 \cdot x$$
$$15 = 2x$$
$$\frac{15}{2} = \frac{2x}{2}$$
$$7.5 = x$$
5 pie crusts require 7.5 cups of flour.

25. $\frac{17}{100} = 17\%$

17% of the people surveyed drive blue cars.

26. $\frac{38}{100} = 38\%$

38% of the shoppers used only cash.

27.
$$13 = 6\frac{1}{2}\% \cdot x$$
$$13 = 0.065x$$
$$\frac{13}{0.065} = \frac{0.065x}{0.065}$$
$$200 = x$$

13 is $6\frac{1}{2}\%$ of 200.

28.
$$54 = 4\frac{1}{2}\% \cdot x$$
$$54 = 0.045x$$
$$\frac{54}{0.045} = \frac{0.045x}{0.045}$$
$$1200 = x$$

54 is $4\frac{1}{2}\%$ of 1200.

29. $x = 30\% \cdot 9$
$x = 0.3 \cdot 9$
$x = 2.7$
2.7 is 30% of 9.

30. $x = 42\% \cdot 30$
$x = 0.42 \cdot 30$
$x = 12.6$
12.6 is 42% of 30.

31. percent increase $= \dfrac{\text{amount of increase}}{\text{original amount}}$
$$= \frac{45 - 34}{34}$$
$$= \frac{11}{34}$$
$$\approx 0.32$$
The scholarship applications increased by 32%.

32. percent increase $= \dfrac{\text{amount of increase}}{\text{original amount}}$

$$= \frac{19 - 15}{15}$$

$$= \frac{4}{15}$$

$$\approx 0.27$$

The price of the paint increased by 27%.

33. sales tax $= 85.50 \cdot 0.075 = 6.4125$
The sales tax is $6.41.
$85.50 + 6.41 = 91.91$
The total price is $91.91.

34. sales tax $= 375 \cdot 0.08 = 30$
The sales tax is $30.
$375 + 30 = 405$
The total price is $405.

35. Point *A* is 4 units left of the *y*-axis and 2 units above the *x*-axis. The coordinates are $(-4, 2)$.
Point *B* is 1 unit right of the *y*-axis and 2 units above the *x*-axis. The coordinates are $(1, 2)$.
Point *C* is on the *y*-axis and 1 unit above the *x*-axis. The coordinates are $(0, 1)$.
Point *D* is 3 units left of the *y*-axis and on the *x*-axis. The coordinates are $(-3, 0)$.
Point *E* is 5 units right of the *y*-axis and 4 units below the *x*-axis. The coordinates are $(5, -4)$.

36. Point *A* is 2 units right of the *y*-axis and 3 units below the *x*-axis. The coordinates are $(2, -3)$.
Point *B* is 5 units left of the *y*-axis and on the *x*-axis. The coordinates are $(-5, 0)$.
Point *C* is on the *y*-axis and 4 units above the *x*-axis. The coordinates are $(0, 4)$.
Point *D* is 3 units left of the *y*-axis and 2 units below the *x*-axis. The coordinates are $(-3, 2)$.

37. No matter what *x*-value we choose, *y* is always 4.

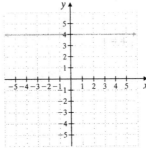

38. No matter what *x*-value we choose, *y* is always −2.

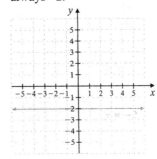

39. The seven numbers are listed in order. The median is the middle number, 57.

40. The five numbers in order are:
60, 72, 83, 89, 95.
The median is the middle number, 83.

41. There is 1 red marble and $1 + 1 + 2 = 4$ total marbles. The probability is $\dfrac{1}{4}$.

42. There are 2 nickels and $2 + 2 + 3 = 7$ total coins. The probability is $\dfrac{2}{7}$.

43. The complement of a 48° angle is an angle that has measure $90° - 48° = 42°$.

44. The supplement of a 137° angle is an angle that has measure $180° - 137° = 43°$.

45. $8 \text{ ft} = \dfrac{8 \text{ ft}}{1} \cdot \dfrac{12 \text{ in.}}{1 \text{ ft}} = 8 \cdot 12 \text{ in.} = 96 \text{ in.}$

46. $7 \text{ yd} = \dfrac{7 \text{ yd}}{1} \cdot \dfrac{3 \text{ ft}}{1 \text{ yd}} = 7 \cdot 3 \text{ ft} = 21 \text{ ft}$

47.

8 tons 1000 lb	7 tons 3000 lb
− 3 tons 1350 lb	− 3 tons 1350 lb
	4 tons 1650 lb

48.

$$
\begin{array}{r}
8 \text{ lb } 15 \text{ oz} \\
+\ 9 \text{ lb }\ \ 3 \text{ oz} \\
\hline
17 \text{ lb } 18 \text{ oz} = 17 \text{ lb} + 1 \text{ lb } 2 \text{ oz} = 18 \text{ lb } 2 \text{ oz}
\end{array}
$$

49. $C = \dfrac{5}{9}(F - 32) = \dfrac{5}{9}(59 - 32) = \dfrac{5}{9}(27) = 15$

59°F is 15°C.

50. $C = \dfrac{5}{9}(F - 32) = \dfrac{5}{9}(86 - 32) = \dfrac{5}{9}(54) = 30$

86°F is 30°C.

Chapter 10

Section 10.1

Practice Problems

1. $(3y+7)+(-9y-14) = (3y-9y)+(7-14)$
$$= (-6y)+(-7)$$
$$= -6y-7$$

2. $(x^2-4x-3)+(5x^2-6x)$
$$= x^2+5x^2-4x-6x-3$$
$$= 6x^2-10x-3$$

3. $(-z^2-4.2z+11)+(9z^2-1.9z+6.3)$
$$= -z^2+9z^2-4.2z-1.9z+11+6.3$$
$$= 8z^2-6.1z+17.3$$

4.
$$\begin{array}{r} x^2\ -x+1.1 \\ +\ -8x^2\ -x-6.7 \\ \hline -7x^2-2x-5.6 \end{array}$$

5. $-(3y^2+y-2)$
$$= -1(3y^2+y-2)$$
$$= -1(3y^2)+(-1)(y)+(-1)(-2)$$
$$= -3y^2-y+2$$

6. $(9b+8)-(11b-20) = (9b+8)+(-11b+20)$
$$= 9b-11b+8+20$$
$$= -2b+28$$

7. $(11x^2+7x+2)-(15x^2+4x)$
$$= (11x^2+7x+2)+(-15x^2-4x)$$
$$= 11x^2-15x^2+7x-4x+2$$
$$= -4x^2+3x+2$$

8. $(-3y^2+5y)-(-7y^2+y-4)$
$$= (-3y^2+5y)+(7y^2-y+4)$$
$$= -3y^2+7y^2+5y-y+4$$
$$= 4y^2+4y+4$$

9.
$$\begin{array}{l} -4x^2+20x+17 \\ -\ (3x^2-12x) \\ \hline \end{array} \qquad \begin{array}{r} -4x^2+20x+17 \\ -3x^2+12x \\ \hline -7x^2+32x+17 \end{array}$$

10. $2y^3+y^2-6 = 2(3)^3+(3)^2-6$
$$= 2(27)+9-6$$
$$= 54+9-6$$
$$= 57$$
The value of $2y^3+y^2-6$ when $y=3$ is 57.

11. $-16t^2+530 = -16(1)^2+530$
$$= -16+530$$
$$= 514$$
The height of the object at 1 second is 514 feet.
$$-16t^2+530 = -16(4)^2+530$$
$$= -16(16)+530$$
$$= -256+530$$
$$= 274$$
The height of the object at 4 seconds is 274 feet.

Vocabulary and Readiness Check

1. The addends of an algebraic expression are the <u>terms</u> of the expression.

3. A polynomial with exactly 2 terms is called a <u>binomial</u>.

5. To <u>add</u> polynomials, combine like terms.

Exercise Set 10.1

1. $(2x+3)+(-7x-27) = 2x-7x+3-27$
$$= -5x-24$$

3. $(-3z^2+5z-5)+(-8z^2-8z+4)$
$$= -3z^2-8z^2+5z-8z-5+4$$
$$= -11z^2-3z-1$$

413

5. $(12y - 20) + (9y^2 + 13y - 20)$
$= 9y^2 + 12y + 13y - 20 - 20$
$= 9y^2 + 25y - 40$

7. $(4.3a^4 + 5) + (-8.6a^4 - 2a^2 + 4)$
$= 4.3a^4 - 8.6a^4 - 2a^2 + 5 + 4$
$= -4.3a^4 - 2a^2 + 9$

9. $-(9x - 16) = -1(9x - 16)$
$= -1(9x) + (-1)(-16)$
$= -9x + 16$

11. $-(-3z^2 + z - 7)$
$= -1(-3z^2 + z - 7)$
$= (-1)(-3z^2) + (-1)(z) + (-1)(-7)$
$= 3z^2 - z + 7$

13. $(8a - 5) - (3a + 8) = (8a - 5) + (-3a - 8)$
$= 8a - 3a - 5 - 8$
$= 5a - 13$

15. $(3x^2 - 2x + 1) - (5x^2 - 6x)$
$= (3x^2 - 2x + 1) + (-5x^2 + 6x)$
$= 3x^2 - 5x^2 - 2x + 6x + 1$
$= -2x^2 + 4x + 1$

17. $(10y^2 - 7) - (20y^3 - 2y^2 - 3)$
$= (10y^2 - 7) + (-20y^3 + 2y^2 + 3)$
$= -20y^3 + 10y^2 + 2y^2 - 7 + 3$
$= -20y^3 + 12y^2 - 4$

19. $\begin{array}{r} 2x + 12 \\ -(9x^2 + 3x \ -4) \\ \hline \end{array}$ $\begin{array}{r} 2x + 12 \\ + -9x^2 + 3x \ -4 \\ \hline -9x^2 \ -x + 16 \end{array}$

21. $\begin{array}{r} 13y^2 - 6y - 14 \\ -(5y^2 + 4y - 6) \\ \hline \end{array}$ $\begin{array}{r} 13y^2 \ -6y - 14 \\ + -5y^2 \ -4y \ +6 \\ \hline 8y^2 - 10y \ -8 \end{array}$

23. $(25x - 5) + (-20x - 7) = 25x - 20x - 5 - 7$
$= 5x - 12$

25. $(4y + 4) - (3y + 8) = (4y + 4) + (-3y - 8)$
$= 4y - 3y + 4 - 8$
$= y - 4$

27. $(3x^2 + 3x - 4) + (-8x^2 + 9)$
$= 3x^2 - 8x^2 + 3x - 4 + 9$
$= -5x^2 + 3x + 5$

29. $(5x + 4.5) + (-x - 8.6) = 5x - x + 4.5 - 8.6$
$= 4x - 4.1$

31. $(a - 5) - (-3a + 2) = (a - 5) + (3a - 2)$
$= a + 3a - 5 - 2$
$= 4a - 7$

33. $(21y - 4.6) - (36y - 8.2)$
$= (21y - 4.6) + (-36y + 8.2)$
$= 21y - 36y - 4.6 + 8.2$
$= -15y + 3.6$

35. $(18t^2 - 4t + 2) - (-t^2 + 7t - 1)$
$= (18t^2 - 4t + 2) + (t^2 - 7t + 1)$
$= 18t^2 + t^2 - 4t - 7t + 2 + 1$
$= 19t^2 - 11t + 3$

37. $(2b^3 + 5b^2 - 5b - 8) + (8b^2 + 9b + 6)$
$= 2b^3 + 5b^2 + 8b^2 - 5b + 9b - 8 + 6$
$= 2b^3 + 13b^2 + 4b - 2$

39. $\begin{array}{r} 6x^2 \qquad -7 \\ + -11x^2 - 11x + 20 \\ \hline -5x^2 - 11x + 13 \end{array}$

41. $\begin{array}{r} 3z + \dfrac{6}{7} \\ -\left(3z - \dfrac{3}{7}\right) \\ \hline \end{array}$ $\begin{array}{r} 3z + \dfrac{6}{7} \\ + -3z + \dfrac{3}{7} \\ \hline \dfrac{9}{7} \end{array}$

43. $-2x + 9 = -2(2) + 9 = -4 + 9 = 5$

45. $x^2 - 6x + 3 = (2)^2 - 6(2) + 3$
$$= 4 - 12 + 3$$
$$= -5$$

47. $\dfrac{3x^2}{2} - 14 = \dfrac{3(2)^2}{2} - 14$
$$= \dfrac{12}{2} - 14$$
$$= 6 - 14$$
$$= -8$$

49. $2x + 10 = 2(5) + 10 = 10 + 10 = 20$

51. $x^2 = 5^2 = 25$

53. $2x^2 + 4x - 20 = 2(5)^2 + 4(5) - 20$
$$= 2(25) + 20 - 20$$
$$= 50 + 20 - 20$$
$$= 50$$

55. $16t^2 = 16(6)^2 = 16(36) = 576$
In 6 seconds, the object travels 576 feet.

57. $3000 + 20x = 3000 + 20(10)$
$$= 3000 + 200$$
$$= 3200$$
The cost for 10 file cabinets is \$3200.

59. $1053 - 16t^2 = 1053 - 16(3)^2$
$$= 1053 - 16(9)$$
$$= 1053 - 144$$
$$= 909$$
After 3 seconds, the height of the object is 909 feet.

61. 2007 corresponds to $x = 7$.
$17x^2 + 210x + 700 = 17(7)^2 + 210(7) + 700$
$$= 17(49) + 1470 + 700$$
$$= 3003$$
3003 million cellular subscribers are expected in 2007.

63. $3^4 = 3 \cdot 3 \cdot 3 \cdot 3 = 81$

65. $(-5)^2 = (-5)(-5) = 25$

67. $x \cdot x \cdot x = x^3$

69. $2 \cdot 2 \cdot a \cdot a \cdot a \cdot a = 2^2 a^4$

71. $P = (5x - 10) + (2x + 1) + (x + 11)$
$$= 5x + 2x + x - 10 + 1 + 11$$
$$= 8x + 2$$
The perimeter is $(8x + 2)$ inches.

73. $(7x - 10) - (3x + 5) = (7x - 10) + (-3x - 5)$
$$= 7x - 3x - 10 - 5$$
$$= 4x - 15$$
The missing length is $(4x - 15)$ units.

75.
$$
\begin{array}{r}
3x^2 + \underline{}x - \underline{} \\
+ \quad \underline{}x^2 - 6x + 2 \\
\hline
5x^2 + 14x - 4
\end{array}
$$
Since $3x^2 + 2x^2 = 5x^2$, $20x - 6x = 14x$ and $-6 + 2 = -4$, the missing numbers are 20, 6, and 2.
$(3x^2 + \underline{20}x - \underline{6}) + (\underline{2}x^2 - 6x + 2) = 5x^2 + 14x - 4$

77. $7a^4 - 6a^2 + 2a - 1$
$$= 7(1.2)^4 - 6(1.2)^2 + 2(1.2) - 1$$
$$= 7.2752$$

79. $1053 - 16t^2 = 1053 - 16(8)^2$
$$= 1053 - 1024$$
$$= 29$$
$1053 - 16t^2$ when $t = 8$ is 29 feet.
$1053 - 16t^2 = 1053 - 16(9)^2$
$$= 1053 - 1296$$
$$= -243$$
$1053 - 16t^2$ when $t = 9$ is -243 feet.
answers may vary

Section 10.2

Practice Problems

1. $z^5 \cdot z^6 = z^{5+6} = z^{11}$

2. $8y^5 \cdot 4y^9 = (8 \cdot 4)(y^5 \cdot y^9) = 32y^{5+9} = 32y^{14}$

3. $(-4r^6 s^2)(-3r^2 s^5) = (-4 \cdot -3)(r^6 \cdot r^2)(s^2 \cdot s^5)$
$$= 12r^{6+2}s^{2+5}$$
$$= 12r^8 s^7$$

4. $11y^5 \cdot 3y^2 \cdot y = (11 \cdot 3)(y^5 \cdot y^2 \cdot y^1) = 33y^8$

5. $(z^3)^6 = z^{3 \cdot 6} = z^{18}$

6. $(z^4)^5 \cdot (z^3)^7 = (z^{20})(z^{21}) = z^{20+21} = z^{41}$

7. $(3b)^4 = 3^4 \cdot b^4 = 81b^4$

8. $(4x^2 y^6)^3 = 4^3 (x^2)^3 (y^6)^3 = 64x^6 y^{18}$

9. $(2x^2 y^4)^4 (3x^6 y^9)^2$
$$= 2^4 (x^2)^4 (y^4)^4 \cdot 3^2 (x^6)^2 (y^9)^2$$
$$= 16x^8 y^{16} \cdot 9x^{12} y^{18}$$
$$= (16 \cdot 9)(x^8 \cdot x^{12})(y^{16} \cdot y^{18})$$
$$= 144x^{20} y^{34}$$

Vocabulary and Readiness Check

1. In $7x^2$, the 2 is called the <u>exponent</u>.

3. To simplify $(x^4)^3$, we <u>multiply</u> the exponents.

Exercise Set 10.2

1. $x^5 \cdot x^9 = x^{5+9} = x^{14}$

3. $a^3 \cdot a = a^{3+1} = a^4$

5. $3z^3 \cdot 5z^2 = (3 \cdot 5)(z^3 \cdot z^2) = 15z^5$

7. $-4x \cdot 10x = (-4 \cdot 10)(x \cdot x) = -40x^2$

9. $2x \cdot 3x \cdot 7x = (2 \cdot 3 \cdot 7)(x \cdot x \cdot x) = 42x^3$

11. $a \cdot 4a^{11} \cdot 3a^5 = (4 \cdot 3)(a^1 \cdot a^{11} \cdot a^5) = 12a^{17}$

13. $(-5x^2 y^3)(-5x^4 y)$
$$= (-5 \cdot -5)(x^2 \cdot x^4)(y^3 \cdot y^1)$$
$$= 25x^6 y^4$$

15. $(7ab)(4a^4 b^5) = (7 \cdot 4)(a^1 \cdot a^4)(b^1 \cdot b^5)$
$$= 28a^5 b^6$$

17. $(x^5)^3 = x^{5 \cdot 3} = x^{15}$

19. $(z^3)^{10} = z^{3 \cdot 10} = z^{30}$

21. $(b^7)^6 \cdot (b^2)^{10} = b^{7 \cdot 6} \cdot b^{2 \cdot 10}$
$$= b^{42} \cdot b^{20}$$
$$= b^{42+20}$$
$$= b^{62}$$

23. $(3a)^4 = 3^4 \cdot a^4 = 81a^4$

25. $(a^{11} b^8)^3 = (a^{11})^3 (b^8)^3 = a^{11 \cdot 3} b^{8 \cdot 3} = a^{33} b^{24}$

27. $(10x^5 y^3)^3 = 10^3 (x^5)^3 (y^3)^3$
$$= 1000x^{5 \cdot 3} y^{3 \cdot 3}$$
$$= 1000x^{15} y^9$$

29. $(-3y)(2y^7)^3 = (-3y) \cdot 2^3 (y^7)^3$
$$= (-3y) \cdot 8y^{21}$$
$$= (-3)(8)(y^1 \cdot y^{21})$$
$$= -24y^{22}$$

31. $(4xy)^3 (2x^3 y^5)^2 = (4^3 x^3 y^3)[2^2 (x^3)^2 (y^5)^2]$
$= (64x^3 y^3)(4x^6 y^{10})$
$= (64 \cdot 4)(x^3 \cdot x^6)(y^3 \cdot y^{10})$
$= 256 x^9 y^{13}$

33. $7(x-3) = 7 \cdot x - 7 \cdot 3 = 7x - 21$

35. $-2(3a + 2b) = -2 \cdot 3a + (-2)(2b) = -6a - 4b$

37. $9(x + 2y - 3) = 9 \cdot x + 9 \cdot 2y - 9 \cdot 3$
$= 9x + 18y - 27$

39. area $= s^2$
$= (4x^6)^2$
$= 4^2 (x^6)^2$
$= 16x^{12}$
The area is $16x^{12}$ square inches.

41. area $= \dfrac{1}{2} bh$
$= \dfrac{1}{2} \cdot (6a^3 b^4) \cdot (4ab)$
$= \left(\dfrac{1}{2} \cdot 6 \cdot 4\right)(a^3 \cdot a^1)(b^4 \cdot b^1)$
$= 12a^4 b^5$
The area is $12a^4 b^5$ square meters.

43. $(14a^7 b^6)^3 (9a^6 b^3)^4$
$= 14^3 (a^7)^3 (b^6)^3 \cdot 9^4 (a^6)^4 (b^3)^4$
$= 2744 a^{21} b^{18} \cdot 6561 a^{24} b^{12}$
$= (2744 \cdot 6561)(a^{21} \cdot a^{24})(b^{18} \cdot b^{12})$
$= 18,003,384 a^{45} b^{30}$

45. $(8.1x^{10})^5 = 8.1^5 (x^{10})^5 = 34,867.84401 x^{50}$

47. answers may vary

49. $(a^{20} b^{10} c^5)^5 \cdot (a^9 b^{12})^3 = a^{100} b^{50} c^{25} \cdot a^{27} b^{36}$
$= a^{127} b^{86} c^{25}$

Integrated Review

1. $(3x + 5) + (-x - 8) = 3x - x + 5 - 8 = 2x - 3$

2. $(15y - 7) + (5y - 4) = 15y + 5y - 7 - 4$
$= 20y - 11$

3. $(7x + 1) - (-3x - 2) = (7x + 1) + (3x + 2)$
$= 7x + 3x + 1 + 2$
$= 10x + 3$

4. $(14y - 6) - (19y - 2) = (14y - 6) + (-19y + 2)$
$= 14y - 19y - 6 + 2$
$= -5y - 4$

5. $(a^4 + 5a) + (3a^4 - 3a^2 - 4a)$
$= a^4 + 3a^4 - 3a^2 + 5a - 4a$
$= 4a^4 - 3a^2 + a$

6. $(2a^3 - 6a^2 + 11) - (6a^3 + 6a^2 + 11)$
$= (2a^3 - 6a^2 + 11) + (-6a^3 - 6a^2 - 11)$
$= 2a^3 - 6a^3 - 6a^2 - 6a^2 + 11 - 11$
$= -4a^3 - 12a^2$

7. $(4.5x^2 + 8.1x) + (2.8x^2 - 12.3x - 5.3)$
$= 4.5x^2 + 2.8x^2 + 8.1x - 12.3x - 5.3$
$= 7.3x^2 - 4.2x - 5.3$

8. $(1.2y^2 - 3.6y) + (0.6y^2 + 1.2y - 5.6)$
$= 1.2y^2 + 0.6y^2 - 3.6y + 1.2y - 5.6$
$= 1.8y^2 - 2.4y - 5.6$

9.
$$\begin{array}{r} 8x+1 \\ -(2x-6) \\ \hline \end{array} \qquad \begin{array}{r} 8x+1 \\ +-2x+6 \\ \hline 6x+7 \end{array}$$

10.
$$\begin{array}{r} 5x^2+2x-10 \\ -(3x^2 \ -x \ +2) \\ \hline \end{array} \qquad \begin{array}{r} 5x^2+2x-10 \\ +-3x^2 \ +x \ -2 \\ \hline 2x^2+3x-12 \end{array}$$

11. $2x - 7 = 2(3) - 7 = 6 - 7 = -1$

12. $x^2 + 5x + 2 = 3^2 + 5(3) + 2 = 9 + 15 + 2 = 26$

13. $x^9 \cdot x^{11} = x^{9+11} = x^{20}$

14. $x^5 \cdot x^5 = x^{5+5} = x^{10}$

15. $y^3 \cdot y = y^{3+1} = y^4$

16. $a \cdot a^{10} = a^{1+10} = a^{11}$

17. $(x^7)^{11} = x^{7 \cdot 11} = x^{77}$

18. $(x^6)^6 = x^{6 \cdot 6} = x^{36}$

19. $(x^3)^4 \cdot (x^5)^6 = x^{3 \cdot 4} \cdot x^{5 \cdot 6} = x^{12} \cdot x^{30} = x^{42}$

20. $(y^2)^9 \cdot (y^3)^3 = y^{2 \cdot 9} \cdot y^{3 \cdot 3} = y^{18} \cdot y^9 = y^{27}$

21. $(5x)^3 = 5^3 x^3 = 125x^3$

22. $(2y)^5 = 2^5 y^5 = 32y^5$

23. $(-6xy^2)(2xy^5) = (-6 \cdot 2)(x \cdot x)(y^2 \cdot y^5)$
$\quad = -12x^2 y^7$

24. $(-4a^2 b^3)(-3ab) = (-4 \cdot -3)(a^2 \cdot a^1)(b^3 \cdot b^1)$
$\quad = 12a^3 b^4$

25. $(y^{11} z^{13})^3 = (y^{11})^3 (z^{13})^3$
$\quad = y^{11 \cdot 3} z^{13 \cdot 3}$
$\quad = y^{33} z^{39}$

26. $(a^5 b^{12})^4 = (a^5)^4 (b^{12})^4$
$\quad = a^{5 \cdot 4} b^{12 \cdot 4}$
$\quad = a^{20} b^{48}$

27. $(10x^2 y)^2 (3y) = 10^2 (x^2)^2 y^2 \cdot 3y$
$\quad = 100x^4 y^2 \cdot 3y$
$\quad = (100 \cdot 3)x^4 (y^2 \cdot y^1)$
$\quad = 300x^4 y^3$

28. $(8y^3 z)^2 (2z^5) = 8^2 (y^3)^2 z^2 \cdot 2z^5$
$\quad = 64y^6 z^2 \cdot 2z^5$
$\quad = (64 \cdot 2)y^6 (z^2 \cdot z^5)$
$\quad = 128y^6 z^7$

29. $(2a^5 b)^4 (3a^9 b^4)^2$
$\quad = 2^4 (a^5)^4 b^4 \cdot 3^2 (a^9)^2 (b^4)^2$
$\quad = 16a^{20} b^4 \cdot 9a^{18} b^8$
$\quad = (16 \cdot 9)(a^{20} \cdot a^{18})(b^4 \cdot b^8)$
$\quad = 144a^{38} b^{12}$

30. $(5x^4 y^6)^3 (x^2 y^2)^5$
$\quad = 5^3 (x^4)^3 (y^6)^3 \cdot (x^2)^5 (y^2)^5$
$\quad = 125x^{12} y^{18} \cdot x^{10} y^{10}$
$\quad = 125(x^{12} \cdot x^{10})(y^{18} \cdot y^{10})$
$\quad = 125x^{22} y^{28}$

Section 10.3

Practice Problems

1. $4y(8y^2 + 5) = 4y \cdot 8y^2 + 4y \cdot 5 = 32y^3 + 20y$

2. $3r(8r^2 - r + 11) = 3r \cdot 8r^2 - 3r \cdot r + 3r \cdot 11$
$\quad = 24r^3 - 3r^2 + 33r$

3. $(b + 3)(b + 5) = b(b + 5) + 3(b + 5)$
$\quad = b \cdot b + b \cdot 5 + 3 \cdot b + 3 \cdot 5$
$\quad = b^2 + 5b + 3b + 15$
$\quad = b^2 + 8b + 15$

4. $(7x-1)(5x+4) = 7x(5x+4) - 1(5x+4)$
$$= 7x \cdot 5x + 7x \cdot 4 - 1 \cdot 5x - 1 \cdot 4$$
$$= 35x^2 + 28x - 5x - 4$$
$$= 35x^2 + 23x - 4$$

5. $(6y-1)^2 = (6y-1)(6y-1)$
$$= 6y(6y-1) - 1(6y-1)$$
$$= 6y \cdot 6y + 6y(-1) - 1 \cdot 6y - 1(-1)$$
$$= 36y^2 - 6y - 6y + 1$$
$$= 36y^2 - 12y + 1$$

6. $(10x-7)(2x+3)$
$$= 10x \cdot 2x + 10x \cdot 3 + (-7)(2x) + (-7)(3)$$
$$= 20x^2 + 30x - 14x - 21$$
$$= 20x^2 + 16x - 21$$

7. $(3x+2)^2 = (3x+2)(3x+2)$
$$= 3x \cdot 3x + 3x \cdot 2 + 2 \cdot 3x + 2 \cdot 2$$
$$= 9x^2 + 6x + 6x + 4$$
$$= 9x^2 + 12x + 4$$

8. $(2x+5)(x^2+4x-1) = 2x(x^2+4x-1) + 5(x^2+4x-1)$
$$= 2x \cdot x^2 + 2x \cdot 4x + 2x(-1) + 5 \cdot x^2 + 5 \cdot 4x + 5(-1)$$
$$= 2x^3 + 8x^2 - 2x + 5x^2 + 20x - 5$$
$$= 2x^3 + 13x^2 + 18x - 5$$

9.

$$
\begin{array}{r}
x^2 + 4x - 1 \\
\times \quad\quad 2x + 5 \\
\hline
5x^2 + 20x - 5 \\
2x^3 + 8x^2 \ - 2x \quad\quad \\
\hline
2x^3 + 13x^2 + 18x - 5
\end{array}
$$

Exercise Set 10.3

1. $3x(9x^2-3) = 3x \cdot 9x^2 - 3x \cdot 3 = 27x^3 - 9x$

3. $-3a(2a^2-3a-5) = -3a \cdot 2a^2 - (-3a)(3a) - (-3a)(5)$
$$= -6a^3 + 9a^2 + 15a$$

5. $7x^2(6x^2 - 5x + 7) = (7x^2)(6x^2) - (7x^2)(5x) + (7x^2)(7)$
$$= 42x^4 - 35x^3 + 49x^2$$

7. $(x+3)(x+10) = x(x+10) + 3(x+10)$
$$= x \cdot x + x \cdot 10 + 3 \cdot x + 3 \cdot 10$$
$$= x^2 + 10x + 3x + 30$$
$$= x^2 + 13x + 30$$

9. $(2x-6)(x+4) = 2x(x+4) - 6(x+4)$
$$= 2x \cdot x + 2x \cdot 4 - 6 \cdot x - 6 \cdot 4$$
$$= 2x^2 + 8x - 6x - 24$$
$$= 2x^2 + 2x - 24$$

11. $(6a+4)^2 = (6a+4)(6a+4)$
$$= 6a(6a+4) + 4(6a+4)$$
$$= 6a \cdot 6a + 6a \cdot 4 + 4 \cdot 6a + 4 \cdot 4$$
$$= 36a^2 + 24a + 24a + 16$$
$$= 36a^2 + 48a + 16$$

13. $(a+6)(a^2 - 6a + 3) = a(a^2 - 6a + 3) + 6(a^2 - 6a + 3)$
$$= a \cdot a^2 + a(-6a) + a \cdot 3 + 6 \cdot a^2 + 6(-6a) + 6 \cdot 3$$
$$= a^3 - 6a^2 + 3a + 6a^2 - 36a + 18$$
$$= a^3 - 33a + 18$$

15. $(4x-5)(2x^2 + 3x - 10) = 4x(2x^2 + 3x - 10) - 5(2x^2 + 3x - 10)$
$$= 4x \cdot 2x^2 + 4x \cdot 3x + 4x(-10) + (-5)(2x^2) + (-5)(3x) - (-5)(10)$$
$$= 8x^3 + 12x^2 - 40x - 10x^2 - 15x + 50$$
$$= 8x^3 - 2x^2 - 55x + 50$$

17. $(x^3 + 2x + x^2)(3x + 1 + x^2) = x^3(3x + 1 + x^2) + 2x(3x + 1 + x^2) + x^2(3x + 1 + x^2)$
$$= x^3 \cdot 3x + x^3 \cdot 1 + x^3 \cdot x^2 + 2x \cdot 3x + 2x \cdot 1 + 2x \cdot x^2 + x^2 \cdot 3x + x^2 \cdot 1 + x^2 \cdot x^2$$
$$= 3x^4 + x^3 + x^5 + 6x^2 + 2x + 2x^3 + 3x^3 + x^2 + x^4$$
$$= x^5 + 4x^4 + 6x^3 + 7x^2 + 2x$$

19. $10r(-3r + 2) = 10r \cdot (-3r) + 10r \cdot 2$
$$= -30r^2 + 20r$$

21. $-2y^2(3y + y^2 - 6) = -2y^2 \cdot 3y + (-2y^2) \cdot y^2 - (-2y^2)(6)$
$$= -6y^3 - 2y^4 + 12y^2$$

23. $(x+2)(x+12) = x(x+12) + 2(x+12)$
$$= x \cdot x + x \cdot 12 + 2 \cdot x + 2 \cdot 12$$
$$= x^2 + 12x + 2x + 24$$
$$= x^2 + 14x + 24$$

25. $(2a+3)(2a-3) = 2a(2a-3) + 3(2a-3)$
$$= 2a \cdot 2a + 2a(-3) + 3 \cdot 2a + 3(-3)$$
$$= 4a^2 - 6a + 6a - 9$$
$$= 4a^2 - 9$$

27. $(x+5)^2 = (x+5)(x+5)$
$$= x(x+5) + 5(x+5)$$
$$= x \cdot x + x \cdot 5 + 5 \cdot x + 5 \cdot 5$$
$$= x^2 + 5x + 5x + 25$$
$$= x^2 + 10x + 25$$

29. $\left(b+\dfrac{3}{5}\right)\left(b+\dfrac{4}{5}\right) = b\left(b+\dfrac{4}{5}\right) + \dfrac{3}{5}\left(b+\dfrac{4}{5}\right)$
$$= b^2 + \frac{4}{5}b + \frac{3}{5}b + \frac{3}{5} \cdot \frac{4}{5}$$
$$= b^2 + \frac{7}{5}b + \frac{12}{25}$$

31. $(6x+1)(x^2+4x+1) = 6x(x^2+4x+1) + 1(x^2+4x+1)$
$$= 6x^3 + 24x^2 + 6x + x^2 + 4x + 1$$
$$= 6x^3 + 25x^2 + 10x + 1$$

33. $(7x+5)^2 = (7x+5)(7x+5)$
$$= 7x(7x+5) + 5(7x+5)$$
$$= 49x^2 + 35x + 35x + 25$$
$$= 49x^2 + 70x + 25$$

35. $(2x-1)^2 = (2x-1)(2x-1)$
$$= 2x(2x-1) + (-1)(2x-1)$$
$$= 4x^2 - 2x - 2x + 1$$
$$= 4x^2 - 4x + 1$$

37. $(2x^2-3)(4x^3+2x-3) = 2x^2(4x^3+2x-3) - 3(4x^3+2x-3)$
$$= 8x^5 + 4x^3 - 6x^2 - 12x^3 - 6x + 9$$
$$= 8x^5 - 8x^3 - 6x^2 - 6x + 9$$

39. $(x^3 + x^2 + x)(x^2 + x + 1) = x^3(x^2 + x + 1) + x^2(x^2 + x + 1) + x(x^2 + x + 1)$
$$= x^5 + x^4 + x^3 + x^4 + x^3 + x^2 + x^3 + x^2 + x$$
$$= x^5 + 2x^4 + 3x^3 + 2x^2 + x$$

41.
$$
\begin{array}{r}
2z^2 - z + 1 \\
\times\; 5z^2 + z - 2 \\
\hline
-4z^2 + 2z - 2 \\
2z^3 - z^2\ \ + z \\
10z^4 - 5z^3 + 5z^2 \\
\hline
10z^4 - 3z^3\qquad + 3z - 2
\end{array}
$$

43. $50 = 2 \cdot 5 \cdot 5 = 2 \cdot 5^2$

45. $72 = 2 \cdot 2 \cdot 2 \cdot 3 \cdot 3 = 2^3 \cdot 3^2$

47. $200 = 2 \cdot 2 \cdot 2 \cdot 5 \cdot 5 = 2^3 \cdot 5^2$

49. $(y - 6)(y^2 + 3y + 2) = y(y^2 + 3y + 2) - 6(y^2 + 3y + 2)$
$$= y^3 + 3y^2 + 2y - 6y^2 - 18y - 12$$
$$= y^3 - 3y^2 - 16y - 12$$

The area is $(y^3 - 3y^2 - 16y - 12)$ square feet.

51. $(x^2 - 1)(x^2 - 1) - (x \cdot x) = x^2(x^2 - 1) + (-1)(x^2 - 1) - x^2$
$$= x^4 - x^2 - x^2 + 1 - x^2$$
$$= x^4 - 3x^2 + 1$$

The area of the shaded figure is
$(x^4 - 3x^2 + 1)$ square meters.

53. answers may vary

Section 10.4

Practice Problems

1. $42 = 2 \cdot 3 \cdot 7$
$\ \ 28 = 2 \cdot 2 \cdot 7$
$\qquad\ \downarrow\qquad\downarrow$
$\qquad\ 2\ \ \cdot\ \ 7$
The GCF is $2 \cdot 7 = 14$.

2. The GCF of z^7, z^8, and $z = z^1$ is $z^1 = z$ since 1 is the smallest exponent to which z is raised.

3. The GCF of 6, 3, and 15 is 3.
The GCF of a^4, a^5, and a^2 is a^2.
The GCF of $6a^4$, $3a^5$, and $15a^2$ is $3a^2$.

4. The GCF of $10y^7$ and $5y^9$ is $5y^7$.
$$10y^7 + 5y^9 = 5y^7 \cdot 2 + 5y^7 \cdot y^2$$
$$= 5y^7(2 + y^2)$$

5. The GCF of the terms is 2.
$$4z^2 - 12z + 2 = 2 \cdot 2z^2 - 2 \cdot 6z + 2 \cdot 1$$
$$= 2(2z^2 - 6z + 1)$$

6. $-3y^2 - 9y + 15x^2$
$$= 3 \cdot -y^2 + 3 \cdot -3y + 3 \cdot 5x^2$$
$$= 3(-y^2 - 3y + 5x^2) \text{ or } -3(y^2 + 3y - 5x^2)$$

Vocabulary and Readiness Check

1. In $-3 \cdot x^4 = -3x^4$, the -3 and the x^4 is each called a <u>factor</u> and $-3x^4$ is called the <u>product</u>.

3. The GCF of a list of variables raised to powers in the variable raised to the <u>smallest</u> exponent in the list.

Exercise Set 10.4

1. $48 = 2 \cdot 2 \cdot 2 \cdot 2 \cdot 3$
$15 = 3 \cdot 5$
GCF $= 3$

3. $60 = 2 \cdot 2 \cdot 3 \cdot 5$
$72 = 2 \cdot 2 \cdot 2 \cdot 3 \cdot 3$
GCF $= 2 \cdot 2 \cdot 3 = 12$

5. $12 = 2 \cdot 2 \cdot 3$
$20 = 2 \cdot 2 \cdot 5$
$36 = 2 \cdot 2 \cdot 3 \cdot 3$
GCF $= 2 \cdot 2 = 4$

7. $8 = 2 \cdot 2 \cdot 2$
$32 = 2 \cdot 2 \cdot 2 \cdot 2 \cdot 2$
$100 = 2 \cdot 2 \cdot 5 \cdot 5$
GCF $= 2 \cdot 2 = 4$

9. $y^7 = y^2 \cdot y^5$
$y^2 = y^2$
$y^{10} = y^2 \cdot y^8$
GCF $= y^2$

11. $a^5 = a^5$
$a^5 = a^5$
$a^5 = a^5$
GCF $= a^5$

13. $x^3 y^2 = x \cdot x^2 \cdot y^2$
$xy^2 = x \cdot y^2$
$x^4 y^2 = x \cdot x^3 \cdot y^2$
GCF $= x \cdot y^2 = xy^2$

15. $3x^4 = 3 \cdot x \cdot x^3$
$5x^7 = 5 \cdot x \cdot x^6$
$10x = 2 \cdot 5 \cdot x$
GCF $= x$

17. $2z^3 = 2 \cdot z^3$
$14z^5 = 2 \cdot 7 \cdot z^3 \cdot z^2$
$18z^3 = 2 \cdot 3 \cdot 3 \cdot z^3$
GCF $= 2z^3$

19. $3y^2 = 3 \cdot y \cdot y$
$18y = 2 \cdot 3 \cdot 3 \cdot y$
GCF $= 3y$
$3y^2 + 18y = 3y \cdot y + 3y \cdot 6 = 3y(y + 6)$

21. $10a^6 = 5a^6 \cdot 2$
$5a^8 = 5a^6 \cdot a^2$
$GCF = 5a^6$
$10a^6 - 5a^8 = 5a^6 \cdot 2 - 5a^6 \cdot a^2$
$\qquad\qquad = 5a^6(2 - a^2)$

23. $4x^3 = 4x \cdot x^2$
$12x^2 = 4x \cdot 3x$
$20x = 4x \cdot 5$
$GCF = 4x$
$4x^3 + 12x^2 + 20x = 4x \cdot x^2 + 4x \cdot 3x + 4x \cdot 5$
$\qquad\qquad\qquad = 4x(x^2 + 3x + 5)$

25. $z^7 = z^5 \cdot z^2$
$6z^5 = z^5 \cdot 6$
$GCF = z^5$
$z^7 - 6z^5 = z^5 \cdot z^2 - z^5 \cdot 6 = z^5(z^2 - 6)$

27. $-35 = -7 \cdot 5$ or $7 \cdot -5$
$14y = -7 \cdot -2 \cdot y$ or $7 \cdot 2 \cdot y$
$-7y^2 = -7 \cdot y^2$ or $7 \cdot -y^2$
$GCF = -7$ or 7
$-35 + 14y - 7y^2$
$= -7 \cdot 5 + (-7)(-2y) + (-7)(y^2)$
$= -7(5 - 2y + y^2)$ or $7(-5 + 2y - y^2)$

29. $12a^5 = 12a^5 \cdot 1$
$36a^6 = 12a^5 \cdot 3a$
$GCF = 12a^5$
$12a^5 - 36a^6 = 12a^5 \cdot 1 - 12a^5 \cdot 3a$
$\qquad\qquad\qquad = 12a^5(1 - 3a)$

31. $30\% \cdot 120 = x$
$0.30 \cdot 120 = x$
$\qquad 36 = x$
30% of 120 is 36.

33. $80\% = \dfrac{80}{100} = \dfrac{4}{5}$

35. $\dfrac{3}{8} = \dfrac{3}{8} \cdot \dfrac{100\%}{1} = \dfrac{300}{8}\% = 37.5\%$

37. area on the left: $x \cdot x = x^2$
area on the right: $2 \cdot x = 2x$
total area: $x^2 + 2x$
Notice that $x(x + 2) = x^2 + 2x$.

39. answers may vary

41. Let $x = 2$ and $z = 7$.
$(xy + z)^x = (2y + 7)^2$
$\qquad\qquad = (2y + 7)(2y + 7)$
$\qquad\qquad = 2y(2y + 7) + 7(2y + 7)$
$\qquad\qquad = 4y^2 + 14y + 14y + 49$
$\qquad\qquad = 4y^2 + 28y + 49$

Chapter 10 Vocabulary Check

1. Factoring is the process of writing an expression as a product.

2. The greatest common factor of a list of terms is the product of all common factors.

3. The FOIL method may be used when multiplying two binomials.

4. A polynomial with exactly 3 terms is called a trinomial.

5. A polynomial with exactly 2 terms is called a binomial.

6. A polynomial with exactly 1 term is called a monomial.

7. Monomials, binomials, and trinomials are all examples of polynomials.

8. In $5x^3$, the 3 is called an exponent.

Chapter 10 Review

1. $(2b + 7) + (8b - 10) = 2b + 8b + 7 - 10$
$\qquad\qquad\qquad\qquad = 10b - 3$

2. $(7s-6)+(14s-9) = 7s+14s-6-9$
         $= 21s-15$

3. $(3x+0.2)-(4x-2.6)$
  $= (3x+0.2)+(-4x+2.6)$
  $= 3x-4x+0.2+2.6$
  $= -x+2.8$

4. $(10y-6)-(11y+6) = (10y-6)+(-11y-6)$
           $= 10y-11y-6-6$
           $= -y-12$

5. $(4z^2+6z-1)+(5z-5)$
  $= 4z^2+6z+5z-1-5$
  $= 4z^2+11z-6$

6. $(17a^3+11a^2+a)+(14a^2-a)$
  $= 17a^3+11a^2+14a^2+a-a$
  $= 17a^3+25a^2$

7. $\left(9y^2-y+\dfrac{1}{2}\right)-\left(20y^2-\dfrac{1}{4}\right)$
  $= \left(9y^2-y+\dfrac{1}{2}\right)+\left(-20y^2+\dfrac{1}{4}\right)$
  $= 9y^2-20y^2-y+\dfrac{1}{2}+\dfrac{1}{4}$
  $= -11y^2-y+\dfrac{3}{4}$

8.

$$\begin{array}{r} x^2-6x+1 \\ -\quad(x-2) \\ \hline \end{array} \qquad \begin{array}{r} x^2-6x+1 \\ +\quad -x+2 \\ \hline x^2-7x+3 \end{array}$$

9. $5x^2 = 5(3)^2 = 5\cdot 9 = 45$

10. $2-7x = 2-7(3) = 2-21 = -19$

11. $(3x+16)+(10x-2)+(3x+16)+(10x-2)$
  $= 3x+10x+3x+10x+16-2+16-2$
  $= 26x+28$
  The perimeter is $(26x+28)$ feet.

12. Perimeter
  $= (4x^2+1)+(4x^2+1)+(4x^2+1)+(4x^2+1)$
  $= 4x^2+4x^2+4x^2+4x^2+1+1+1+1$
  $= 16x^2+4$
  The perimeter is $(16x^2+4)$ meters.

13. $x^{10}\cdot x^{14} = x^{10+14} = x^{24}$

14. $y\cdot y^6 = y^{1+6} = y^7$

15. $4z^2\cdot 6z^5 = 4\cdot 6z^{2+5} = 24z^7$

16. $(-3x^2 y)(5xy^4) = -3\cdot 5x^{2+1}y^{1+4} = -15x^3 y^5$

17. $(a^5)^7 = a^{5\cdot 7} = a^{35}$

18. $(x^2)^4\cdot(x^{10})^2 = x^{2\cdot 4}\cdot x^{10\cdot 2}$
        $= x^8\cdot x^{20}$
        $= x^{8+20}$
        $= x^{28}$

19. $(9b)^2 = 9^2\cdot b^2 = 81b^2$

20. $(a^4 b^2 c)^5 = (a^4)^5\cdot(b^2)^5\cdot(c)^5$
      $= a^{4\cdot 5}\cdot b^{2\cdot 5}\cdot c^5$
      $= a^{20}b^{10}c^5$

21. $(7x)(2x^5)^3 = 7x\cdot 2^3(x^5)^3$
       $= 7x\cdot 8x^{5\cdot 3}$
       $= 7x\cdot 8x^{15}$
       $= 56x^{16}$

22. $(3x^6 y^5)^3(2x^6 y^5)^2$
  $= 3^3(x^6)^3(5^5)^3\cdot 2^2(x^6)^2(y^5)^2$
  $= 27x^{18}y^{15}\cdot 4x^{12}y^{10}$
  $= 108x^{30}y^{25}$

23. $A = s^2 = (9a^7)(9a^7) = 9\cdot 9a^{7+7} = 81a^{14}$
  The area is $81a^{14}$ square miles.

24. $A = lw = 3x^4 \cdot 9x = 3 \cdot 9x^{4+1} = 27x^5$

The area is $27x^5$ square inches.

25. $2a(5a^2 - 6) = 2a \cdot 5a^2 - 2a \cdot 6 = 10a^3 - 12a$

26. $-3y^2(y^2 - 2y + 1) = -3y^2 \cdot y^2 - (-3y^2)(2y) + (-3y^2)(1)$
$$= -3y^4 + 6y^3 - 3y^2$$

27. $(x + 2)(x + 6) = x \cdot x + x \cdot 6 + 2 \cdot x + 2 \cdot 6$
$$= x^2 + 6x + 2x + 12$$
$$= x^2 + 8x + 12$$

28. $(3x - 1)(5x - 9) = 3x \cdot 5x + 3x(-9) - 1 \cdot 5x - 1(-9)$
$$= 15x^2 - 27x - 5x + 9$$
$$= 15x^2 - 32x + 9$$

29. $(y - 5)^2 = (y - 5)(y - 5)$
$$= y \cdot y + y(-5) - 5 \cdot y - 5(-5)$$
$$= y^2 - 5y - 5y + 25$$
$$= y^2 - 10y + 25$$

30. $(7a + 1)^2 = (7a + 1)(7a + 1)$
$$= 7a \cdot 7a + 7a \cdot 1 + 1 \cdot 7a + 1 \cdot 1$$
$$= 49a^2 + 7a + 7a + 1$$
$$= 49a^2 + 14a + 1$$

31. $(x + 1)(x^2 - 2x + 3) = x(x^2 - 2x + 3) + 1(x^2 - 2x + 3)$
$$= x^3 - 2x^2 + 3x + x^2 - 2x + 3$$
$$= x^3 - x^2 + x + 3$$

32. $(4y^2 - 3)(2y^2 + y + 1) = 4y^2(2y^2 + y + 1) - 3(2y^2 + y + 1)$
$$= 8y^4 + 4y^3 + 4y^2 - 6y^2 - 3y - 3$$
$$= 8y^4 + 4y^3 - 2y^2 - 3y - 3$$

33. $(3z^2 + 2z + 1)(z^2 + z + 1) = 3z^2(z^2 + z + 1) + 2z(z^2 + z + 1) + 1(z^2 + z + 1)$
$$= 3z^4 + 3z^3 + 3z^2 + 2z^3 + 2z^2 + 2z + z^2 + z + 1$$
$$= 3z^4 + 5z^3 + 6z^2 + 3z + 1$$

34. $(a+6)(a^2 - a + 1)$
$= a(a^2 - a + 1) + 6(a^2 - a + 1)$
$= a^3 - a^2 + a + 6a^2 - 6a + 6$
$= a^3 + 5a^2 - 5a + 6$
The area is $(a^3 + 5a^2 - 5a + 6)$ square centimeters.

35. $20 = 2 \cdot 2 \cdot 5$
$35 = 5 \cdot 7$
$GCF = 5$

36. $12 = 2 \cdot 2 \cdot 3$
$32 = 2 \cdot 2 \cdot 2 \cdot 2 \cdot 2$
$GCF = 2 \cdot 2 = 4$

37. $24 = 2 \cdot 2 \cdot 2 \cdot 3$
$30 = 2 \cdot 3 \cdot 5$
$60 = 2 \cdot 2 \cdot 3 \cdot 5$
$GCF = 2 \cdot 3 = 6$

38. $10 = 2 \cdot 5$
$20 = 2 \cdot 2 \cdot 5$
$25 = 5 \cdot 5$
$GCF = 5$

39. $x^3 = x^2 \cdot x$
$x^2 = x^2$
$x^{10} = x^2 \cdot x^8$
$GCF = x^2$

40. $y^{10} = y^7 \cdot y^3$
$y^7 = y^7$
$y^7 = y^7$
$GCF = y^7$

41. $xy^2 = x \cdot y \cdot y$
$xy = x \cdot y$
$x^3 y^3 = x \cdot x^2 \cdot y \cdot y^2$
$GCF = x \cdot y$

42. $a^5 b^4 = a^5 \cdot b^2 \cdot b^2$
$a^6 b^3 = a^5 \cdot a \cdot b^2 \cdot b$
$a^7 b^2 = a^5 \cdot a^2 \cdot b^2$
$GCF = a^5 b^2$

43. $5a^3 = 5 \cdot a \cdot a^2$
$10a = 2 \cdot 5 \cdot a$
$20a^4 = 2 \cdot 2 \cdot 5 \cdot a \cdot a^3$
$GCF = 5a$

44. $12y^2 z = 4y^2 z \cdot 3$
$20y^2 z = 4y^2 z \cdot 5$
$24y^5 z = 4y^2 z \cdot 6y^3$
$GCF = 4y^2 z$

45. $2x^2 = 2x \cdot x$
$12x = 2x \cdot 6$
$GCF = 2x$
$2x^2 + 12x = 2x \cdot x + 2x \cdot 6 = 2x(x+6)$

46. $6a^2 = 6a \cdot a$
$12a = 6a \cdot 2$
$GCF = 6a$
$6a^2 - 12a = 6a \cdot a - 6a \cdot 2 = 6a(a-2)$

47. $6y^4 = y^4 \cdot 6$
$y^6 = y^4 \cdot y^2$
$GCF = y^4$
$6y^4 - y^6 = y^4 \cdot 6 - y^4 \cdot y^2 = y^4(6 - y^2)$

48. $7x^2 = 7 \cdot x^2$
$14x = 7 \cdot 2x$
$7 = 7 \cdot 1$
$GCF = 7$
$7x^2 - 14x + 7 = 7 \cdot x^2 - 7 \cdot 2x + 7 \cdot 1$
$\qquad\qquad\qquad = 7(x^2 - 2x + 1)$

49. $5a^7 = a^3 \cdot 5a^4$

$a^4 = a^3 \cdot a^1$

$a^3 = a^3 \cdot 1$

GCF $= a^3$

$5a^7 - a^4 + a^3 = a^3 \cdot 5a^4 - a^3 \cdot a + a^3 \cdot 1$

$= a^3(5a^4 - a + 1)$

50. $10y^6 = 10y \cdot y^5$

$10y = 10y \cdot 1$

GCF $= 10y$

$10y^6 - 10y = 10y \cdot y^5 - 10y \cdot 1 = 10y(y^5 - 1)$

51. $z^2 - 5z + 8$

$\underline{+ 6z - 4}$

$z^2 + z + 4$

52. $8y - 5$ $8y - 5$

$\underline{-(12y - 3)}$ $\underline{+ -12y + 3}$

$-4y - 2$

53. $x^5 \cdot x^{16} = x^{5+16} = x^{21}$

54. $y^8 \cdot y = y^{8+1} = y^9$

55. $(a^3 b^5 c)^6 = (a^3)^6 (b^5)^6 c^6 = a^{18} b^{30} c^6$

56. $(9x^2) \cdot (3x^2)^2 = (9x^2)(9x^4) = 81x^6$

57. $3a(4a^3 - 5) = 3a \cdot 4a^3 + 3a(-5) = 12a^4 - 15a$

58. $(x + 4)(x + 5) = x^2 + 4x + 5x + 20$

$= x^2 + 9x + 20$

59. $(3x + 4)^2 = (3x + 4)(3x + 4)$

$= 9x^2 + 12x + 12x + 16$

$= 9x^2 + 24x + 16$

60. $(6z + 5)(z - 2) = 6z^2 - 12z + 5z - 10$

$= 6z^2 - 7z - 10$

61. $28 = 2 \cdot 2 \cdot 7$

$32 = 2 \cdot 2 \cdot 2 \cdot 2 \cdot 2$

$40 = 2 \cdot 2 \cdot 2 \cdot 5$

GCF $= 2 \cdot 2 = 4$

62. $5z^5 = 5 \cdot z^4 \cdot z$

$12z^8 = 2 \cdot 2 \cdot 3 \cdot z^4 \cdot z^4$

$3z^4 = 3 \cdot z^4$

GCF $= z^4$

63. $z^9 = z^7 \cdot z^2$

$4z^7 = z^7 \cdot 4$

GCF $= z^7$

$z^9 - 4z = z^7 \cdot z^2 - z^7 \cdot 4 = z^7(z^2 - 4)$

64. $x^{12} = x^5 \cdot x^7$

$6x^5 = x^5 \cdot 6$

GCF $= x^5$

$x^{12} + 6x^5 = x^5 \cdot x^7 + x^5 \cdot 6 = x^5(x^7 + 6)$

65. $15a^4 = 15a^4 \cdot 1$

$45a^5 = 15a^4 \cdot 3a$

GCF $= 15a^4$

$15a^4 + 45a^5 = 15a^4 \cdot 1 + 15a^4 \cdot 3$

$= 15a^4(1 + 3a)$

66. $16z^5 = 8z^5 \cdot 2$

$24z^8 = 8z^5 \cdot 3z^3$

GCF $= 8z^5$

$16z^5 - 24z^8 = 8z^5 \cdot 2 - 8z^5 \cdot 3z^3$

$= 8z^5(2 - 3z^3)$

Chapter 10 Test

1. $(11x - 3) + (4x - 1) = 11x + 4x - 3 - 1$

$= 15x - 4$

2. $(11x - 3) - (4x - 1) = (11x - 3) + (-4x + 1)$

$= 11x - 4x - 3 + 1$

$= 7x - 2$

3. $(1.3y^2 + 5y) + (2.1y^2 - 3y - 3) = 1.3y^2 + 2.1y^2 + 5y - 3y - 3$

$$= 3.4y^2 + 2y - 3$$

4. $6a^2 + 2a + 1$ $6a^2 + 2a + 1$

      $\underline{-(8a^2 \quad + a)}$ $\underline{+ \;{-8a^2} \quad -a}$

                                   $-2a^2 \quad + a + 1$

5. $x^2 - 6x + 1 = (8)^2 - 6(8) + 1$

$$= 64 - 48 + 1$$

$$= 17$$

6. $y^3 \cdot y^{11} = y^{3+11} = y^{14}$

7. $(y^3)^{11} = y^{3 \cdot 11} = y^{33}$

8. $(2x^2)^4 = 2^4 \cdot (x^2)^4 = 16 \cdot x^{2 \cdot 4} = 16x^8$

9. $(6a^3)(-2a^7) = (6)(-2)(a^3 \cdot a^7) = -12a^{10}$

10. $(p^6)^7 (p^2)^6 = p^{6 \cdot 7} \cdot p^{2 \cdot 6}$

$$= p^{42} \cdot p^{12}$$

$$= p^{42+12}$$

$$= p^{54}$$

11. $(3a^4 b)^2 (2ba^4)^3 = (3^2 a^{4 \cdot 2} b^2)(2^3 b^3 a^{4 \cdot 3})$

$$= 9a^8 b^2 \cdot 8b^3 a^{12}$$

$$= 9 \cdot 8 a^{8+12} b^{2+3}$$

$$= 72 a^{20} b^5$$

12. $5x(2x^2 + 1.3) = 5x \cdot 2x^2 + 5x \cdot 1.3$

$$= 10x^3 + 6.5x$$

13. $-2y(y^3 + 6y^2 - 4) = -2y \cdot y^3 - 2y \cdot 6y^2 - 2y \cdot (-4)$

$$= -2y^4 - 12y^3 + 8y$$

14. $(x - 3)(x + 2) = x(x + 2) - 3(x + 2)$

$$= x \cdot x + x \cdot 2 - 3 \cdot x - 3 \cdot 2$$

$$= x^2 + 2x - 3x - 6$$

$$= x^2 - x - 6$$

15. $(5x + 2)^2 = (5x + 2)(5x + 2)$
$$= 5x(5x + 2) + 2(5x + 2)$$
$$= 5x \cdot 5x + 5x \cdot 2 + 2 \cdot 5x + 2 \cdot 2$$
$$= 25x^2 + 10x + 10x + 4$$
$$= 25x^2 + 20x + 4$$

16. $(a + 2)(a^2 - 2a + 4) = a(a^2 - 2a + 4) + 2(a^2 - 2a + 4)$
$$= a \cdot a^2 + a(-2a) + a \cdot 4 + 2 \cdot a^2 + 2(-2a) + 2 \cdot 4$$
$$= a^3 - 2a^2 + 4a + 2a^2 - 4a + 8$$
$$= a^3 + 8$$

17. Area:
$$(x + 7)(5x - 2) = x(5x - 2) + 7(5x - 2)$$
$$= 5x^2 - 2x + 35x - 14$$
$$= 5x^2 + 33x - 14$$
The area is $(5x^2 + 33x - 14)$ square inches.

Perimeter: $2(2x) + 2(5x - 2) = 4x + 10x - 4 = 14x - 4$
The perimeter is $(14x - 4)$ inches.

18. $45 = 3 \cdot 3 \cdot 5$
$60 = 2 \cdot 2 \cdot 3 \cdot 5$
$GCF = 3 \cdot 5 = 15$

19. $6y^3 = 2 \cdot 3 \cdot y^3$
$9y^5 = 3 \cdot 3 \cdot y^3 \cdot y^2$
$18y^4 = 2 \cdot 3 \cdot 3 \cdot y^3 \cdot y$
$GCF = 3y^3$

20. $3y^2 = 3y \cdot y$
$15y = 3y \cdot 5$
$GCF = 3y$
$3y^2 - 15y = 3y \cdot y - 3y \cdot 5 = 3y(y - 5)$

21. $10a^2 = 2a \cdot 5a$
$12a = 2a \cdot 6$
$GCF = 2a$
$10a^2 + 12a = 2a \cdot 5a + 2a \cdot 6 = 2a(5a + 6)$

22. $6x^2 = 6 \cdot x^2$

$\qquad 12x = 6 \cdot 2x$

$\qquad 30 = 6 \cdot 5$

$\qquad$ GCF $= 6$

$\qquad 6x^2 - 12x - 30 = 6 \cdot x^2 - 6 \cdot 2x - 6 \cdot 5$

$\qquad\qquad\qquad\qquad = 6(x^2 - 2x - 5)$

23. $7x^6 = x^3 \cdot 7x^3$

$\qquad 6x^4 = x^3 \cdot 6x$

$\qquad x^3 = x^3 \cdot 1$

$\qquad$ GCF $= x^3$

$\qquad 7x^6 - 6x^4 + x^3 = x^3 \cdot 7x^3 - x^3 \cdot 6x + x^3 \cdot 1$

$\qquad\qquad\qquad\qquad = x^3(7x^3 - 6x + 1)$

Cumulative Review Chapters 1–10

1. Area $= 380 \cdot 280 = 106,400$

The area of Colorado is approximately 106,400 square miles.

2. $21 \times 7 = 147$

There are 147 pecan trees.

3. $1 + (-10) + (-8) + 9 = -9 + (-8) + 9$

$\qquad\qquad\qquad\qquad\quad = -17 + 9$

$\qquad\qquad\qquad\qquad\quad = -8$

4. $-2 + (-7) + 3 + (-4) = -9 + 3 + (-4)$

$\qquad\qquad\qquad\qquad\qquad = -6 + (-4)$

$\qquad\qquad\qquad\qquad\qquad = -10$

5. $8 - 15 = 8 + (-15) = -7$

6. $4 - 7 = 4 + (-7) = -3$

7. $-4 - (-5) = -4 + 5 = 1$

8. $3 - (-2) = 3 + 2 = 5$

9. $\qquad 7x = 6x + 4$

$\qquad 7x - 6x = 6x + 4 - 6x$

$\qquad\qquad x = 4$

10. $\qquad 4x = -2 + 3x$

$\qquad 4x - 3x = -2 + 3x - 3x$

$\qquad\qquad x = -2$

11. $\qquad 17 - 7x + 3 = -3x + 21 - 3x$

$\qquad\quad 20 - 7x = -6x + 21$

$\qquad 20 - 7x + 7x = -6x + 21 + 7x$

$\qquad\qquad 20 = x + 21$

$\qquad\quad 20 - 21 = x + 21 - 21$

$\qquad\qquad -1 = x$

12. $\qquad 20 - 6x + 4 = -2x + 18 + 2x$

$\qquad\quad 24 - 6x = 18$

$\qquad 24 - 6x - 24 = 18 - 24$

$\qquad\qquad -6x = -6$

$\qquad\qquad \dfrac{-6x}{-6} = \dfrac{-6}{-6}$

$\qquad\qquad x = 1$

13. $\dfrac{2x}{15} + \dfrac{3x}{10} = \dfrac{2x}{15} \cdot \dfrac{2}{2} + \dfrac{3x}{10} \cdot \dfrac{3}{3} = \dfrac{4x}{30} + \dfrac{9x}{30} = \dfrac{13x}{30}$

14. $\dfrac{5}{7y} - \dfrac{9}{14y} = \dfrac{5}{7y} \cdot \dfrac{2}{2} - \dfrac{9}{14y} = \dfrac{10}{14y} - \dfrac{9}{14y} = \dfrac{1}{14y}$

15. 736.2359 rounded to the nearest tenth is 736.2.

16. 328.174 rounded to the nearest tenth is 328.2.

17. $\quad\begin{array}{r} 23.850 \\ +\ 1.604 \\ \hline 25.454 \end{array}$

18. $\quad\begin{array}{r} 12.762 \\ +\ 4.290 \\ \hline 17.052 \end{array}$

19. $\qquad 3.7y = -3.33$

$\qquad 3.7(-9) \stackrel{?}{=} -3.33$

$\qquad\quad -33.3 = -3.33$ False

No, -9 is not a solution.

20. $2.8x = 16.8$

     $2.8(6) \stackrel{?}{=} 16.8$

       $16.8 = 16.8$ True

   Yes, 6 is a solution.

21. $\dfrac{786.1}{1000} = 0.7861$

22. $\dfrac{818}{1000} = 0.818$

23. $-\dfrac{0.12}{10} = -0.012$

24. $\dfrac{5.03}{100} = 0.0503$

25. $-2x + 5 = -2(3.8) + 5 = -7.6 + 5 = -2.6$

26. $6x - 1 = 6(-2.1) - 1 = -12.6 - 1 = -13.6$

27.

$$
\begin{array}{r}
3.142 \approx 3.14 \\
7\overline{)\,22.000} \\
\underline{-21} \\
1\,0 \\
\underline{-\,7} \\
30 \\
\underline{-\,28} \\
20 \\
\underline{-\,14} \\
6
\end{array}
$$

$\dfrac{22}{7} \approx 3.14$

28.

$$
\begin{array}{r}
1.9473 \approx 1.947 \\
19\overline{)\,37.0000} \\
\underline{-19} \\
18\,0 \\
\underline{-17\,1} \\
90 \\
\underline{-76} \\
140 \\
\underline{-133} \\
70\\
\underline{-57} \\
13
\end{array}
$$

$\dfrac{37}{19} \approx 1.947$

29. $\sqrt{\dfrac{1}{36}} = \dfrac{1}{6}$ since $\left(\dfrac{1}{6}\right)^2 = \dfrac{1}{6} \cdot \dfrac{1}{6} = \dfrac{1}{36}$.

30. $\sqrt{\dfrac{4}{25}} = \dfrac{2}{5}$ since $\left(\dfrac{2}{5}\right)^2 = \dfrac{2}{5} \cdot \dfrac{2}{5} = \dfrac{4}{25}$.

31.   Let x be the height of the tree.

$$\frac{6}{9} = \frac{x}{69}$$
$$6 \cdot 69 = 9x$$
$$414 = 9x$$
$$\frac{414}{9} = \frac{9x}{9}$$
$$46 = x$$

   The height of the tree is 46 feet.

32.   Let x be the height of the hydrant.

$$\frac{1}{2} = \frac{x}{6}$$
$$1 \cdot 6 = 2 \cdot x$$
$$6 = 2x$$
$$\frac{6}{2} = \frac{2x}{2}$$
$$3 = x$$

   The height of the fire hydrant is 3 feet.

33. $1.2 = 30\% \cdot x$

34. $9 = 45\% \cdot x$

35. $x \cdot 50 = 8$

$50x = 8$

$\dfrac{50x}{50} = \dfrac{8}{50}$

$x = 0.16$

$x = 16\%$

16% of 50 is 8.

36. $x \cdot 16 = 4$

$16x = 4$

$\dfrac{16x}{16} = \dfrac{4}{16}$

$x = 0.25$

$x = 25\%$

25% of 16 is 4.

37. $31 = 4\% \cdot x$

$31 = 0.04x$

$\dfrac{31}{0.04} = \dfrac{0.04x}{0.04}$

$775 = x$

There are 775 freshman.

38. $2\% \cdot x = 29$

$0.02x = 29$

$\dfrac{0.02x}{0.02} = \dfrac{29}{0.02}$

$x = 1450$

There are 1450 apples in the shipment.

39. simple interest $= P \cdot R \cdot T$

$= \$2400 \cdot 10\% \cdot \dfrac{8}{12}$

$= \$2400 \cdot 0.10 \cdot \dfrac{2}{3}$

$= \$160$

The interest is $160.

40. simple interest $= P \cdot R \cdot T$

$= \$1000 \cdot 3\% \cdot \dfrac{10}{12}$

$= \$1000 \cdot 0.03 \cdot \dfrac{5}{6}$

$= \$25$

The interest is $25.

41. $47\% + 15\% + 6\% + 3\% = 71\%$

71% of Americans has one or more computers at home.

42. $15\% + 47\% + 29\% = 91\%$

91% of Americans have fewer than 3 computers at home.

43. $P = 2l + 2w = 2(11) + 2(3) = 22 + 6 = 28$

The perimeter is 28 inches.

44. $P = 6 + 8 + 11 = 25$

The perimeter is 25 feet.

45. $A = bh = 1.5 \cdot 3.4 = 5.1$

The area is 5.1 square miles.

46. $A = \dfrac{1}{2}bh = \dfrac{1}{2}(17)(8) = 68$

The area of the triangle is 68 square inches.

47.

$$\begin{array}{r} 8 \text{ tons } 1000 \text{ lb} \\ -\ 3 \text{ tons } 1350 \text{ lb} \\ \hline \end{array} \qquad \begin{array}{r} 7 \text{ tons } 3000 \text{ lb} \\ -\ 3 \text{ tons } 1350 \text{ lb} \\ \hline 4 \text{ tons } 1650 \text{ lb} \end{array}$$

48.

$$\begin{array}{r} 5 \text{ tons } 700 \text{ lb} \\ \times \qquad\qquad 3 \\ \hline 15 \text{ tons } 2100 \text{ lb} \end{array} = 15 \text{ tons} + 1 \text{ ton } 1000 \text{ lb}$$

$$= 16 \text{ tons } 1000 \text{ lb}$$

49. $3210 \text{ ml} = \dfrac{3210 \text{ ml}}{1} \cdot \dfrac{1 \text{ L}}{1000 \text{ ml}} = \dfrac{3210}{1000} \text{ L} = 3.21 \text{ L}$

50. $4321 \text{ cl} = \dfrac{4321 \text{ cl}}{1} \cdot \dfrac{1 \text{ L}}{100 \text{ cl}} = \dfrac{4321}{100} \text{ L} = 43.21 \text{ L}$

Appendices

Practice Problems

1. $\dfrac{y^{10}}{y^6} = y^{10-6} = y^4$

2. $\dfrac{5^{11}}{5^8} = 5^{11-8} = 5^3 = 125$

3. $\dfrac{12a^4 b^{11}}{ab} = 12 \cdot \dfrac{a^4}{a^1} \cdot \dfrac{b^{11}}{b^1}$
$= 12 \cdot (a^{4-1}) \cdot (b^{11-1})$
$= 12a^3 b^{10}$

4. $6^0 = 1$

5. $(-8)^0 = 1$

6. $-8^0 = -1 \cdot 8^0 = -1 \cdot 1 = -1$

7. $7y^0 = 7 \cdot y^0 = 7 \cdot 1 = 7$

8. $5^{-2} = \dfrac{1}{5^2} = \dfrac{1}{25}$

9. $5x^{-2} = 5 \cdot \dfrac{1}{x^2} = \dfrac{5}{x^2}$

10. $4^{-1} + 3^{-1} = \dfrac{1}{4} + \dfrac{1}{3} = \dfrac{3}{12} + \dfrac{4}{12} = \dfrac{7}{12}$

11. $\left(\dfrac{6}{7}\right)^{-2} = \dfrac{6^{-2}}{7^{-2}}$
$= \dfrac{6^{-2}}{1} \cdot \dfrac{1}{7^{-2}}$
$= \dfrac{1}{6^2} \cdot \dfrac{7^2}{1}$
$= \dfrac{7^2}{6^2}$
$= \dfrac{49}{36}$

12. $\dfrac{x}{x^{-4}} = \dfrac{x^1}{x^{-4}} = x^{1-(-4)} = x^5$

13. $\dfrac{y^{-4}}{y^6} = y^{-4-6} = y^{-10} = \dfrac{1}{y^{10}}$

14. $y^{-6} \cdot y^3 \cdot y^{-4} = y^{-6+3} \cdot y^{-4}$
$= y^{-3} \cdot y^{-4}$
$= y^{-3+(-4)}$
$= y^{-7}$
$= \dfrac{1}{y^7}$

15. $(a^6 b^{-4})(a^{-3} b^8) = a^{6+(-3)} \cdot b^{-4+8} = a^3 b^4$

16. $(3y^9 z^{10})(2y^3 z^{-12}) = 3 \cdot 2 \cdot y^{9+3} \cdot z^{10+(-12)}$
$= 6 \cdot y^{12} \cdot z^{-2}$
$= \dfrac{6y^2}{z^2}$

Appendix C Exercise Set

1. $\dfrac{x^3}{x} = \dfrac{x^3}{x^1} = x^{3-1} = x^2$

3. $\dfrac{9^8}{9^6} = 9^{8-6} = 9^2 = 81$

5. $\dfrac{p^7 q^{20}}{pq^{15}} = \dfrac{p^7}{p^1} \cdot \dfrac{q^{20}}{q^{15}} = p^{7-1} \cdot q^{20-15} = p^6 q^5$

7. $\dfrac{7x^2 y^6}{14x^2 y^3} = \dfrac{7}{14} \cdot \dfrac{x^2}{x^2} \cdot \dfrac{y^6}{y^3}$

$\qquad = \dfrac{1}{2} \cdot x^{2-2} \cdot y^{6-3}$

$\qquad = \dfrac{1}{2} \cdot x^0 \cdot y^3$

$\qquad = \dfrac{y^3}{2}$

9. $7^0 = 1$

11. $2x^0 = 2 \cdot x^0 = 2 \cdot 1 = 2$

13. $-7^0 = -1 \cdot 7^0 = -1 \cdot 1 = -1$

15. $(-7)^0 = 1$

17. $4^{-3} = \dfrac{1}{4^3} = \dfrac{1}{64}$

19. $7x^{-3} = 7 \cdot \dfrac{1}{x^3} = \dfrac{7}{x^3}$

21. $3^{-1} + 2^{-1} = \dfrac{1}{3^1} + \dfrac{1}{2^1} = \dfrac{1}{3} + \dfrac{1}{2} = \dfrac{2}{6} + \dfrac{3}{6} = \dfrac{5}{6}$

23. $\dfrac{1}{p^{-3}} = p^3$

25. $\dfrac{x^{-2}}{x} = \dfrac{x^{-2}}{x^1} = x^{-2-1} = x^{-3} = \dfrac{1}{x^3}$

27. $\dfrac{z^{-4}}{z^{-7}} = z^{-4-(-7)} = z^3$

29. $3^{-2} + 3^{-1} = \dfrac{1}{3^2} + \dfrac{1}{3^1} = \dfrac{1}{9} + \dfrac{1}{3} = \dfrac{1}{9} + \dfrac{3}{9} = \dfrac{4}{9}$

31. $\left(\dfrac{5}{y}\right)^{-2} = \dfrac{5^{-2}}{y^{-2}} = \dfrac{5^{-2}}{1} \cdot \dfrac{1}{y^{-2}} = \dfrac{1}{5^2} \cdot \dfrac{y^2}{1} = \dfrac{y^2}{25}$

33. $\dfrac{1}{p^{-4}} = p^4$

35. $a^2 \cdot a^{-9} \cdot a^{13} = a^{2+(-9)} \cdot a^{13}$

$\qquad = a^{-7} \cdot a^{13}$

$\qquad = a^{-7+13}$

$\qquad = a^6$

37. $(x^8 y^{-6})(x^{-2} y^{12}) = x^{8+(-2)} \cdot y^{-6+12}$

$\qquad = x^6 y^6$

39. $x^{-7} \cdot x^{-8} \cdot x^4 = x^{-7+(-8)} \cdot x^4$

$\qquad = x^{-15} \cdot x^4$

$\qquad = x^{-15+4}$

$\qquad = x^{-11}$

$\qquad = \dfrac{1}{x^{11}}$

41. $(5x^{-7})(3x^4) = 5 \cdot 3 \cdot x^{-7+4}$

$\qquad = 15 \cdot x^{-3}$

$\qquad = 15 \cdot \dfrac{1}{x^3}$

$\qquad = \dfrac{15}{x^3}$

43. $y^5 \cdot y^{-7} \cdot y^{-10} = y^{5+(-7)} \cdot y^{-10}$

$\qquad = y^{-2} \cdot y^{-10}$

$\qquad = y^{-2+(-10)}$

$\qquad = y^{-12}$

$\qquad = \dfrac{1}{y^{12}}$

45. $(8m^5 n^{-1})(7m^2 n^{-4}) = 8 \cdot 7 \cdot m^{5+2} n^{-1+(-4)}$
$$= 56 \cdot m^7 \cdot n^{-5}$$
$$= 56 \cdot m^7 \cdot \frac{1}{n^5}$$
$$= \frac{56m^7}{n^5}$$

47. $\dfrac{x^{15}}{x^8} = x^{15-8} = x^7$

49. $\dfrac{a^9 b^{14}}{ab} = \dfrac{a^9}{a^1} \cdot \dfrac{b^{14}}{b^1} = a^{9-1} \cdot b^{14-1} = a^8 b^{13}$

51. $\dfrac{x^3}{x^9} = x^{3-9} = x^{-6} = \dfrac{1}{x^6}$

53. $3z^0 = 3 \cdot z^0 = 3 \cdot 1 = 3$

55. $5^{-3} = \dfrac{1}{5^3} = \dfrac{1}{125}$

57. $8x^{-9} = 8 \cdot \dfrac{1}{x^9} = \dfrac{8}{x^9}$

59. $5^{-1} + 10^{-1} = \dfrac{1}{5^1} + \dfrac{1}{10^1}$
$$= \dfrac{1}{5} + \dfrac{1}{10}$$
$$= \dfrac{2}{10} + \dfrac{1}{10}$$
$$= \dfrac{3}{10}$$

61. $\dfrac{z^{-8}}{z^{-1}} = z^{-8-(-1)} = z^{-7} = \dfrac{1}{z^7}$

63. $x^{-7} \cdot x^5 \cdot x^{-7} = x^{-7+5} \cdot x^{-7}$
$$= x^{-2} \cdot x^{-7}$$
$$= x^{-2+(-7)}$$
$$= x^{-9}$$
$$= \dfrac{1}{x^9}$$

65. $(a^{-2} b^3)(a^{10} b^{-11}) = a^{-2+10} \cdot b^{3+(-11)}$
$$= a^8 \cdot b^{-8}$$
$$= a^8 \cdot \dfrac{1}{b^8}$$
$$= \dfrac{a^8}{b^8}$$

67. $(3x^{20} y^{-1})(10x^{-11} y^{-5})$
$$= 3 \cdot 10 \cdot x^{20+(-11)} y^{-1+(-5)}$$
$$= 30 \cdot x^9 \cdot y^{-6}$$
$$= 30x^9 \cdot \dfrac{1}{y^6}$$
$$= \dfrac{30x^9}{y^6}$$

Appendix D

Practice Problems

1. a. $760{,}000 = 7.6 \times 10^5$

   b. $0.00035 = 3.5 \times 10^{-4}$

2. a. $9.062 \times 10^{-4} = 0.0009062$

   b. $8.002 \times 10^6 = 8{,}002{,}000$

3. a. $(8 \times 10^7)(3 \times 10^{-9}) = 8 \cdot 3 \cdot 10^7 \cdot 10^{-9}$
$$= 24 \times 10^{-2}$$
$$= 0.24$$

b. $\dfrac{8\times10^4}{2\times10^{-3}} = \dfrac{8}{2}\times10^{4-(-3)}$

$= 4\times10^7$

$= 40,000,000$

Appendix D Exercise Set

1. $78,000 = 7.8\times10^4$

3. $0.00000167 = 1.67\times10^{-6}$

5. $0.00635 = 6.35\times10^{-3}$

7. $1,160,000 = 1.16\times10^6$

9. $13,600 = 1.36\times10^4$

11. $8.673\times10^{-10} = 0.0000000008673$

13. $3.3\times10^{-2} = 0.033$

15. $2.032\times10^4 = 20,320$

17. $7.0\times10^8 = 700,000,000$

19. $940,000,000 = 9.4\times10^8$

21. $1.23\times10^{12} = 1,230,000,000,000$

23. $23,000,000,000 = 2.3\times10^{10}$

25. $(1.2\times10^{-3})(3\times10^{-2}) = 1.2\cdot3\times10^{-3+(-2)}$

$= 3.6\times10^{-5}$

$= 0.000036$

27. $(4\times10^{-10})(7\times10^{-9})$

$= 4\cdot7\cdot10^{-10+(-9)}$

$= 28\times10^{-19}$

$= 2.8\times10^1\times10^{-19}$

$= 2.8\times10^{-18}$

$= 0.0000000000000000028$

29. $\dfrac{8\times10^{-1}}{16\times10^5} = \dfrac{8}{16}\times10^{-1-5}$

$= 0.5\times10^{-6}$

$= 5\times10^{-1}\times10^{-6}$

$= 5\times10^{-7}$

$= 0.0000005$

31. $\dfrac{1.4\times10^{-2}}{7\times10^{-8}} = \dfrac{1.4}{7}\times10^{-2-(-8)}$

$= 0.2\times10^6$

$= 200,000$

33. $(7.5\times10^5)(3600) = (7.5\times10^5)(3.6\times10^3)$

$= 7.5\cdot3.6\times10^{5+3}$

$= 27\times10^8$

$= 2.7\times10^1\times10^8$

$= 2.7\times10^9$

On average, 2.7×10^9 gallons of water flow over Niagara Falls each hour.